BURLINGTON NORTHERN
A Great Adventure

BURLINGTON NORTHERN

A Great Adventure
1970–1979

BY EARL J. CURRIE

• RAILS NORTHWEST •

(*hardcover*) ISBN-13: 978-0-9961225-6-6
(*paperback*) ISBN-13: 978-0-9961225-7-3

• • •

Library of Congress Control Number: 2019938278

• • •

Copyright © 2019 by Earl J. Currie

• • •

Manufactured in the U.S.A.
Design by Ariane C. Smith
CAPITAL A PUBLICATIONS, LLC

• • •

• • •

FRONTISPIECE

An illustration from an informational brochure issued in 1961 by the four railway companies making application to consolidate and form Great Northern Pacific and Burlington Lines, Inc. (later named Burlington Northern, Inc.). BURLINGTON NORTHERN.

ON THE COVER

(*front*) An aerial view of the hump classification yard. Minneapolis. BURLINGTON NORTHERN photo. (*back*) Repairing the line. JOHN PHILLIPS III COLLECTION. (*front flap, dust jacket, top*) In the 1970s, Burlington Northern provided freight transportation service by rail, truck, or air. Each mode was represented on its wall calendar for 1973. (*front flap, dust jacket, bottom*) A 250-ton wrecking derrick was based at a few strategic locations on Burlington Northern's rail system. These large, powerful machines were used to clear the track after a derailment or other type of accident. In later years, the derricks were replaced by off-track Caterpillar tractors equipped with side booms and counterweights. BURLINGTON NORTHERN. (*back flap, dust jacket*) BN News, no. 16 (February 26, 1970), announces the merger.

CONTENTS

ILLUSTRATIONS

• • •

UNLESS OTHERWISE CREDITED, MAPS ARE BY MIKE BARTENSTEIN.

PREFACE

SEVERAL BOOKS HAVE BEEN WRITTEN IN THE last twenty-five years or so on Burlington Northern and its predecessor companies, the most recent, being *Leaders Count,* a history of BN written by Lawrence Kaufman. Other books about BN are made up mainly of photographs of trains, freight cars, and locomotives, with a limited amount of text covering its history and legacy. Kaufman's book opened the door for additional writings on BN by qualified researchers and business historians, or in my case, by people who worked for BN who are willing and able to document and interpret their experiences, and also reveal the evaluations they have formed of events and decisions made since the company was formed.

My writings are not intended to be a history or chronology of landmark events in BN's history. An effort has been made to avoid repeating much of what has been covered by others and, instead, to write from the perspective of an operating officer who went "through all the chairs" of the Operating Department at BN and two of its predecessor roads to Senior Vice President–Maintenance and Transportation. My writings cover the series of changes in policies and directives, the challenging cases and conflicts, and the major undertakings that set the course. There is particular emphasis on the roles and contributions made by representative individuals and groups of employees and managers who, from all levels of the organization, came to contribute to the success BN built in its early years.

My first book, *Burlington Northern: A Great Adventure, 1970–1979,* concentrates mainly on the years 1970 through 1979, when the foundation was laid for the achievements the company made in those years. My second book, *Transformation of a Railroad Company,* covers the years 1980 through about 1995, years of transformation and innovation in which BN broke away and differentiated itself from the alliances, practices, and norms that had been built up in the railroad industry over many decades. A period of high stress for the people of BN, it was worthy of interpretation and evaluation of the efforts they applied while working under a much different culture, with new objectives, strategies, and directives, while having to cope with a fundamental change in leadership style.

The statements, recollections, and opinions I have expressed are from first-hand knowledge I obtained in the years I was employed with BN, from research material I located or gathered over the years, and from conversations with fellow officers and employees with whom I have maintained contact over the years. Every effort has been made to accurately convey our recollections of events and the spirit of the times. My writings are centered on the work, talents, and commitment applied by the people of BN who moved it into a position of success in its first fifteen years.

It is hoped that those who had an association with BN in those years—whether as an executive, manager, employee, customer, business historian, or resident living on or near one of BN's lines—will find empathy and satisfaction in reading of the progress the company made through periods with mixtures of blessings, opportunities, turbulence, trial, and tension on its way to becoming a leader in the rail industry.

To avoid this manuscript becoming encumbered with footnotes, the bibliography at the end of the book lists the sources that have been consulted for this project. For direct quotes and certain key data, the sources are stated in footnotes. A list of "primary sources," i.e. references used throughout the book and not limited to one chapter, are listed at the end of the book.

Although I am fortunate to have had access to many highly knowledgeable and capable people, and excellent research material for these writings, I assume responsibility for errors or omissions in the text, tables, and drawings, for facts and opinions stated, and for interpretations of facts and events.

The pronouns "he" and "him" are used for clarity, to reduce the number of pronouns needed, and for grammatical convenience. They are not intended to limit the ideas or thoughts in this book to one gender.

ACKNOWLEDGEMENTS

THE POSSIBILITY OF WRITING THIS history of Burlington Northern's first two decades originated with Robert W. (Bob) Downing, who held positions as Executive Vice President, President, and Vice Chairman–Chief Operating Officer from BN's formation in 1970 to his retirement in 1976. He felt it was important that the company's history be reviewed and documented by employees and managers who contributed to the success BN had in merging the properties and organizations of four railroad companies, and then taking on the challenge of upgrading a large part of the railroad's network to handle the unprecedented tonnages of low-sulfur coal mined in the Powder River Basin. I was privileged to spend many days with Bob Downing while serving as an operating officer on BN, as well as after his retirement as he shared his recollections, insights, and evaluations of many aspects of BN's history.

I was also fortunate that Norman Lorentzsen, who followed L. W. (Lou) Menk as President and CEO, contacted me when he learned I had started to write a history of BN. Norman encouraged me to look over and include some of his writings on specific issues in the late 1970s, when he was responsible for leadership and direction of both the Transportation and Resources divisions.

I am grateful to Francis J. (Fran) Coyne, retired Vice President–Human Resources, for putting me in touch with C. Robert (Bob) Binger, who headed BN's Resources Division from 1970 through 1980. Bob gave generously of his time and provided insights to the strategies he employed to generate the ever increasing level of operating income that was sorely needed to finance the high level of operating expense and capital expenditures to upgrade and expand the capacity of its railroad.

I wish to thank two senior officers in BN's Finance Department, Robert F. (Bob) Garland and Raymond C. (Ray) Burton, who contributed generously to these books by each writing a chapter on experiences that were highlights in the years they held positions of leadership. Other vice presidents who took time to meet with me on one or more occasions to provide information or simply to refresh my memory of certain events and challenges were David Ylkanen (Human Resources), Edward Bauer (Mechanical), Wayne Hatton (Regional Vice President and, later, Vice President–Transportation), Fran Coyne, Norm Doerr (Purchasing and Material), and Roger Nelson (Safety).

Several officers in the Engineering Department (the function responsible for maintenance and construction of track, bridges, buildings, and signals) have been particularly helpful, among them Mike Armstrong, George Lamphier, Dave Hesterman, Bill Ferryman, Jim Daume, Cliff Inmon, Chuck McCormick, Don Rogers, Eldon Ficke, Mark Sprattler, Carl Peglow, Sam Melonas, Ralph Knutson, and Bill Bergmeier. In the Mechanical function (responsible for maintenance of cars and locomotives), I am especially grateful to Ed Bauer for sharing the strategies and implementation plans for the innovations he developed so successfully in the 1980s. Other Mechanical officers who have been helpful include Bob Seeley, Max Lary, Tom Hackney, and John Hilliard.

A number of operating officers who served as division or terminal superintendents in those years shared a great deal about the challenges to which they applied their leadership skills. Among them were Tom Lynch, Cliff Tye, Mike Martin, Bob Seeley, Mike Holsteen, Dave Burns, Bill Reilly, Dan Watts, Max Steele, Wayne Eisenman, and Don Maze. Other operating officers whose work contributed greatly to the company's success and were very helpful to me were Bob Hanson, Jerry Wick, Jake Greeling, Forester Dusell, Pat Keim, Cal Evans, Jerry Pinkepank, Al Johnson, and Greg Mangieri.

Jim Hudson, a retired Conductor for the Spokane, Portland and Seattle (SP&S) Railway and BN at Vancouver, Washington, organized a meeting of six former SP&S employees who also served as general chairmen or held other positions of leadership in the two unions that represented conductors and brakemen in the United Transportation Union (UTU) and engineers who belonged to the Brotherhood of Locomotive Engineers (BLE). This group of union officers played a major role in developing the unprecedented agreements that BN and the unions made for wage guarantees and lifetime job protection. Reaching agreement with the unions was vital in the efforts undertaken by the applicant railroads to get the merger approved upon reconsideration by the Interstate Commerce Commission. Meeting with these union officials provided a unique opportunity for me to learn of the contribution they made in working with the railroad companies' Labor Relations officers to make the proposed merger a reality. Jim Hudson also was helpful in providing insights to the new work environment that employees of the former SP&S Railway had to adjust to in the early years of the consolidated operation.

Merle Geiger, a Locomotive Engineer in Vancouver who advanced to president of the BLE in the 1980s, provided insights on the implementation of merger agreements and on how many new and vital issues were handled several years after the merger was consummated.

Wendell Bell, Assistant Vice President and Attorney in BN's Labor Relations Department, provided excellent assistance to me in working through details of landmark events in the development of new or revised agreements governing the work of the operating crafts from 1980 and into the 1990s. Especially valuable was the opportunity to draw on Wendell's participation in preparing the strategies applied in the effort to reach agreement with the UTU for reducing the number of employees required on train crews on a major part of the BN system.

I am grateful to Robert L. (Bob) Portsche, Operating and Safety Officer for BN and, in later years, for the Federal Railroad Association (FRA), for his assistance in making sure I was drawing upon a consistent set of numbers and facts relative to BN's record in employee safety.

I am fortunate to have had the interest, time, and support of several BN employees in my effort to reveal details about the changes in work environment and other cultural adjustments they had to work through as the merger was implemented. Among the employees who deserve particular recognition are John Langlot (Conductor at Spokane), Gary Nelson (Conductor at Minneapolis), Dale Wright and Joe Donovan (Locomotive Engineers at Minneapolis), Alex (Bucky) Hasson (Engineer at Whitefish), and Maury Godsil (Yardmaster at Galesburg).

I particularly enjoyed putting together the anecdotes of a selected group of operating and maintenance employees and officers who worked mainly on the Alliance and Nebraska divisions in the years when BN was going all-out to hire and train the large number of people needed to upgrade and expand the capacity of its railroad, and to operate the increased number of trains needed to handle the heavy tonnage of coal being mined in the Powder River Basin. I was able to meet with most of these railroaders or talk with them by phone, but for a few of them I had to rely on information from their fellow workers or reports about their work in company publications. In any event, it was rewarding to prepare brief summaries of work histories and remembrances of this representative group of railroad people in the chapter entitled "How and Why They Came."

I wish to give a great amount of thanks and credit for helping me get these two manuscripts finalized and ready for publishing to Dave Burns, a longtime friend and fellow operating officer. Dave generously applied his fine skills as a leader, teacher, and communicator to these tasks over the five years this manuscript has been under development, review, and reconstruction. Through the entire process, Dave's talents have been very valuable to me.

There are many members of what I would call the "supporting cast" who made it possible to complete my writings on BN history. Amy Jo Stockinger and I have worked together for many hours in assembling the various drafts and making an untold number of revisions to finally get the text in shape. I am fortunate she has been available to do similar work on other books and essays I have written about railroad operations in the past ten years.

I have been fortunate to have had the service of

Ariane C. Smith of Capital A Publications for editing and layout. I am well pleased with the exceptional skills and experience Ariane demonstrated in editing, page layout, design of the dust jackets, and the general advice she provided on creating attractive and professional publications. I became acquainted with the high quality of work done by Ariane in my reading of *The Rusty Dusty*, written by John Langlot and Mac McCulloch about their work as employees on the Great Northern's Wenatchee–Oroville line. I was so favorably impressed that I asked John and Mac for Ariane's e-mail address. I am fortunate that she was willing to edit the two books I had nearly completed on Burlington Northern history.

Mike Bartenstein, a retired engineer for Alcoa at Malaga, Washington, drew all of the maps I have used. Having a person with Mike's skills at graphics, together with a good knowledge of railroad history and operations, has been fortunate.

I am grateful to Don Hofsommer for his willingness to review the entire manuscript when it was still in the early stages of development. With the success Don has had in writing several widely acclaimed books on railroad history, I was fortunate that a person of his standing would take the time needed to advise me on how the manuscript should be constructed and prepared. Don evaluated the content as to the significance it would have to potential readers, and offered advice as to the style of writing that should be applied in reviewing the history of a railroad company.

The personnel at several libraries were very helpful in locating materials of interest in various activities and local issues involving BN. Among those libraries were the John W. Barriger III National Railroad Library in St. Louis, the libraries of the Minnesota and Montana historical societies, public libraries in Alliance, Nebraska; Gillette and Douglas, Wyoming; Spokane, Washington; and Missoula, Montana; the Galesburg (Illinois) Railroad Museum; the Newberry Library in Chicago; and the libraries of the Great Northern and Northern Pacific historical societies.

In looking back over my career of forty-two years in the rail industry, I reflect on the large number of fine people who had a hand in my development and the progress I made in my career as a railroad operating officer. It has been through that kind of support that I developed the spirit, interest, and background needed in later years to undertake the writing of a history of the company and its predecessor railroads that opened opportunities for me to advance and contribute toward their success. These railroaders became part of a network I could rely on when for coaching, advice, or other input at important junctures. For all of the "rails" I worked with who applied professionalism in their work and upheld the standards, values, and traditions that helped make BN and the other freight railroads of North America that are ascribed by many as "the envy of the world," I am most grateful and respectful.

EARL J. CURRIE
December 4, 2016

BURLINGTON NORTHERN

A GREAT ADVENTURE

Our basic goal is to provide better service to our shippers from the first day of the merger.
Integration of activities will be made on a carefully planned, step-by-step basis
to ensure success and minimize the possibilities of service deficiencies.
In the long run, Burlington Northern will be further strengthened by this careful approach.
BURLINGTON NORTHERN, INC., ANNUAL REPORT 1969, P. 2

With the cooperation of the Union organizations, labor agreements
negotiated prior to merger removed all impediments to an integrated operation.
BURLINGTON NORTHERN 1970 ANNUAL REPORT, P. 13

The Commission (ICC) found that . . . $40 million per year in savings
would be realized by the tenth year after merger.
UNITED STATES V. ICC 396 U.S. 491 (1970), P. 11

In 10 years Burlington Northern's consolidated net income
has grown at an average annual rate of 22 percent.
BURLINGTON NORTHERN ANNUAL REPORT FOR 1979, P. 3

RIGHT FROM THE START, THE MERGER THAT created Burlington Northern (BN) in 1970 was a success. BN did not have the initial struggles that have occurred in other railroad mergers, especially the one that created the Penn Central or the mergers several years later of Union Pacific with Chicago and NorthWestern (C&NW) and Southern Pacific. A number of important factors made that possible. First and foremost was the quality and depth of planning that was done under the leadership of Robert W. (Bob) Downing of Great Northern Railway and D. E. (Doug) Shoemaker of Northern Pacific, A. R. (Art) McDonald of Burlington, N. S. (Jim) Westergard of the Spokane, Portland and

Seattle (SP&S). Managers in all departments across each of the four railroad systems participated in the planning process over a period of fifteen years.

BN was blessed with outstanding leadership at the startup of the consolidated operation. John Budd, its first chief executive officer and chairman, had revived the plan for merger developed forty years earlier under his father, Ralph Budd, and, before his time, by James J. Hill. Louis W. Menk, BN's first president and a relative newcomer to the "Hill family of railroads," had the qualities of strength, perseverance, leadership, and communication skills needed to provide direction in a newly formed enterprise. As one of the leaders of the planning effort for the merged company, Downing was ideally suited for the role of an executive vice president at the time of merger, and ready to become president upon Menk's move up to chairman and CEO when Budd retired.

Through joint ownership of the Burlington and the SP&S railways, as well as the operation of several large jointly owned facilities, a large number of managers and employees of each of the four component railroads already were well acquainted. During their careers, they had become familiar with the standards and operating practices in place on one or more of the other railroads that would make up BN. The many years of affiliation they had experienced helped pave the way for success under a consolidated operation. Managers at all levels in each department had been involved in the planning effort, giving them ownership and familiarity with the plans they would be responsible and accountable for implementing upon merger.

In a presentation Downing made on the planning process, he stated, "We went out of our way to ensure that everybody in our companies understood that this was not a takeover. It was a merger of equals, that everybody would be treated equally and that we respected the way they were doing things."[1] In this way, BN did not start out with "red teams" and "green teams," as had been the case in some other corporate mergers, both in and out of the rail industry.

There are some who say BN fell into its success, that it was an accident of history, and that it was just plain lucky by having ownership of lines already built in the area of the Powder River Basin. This definitely was not the case: as BN had to take a major risk in which it "bet the company" in making huge investments to handle the large tonnages of coal that were demanded by the electric power industry starting in the early 1970s. In addition, BN's leadership had to transcend the long-standing mindset in the rail industry that a railroad could survive only by cutting costs and reducing the extent of operations. The culture had to be changed into one of managing a growing enterprise with unprecedented challenge and opportunity. Many companies in the railroad industry (including the Union Pacific, its arch competitor) have benefited from the risks BN took in its early years

to gear up to handle large tonnages of coal that would move to power plants located on their lines, as have the consumers of low-cost electric power.

There also were large demands for capital for the projects needed to fully integrate the operations of the component roads. BN paid a high price at the outset by granting job protection to all of its employees, as well as in the concessions it made to its competitors, particularly Chicago and NorthWestern Railway (C&NW) and Milwaukee Road. There were some observers who said the newly merged company was so big it was unmanageable, including executives on some of the railroads that connected with BN. However, BN was successful in the years that some railroads failed, went out of business, or were taken over by larger railroad companies.

The growth in low-sulfur coal mined in the Powder River Basin built the density (gross ton-miles per mile of road) of traffic and the level of utilization higher on large parts of BN's roadway than on any other freight railroad in the world. With such high density, the economies of scale that are so inherent in the railroad business came into play. The operating ratio declined and operating income increased. To meet the challenge of handling this extremely high tonnage, BN's Engineering and Mechanical departments had to develop and apply significant advances in maintenance and inspection practices within a short time span. Under the direction and leadership BN applied to these technical areas, impressive advances were made in the design of track components and in the maintenance of track, locomotives, and cars.

The opportunity to move heavy tonnages out of the Powder River Basin transformed a large part of the BN system from a light or medium level of density into a heavy-tonnage railroad, far exceeding the projections for coal tonnage made by the Marketing and Planning departments in the mid to late 1970s. The upgrading of most of the lines in Wyoming and throughout the Denver and Chicago regions of the BN system made it possible to overcome the deferred maintenance of previous years. Without the revenue from the heavy tonnage of coal, such major investments could not have been justified and financed. This book contains a discussion of the innovations and major capital projects undertaken to expand the capacity of the network and improve the quality of track of hundreds of miles of road.

---

1   Solomon, *Burlington Northern Santa Fe Railway*, 118–119.

In this book I have expanded on the leadership, planning, and other factors that got the new company off to a good start and laid the foundation for long term success. Those of us who were there in those years experienced the kind of success we felt the rail industry was capable of achieving, if it could gain some relief from undue government regulation and if a "can do" spirit could be developed at all levels and in all disciplines within the company. The unprecedented opportunities we had at BN produced a generation of leadership ready for the task of managing in the competitive, market-driven environment we would be facing under deregulation. We became a respectable, competent business enterprise as opposed to an organization that might have been seen by observers from outside the industry as a constrained, declining institution forced to operate on a survival mentality. BN's success created employment opportunity for thousands of people with well-paying, rewarding jobs, far better than the jobs they may have been able to get in other sectors of the economy.

FOREWORD

by David H. Burns

I'm grateful to Earl Currie for devoting more than five years of his life to make this saga
available for others. It is an important contribution to an important industry.

DAVID H. BURNS, NOVEMBER 2016[1]

THE THEME AND CULTURE FOR *Burlington Northern: A Great Adventure* grew out of a whirlwind tour by President Louis W. Menk shortly after the 1970 merger that created Burlington Northern (BN). Referring to the new company as "our magnificent enterprise," he solicited everyone's participation in a "great adventure" to create a "shining railroad on a hill." Some scoffed, but many took hold of that powerful imagery. While BN's inauguration was not without some surprises in spite of very careful planning, none were without quick remedy. Within a few years, BN was being heralded as "the merger that worked."

While there have been a number of good books on BN, none have been written from the perspective of actually being there. This author was present at both the line operating and senior management levels, which affords rich and unique insights to share.

Indeed, you will meet many people who will share their personal involvement at all levels. One chapter is "Financing for the Nation's Largest and Newest Railroad" by Robert F. Garland, senior vice president–Finance. Another is "Finding Billions for New Ventures." Others will be from the critical level of making the railroad run in Part VII, "The People of the Railroad."

This project is in two volumes because there were vastly different adventures in the first and second decades of BN. The 1970s saw the sudden explosion in demand for vast amounts of low-sulfur coal available for development from the Powder River Basin of Montana and Wyoming. To its credit, BN took the lead in this unknown venture rather than waiting to be pushed into it. But to do so required "betting the farm" on such an unprecedented expansion. Doing so required heroic and innovative efforts at securing frightening amounts of capital, followed by heroic and innovative efforts in constructing new trackage, rebuilding most of the physical plant elsewhere, acquiring staggering amounts of new equipment, and hiring and training thousands of new employees, all while handling our existing traffic. This

---

1   David H. Burns began his career as a student officer on the Great Northern Railway in 1963. He advanced rapidly in the Operating Department to division superintendent of the GN's Dakota Division, headquartered at Grand Forks, North Dakota, in 1968. After the 1970 merger that formed Burlington Northern, Dave served as manager of Operations for the Chicago Region, and as superintendent of the Rocky Mountain, Pacific, and Colorado divisions. He advanced to general superintendent–Transportation, then general manager–Intermodal and director–Quality Control at the headquarters level. In his retirement, Dave and his wife, Barbara, teach courses in Texas history to adults, and remedial math for high school students in the Fort Worth area.

was indeed an adventure, although not one that some may not want to repeat. But BN survived and thrived. And that's the overarching theme of the first book.

The second book, *Transformation of a Railroad Company: Burlington Northern, 1980–1995,* covers the 1980s and early 1990s and details three other major "life changing" events: the merger with the Frisco Railroad, the hiring of Richard M. Bressler from Atlantic Richfield to replace Norman M. Lorentzsen as president and CEO, and the advent of deregulation of the rail industry that came with the Staggers Act of 1980. As the author wryly notes, "Having even one of these three events would have been enough to cause a transformation of the company, but having all three within a short period of time really set things astir in every department and function." He then describes each in fascinating detail. The second book ends with the merger of BN and the Santa Fe Railway and fading into history of the BN emblem.

Both books have the added advantage of well-written introductions and conclusions, which deftly summarize the major events in each and why they were for us a great adventure. The conclusion for the second book is perhaps the most compelling reason to read both books— because they reflect the BN so many of us knew, loved, fought for, and admired:

> Perhaps the highest level of recognition that could be shown . . . has been the decision Warren Buffett made (shortly after the merger that created BNSF) to acquire ownership of the company. Buffett probably could have made an offer to purchase 100 percent of the stock of any of the large North American Railroad Companies, or a large company in another industry. The fact he chose BNSF clearly indicates he evaluated it overall as the railroad company having the best prospects for the long term. In announcing his decision Buffett provided the basis for his evaluation as, "I think . . . it's a great business in that you know it's going to be here forever. . . . We bought it because it is well-managed."

I'll wager Mr. Buffett also appreciates being part of a magnificent enterprise, and continuing a great adventure.

LIST OF ACRONYMS, ABBREVIATIONS, AND SHORTENED NAMES

AAR	Association of American Railroads
ABS	Automatic Block System
AREA	American Railway Engineering Association
BLE	Brotherhood of Locomotive Engineers
BRC	Belt Railway Company of Chicago
BN	Burlington Northern
BNI	Burlington Northern, Inc.
BNRR	Burlington Northern Railroad
CB&Q	Chicago, Burlington and Quincy (Burlington Route)
CGW	Chicago Great Western
CSTPM&O	Chicago, St. Paul, Minneapolis and Omaha ("the Omaha")
CMSTP&P	Chicago, Milwaukee, St. Paul and Pacific ("the Milwaukee Road")
CN or CNR	Canadian National Railway
C&NW	Chicago and North Western Railway ("the North Western")
CP or CPR	Canadian Pacific Railway
C&S	Colorado and Southern
CTC	Centralized Traffic Control
DM&IR	Duluth, Missabe and Iron Range ("the Missabe")
DOT	Department of Transportation (may be federal or state)

FRA	Federal Railroad Administration
FW&D	Fort Worth and Denver
GN	Great Northern
GTW	Grand Trunk Western
IC	Illinois Central
ICC	Interstate Commerce Commission
Manitoba Road	St. Paul, Minneapolis and Manitoba
Metra	An operating organization established by the Regional Transportation Authority for the greater Chicago area for rail commuter train service on lines acquired by Metra, or to negotiating purchase of service agreements for service on lines that continued to be in the ownership of privately-owned railroads, among them BNSF and UP.
Milwaukee Road	Chicago, Milwaukee, St. Paul and Pacific
Missabe Road	Duluth, Missabe and Iron Range
MP	Milepost or Missouri Pacific
MRL	Montana Rail Link
MSTP&SSM	Minneapolis, St. Paul and Sault Ste. Marie ("the Soo Line")
M&STL	Minneapolis and St. Louis

North Western	Chicago and North Western	SI	Spokane International
NP	Northern Pacific	Soo Line	Minneapolis, St. Paul and Sault Ste. Marie
NS	Norfolk Southern		
		SP	Southern Pacific
OE	Oregon Electric	SP&S	Spokane, Portland and Seattle
OT	Oregon Trunk	SSW	St. Louis Southwestern (also known as the "Cotton Belt")
PC	Penn Central	STB	Surface Transportation Board
PCRR	Pacific Coast Railroad		
PRR	Pennsylvania Railroad	UMTA	Urban Mass Transit Authority
		UP	Union Pacific
RRV&W	Red River Valley and Western	UTU	United Transportation Union
RTA	Regional Transportation Authority		
		WC	Wisconsin Central
		WP	Western Pacific

PART I

MERGER PLAN
AND IMPLEMENTATION

CHAPTER 1

THE NEED TO MERGE

• the importance of having James J. Hill's vision realized •
• to build density on major corridors •
• to take full advantage of the most efficient routes •
• to eliminate duplicate line segments •
• to rationalize yards and support facilities •

BECAUSE RAILWAYS ARE A DECREASING-cost industry, their economic success depends to a great extent on the level of density they are able to develop and maintain. This is true primarily because of the high labor costs and high fixed costs railways have in operating and maintaining their fixed plant. Without the high density that railways have built up in recent years, they would have even more difficulty in achieving a satisfactory rate of return on the capital they have invested in plant and equipment than they had since the 1920s. Their operating margins would not be high enough to justify the amount of reinvestment needed to keep their property in a satisfactory level of maintenance, to attract capital for new business opportunities, or to invest in a major renewal of roadway and facilities when these expensive assets are worn out.

The railroads that have the highest rates of return are those with a high density of traffic. An example would be the Union Pacific, with the high-density main line it operated between the east end of its system at the Missouri River and Ogden, Utah. Other examples of high-density operation at the time the northern lines' merger proposal was developed were on the coal-hauling eastern railways: the Norfolk and Western and the Chesapeake and Ohio.

For the writing of this chapter, the following tonnages are used in defining a levels of density:

high density	30 million gross tons per mile (GTM) per year
medium density	from 5 million to 29 million GTM per year
low density	less than 5 million GTM per year

Because James J. Hill understood how essential it was to build up density on a railway, he pushed hard to integrate the operations of the Great Northern (GN) and Northern Pacific (NP) and, in time, to merge those properties. Several years later, in the 1920s, Ralph Budd made the same attempt, but again without success. By the mid-1950s, the need to reduce costs by concentrating traffic on fewer line segments became increasingly urgent. John Budd of the GN and Robert Macfarlane of the NP agreed that, more than ever, their railroads had to be merged.

Before the opening of the Panama Canal in 1914, the volume of traffic moving across the northern tier of states to and from the Pacific Northwest became so high that the NP built either double-track or alternate routes between Paradise and Laurel, Montana (411 miles), and added a second main track between Buffalo, North Dakota, and Minneapolis (277 miles). The GN increased its capacity by building an alternate main line between Fargo and Minot (the Surrey cut-off), and started to build a second main line across Montana, east of the Rocky Mountains between Great Falls and New Rockford in 1912.

The opening of the Panama Canal and completion of the Milwaukee's Puget Sound extension at about the same time caused a major reduction in business moving over the GN and NP. After World War I, traffic on the northern lines recovered to some extent for about ten years, to the beginning of the Great Depression. After World War II, with the great improvements made in highways throughout the country, railroad traffic declined and caused the GN and NP again to organize a strong effort to merge and combine their operations. Consolidation would increase density by concentrating traffic on fewer routes, eliminating duplicate terminals and shops at common points, and allowing the abandonment of some parallel or duplicating lines.

Density was somewhat higher on parts of the Burlington, mainly between Chicago and Lincoln, and on the line serving the coal fields in southern Illinois. Until World War II and the economic boom in the Pacific Northwest after the war, density never built up on the Spokane, Portland & Seattle (SP&S), which had been completed in 1908. On all four of the "Hill roads" (the GN, NP, Burlington, and SP&S), the decline in local business handled by railroads—i.e., consumer goods, general merchandise, machinery, and construction materials—reduced the level of traffic density on both main lines and branch lines. The principle competition was no longer between railroad companies but was instead between railroads and trucking companies. There was nothing on the horizon for economic growth that would provide much opportunity to use the excess capacity owned and maintained by any of the four railroads in the merger proposal.

A number of shared lines and facilities were set up on the Hill lines over the years, but this took place mainly in Mr. Hill's day. One example was the General Office building in St. Paul, shared by the GN and NP. Another was the yard at Laurel, Montana, owned and operated by the NP but also used by the GN and the Burlington. Both the GN and NP used the King Street passenger station in Seattle. In addition to the GN and NP, the Burlington used the GN passenger station in Minneapolis and the GN's route between St. Paul and Minneapolis. The GN had trackage rights on the NP's main line between Seattle and Vancouver, Washington—173 miles. With these jointly owned or shared facilities in service, density and efficiency were much higher than if each road had built and operated its own lines and facilities. For the most part, however, there were duplicate yards and support facilities at common points throughout the GN and NP systems. In major terminals—the Twin Cities, Spokane, Seattle, and Duluth–Superior—both railroads operated independent yards and equipment maintenance facilities.

Each company brought particular strengths and weaknesses to the merger "table." The team established to prepare the plan for a consolidated operation had to evaluate these differences and prepare a plan that would take full advantage of the best-located, most-efficient, or best-engineered route and facilities. Such evaluation would determine which line segments would make up the preferred route for moving long-haul, transcontinental business. It would also determine what capital projects would be needed to integrate the operation.

The operating plan would bring about increases in the density of traffic on the line segments selected for the preferred route, and allow a lower standard of maintenance on line segments that would be relegated to a secondary position in the network. Supporting facilities such as classification yards, station and office buildings, facilities for servicing and maintaining cars and locomotives would be consolidated, thereby allowing many of them to be closed. The headquarters functions carried out in the General Office of the Burlington in Chicago would be consolidated with those in the headquarters building in St. Paul. The overall result would be large reductions in operating and maintenance costs, a reduced asset base, and an increase in density on lines to be retained between common points or junctions.

Following is a brief, general description of the density (gross ton-miles per mile of road) handled on each of the four railroads that made up BN.

GREAT NORTHERN   The GN benefited from the movement of high tonnages of iron ore over much of its history. The line between Superior and the Mesabi Iron Range in northeastern Minnesota formed the GN's only high-density operation. However, by the mid-1950s the "natural ore" had been depleted in mines served by the GN. The taconite pellets produced from low-grade ore in subsequent years provided the GN only a fraction of the tonnage of ore it had been moving since the 1890s. The GN had the advantage of lower grades on its transcontinental line than those on the NP, Milwaukee, and Union Pacific. Until the late 1960s, the GN produced a low operating ratio, below 80. However, the ratio was exceeding 80 in the last few years before the 1970 merger. Clearly, the economics of merger were vital and in the long-term interest of its shareholders, employees, and the territory it served.

NORTHERN PACIFIC   The NP brought the advantages of a high-capacity, well-maintained double-track route between Minneapolis and Buffalo, North Dakota (277 miles). While the NP did not have density high enough to require the capacity of this line, it is fortunate the NP did not reduce to a single-track operation before the merger. It also had a high-capacity, two-main-track line between Seattle and Portland. With the GN and UP having trackage rights on this line, the density was fairly high. On its transcontinental line, the NP had the disadvantage of heavier grades and longer distances between major terminals than the GN. As was the case with the GN, the NP had only moderate to light density of traffic on nearly all of its system.

BURLINGTON   More than the GN and NP, the Burlington was hit with traffic losses as the interstate highway network was expanded. It had a fairly high-density double-track line on most of the 525-mile corridor between Chicago and Lincoln, but the rest of its system had only light to moderate density. The Burlington had the advantage of a fairly large traffic base through interchange with the GN and NP at the Twin Cities and Laurel, and with the GN at Sioux City. Also, the largely double-track line between Savanna, Illinois, and the Twin Cities (282 miles) had far more capacity than was needed, although it is fortunate that capacity was not reduced by converting the line to single track with Centralized Traffic Control (CTC), in view of the large increase in business that was built up shortly after the 1970 merger.

SPOKANE, PORTLAND & SEATTLE   The SP&S was superbly engineered with low grades and minimal curvature. As a result, it had excellent operating capability. With the economic expansion that occurred in the Pacific Northwest during and after World War II, density built up to a moderate level. Soon after the 1970 merger, with more traffic being routed on the most-efficient line segments, the volume of business increased significantly

on the SP&S main line between Vancouver, Washington, and Spokane. Business remained quite light on the north-south line through central Oregon until the formation of the BNSF Railway. It has been fortunate that the east-west main line of the SP&S was built with low grades and minimum curvature, which enable it to handle heavy-unit trains of grain and coal very efficiently. Together with the trains that handle containers with imported merchandise and automobiles, the former SP&S line has become a high-density operation nearing capacity.

PACIFIC COAST RAILROAD   A fifth but seldom-mentioned railroad company that formed BN was the Pacific Coast Railroad (PC), a small company with twenty-two miles of track between Seattle and Maple Valley, Washington. It was acquired by the GN in 1951 but never merged with the GN. The track of the PC provided the Milwaukee Road access from Maple Valley to Seattle under trackage rights. It was used by the GN only to serve shippers in the Renton industrial area (refer to chapter 2 for details on the operation and use of the PC).

Table 1.1 contains a comparison of the density on main line segments on each of the four railroads in 1955 when a study of the economics of the proposed consolidation was undertaken. The table also contains "snapshots" of the density of those line segments in later years, to show the extent to which density increased in the years after the 1970 merger. Some of this build-up resulted from consolidation of traffic on the most efficient routes, but it also is the result of effective marketing programs to develop international trade in grain and consumer goods, and on many lines from the movement of large tonnages of coal mined in the Powder River Basin.

If the merger had not occurred, it is likely that in a few years, GN and NP would have had to start reducing the quality of track maintenance, even on their main lines. Over the long run, the condition of their track would have followed the path of general deterioration that happened on the Milwaukee, C&NW, and Rock Island. Already on the Burlington, the condition of the track was deteriorating from the squeeze on earnings and the requirement to generate enough cash to maintain dividend payments to its owners.

The three large roads that combined to become BN were no longer in a position strong enough to "go it

Table 1.1.

COMPARISON OF DENSITY, 1955, 1981, AND 2009
NORTHERN CORRIDOR

	1955*	1981	2009
Everett–Spokane	9	24	20
Spokane–Shelby	16	51**	69**
Shelby–Minot	13	31	66
Minot–Casselton	15	42	61
Casselton–Staples	15	64†	102†
Staples–Minneapolis	16	62†	77†
Vancouver, Wash.–Wishram	13	29	85
Wishram–Pasco	11	25	83
Pasco–Spokane	18‡	33‡	85

*   Contains data filed with the Interstate Commerce Commission (ICC) in application for merger that was effective in 1970. The numbers shown are for freight only and do not include tonnage of passenger trains.

**Does not include tonnage handled by Montana Rail Link under trackage rights between Spokane and Sandpoint beginning in 1987.

† Includes tonnage (primarily coal) moving east from former NP line between Casselton and Huntley, Montana.

‡ Includes combined tonnage for lines of former SP&S and NP in the years while the lines of both companies were still in service.

Table 1.2.

COMBINED AVERAGE EARNINGS (FOR GN, NP, BURLINGTON, AND SP&S)

    1951–55 $121.6 million
    1956–60   86.9 million
    1961–65   73.9 million
Average rate of return, 1955–61: from a high of 3.73% in 1955 to
    a low of 2% in 1961.

alone." The merger was vital if they were to continue to maintain their plant and equipment to a high standard, achieve higher earnings, and maintain dividend payments to their shareholders. Since no major opportunities for growth in traffic were on the horizon at that time, the ability to at least maintain (and, hopefully, to increase) earnings would have to come mainly through the cost-reduction opportunities that a merger would afford by elimination of duplicate facilities, achieving economies of scale by concentrating traffic on the most-efficient routes, obtaining a higher rate of utilization of cars and locomotives, and making a large reduction in the labor force.

In the several years preceding the merger, the financial position of the northern lines and the Burlington deteriorated, due primarily to loss of business to transportation modes using publicly supported highways and waterways, and by delays in getting rate increases approved by the Interstate Commerce Commission (ICC). In contrast to truckers and inland water carriers, railroad companies had to maintain their own roadway and were dependent on private capital for financing. A look at the earnings of the applicant companies in five-year intervals in Table 1.2 shows the trend of earnings and the extent of their decline up to the mid-1960s when they prepared their case for merger:

It was likely that some or all of the Hill roads' competitors already were or would soon be looking for opportunities to extend their systems through mergers and acquisitions. Those initiatives would be sure to deprive the Hill roads of traffic—especially the large volume of highly competitive bridge traffic the Burlington handled to and from its connections—and further reduce its density of traffic.

The economies gained through the 1970 merger, plus earnings from the increased tonnages of coal moved out of the Powder River Basin starting in 1974, enabled BN to become much stronger. In the early 1980s, with density increasing on most of the BN's system, and by reducing costs, its financial results improved. BN became able to move beyond coal and make the investments in terminals and equipment needed to handle consumer goods shipped from the Pacific Rim and export grain.

We are indeed fortunate that Budd and Macfarlane

decided to move against strong odds and accomplish the merger of the Hill roads. There was strong resistance at the outset from labor unions; competing and neighboring railroads; the communities in which one or more of the companies had offices, shops, or yards; and customers who feared the loss of competition. Most elected representatives and the U.S. Department of Justice also lined up in opposition. Other than the financial community and the rare customer and others who understood railroad economics, there were only a few groups or individuals who favored the merger during the years in which it was under review by the ICC. Budd and Macfarlane knew they had an opportunity to protect the interests of all of these constituents over the long run, if they could prevail over those who saw the merger as destructive and unnecessary and who believed its basis amounted to no more than corporate greed.

During the application process, not too many factions seemed to understand or appreciate the benefits the merger could produce. They failed to recognize the need and benefits to be gained by reducing operating costs and eliminating expensive overhead functions, and that service could be improved as a result of reducing terminal delay and having enough density to set up more "through" train operations. By eliminating duplicate or overlapping routes and terminal facilities, the size of the underutilized network would be reduced to a more affordable level, and more in line with foreseen levels of business. With better earnings, the railroads would make a higher rate of return on invested capital, and their capability to finance much needed capital improvements would improve.

Accomplishing the merger would strengthen railroading as a viable transportation service for a sizable area of the United States. The consolidated company would have a better chance of preserving railroad jobs—some of America's best-paying and most secure jobs for thousands of families—and would help keep trucks off highways and rural roads. It would enable the agricultural business community to reach new overseas markets. Moreover, having a large network created by the merger has made it possible for BN to move high tonnages of coal on a single system for long distances at very low rates. That capability has done more than anything else to hold down the price of electricity for millions of people throughout America. The use of low-sulfur coal throughout the United States has helped greatly in improving the quality of the air we breathe. Without the merger that formed BN, the road to gaining these many economic and social benefits might not have been available.

CHAPTER 2

THE PREDECESSOR ROADS:
THE BURLINGTON ROUTE

EVERYWHERE WEST

A SLOGAN USED BY THE BURLINGTON

ON MANY OF ITS LOCOMOTIVES AND FREIGHT CARS

• *strengths and weaknesses* •
• *characteristics: route structure, traffic, culture* •
• *condition of roadway* •

MPLOYEES OF THE BURLINGTON EXPERI-
enced less tension and uncertainty at the time
of the merger than those who worked on GN,
NP, or SP&S. The only common points for Burlington
and the northern lines were the Twin Cities, Sioux City,
and Laurel. In the Twin Cities terminal, operations at
the Burlington's Daytons Bluff yard were integrated on
March 2, 1970—M-Day—with the other ten yards in
that complex. Daytons Bluff was kept in service until the
new hump yard at Northtown was opened in 1975. The
biggest change for Burlington employees at the time of
the merger was the agreement allowing their road crews
from La Crosse to operate trains directly into the other
yards. Before the merger, deliveries to and from other
yards were handled by yard transfer crews based at Daytons Bluff. Because BN was able to provide a single-line
haul from origin points to Chicago, eastbound business increased right from the start following M-Day.
The number of pool crews based at La Crosse increased
from twenty-five to about fifty by late 1972.

At such common points (common to GN and NP) as
Spokane, Seattle, and the Twin Cities, the merger caused

some pain through difficulties in combining seniority on
an equitable basis that would satisfy the concerns and
preferences of everyone. There were cases of the main
flow of traffic being shifted, which forced some employees to relocate in order to follow their work. Home terminals for some crews were changed. Some employees
endured cases of harassment and intimidation. Employees of the Burlington experienced no disruption of such
magnitude except for those in the General Office in Chicago whose work was transferred to the new company's
headquarters office in St. Paul.

Sioux City, Iowa, was a common point for Burlington and GN. Upon merger the operation was changed
to allow road crews bringing trains in from Lincoln,
Nebraska, and Pacific Junction, Iowa, to operate into the
GN's yard in Sioux City. Previously, transfer crews had to
be used to move cars between GN's yard and Ferry, on the
Nebraska side of the Missouri River. An implementing
agreement was necessary to allow this change.

The only location besides the Twin Cities that was
served by all three of the major railroads in the merger
was Laurel, Montana. The yard there was operated by

NP, which handled train makeup and general switching for GN and Burlington. Trains operated on Burlington's Sheridan line used twenty-six miles of trackage rights (Huntley–Laurel) on NP to reach the yard at Laurel. With the joint facility and usage payments made by Burlington, NP was able to maintain this line segment as a double-track railroad, in excellent condition. Burlington generally operated two trains per day on this line, compared to NP's six through-freight trains and four passenger trains. Trains moving to and from Burlington's Casper line used NP's Fromberg branch for twenty-two miles to reach Laurel.

Although 98 percent of the stock of Burlington was held by GN and NP, the owner lines did not micro-manage or try to dominate in managing its day-to-day activities. The Burlington was larger than either of its owners by any measure. Also, because Burlington served more large metropolitan areas, it generally was better known than either GN or NP. This was a factor in the decision to use the name "Burlington" in the name of the new company. The board of the Burlington consisted mainly of top executives of GN and NP. Their main concern was for Burlington to have earnings high enough every year to maintain the payment of the same level of dividends to the owning roads. Dividends received from Burlington made up a large part of the earnings reported by GN and NP. In the last few years before the merger, Burlington had to cut its expenses for roadway maintenance in order to keep earnings high enough to support the dividend.

To be a competitive leader in service, Burlington had to run a fast-paced operation for the merchandise and perishable business moving to and from its major western connections: Rio Grande at Denver; Union Pacific at Grand Island, Nebraksa; and Santa Fe and Rock Island at Kansas City. That was also true for connections with the GN and NP in Minneapolis. The owner roads did not simply turn all of their eastbound business over to the Burlington at the Twin Cities. A large volume of business also was interchanged with the Burlington's competitors, mainly the C&NW, as well as the Rock Island, Soo Line, and Milwaukee. Some tension built up in the early days of BN over the priorities it should have for its routes and connections, in the face of a shortage of motive power.

Burlington faced strong competition from the seven other railroads that connected with UP at Council Bluffs,

Iowa; Fremont, California; or Kansas City, Missouri, but it managed to build up year-to-year increases in its volume with UP through Grand Island. Most of the business interchanged with UP was moving to or from California, but a fair amount connected with UP's NWF (Northwest Forwarder) that served Portland and Seattle in competition with GN and NP. Shortly after the 1970 merger, business increased on many of BN's lines, which caused some difficulty in providing enough motive power to run all westbound priority trains out of Cicero, Illinois, on schedule as Burlington had done with success, year in and year out, before the merger.

In the merger, Burlington and its people were not hurt. Instead they benefited from an immediate, large increase in business, with all priority trains powered by the best locomotive types in the combined BN fleet and—very importantly—with the money needed to begin to overcome developing deficiencies in the track. Rather than being disadvantaged, the Burlington benefited greatly from the merger.

The Burlington ran a high-capacity railroad. On its main line between Cicero and across Iowa to the Missouri River, double track was in place on all but the first 33 miles east of the river. Three main tracks were in service for the 31 miles between Cicero and Aurora, and four main tracks east of Cicero. Except for four short stretches of single track on bridges over major rivers, there were two main tracks on all 282 miles of the Burlington's north line between Savanna and St. Paul. CTC was in service on all but 34 miles of the single-track line of 300 miles between Galesburg and Kansas City. CTC was also in service on the 93 miles of main line north from Kansas City to Napier, Missouri, and for 541 miles between the Missouri River and Denver.

Fortunately for the customers and employees, BN started major upgrading programs within about two months of the effective date of the merger, almost as soon as the ground had thawed out after winter. Tie gangs, rail gangs, and surfacing gangs were established to work on lines and in yards that had seen little major work for several years. In addition, large yard-cleaning machines were sent to the Chicago and Omaha regions to remove the debris, spilled bulk commodities, and trash that had accumulated for many years due to not having the labor and machinery needed to keep the yards clean and free of tripping hazards. Several small maintenance gangs were set up to make repairs to weak spots in the track that were under slow orders or that were causing minor yard derailments.

All of this work was a clear demonstration to customers, elected representatives, regulatory agencies, and employees that the newly formed company was determined to restore the track in those areas to the condition needed for safety and reliability of service, as well as to handle the projected growth in business. These efforts to improve the property continued without letup for the next ten years. The amount of money spent to improve the lines of the former Burlington clearly showed a commitment of the new company to overcome the problems of insufficient maintenance due at least in part to the amount of dividends Burlington had been paying to GN and NP. The investment BN made in the territory of the former Burlington was many times higher than any amount of "excess" dividends it had been paying in the 1960s. The main lines and yards were brought up to a standard of "the best in the business." Finally, it was an inspiration to the people of the railroad to see the property cleaned up, in addition to such simple but meaningful things done as to paint buildings, get rid of scrap material, improve locker- and lunch-room facilities, eradicate unwanted vegetation, and generally make things look like we were an up-and-coming business.

CHAPTER 3

THE GREAT NORTHERN

Only 55 Miles of the Line Above 4,000 Feet

A STATEMENT GN PLACED ON MANY OF ITS PUBLICATIONS

T HE GREAT NORTHERN (GN) WAS KNOWN to many as "an operating man's railway."[1] It ran an efficient operation with an operating ratio below 79 up to the last few years before the 1970 merger. The GN knew how to take advantage of its low grades over the Rocky Mountains. Even in North Dakota and eastern Montana, the GN was located and built with lower grades than its competitors—the Milwaukee, NP, and the Union Pacific—on its line to the Pacific Northwest.

Until the mid-1950s, the GN profited greatly from handling large tonnages of iron ore from the Mesabi Range in northeastern Minnesota to its docks near Superior, Wisconsin. The GN made moving the ore a low-cost operation by locating its line on a grade of only 0.4 percent, descending in the loaded direction. The other major hauler of ore from the Mesabi Range, the Duluth, Missabe and Iron Range (DM&IR), had to contend with

grades of 2.2 percent and 2.0 percent to reach its ore docks in Duluth and Two Harbors. The GN was fortunate to serve highly productive grain-growing areas in Minnesota, North Dakota, and Montana. Together with iron ore, grain, and forest products, the GN had a very strong base of revenue.

Having the lowest grades and lowest elevation in crossing the Continental Divide of all the northern transcontinental railways was a source of pride to the GN over the years. Another claim to fame was that the GN was built without the benefit of land grants. Continuing forward from its early days under James J. Hill, the GN carried out programs every year to improve the alignment on its main lines. With these projects, curvature, grades, and the length of the line were reduced. Other than during the Great Depression, line improvements were made every year, right up to the 1970 merger. The result of these year-to-year line improvements made in those eighty years following completion of its transcontinental line was a vastly improved property. Each such project, if looked at individually, may not always have had a high return on investment. However, when looking at the cumulative effect of all of the improvements made over a number of years, the reduced track-maintenance cost and improved train performance that resulted from these line changes is impressive. This approach to improving the railroad on a sustained basis was one of the most significant and enduring parts of Hill's legacy.

The GN also was seen as a "class railroad." It maintained its property to a high standard, with high-quality track on all of its main lines. Its buildings and other properties

---

1   The expression "an operating man's railroad" was often used in years past as a compliment or indication of respect for particular railroads that seemed to have the physical characteristics and other attributes that formed a strong foundation for operating at a high level of efficiency. In my experience, it was a characteristic often credited to the former Great Northern Railway. In what I learned about the lines of the former SP&S Railway while serving as vice president–general manager of BN's Seattle Region in the early 1980s, I found that the SP&S also deserved such recognition. In more recent times, we might express it as "an operating *officer's* railway," but for an essay I wrote about the SP&S Railway covering the years 1908 through 1969, I decided it would be appropriate to give it the traditional name: "An Operating Man's Railway."

were clean and neat in appearance, in contrast to what one would observe on the property of many railroads of that era. In many small communities, the GN's buildings looked better than other commercial buildings in the town. Because of the low grades and shorter distance of its transcontinental route, the GN had a competitive advantage over the NP for time-sensitive merchandise business. In competing for business, there was good old-fashioned rivalry between the GN and NP.

In the process of planning for the merger each of the presidents of the component lines resolved that this was to be a merger of equals: none of the three companies would dominate the planning process, nor would the senior management of one company be awarded the majority of the most "dominant" positions in the new company, such as Operations, Sales-Marketing, Finance, or Law. In spite of that declaration, there was said to be a feeling among some GN people that in time, they would dominate the company, because the GN had a better property and a better record of earnings and operating performance than the NP or Burlington. Of course, that did not happen. If anything, the people of the NP held more of the commanding positions in BN by the mid-1970s.

Of the seven top positions on BN in the mid- to late 1970s, five were held by officers of the former NP. Only one was filled by a GN person, and one by a person with roots in the Burlington. On the five operating regions, three were headed by officers of the former NP. It could be argued that by then, the time had come to set aside the early effort to have a "balance of power" and, instead, to place the best and most capable persons on job vacancies, at any level or within any department in the company regardless of heritage. After all, by that time we had been Burlington Northern for about five years. We were not three separate companies, each having officers vying for power in the new company.

However, it should be noted that two of BN's "rising stars" with roots on the GN left BN in the early 1970s: W. L. Smith became a president of the Milwaukee, and P. F. Cruikshank went to the United States Railroad Administration, and then a short time later became vice president–Operations on the Milwaukee. Both Smith and Cruikshank would likely have been contenders for positions filled by former NP people shortly after they left BN. Whether the decisions made on management succession were a deliberate effort on the part of the board and the chairman and CEO of BN to favor NP people in preference to those of the GN or Burlington could be the subject of endless debate.

When the application for merger was turned down by the ICC in 1966, Budd decided the GN needed to prepare to "go it alone," if need be. An outcome of Budd's thinking was a complete overhaul of the Traffic Department into a new marketing organization in 1967. The resources for market research, service innovation, and equipment planning, plus developing better understanding of customer needs, were integrated into a new market-planning group. It was staffed in large part by new market managers hired from industries that used rail, truck, and barge service. The new GN structure for marketing was retained and expanded in the organization set up for BN. Another part of the program of reorienting the GN for the future was to adopt the "big sky blue" paint scheme for cars, locomotives, and corporate publications. The GN's famous "goat" logo was retained, but symbolically made "trimmer and more agile" in keeping with the new image.[2]

The GN produced the best operating and financial results of the three companies in the fifteen years of planning and review by the ICC and courts leading up to the merger. However, by 1967 the GN's historically low operating ratio exceeded the threshold level of 80. By the time of the merger, there was no question that each of the three companies needed to merge, build on the synergy that would result, and structure operations, mission, and priorities for the new company in ways that would make it more competitive and able to overcome the problem of having light density on much of the system.

---

2   Great Northern Railway Annual Report for 1967, page 2.

CHAPTER 4

THE NORTHERN PACIFIC

MAIN STREET OF THE NORTHWEST

A SLOGAN USED BY THE NORTHERN PACIFIC

ON MANY OF ITS FREIGHT CARS AND LOCOMOTIVES

THE NP MAINTAINED HIGH-QUALITY LINES of railway, to standards of the same level as those used on GN. Even with the disadvantages that it had in grades and distance compared to GN, NP did what was necessary to overcome these disadvantages and provide service that was fully competitive. However, these disadvantages were reflected to some extent in the higher operating ratio and generally lower level of productivity NP had compared to GN.

NP did not have the advantage of the enormously large and very profitable segment of business GN had through its service to iron mines on the Mesabi Range in Minnesota. NP served only the Cuyuna Range, which was a much smaller operation that had wound down its operations by the early 1950s. However, unlike its merger partners, NP owned millions of acres of land, much of it forested. It also held the mineral rights over extended areas of North Dakota, Montana, and Washington. In 1968 NP acquired Plum Creek Lumber Company, a forest-products-manufacturing company. These non-rail assets generated substantial income for NP and added great value to its balance sheet. Development of the resources became an item of contention for the boards of NP and subsequently BN in setting policies and priorities.

Since GN did not receive land grants to help finance its construction, it did not have the benefit of owning the kind of timber and mineral resources that NP brought to the merged company. The earnings from NP's non-rail properties were an item of contention in determining the rate for the exchange of NP and GN stock in the merged company. Not only did NP benefit from current earnings from sales and royalties from its resources, but those properties were anticipated to have a much higher value in years to come.

NP served many large lumber and paper mills, coal mines in southeastern Montana, and highly productive agricultural areas in many parts of its system. A number of major line changes and other roadway-improvement projects made over the years greatly improved the efficiency of NP's main line.

Fortunately for the merged company, NP had retained double track on its main line between Minneapolis and Casselton, North Dakota. That amount of capacity was needed for the combined traffic to be moved over the east end of the preferred route between Minneapolis and Seattle. NP also had a high-capacity double-track line of 176 miles between Seattle and Vancouver, Washington. This line handled trains of GN and UP in addition to those of NP. NP's operating ratio was always a few points higher than GN's, due in large part to its greater distance between the Twin Cities and the West Coast, and having to cross five mountain ranges compared to only two on GN.[1] The advantages of GN's route became

---

1   NP's main lines crossed Mullan Pass and Homestake Pass over the Continental Divide; Bozeman Pass; the Mission

apparent in the designation of the preferred route for the merged company between the Twin Cities and the Pacific Northwest.

Under the plans and direction set by James J. Hill when he controlled both NP and GN, the two northern lines operated under several joint-facility arrangements. The most prominent jointly owned and operated entities or facilities were ownership of Burlington and SP&S, joint use of the passenger stations in Seattle and Minneapolis, the "east-side line" between Minneapolis and St. Cloud, the line between Seattle and Vancouver, Washington, and the large classification yard at Laurel, Montana.

Some resentment over James J. Hill's dominance of NP

Range (Corican Defete) at Evaro, Montana; and Stampede Pass on the Cascade Range. GN by comparison had two mountain crossings, Marias Pass over the Continental Divide and Stevens Pass over the Cascade Range. (See chapter 8 for details on the percent grade on each line.)

in the early 1900s continued for a long time, even into the early days of the 1970 merger. Even though Hill's attempt to merge the northern lines was unsuccessful, he still had considerable influence on the NP. He established operating rights for GN on NP's line between Seattle and Vancouver, Washington, as well as joint ownership or use in large facilities such as the passenger stations in Minneapolis and Seattle, construction of the SP&S Railway, and the expanded yard at Laurel, Montana. Hill brought Howard Elliott from the Burlington to serve as president of NP in 1903. Under Elliott's leadership, NP enjoyed some of its best years. Many line improvements were made, among them line changes to reduce grade and curvature and construction of several miles of additional second main track.

In the late 1960s, NP benefited from the development of low-sulfur coal in mines in southeastern Montana, which was moved to electric power plants in Minnesota and

North Dakota. This was the beginning of the movement of large tonnages of coal over NP's main line between Huntley, Montana, the Twin Cities, and Duluth–Superior. In the operating plan for the merged company, the entire NP line of 1,156 miles between Casselton, North Dakota, and Sandpoint, Idaho, was to become a secondary line. However, the market for low-sulfur coal that developed in the 1970s made a large part of that line a major high-tonnage coal corridor.

Under the consolidated operation, only 339 miles of NP's transcontinental line was designated as part of the 1,763 mile "preferred route" (261 miles, Minneapolis–Casselton and 68 miles, Sandpoint–Spokane). For traffic moving between Kansas City and the Pacific Northwest, NP's line between Spokane and Huntley (626 miles) was designated the preferred route rather than GN's line via Whitefish, Shelby, and Great Falls.

Shortly before the 1970 merger, NP completed a major line change at Granite Lake, Idaho, on the Sandpoint–Spokane segment of the preferred route. Reducing curvature in this area made it possible to raise train speeds from 25 miles per hour to 60. The length of the line was reduced by 1.82 miles. A major line change of 9.3 miles was made at New Salem, North Dakota, in the 1950s to reduce grade and curvature. NP constructed a large hump classification yard—where cars are shunted onto one of several tracks, with an artificial hill in order to use gravity for moving the cars onto their track—at Pasco, Washington, in the 1960s. The new hump yard at Northtown (Minneapolis) was built on property owned by NP. Completion of the Northtown Yard made it possible to consolidate the operations in the Twin Cities and achieve a major part of the savings projected in the plan for the merger.

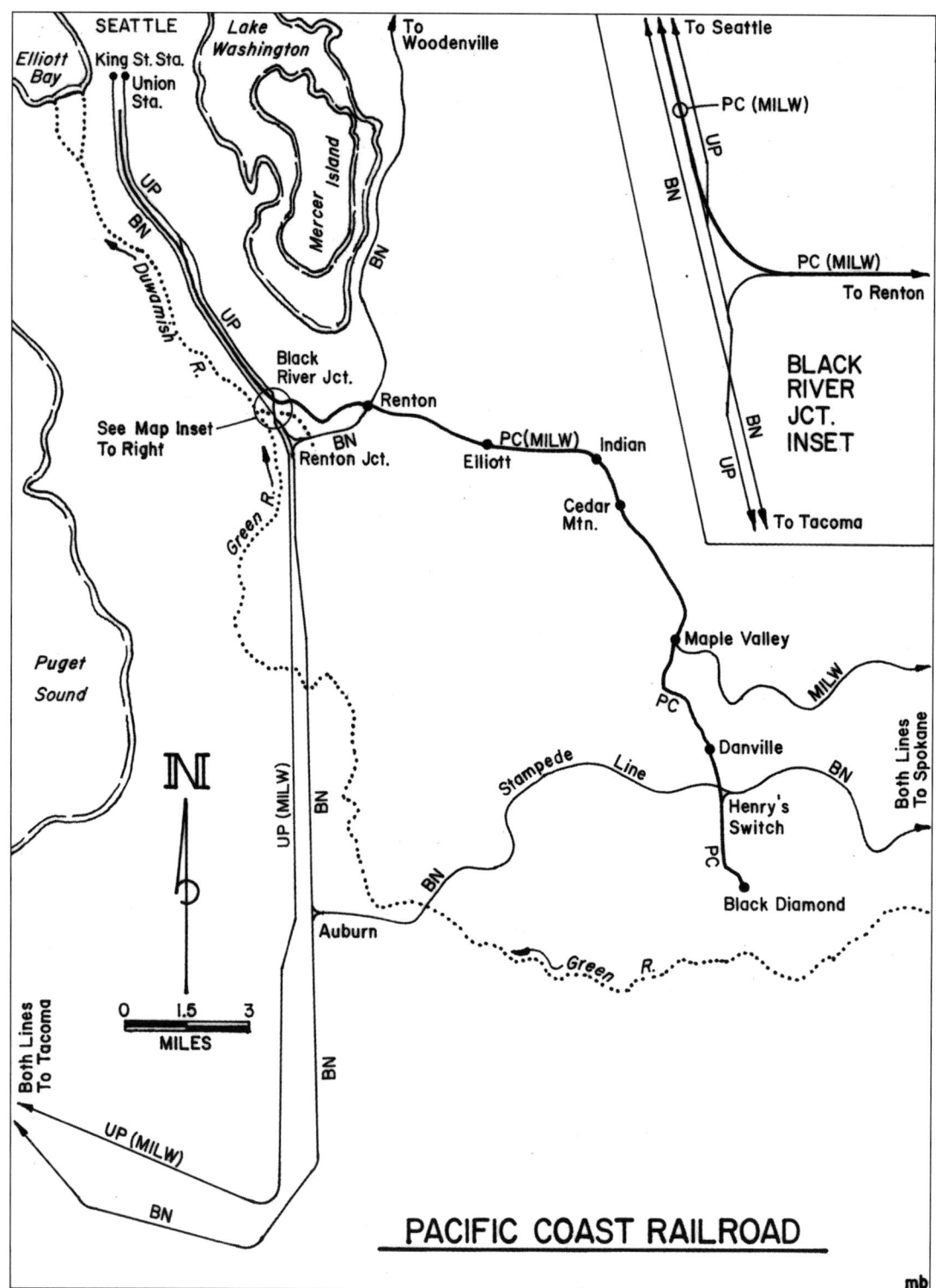

Figure 1.1.

CHAPTER 5

THE PACIFIC COAST R.R. CO.

⸰ a small railroad of thirty-two miles acquired by GN in 1951
nineteen miles used by Milwaukee to reach Seattle ⸰

THE MERGER OF THE THREE LARGE RAIL-road companies that formed BN also included the Pacific Coast Railroad (PC), a small company with thirty-two miles of road in western Washington, which was wholly owned by GN (refer to figure 4.1). PC was acquired by GN in 1951 for $1.7 million. It was attracted to PC because of the industrial property PC owned at Renton and the terminal trackage it had built in Seattle.

Milwaukee Road gained access to Seattle through trackage rights on nineteen miles on PC between Maple Valley and Argo, located 3.4 miles from the UP station in Seattle where the Milwaukee's passenger trains originated and terminated. PC maintained a double-track railroad with mainly 85-pound rail with an automatic block signal system where Milwaukee had trackage rights. Virtually all of the tonnage moved over PC was generated by the trains Milwaukee operated under trackage rights. That placed nearly all of the burden for operating and maintenance expense on Milwaukee. Since neither GN nor BN moved much tonnage on PC, there was no

incentive for them to upgrade PC's track with heavier rail or to maintain it at a standard other than for operation safely at a low speed. It is interesting that Milwaukee was dependent on GN, its arch rival, to reach the vital end point of its Pacific Coast extension.

It is even more surprising that Milwaukee did not even attempt to gain ownership and control of that important link when it had the opportunity, before the GN was given the chance to acquire PC. Apparently Milwaukee did not suspect GN or possibly NP would make a move to acquire control of PC, or perhaps it was because of the financial pressures it was under at that time to upgrade and modernize its railroad after coming out of bankruptcy. Also, Milwaukee still carried a heavy debt load for construction of its West Coast extension. Not long after Milwaukee it shut down in 1980, BN removed the portion of PC's double-track line where it closely paralleled the former NP main line for the nine miles between Black River and Seattle. The remainder of PC was either abandoned or reduced to single track for access to local industries.

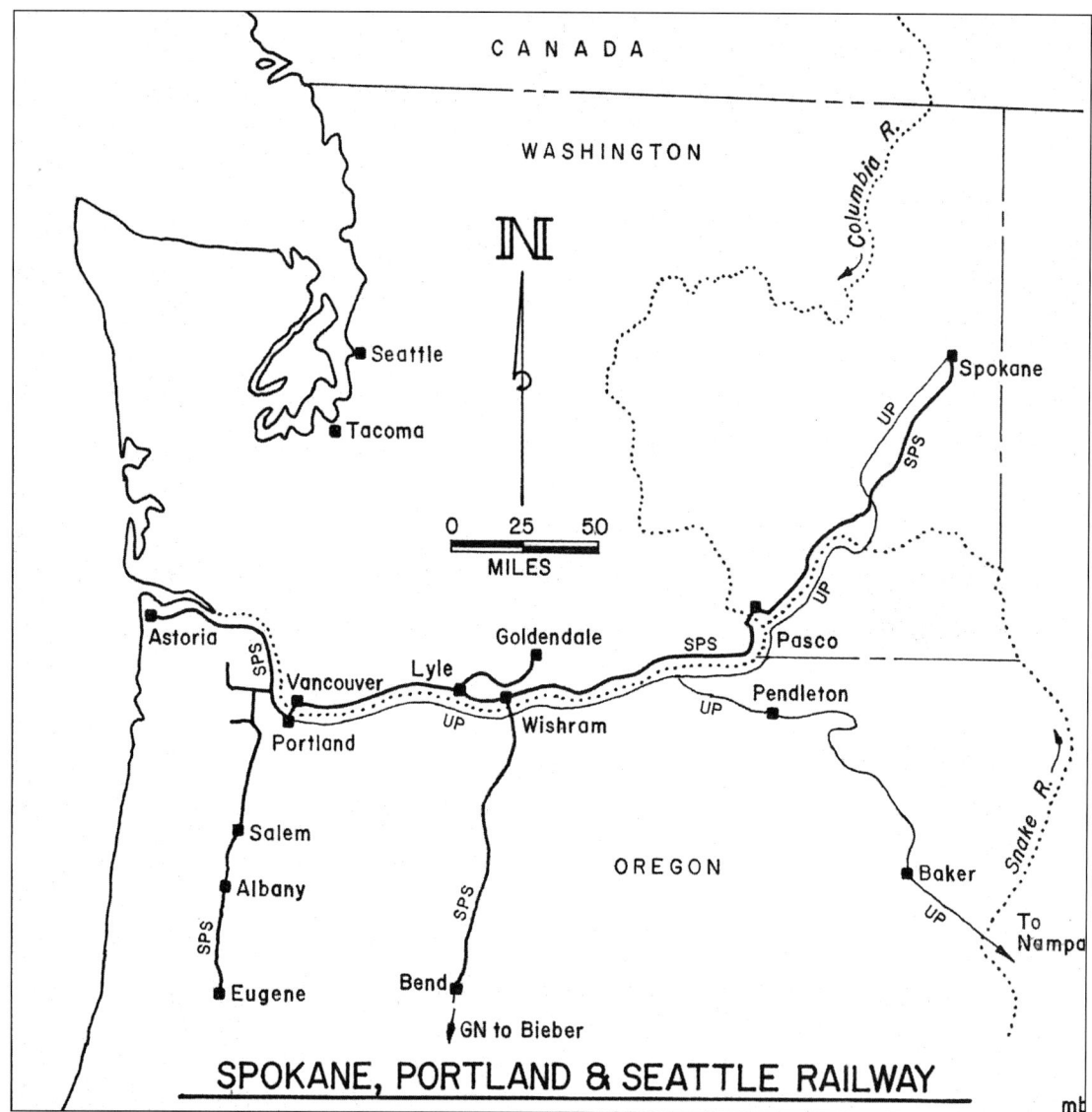

CANADA

WASHINGTON

Seattle

Tacoma

N

0   25   50
MILES

Columbia R.

Spokane

UP
SPS

UP

Pasco

Astoria

SPS

Goldendale

Lyle

Vancouver

SPS

Portland

UP

Wishram

SPS

Pendleton

UP

Snake R.

Salem

Albany

SPS

OREGON

Baker

UP

To Nampa

Eugene

SPS

Bend

GN to Bieber

SPOKANE, PORTLAND & SEATTLE RAILWAY

mb

Figure 6.1.

CHAPTER 6

THE SPOKANE, PORTLAND
AND SEATTLE RAILWAY

"THE NORTHWEST'S OWN RAILWAY"
A SLOGAN DISPLAYED ON MANY SP&S
LOCOMOTIVES, FREIGHT CARS, AND CABOOSES

THE RAIL INDUSTRY IS REPLETE WITH subsidiaries that are jointly owned by two or more railway companies. Many of them are terminal companies that were created to avoid the need for each railway to build its own yard. With a terminal company providing service to each owner (and in some cases, two non-owners as well), the expense of building and operating many smaller duplicate yards and routes was avoided. Two of the subsidiaries of GN and NP came under their ownership in a much different and perhaps unique way. Ninety-seven percent of the stock of Burlington was acquired by GN and NP at the time James J. Hill controlled both companies. Burlington had been a strong, well-managed company and needed no intervention by its owners in managing its day-to-day activities. Also, Burlington was a large company, with more mileage and a greater volume of business than either the GN or NP handled.

The situation with the Spokane, Portland and Seattle Railway was quite different. It was created and built under the direction of James J. Hill, who decided it would be wholly owned in equal shares by the GN and NP. The SP&S struggled in its first forty years to generate enough revenue to even make the interest payments on its debt.

The line SP&S built between Spokane and Vancouver, Washington, marked the fulfillment of a major part of Hill's vision and his ambitious plan for building a low-grade, efficient line across the northern tier of states. In announcing his plan to a group of business leaders in Portland in 1905, Hill declared, "We will build a road of such character of construction as has not before been built west of the Rocky Mountains, of low gradients and slight curvature, and it will be completely ready for trains to enter Portland by a year from next Christmas. It is finishing the work that we started to do and that we must carry to completion."[1] Thus the foundation and standards to which SP&S would be built were clearly laid out.

The GN line over the Cascades on Stevens Pass had been completed, but even after a tunnel was built at the summit to eliminate switchbacks and grades steeper than 2.2 percent, the line still was plagued with problems. Later, Hill decided to proceed with construction of a much better engineered route that would reach the Pacific, on the north bank of the Columbia River. Even though NP owned and operated a main line between Spokane and Pasco, Hill decided to build a generally parallel line for SP&S that would have much lower grades and less curvature. There were several locations with grades of one percent and sharp curvature on NP's line that Hill considered excessive for any main line built outside of mountain territory. With some heavy grading and construction of five long, high steel-trestle bridges over coulees and canyons, Hill accomplished his objective of having a line that required no helper locomotives,

---

1   Hill speaking at the Lewis and Clark Exposition, Portland, Oregon, October 3, 1905, printed in the *Oregonian*, October 4, 1905.

no speed restrictions due to curvature, shorter running time, less motive power, lower fuel consumption and lower track-maintenance costs.

The SP&S has been evaluated by many as the best engineered line of railway in North America. It has no grades in excess of 0.4 percent (ascending eastward) in the first 137 miles west of Spokane[2] and 0.2 percent (also ascending eastward) in the next 221 miles to Vancouver. Curvature is minimal, with a maximum of 3 degrees. With such low grades, heavier trains could be run, and they required less motive power than either GN or NP lines across Washington, both of which had mountain grades of 2.2 percent. The track was laid ten feet above the previous high-water mark of the Columbia River that was established in 1894.

Rather than building a new river grade line to Portland solely for GN, Hill believed that because both GN and NP would benefit from the capability of the new line, it should be financed and owned in equal shares by both companies. Construction of SP&S cost over $45 million, far more than the $5 million he had projected. Hill was roundly criticized by his bankers for this huge overrun. He blamed himself for not giving enough attention to the details during construction, calling it "the most unsatisfactory thing that has occurred in my experience."[3] Even with this shaky start, BN was indeed fortunate that Hill had had the instinct, courage, and knowledge to establish the confidence and determination needed to build SP&S. With the 1970 merger, SP&S evolved into a vital part of the BN system. On today's BNSF Railway, the main line of the former SP&S between Pasco and Vancouver often operates at or very near its capacity.

Under the plan for the 1970 merger, SP&S was leased by BN. In 1981 it was formally merged into BN. The delay was due to SP&S holding $54.7 million of debt that would mature in 1971. The maturity date had already been extended by ten years, and GN and NP received approval from the ICC to extend it for another ten. The owners had hoped to draw the amount of debt down to $40 million and then sell the bonds to the public by their

maturity date in 1971. Accordingly, a lease of SP&S was set up for ten years. With the bonds in the hands of the public, SP&S would then be merged into BN. In 1981, the bonds were cancelled and SP&S was merged with BN.

In addition to its main line between Spokane and Portland/Vancouver, SP&S also owned and operated a major link in the "Inside Gateway," a service route to and from northern California operated jointly by GN and SP&S. The 150-mile segment of that corridor, between Wishram and Bend, was built by Oregon Trunk Railway, a wholly owned SP&S subsidiary. Early in its history, SP&S had acquired rail lines in western Oregon, including the Oregon Electric, which served Salem and Eugene, and the Astoria and Columbia River Railroad, which owned the line between Portland and Astoria.

A new item of contention on the building of the SP&S arose in later years when GN wanted to extend the line of the Oregon Trunk (an SP&S subsidiary) from Bend into California. When NP decided not to invest the 50 percent that would represent its ownership, GN proceeded to build the line on its own. There are some who say SP&S became more of a GN-only subsidiary as a result.

Through the years up to the creation of BN, the level of business on the SP&S remained low relative to its capacity and the standard to which it was built and maintained. For the Inside Gateway, GN and Western Pacific worked hard to develop business moving to and from California. The service offered by this "chain" of railroads (GN, SP&S, WP, and Santa Fe) was competitive with the schedules and delivery times offered by Southern Pacific. However, because SP had the advantage of being the only railroad with direct access to the facilities and plants of most of the rail-served customers in California, it was difficult to pull business away from it.

In the decade leading up to the 1970 merger, SP&S maintained a respectable operating ratio in the low to mid-70s through 1967. In 1968 and 1969, the ratio came very close to 80 but was still lower than either of its parents. The numbers clearly showed the need for the northern lines, plus SP&S and Burlington, to improve their performance. The proposed merger held the fastest and strongest opportunity for them to improve their earnings.

Soon after the merger occurred, the amount of business moving on SP&S's main line between Vancouver and Spokane increased considerably. The operating plan

---

2   For westbound trains, there was an ascending grade of 0.8 percent for 5.2 miles between the crossing of Latah Creek and Marshall, just west of Spokane.

3   Letter from Hill to George Baker, president of the First National Bank of New York, June 14, 1910.

provided that tonnage would move on the designated preferred routes, rather than on routes owned by GN and NP, which would give them the longest possible haul and thereby maximize their revenue. Moving more business on the low-grade, economical route of SP&S instead of moving it to and from Vancouver over the Cascade Mountains reduced the cost of moving it and increased the margin of profitability. Also at this time, international business moving to and from the Port of Portland began to increase: mainly export grain, imported automobiles, and consumer goods moving in containers. By the 1990s the main line of the former SP&S was often running close to its capacity of thirty to thirty-five trains per day. The river grade is so favorable for the operation of loaded unit trains of grain, coal, and oil that it is more economical to run them "the long way around" via Vancouver than to use the shorter routes over the Cascades to reach the grain elevators and Kalama, Tacoma, and Seattle, and the coal terminal at Roberts Bank, British Columbia, on the grades 2.2 percent. Thus, James J. Hill's vision and his understanding of the economics of railroading on well engineered lines has been validated. The potential he saw from construction of the SP&S has been realized.[4]

It is interesting to note the large number of SP&S employees who have been elected over the years to system-level positions in the two unions that represented operating employees on all or a large part of the BN. Some of these officers had a major role in developing the implementing agreements for the 1970 merger that made it possible for BN to begin to integrate its operations right from the day the merger became effective. Jack Hurley, as the general chairman for the yard crews, and Jack Carstens, for road crews on the SP&S, were members of the team of union officers and officials from the Labor Relations and Operating Departments who formulated those agreements, together with ones that provided for job protection and wage guarantees. Hurley also was instrumental in setting up what became known as the Hanlon Board for making decisions on how the implementing agreements should be applied to specific situations that arose during the early years of implementation. Decisions made by the Hanlon Board were applicable on all parties over the entire system.

Having this board served to prevent the build up of a large number of claims through quick resolutions of the issues involved. This turned out to be a very effective way of handling merger-related issues for all concerned. In its first ten years, the Hanlon Board issued over five

---

4   Operating officers on the SP&S had the advantage of exceptionally favorable grades for the full length of the main line on their railway. It was laid out with curvature that required no permanent speed restrictions. Safety in the operation of trains was greatly enhanced by the installation of ABS (an automatic block-signal system) on all of the main line, plus the 152-mile line of the Oregon Trunk line into central Oregon. CTC was extended in increments eastward from Vancouver in the 1960s.

The track was well maintained, to a standard of excellence for safety and for consistency in running trains on schedule. The number of miles of main track with continuous welded rail was increased in rail-replacement programs carried out every year. Preventive maintenance of track, bridges, and signals was diligently practiced, to avoid the need to impose slow orders (temporary speed restrictions) except in cases of "work in progress" on scheduled or planned maintenance or an upgrading project.

Together, these factors helped make it possible to achieve a low operating ratio. A high level of operating efficiency could be achieved in such performance indicators as the gross ton-miles per train hour, cost per train-mile, and average tons per train.

SP&S operating officers were fortunate that over the years, the top executives of its owners, the NP and GN, had experience and successful careers in maintenance and operations. As a result, the owners favored strong programs for maintenance of the roadway and equipment. They supported and advocated the application of high grade materials and maintenance equipment on the SP&S, and in applying new technology and maintenance practices. The owning roads allowed strong maintenance programs to be continued even through times of a downturn in business, rather than deferring maintenance work until normal business levels resumed.

As a result of these and other highly favorable factors, the Operating Department of the SP&S was in position to run a safe and very efficient operation that would meet the service requirements of both on-line shippers and those whose business originated on a connecting railway.

THE SIGNIFICANCE OF THE OPERATING RATIO

NEXT TO MEASURES OF SAFETY, THE OPERATING ratio is considered the best overall indicator of a railway company's performance. Executives, managers, and employees in all departments and in all specialties have both a stake and an influence on the level that is achieved. All of the company's operating expenses fall into one of the categories that make up the ratio—maintenance-of-way and structures, maintenance of equipment, transportation, and general and administrative—regardless of which department incurs expenses defined under standard railway accounting as operating expense (as opposed to capital expenditures).

Operating expenses make up the denominator of the ratio. The numerator is based on the gross revenue taken into account. The Operating Department is responsible for a very high percentage of the expenses, with the Sales and Marketing Department largely accountable for the amount of revenue taken in. The "quality" of revenue has a great bearing on the level of operating ratio that is achieved.

It is vital that the Sales and Marketing team succeeds in capturing business with a good margin of profit. Their success depends on the company having suitable equipment available for loading, plus reliable and competitive transit times that meet the needs of shippers. Sales and Marketing must be skilled in negotiating and adjusting rates according to changes in the marketplace. Some of the rail industry's success in reducing its operating ratio in the last thirty years has been due to the benefit of deregulation in pricing. Railways were finally given the freedom to price competitively without the burden and restrictions of government oversight.

On the cost side of the ratio, railways have benefited from new labor agreements to allow reduction of the size of train and yard switching crews. The improvement in earnings has made it possible to develop and apply new technology to further reduce costs and improve service in many ways.

At the time of the merger that formed Burlington Northern, a ratio of 79 was considered a threshold level for a rating of "good" in operating performance. In the past ten years or so, all of the large railway systems in North America have reduced their operating ratios down to the low to mid-60s. There no longer are any large railway companies performing at a level of operating ratio that would be judged mediocre or unsatisfactory.

Referring to table 6.1, the Burlington (CB&Q) produced a ratio above 80 in all but one of the five years before the merger took place. In 1965 there was an

hundred awards, and its structure became a pattern for use on other railroads involved in mergers.[5]

Another leader with roots on the SP&S was Bill Dunegan, who served as general chairman for the Locomotive Engineers. In those years, he provided leadership on such major issues as drug and alcohol testing, setting up interdivisional runs, merger with the Frisco, Engineer certification, and Locomotive Engineer training, as well as in making the changes dictated in the revised Hours of Service Law.

5   This tribunal was officially designated the Appendix E Board, named after a section in the agreement that set it up. Paul Hanlon was the first of the three neutral members to serve during the board's tenure of just over twenty years. It was formed in 1967 to examine the various protective agreements at the level of detail needed to memorialize the negotiating parties' intent while memories were fresh. The board had exclusive jurisdiction to hear any disputes on the interpretation or implementation of the protective agreements, or to provide the arbitration needed to reach any

new implementing agreements needed as a result. (Source: Wendell Bell, retired attorney in the BNSF and BN Labor Relations departments.)

Table 1.1. OPERATING RATIOS, 1960–1968*

Year	CB&Q	GN	NP	SP&S**
1960	81.2	78.9	84.5	77.8
1961	79.4	79.9	86.6	80.1
1962	78.9	78.6	86.0	78.9
1963	79.7	76.8	84.9	75.2
1964	81.3	77.9	85.3	76.9
1965	82.0	75.6	82.7	73.3
1966	78.6	73.8	80.3	71.8
1967	81.4	81.8	86.9	74.8
1968	80.1	81.6	84.5	79.1

* The results for 1969 were shown on a consolidated basis as though the merger had already occurred.
** Includes SP&S, OT, and OE railways.

intercity passenger service and the commuter service it provided in the Chicago area. It also had a large network of marginal branch lines.

For many years, the Great Northern achieved a favorable ratio of 79 or less. The GN benefited from the low grades on which it was built. However, economic pressures, truck competition, and not having great success or opportunities to grow its business caused the ratio to move above 80 in the last two years before merger.

The Northern Pacific had the disadvantage of heavier mountain grades than the GN and longer hauls between major terminals. Density was quite low on even some of its main line segments. As a result, the NP ran well above the threshold level of 79 for an operating ratio.

Of the four railway companies to be merged, the SP&S produced the best operating ratios, year after year. Having been located on a water-level grade on its entire main line gave it an advantage over all of the other western railways. By the late 1960s, even that favorable characteristic barely kept its ratio below 80.

All of these factors made it vital that the four railways merge, thereby concentrating traffic, building density on the most favorable routes, and eliminating the cost of operating duplicate support facilities.

unusually heavy volume of export grain, and major cuts in expenses had been made in L. W. Menk's first year as president. Heavy truck competition and having several railways competing for business on its major corridors forced the Burlington to hold its rates down. The Burlington also endured losses in its

Jim Hudson, as local chairman for the road crews, served for two full years on a task force set up by the Federal Railroad Administration (FRA) to evaluate ways for the railway companies to improve operations in the Portland terminal. Jack Carstens, general chairman for SP&S road crews, was able to come up with an agreement for a reduced crew consist that was implemented well before the national agreement was finalized. Jack was ahead of his time. Merle Geiger advanced through the Brotherhood of Locomotive Engineers to become an international vice president. He began working for

the SP&S in 1966 as a fourth-generation railroader. John Fitzgerald hired out as a brakeman with BN in 1970 at Vancouver. In the early 1990s, he became assistant general chairman representing the Conductors on the former GN and SP&S lines. Later, John replaced Mel Winter as general chairman. In those years he was involved in bringing the highly controversial crew consist agreement to a successful conclusion.

After Sherman Holliday served as general chairman and vice president of the Order of Railway Conductors and Brakemen (ORC&B) (later merged into the United

Transportation Union), the SP&S hired him for its Labor Relations Department. Under BN, "Sherm" advanced to one of the three assistant vice presidents reporting to T. C. (Tom) de Butts, vice president for Labor Relations for the entire BN system. Sherm's main responsibility was to oversee implementation of the agreements made to facilitate full integration and the consolidated operating plan. In the mid-1970s, he also negotiated the agreements needed for operation of the rapidly expanding coal business on BN's lines in Wyoming and Nebraska. Early in his career as an international officer in the ORC&B, Sherm had the honor of serving on the Presidential Railroad Commission under President John F. Kennedy, which was established to examine controversial issues involving the manning of crews, including the employment of Firemen on diesel locomotives.

All in all, a great deal of leadership talent and skills came from the ranks of SP&S employees. Together, they established a real legacy for that part of BN.

CHAPTER 7

LEADERSHIP IN THE 1970S

THE PRESIDENTS

IN ITS FIRST TEN YEARS, BN AND ITS TWO divisions, Transportation and Resources, were served by seven leaders with the title of president, CEO, chairman, or vice chairman. This section contains a brief background check for each of these leaders, together with some description and evaluation of particular initiatives undertaken in the years they held their senior leadership positions.

• J. M. BUDD •

John Budd gave the railroad a balanced, unpretentious style of leadership. . . . Those who knew him considered John Budd extremely ethical, thoughtful, sound of judgment, loyal, reserve in manner, fair-minded. . . . Not a publicity seeker, Budd avoided drama simply for effect, but his positions reflected careful analysis.

RALPH W. HIDY, MURIEL E. HIDY, AND ROY V. SCOTT
WITH DON L. HOFSOMMER, *GREAT NORTHERN RAILWAY: A HISTORY*, P. 249

PERHAPS MORE THAN ANYONE, JOHN BUDD WAS the principal architect of the 1970 merger and deserves credit for finally achieving a merger of the "Hill family" of railroad companies. Together with Robert Macfarlane of NP, John Budd fulfilled James J. Hill's dream to consolidate the ownership and management of GN and NP and the two railroads they controlled, Burlington and SP&S. Accomplishing this merger had also been a goal of Budd's father, Ralph Budd, in the 1920s. Because the conditions prescribed by the ICC at that time were so onerous (including divestiture of Burlington), GN and NP dropped the proposal. By 1955 John Budd and Macfarlane agreed it was time for them to raise the torch of their forebears and, this time, get it done. After fifteen years of planning, presenting their case to the ICC, fighting their way through the courts, and establishing agreements with the unions, along with Milwaukee and C&NW agreeing to withdraw their opposition, the merger was finally approved.

By any reasonable measure and means of evaluation, the merger that created BN has been a resounding success. In fact, the foundation on which BN was created was so strong that even the restructuring, highly disruptive changes in culture and values, and lack of faith or enthusiasm for the future of railroad transportation would be inflicted on BN in the 1980s did not "kill" the company.

Following his graduation from Yale, Budd took a junior position in GN's Engineering Department. He soon advanced to division superintendent and assistant general manager. In 1947, Budd took the opportunity to become president of the C&EI (Chicago and Eastern Illinois) Railroad, a fairly small property with lines between Chicago, St. Louis, and Evansville, Indiana. Two years later, Budd returned to the GN as vice

MANAGEMENT SUCCESSION IN THE 1970S

IN THE FIRST EIGHT YEARS AFTER THE 1970 MERGER, THE LARGE NUMBER OF CHANGES IN titles of senior executives may be a little difficult to follow for those who were not with BN as that time. Hopefully, the outline below will be helpful in following the succession that occurred.

JOHN M. BUDD

1970 Chairman and Chief Executive Officer

1971 Chairman of the Finance Committee, Board of Directors

1976 Retired from Board of Directors

LOUIS W. MENK

1970 President and Chief Operating Officer

1971 Chairman and Chief Executive Officer

1978 Chairman

1981 Retired

ROBERT W. DOWNING

1970 Executive Vice President

1971 President and Chief Operating Officer

1973 Vice Chairman and Chief Operating Officer

1976 Retired

NORMAN M. LORENTZSEN

1970 Vice President–Operations

1971 Executive Vice President

1973 President, Transportation Division

1976 President, Burlington Northern, Inc.

1978 President and Chief Executive Officer

1980 Chairman of the Executive Committee, Board of Directors

1981 Retired

THOMAS J. LAMPHIER

1970 Vice President–Planning

1971 Vice President–Management Services and Planning

1972 Vice President–Executive Department

1973 Executive Vice President

1976 President, Transportation Division

1980 Senior Vice President–Transportation Policy and Analysis

1981 Retired

C. ROBERT BINGER

1970 Vice President–Resources Development

1973 President, Resources Division

1980 Retired

president–Operations, and after another two years he was named GN's president.

As president of the GN for nineteen years, Budd had time to build the kind of culture and organizational character he wanted the GN to have. By most (but not all) measures, GN was the strongest of the four railroads that were combined to form BN. A look at the operating ratios, safety record, and improvements made to the property over the years would bear this out.

There were many things that made the GN a "class railroad," as ascribed by a number of writers. The GN kept its property clean and orderly. Shop buildings and the small buildings used by track and signal maintenance crews along the railroad always were kept painted and in good condition. Shrubs and flowers were planted adjacent to station buildings. In many small towns in remote areas along the GN, the depot and maintenance buildings often were among the best-looking structures in town. Employees could take pride in their company. The GN was a credit to the communities it served, not a disgrace or a junk yard as sometimes was the case with the properties of other railroads. Fortunately for BN, these standards were applied across the merged company. Having clean, respectable property and equipment made a difference in the attitudes of employees, customers, and the public in their impression of the new company.

At the time of the BN merger in 1970, Budd assumed the position of chairman and CEO. L. W. Menk was appointed president. In 1972 Budd retired, but he remained on the board of BN until 1976. Budd and Menk were different kinds of people, although they were able to work well together in selecting the heads of departments at the time of merger and finalizing the operating plan. They communicated with consistency on the need for the merger in statements made to employees, customers, and the public on the policies and plans under which the new company would be operated. They stayed the course and implemented changes conservatively.

Budd was seen by employees as a leader who was consistent; he "walked the talk" and knew the business of railroading. He was approachable and a good listener. He demonstrated strong knowledge and interest in the work of the maintenance-of-way people who have to do much of the hard labor and dirty work of a railroad, an attribute not always seen in senior managers in the rail industry.

J. M. BUDD WORK HISTORY

GREAT NORTHERN RAILWAY CO.

1926	Assistant Chainman (Engineering Dept.)
1930–1931	Assistant to Electrical Engineer
1932–1933	Yale Transportation Program
1933–1934	Assistant Trainmaster
1934–1940	Trainmaster
1940–1942	Division Superintendent
1942–1945	Military Service
1945–1947	Assistant General Manager, Lines East

CHICAGO AND EASTERN ILLINOIS RAILROAD CO.

1947–1949	President

GREAT NORTHERN RAILWAY CO.

1949–1951	Vice President–Operations
1951–1970	President

BURLINGTON NORTHERN, INC.

1970–1971	Chairman and Chief Executive Officer
1971–1976	Member, Board of Directors (Chairman of the Finance Committee, 1971–1972)

A very noteworthy project initiated by Budd was the recognition in the early 1950s of the need for a program to provide counseling and treatment for alcoholism for employees. The GN set up professional services to help restore experienced, capable employees in need of treatment to overcome the problem of addiction and thereby maintain useful lives on the railroad. The GN set up this program many years before other railroad companies and other industries decided to commit resources for such a program. The GN's program set a standard and example for the rail industry to adopt in later years. Best of all, it transformed the work lives and personal lives of many employees who needed such assistance. Needless to say, this counseling program made the operation of the railroad safer.

Budd maintained and extended the vision of James J. Hill; Budd's father, Ralph Budd; and their successors to never stop improving its roadway. Every year, outside of the worst years in the Great Depression of the 1930s, GN

committed funds for line changes that would reduce curvature, grades, and the length of the line. Over the long run, these improvements produced a vastly improved property. Maintenance costs were reduced, and trains could operate with less delay from speed restrictions due to curvature. Budd took a strong interest in the selection of such projects and saw to it that funds were provided in the annual budget to carry them out.

By the mid to late 1960s, GN was struggling to maintain its historic low operating ratio. Outside of the western part of the state of Washington, business remained relatively flat for many years, with not much opportunity for growth in the kinds of freight most conducive to haul by rail. Certainly GN, and nearly all of the rest of the rail industry at that time, needed to reduce costs by eliminating duplicate, underutilized routes and facilities, and to achieve economies of scale by increasing density on a smaller network. The creation of BN made that possible.

· L. W. MENK ·

The transportation revolution (referring to the large increase in rail business in the western states) could spread and bring great benefits to the public and to all major railroads in the nation. Even our industry itself has not yet been able to calculate with accuracy the extent to which the nation's railroads would be transformed by a doubling of coal revenues over the next 10 years.

L. W. MENK TESTIMONY BEFORE THE
COMMITTEE ON INTERIOR AND INSULAR AFFAIRS,
U.S. HOUSE OF REPRESENTATIVES, WASHINGTON, D.C., NOVEMBER 7, 1975.
FROM MENK, *A RAILROAD MAN LOOKS AT AMERICA:
EXCERPTS FROM THE SPEECHES OF LOUIS W. MENK,* 65

LOUIS W. MENK WAS BROUGHT INTO THE "Hill family of railroads" in 1965 when he was hired to become president of Burlington. His predecessor, H. C. Murphy, was to become vice chairman of the merged company, but as delays mounted up in the long process of getting the merger evaluated and approved through the ICC and the courts, Murphy decided to retire ahead of the merger. Certain members of the board of Burlington (i.e., those who represented NP's ownership) apparently saw this as an opportunity to satisfy concerns that some members of their own board (on NP) that some new blood be brought

in. Some former executives who worked for the northern lines in that era allege that Norton Simon of the NP board had been told by Henry Crown, a major shareholder of the SLSF (Frisco) Railroad, that Menk was the type of young executive who would be a good candidate to head a large railroad and that he should be considered when such an opportunity came.

After serving as president of Burlington for only a year, Menk was brought to NP to replace Robert S. Macfarlane upon his retirement as president. Having served as president of one of the "owning" roads in the merger in addition to the Burlington put Menk in position to head the merged company as president, reporting to J. M. Budd, who would become its chairman and CEO. When Budd retired (only two years after the merger took place), Menk was named president and CEO of BN. He became chairman of the board in addition to those responsibilities in 1972. In 1982, shortly after Richard Bressler replaced Norman Lorentzsen as president and CEO of BNI, Menk would retire as chairman as well. He closed out his career in business as CEO of International Harvester, then a troubled company for which he had served as a board member.

Menk and Budd worked well together in implementing the merger that formed BN. Overall, their leadership is judged by most as a resounding success. Certainly in comparison with the merger that created Penn Central in 1968, the BN merger did not result in severe disruptions to service or chaos in operations, or produce a deficit. The plans for transition to the consolidated operation had been well thought out over a period of about fifteen years, and they were implemented in an orderly fashion. The "balance of power" among executives appointed as department heads and as regional vice presidents prevented any one of the three component roads from "taking over" the company. In the minds of most managers and employees, there was "one team," and it was the "BN team." It took strong leadership at the top to establish and maintain this mantra at the outset, as well as until personnel decisions could be made on the basis of which person was the best for the job, rather than on his or her pre-merger pedigree. Any attempt at corporate gamesmanship was thwarted during the company's first ten years, indeed a notable achievement.

Menk and Downing demonstrated great courage and foresight when they convinced the BN board to invest $2 billion in the expansion of capacity and general upgrading of hundreds of miles of track to handle the unprecedented tonnage of coal that was to be mined in Wyoming and Montana. This was only four years after the company had been formed. Major capital investments were still being made to complete consolidation of the operation as provided in the plans for the merger. Menk and Downing had to face skepticism and a general lack of faith from the financial community that an investment this large in a railroad company would pay off.

L. W. MENK WORK HISTORY

UNION PACIFIC RAILROAD COMPANY

1936–1940	Telegraph messenger

ST. LOUIS–SAN FRANCISCO (FRISCO) RAILROAD COMPANY

1940–1945	Telegrapher, Dispatcher, Chief Dispatcher
1945–1950	Terminal Trainmaster, Assistant Superintendent
1950–1953	Superintendent
1954–1956	Assistant General Manager
1956–1958	General Manager
1958–1962	Vice President–Operations
1962–1965	Chairman and President

CHICAGO, BURLINGTON AND QUINCY RAILROAD (CB&Q) COMPANY

1965–1966	President

NORTHERN PACIFIC RAILWAY COMPANY

1966–1970	President

BURLINGTON NORTHERN, INC.

1970–1971	President and Chief Operating Officer
1971–1978	Chairman and Chief Executive Officer
1978–1981	Chairman

INTERNATIONAL HARVESTER COMPANY

1982–1983	President and Chief Executive Officer

Together with most board members, Menk, Downing, and Lorentzsen stayed the course through eight years of heavy capital investment and increased maintenance expenses to bring the property up to the standard where it could efficiently handle the coal. The business of moving heavy tonnages of coal over a large part of the system would transform BN.

When Menk had Burlington buy a 10 percent interest in Frisco, his desire for a merger with his former railroad became obvious. At that time, Burlington was, of course, already included in the plans for merger with its owners, GN and NP. The decision was soon made to sell the stock in Frisco, but it seemed there was no question that at some time in the future, merger with Frisco would again be in the offing. In 1977 BN and Frisco would announce their plan to seek authority from the ICC to merge.

Menk seemed to want his long-term associate and president of Frisco, Richard Grayson, to head BN. Menk was under increasing pressure to improve the return on the capital that had been invested in BN from the time of the 1970 merger. He thought that under Grayson's leadership, improvement in the performance of the railroad could be accelerated. Menk had also been criticized for not doing more to exploit the value of BN's vast holdings in timberlands, oil, gas, coal, and other minerals. While there had been some year-to-year improvement in the financial results from BN's Resource Division, it was seen as much too slow and inadequate by some board members and the financial community. Although BN was well positioned for improved results in its second decade, Menk apparently did not see that level of expectation or potential as sufficient for his legacy. In 1980 Richard Bressler would be brought in from Atlantic Richfield as president and CEO of BNI, to be over both the Transportation and Resource divisions. Menk continued as chairman until 1982 and as a member of the board until 1986.

Menk told Jay Lorsch of the Harvard Business School:

It was after those talks that I came to realize Burlington Northern needed a generalist to run the place: someone who knew finance, who had some operating experience, and who knew how to develop the resource end of the business. I realized the company had a lot of good railroad men, but the company needed someone with a little something more. . . . he [Daniel Davisson, a member of BN's board and president of United States Trust Company of New York] explained to me why he felt Burlington Northern needed a man who could manage a diverse corporation in a changing industry.[1]

Lorsch quotes Pemberton Hutchinson, another BN board member, as suggesting that in hiring an executive from outside the company, Menk recognized he was not leaving a legacy of trained people to take over when he left. I do not believe that was the case, at least not for the position of CEO of the railroad company. In making the decision to merge with the Frisco Railroad and bringing Richard Grayson to head the railroad, Menk believed it would be in the hands of a very capable executive. Further, BN had developed a strong cadre of vice presidents in charge of the railroad's major functions, and a very capable group of younger executives at the next level to succeed the vice presidents in the next few years. The same could be said for the management team in the Resources Division, most of whom had been recruited under Menk's direction. In summary, BN was not lacking in management talent for the future.

---

1   Jay Lorsch with Elizabeth MacIver, *Pawns or Potentates: The Reality of America's Corporate Boards*, 1989.

· W. J. QUINN ·

*When I was asked to go to the Burlington, I thought I was taken care of
for years, for life! The Burlington was a very prestigious railroad . . . was very profitable.*

. . .

*I was approached by a committee of the Board of the Milwaukee, asking me if I would
consider coming back. The Milwaukee had been under merger negotiations
with the North Western . . . and it had reached a hitch.*

"AN INTERVIEW OF WILLIAM QUINN,"

RAILROAD EXECUTIVE ORAL HISTORY PROGRAM,

JOHN W. BARRIGER III NATIONAL RAILROAD LIBRARY,

MARCH 21 AND 25, 2002

· *recruited from the Milwaukee Road; returned to the Milwaukee three years later* ·

WILLIAM J. QUINN WAS HIRED TO REPLACE L. W. Menk as president of Burlington in 1966 when Lou Menk became president of NP. Quinn had served as president of Milwaukee Road since 1958. He was to become vice chairman upon merger of the northern lines and Burlington, which all who were concerned anticipated would be approved by the ICC within a short time. As was the case with Menk coming to Burlington less than a year before, it was not entirely clear why the board of Burlington decided to recruit a president from outside the company. Apparently, with retirement of several of Burlington's senior executives in the offing, the board felt none of its younger senior executives were ready for the presidency at that time.

In the oral history Quinn gave the Barriger Library, he stated, "It was surprising to me that I would be asked." He was visited by Lou Menk and John Budd at the time Menk was about to accept the presidency of NP. Quinn said that since the application of the northern lines and Burlington had been turned down by the ICC, he felt Burlington would remain independent for the long term. In addition, it was a very strong company that paid a dividend high enough "to keep the Northern Pacific going. And, they [NP] waited for the dividend day like you waited for your birthday. I thought this is the way I'm going to end my life in peace and happiness." Quinn thought the appeal to the ICC for reconsideration of the merger application was "a long shot because they had

been working on attempting to accomplish this for 75 to 100 years, and it didn't look like there was much chance. And, this time it worked. . . . And, things didn't turn out as I had thought they were going to."[2]

Quinn was generally liked and held in high regard by his peers in the industry. There were BN people who felt he was brought to Burlington as a means of removing some of the strength of Milwaukee's opposition to the northern lines merger. That did not turn out to be the case, as Milwaukee's opposition intensified, if anything, in the days and months leading to the merger, and by the fact that the ICC turned down the merger application on April 27, 1966.

When the merger was impending, Quinn bought a house in the Twin Cities and was ready to move and begin his new role as vice chairman of the board of BN. He was to have responsibility for the Law Department and the Government Affairs Department and also serve as the public spokesman for the company. However, Quinn had second thoughts about moving his family, and he thought the organization structure of BN might be top heavy. Also at that time, negotiations for a merger of Milwaukee and C&NW had just been terminated. Quinn related in his oral history that a committee of

---

2   "An Interview with William Quinn," Railroad Executive Oral History Program, John W. Barriger III National Railroad Library, 29–43.

W. J. QUINN
WORK HISTORY

1935–1937	Attorney in private practice
1937–1939	Assistant U.S. District Attorney for Minnesota
1939–1942	Attorney, Soo Line Railroad
1942–1945	Special Agent, Federal Bureau of Investigation

SOO LINE RAILROAD

1945–1953	Assistant Commerce Counsel, Commerce Counsel and General Counsel, Soo Line Railroad
1953–1954	Vice President and General Counsel

MILWAUKEE ROAD

1954–1958	Vice President–Law
1958–1966	President

BURLINGTON ROUTE (CB&Q)

1966–1970	President

BURLINGTON NORTHERN, INC.

1970	Vice Chairman

MILWAUKEE ROAD

1970–1978	Chairman and Chief Executive Officer

the board of Milwaukee asked him if he would consider coming back to Milwaukee. He was offered a contract to stay with Milwaukee until he reached age 70. Only two weeks after the effective date of the BN merger, Quinn resigned his position and resumed his affiliation with Milwaukee.

In his ensuing years with Milwaukee, Quinn tried without success to merge Milwaukee with BN or another merger partner. Before long, Milwaukee decided to shut down its West Coast extension, and a few months later it declared bankruptcy.[3]

---

3   Speculation continues to this day on what was behind the decision of the board of the Burlington (that is, the Chicago, Burlington and Quincy Railroad) to offer the job of president to Quinn. Based on his oral history and conversations with Robert Downing, it appears it was because Quinn was well liked by his peers on the GN and NP, and that the board felt he was well suited to serve as president in whatever time remained until the ICC would reverse its decision and approve the merger. Contrary to the thoughts of some rail industry observers, it appears Quinn was not hired to remove or settle the objections the Milwaukee had taken to the merger. Further, nothing has been found to indicate there was a plan to have Quinn return to the Milwaukee after the merger and then orchestrate a move to have it taken over by BN.

· R. W. DOWNING ·

*We went out of our way to ensure everybody in our companies understood that
this was not a takeover. It was a merger of equals, that everybody would be
treated equally, and that we respected the way we were doing things.*

R. W. DOWNING, QUOTED IN BRIAN SOLOMON,
BURLINGTON NORTHERN SANTA FE RAILWAY, 118–119

*The planners, Downing points out, say that the railroad industry
can expect an average of four percent increase in tonnage each year.
"Figure that out for 10 years," Downing notes, "and you can see that we
will be handling about half again as much traffic as we're handling today."[4]*

INSIDE STORY, SPRING/SUMMER 1971,
QUOTED IN *THE BN EXPEDITER*, JULY 1014, PP. 11 AND 13

*He [Bob Downing] may not have all the charisma of some of the better known presidents,
but he's the best technical railroad man in the industry and he's
equally good at understanding railroad financing.*

L. W. MENK, QUOTED IN "A MERGER THAT STARTED ON THE RIGHT TRACK,"
BUSINESS WEEK, JUNE 12, 1971

THE VICE CHAIRMAN AND CHIEF OPERATING officer in the years 1971–1976, Robert (Bob) Downing is one of the most revered and admired leaders of BN. Downing served as president of BN from 1972 through 1974 and as cice chairman of the board and chief operating officer from 1974 until his retirement in 1976. At the time of the 1970 merger, Downing held the position of executive vice president in the Great Northern Railway. He worked directly under John Budd for fifteen years and often referred to him as his mentor.

Budd first became familiar with Downing's capabilities during inspection trips they made together when Bob was working as trainmaster on the Minnesota Iron Range and Budd was serving as vice president–Operations. It was at the time of these visits that Downing climbed the ladder from trainmaster to division superintendent of the GN's Minot Division, a big step in his career. After serving two years as division superintendent, Budd brought Downing to the General Office in

St. Paul to work under him as assistant to the president. This promotion was made in 1956, shortly after Budd and Robert Macfarlane of the NP revived efforts to merge the GN, NP, and the Burlington.

Right from the outset of the planning process, Downing had an important role in preparing the operating plan for the merged company. The early success BN had in consolidating its operations was due to the quality and level of detail in the plan. Having a thorough and well-thought-out plan in place right from the start is considered by most observers as the most important factor in the early success of the merger. It was prepared with broad participation of officers from all levels of the roads that would make up BN. In addition to the plan for operations, Downing and the other officers who worked on the planning team prepared the plans for the organization structure and for consolidating the functions and activities of all departments.

A large part of the planning for the combined operation was the determination of which line segments of the GN and NP should comprise the preferred route for transcontinental service and which main line routes would be relegated to secondary status. The "preferred route" that evolved from this process was superior to the routes of either of the predecessor roads.

---

4   When Mr. Downing stated in 1971 that BN's business volume would double by 1980, it seemed hard to believe that would be possible, given the problems railroads were having from the diversion of business to other modes. However, even without the growth in coal tonnage, that did happen.

R. W. DOWNING
WORK HISTORY

PENNSYLVANIA RAILROAD COMPANY

1935–1938	Maintenance of Way Department

GREAT NORTHERN RAILWAY COMPANY

1938	Assistant to Superintendent, Whitefish, Montana
1938–1941	Acting District Roadmaster, Whitefish, Montana
[1941–1945	Service in U.S. Navy]
1945–1946	District Roadmaster, Great Falls, Montana
1946–1947	Trainmaster, Great Falls, Montana
1947–1949	Trainmaster, Glasgow, Montana
1950–1951	Trainmaster, Spokane, Washington
1951–1954	Trainmaster, Kelly Lake, Minnesota
1954–1956	Division Superintendent, Minot, North Dakota
1956–1958	Assistant to the President, St. Paul, Minnesota
1958–1967	Vice President, Executive Department
1967–1971	Executive Vice President

BURLINGTON NORTHERN, INC.

1970–1971	Executive Vice President
1971–1973	President and Chief Operating Officer

Together with Lou Menk, Downing convinced the board of directors in 1974 to commit $2 billion to expand the capacity of the railway to move the vast increase in coal tonnage that would be mined in Wyoming and Montana. During the next six years, while that very large capital investment was being made, heavy operating expenses were incurred as well, which of course reduced net income. That in turn increased the operating ratio and reduced the return on investment from the large program for capital expenditures. Fortunately, Menk and Downing stayed the course that had been set in 1974 and did not yield to pressures to stop or reduce the amount of investment being made in roadway and equipment. In remarks made in 2010 before the Lexington Group, Bob Downing stated, "I can personally testify that when we started work on upgrading hundreds of miles of lines to handle the Powder River Basin coal traffic in the 1970s, we were criticized for spending far too much money on which we could never expect a return. I haven't run into anyone in the last 20 years who even suggests that we should have done nothing."[5]

Bob is best remembered for his character, his vast knowledge of railroading, and his willingness to share it, even during the thirty-four years he lived after his retirement. Bob continued to mentor people of BN, BNSF, and other railway companies. He wrote extensively on many aspects of railroading for the Great Northern Railway Historical Society and the Lexington Group. He provided written oral histories for many rail and business historians and other "students" of railroading. Some of my most rewarding experiences in railway work have been the trips I have made with Bob—both before and after his retirement—and our frequent conversations on a variety of subjects. The rail industry and its people are indeed fortunate to have had the benefit of his leadership, intelligence, and knowledge for so many years.

Bob found some of the decisions made by the management in place at BN in the early to mid-1980s disturbing and in violation of the standards and principles that had been established in its first ten years. Bob felt the management in place in those years lacked understanding

5 Panel discussion on the Powder River Basin, annual meeting of the Lexington Group in Transportation, Sheridan, Wyoming, September 23, 2009. I was present and also served on the panel.

and knowledge of the basic principles of running a rail-road, and they did not have the level of commitment they should have had in order for BN to be successful in the long term. He strongly believed that some decisions made in those years were harmful to the company both for the present and the future. While Bob was a mild-mannered, non-emotional person, he would become exercised when discussing some of the decisions made at the top level of management in those years. He saw nothing in BN's financial or operating performance, in either the short term or for the long term, that would vindicate those leaders for the course on which they took the company.

Among the decisions that concerned Bob the most was the abandonment of the efficient, well-engineered line built by the former SP&S Railway between Spokane and Pasco. Another was the decision to give up the route of the former Milwaukee Road over Snoqualmie Pass that BN had acquired in 1980 as a more efficient route over the Cascade Range. Bob did not see any merit or benefit to the company in the sale and lease of the lines of the for-mer NP between Huntley (near Billings) and Sandpoint, Idaho, to the third party that established Montana Rail Link. Another issue Bob regretted was the loss of several executives whom he believed were in position to lead the company in its second decade, but who left the company because of the kind of culture established by the new management. Bob was never convinced that merger with the Frisco produced benefits of any magnitude for the shareholders or that the Frisco had any strategic value for BN. Bob felt that progressing plans to merge with the Missouri Pacific before the Union Pacific made its move would have produced a far better return on the investment of management time and energy needed to make a merger succeed. When discussing such contro-versial issues, Bob Downing was neither threatening nor contemptuous when he expressed displeasure over such decisions and their results.

Bob never became unpleasant during inspection trips when he observed work that was substandard or when he felt that insufficient progress was being made in complet-ing projects. Instead, his reaction was mild mannered, and his criticism was constructive. By no means was he a pushover, because it was always clear to those present that he expected results, and that he wanted to see that improvements had been made since his last trip over that part of the railroad.

One of Bob Downing's many accomplishments on both the GN and BN was for the company to have a policy in support of making year-to-year improvements in the alignment of the roadway. He also made sure that the necessary financial resources were designated to carry out such projects. This policy was the continuation of James J. Hill's directive that every year, the railroad would identify its weakest segment—where the opera-tion was the most impaired by restrictions in train speed, or where congestion was occurring the most often. By identifying those locations and then committing funds for improvements such as line changes to reduce curva-ture, to extend one or more sidings, or to add a few miles of second main track, an area of weakness or impairment could be brought up to a standard "as good as the best."

Carrying out such projects annually on a particular corridor would produce a vast improvement in the opera-tion when looked at over a period of time. Each such project, if looked at singly, may not always have had a high return on investment. However, when looking at all of the improvements made over an entire main line corridor over a number of years, the aggregate amount of reduced maintenance cost and the improvement in train performance from line changes and capacity added to the infrastructure was impressive. Bob deserves credit for keeping this policy in place over the years.

Overall, Bob Downing was one of the best and bright-est of BN's senior executives. The company and its people are fortunate to have had his services for so many years.

· C. ROBERT BINGER ·

[BN] company lands are logged to provide a regular, sustained harvest
through the orderly removal of mature and diseased timber, and
with an eye to preserving views and protecting streams and watersheds.
BURLINGTON NORTHERN ANNUAL REPORT FOR 1973, PAGE 13

These [financial] results for [1976] gave the [Resources] division
seven percent of Burlington Northern's consolidated revenues for the year
and 34 percent of consolidated net operating income.
BURLINGTON NORTHERN ANNUAL REPORT FOR 1976, PAGE 16

His [Bob Binger's] early wilderness experiences nurtured
a deep devotion to the land and Bob went on to make a distinguished career of land management.
"IN MEMORIAM," YMCA CAMP WIDJIWAGAN NEWSLETTER, PAGE 10

C. ROBERT (BOB) BINGER WAS A GRADUATE OF the School of Forestry at the University of Minnesota and also the Graduate School of Forestry at Yale University. After serving in the U.S. Navy in World War II, Binger worked for the Minnesota and Ontario Paper Company from 1946 to 1968. In 1967 he was named vice president of Operations in Canada and the United States.

In 1968 NP Railway hired Binger to manage its resources unit, which consisted of large timber holdings, oil and gas producing properties, and millions of acres of land on which NP had retained mineral rights. He was also a member of NP's board of directors. When NP, GN, and their affiliates merged to form BN in 1970, Binger was named vice president–Resources Development.

When BN was reorganized into two major units, Resources and Transportation, in 1974, Bob Binger was named president of the Resources Division. At that time, he also served as president and chief executive officer of Plum Creek Lumber Company, a forest-products manufacturing company that NP had acquired in 1968. Under his leadership, Plum Creek was expanded to become one of the country's largest and most efficient lumber companies.

Bob Binger delighted in recalling his early exposure to the rail industry when he worked on a "steel gang" (rail-laying gang) for the GN in North Dakota as summer employment while attending the University of Minnesota. He talked about the time when C. O. Jenks, vice president of Operations, stopped the train handling his business car and then walked along the track to inspect the work being done by the gang. The Jenks and Binger families were acquainted through social activities in St. Paul, but Jenks made sure he did not acknowledge young Bob at work on that gang. Bob was glad Jenks spared him of any special attention of even acknowledging he was acquainted with him.

Binger extolled the practices, values, and vision of a professional forester. In managing the timber resources of NP and then BN, he had responsibility for both the forest-products manufacturing activities and stewardship of those companies' forested lands. He directed large reforestation projects on land owned by NP or BN where timber had been cut, or when a forest needed to be renewed to become more valuable and productive in the years ahead. In the manufacture of forest products, plants owned by BN were set up to make use of 100 percent of every tree cut, including the bark and small branches. Although Binger had no experience in any of the resource business other than managing the timberlands and forest-projects-manufacturing units, he also had responsibility for managing BN's vast holdings of oil, gas, and minerals.

The principles Binger held in forest management brought him into conflict with some members of the BN board who pushed for a shift from managing forested lands from the sustained-yield basis to clear-cutting, as the means to maximize revenue in the short run. He never relented on the beliefs that were inculcated in him as a professional forester and a person who believed in managing forest resources for the long term. Binger wanted to insure that a good and predictable supply of timber would be available for future generations.

The numbers show that C. Robert Binger was successful in increasing earnings from the Resources Division in the 1970s, during the years in which earnings from the Transportation Division (primarily BN's railroad company) were held back because of the large amount of operating expense incurred in upgrading thousands of miles of track. He was respected by most members of the financial community and admired by the entire BN family for his leadership he gave in producing such favorable results at a critical time in the company's history.

• N. M. LORENTZSEN •

Burlington Northern's mission is "to fulfill the transportation and distribution needs
of the shipping public it serves and to develop a sustained, profitable expansion
from all of the company's physical and financial resources."
1974 ANNUAL REPORT, BURLINGTON NORTHERN, INC., PAGE 9

The advance of Burlington Northern toward the longer-range goals we have set
for our company has not been deterred by temporary adversities, economic or otherwise.
1977 ANNUAL REPORT, BURLINGTON NORTHERN, INC., PAGE 4

Unlike a lot of large corporate executives he's got a lot of feeling for
not only the area he came from, but for the people who are customers of his railroad.
DICK GOLDBERG, PRESIDENT OF GOLDBERG FEED AND GRAIN,
WEST FARGO, NORTH DAKOTA, QUOTED IN *BN NEWS*, JANUARY 1978

NORMAN LORENTZSEN HELD SENIOR-LEVEL positions right from the time of the 1970 merger to 1980, the year of the merger of BN and the SLSF (Frisco) and when R. M. Bressler was brought in to head the company. Norman played a major part in setting the course for the company in its first decade, in developing leaders for the second decade and beyond, and in overseeing the expansion projects underway in the Transportation and Resources divisions.

As vice president–Operations of the new company on M-Day, Norman was responsible for putting the consolidated operating plan in place, getting major construction projects underway, and setting the pace for the field and headquarters operations and maintenance officers. The network of the new company was made up of 23,609 miles of railroad in nineteen states and two Canadian provinces. BN's early success in getting the new operation established can be attributed, in large part, to Norman's skills in leadership.

In designating the heads of the units in the Operating Department, every effort was made to prevent any one of the three component lines from dominating the new company. A year or two into the merger, promotions were made on the basis of "the best person for the job," regardless of "heritage," but at the time of the merger it was important to have a "balance of power" and avoid a situation of "in groups and out groups," as happens so often in mergers of any kind. The position of VPO was especially sensitive since by far the largest number of people in a railroad company are in the Operating Department, and because the early success of the merger depended mainly on how well the operations could be integrated.

In deciding which of the three VPOs should head the department at the time of merger, the choice was

N. M. LORENTZSEN WORK HISTORY

NORTHERN PACIFIC RAILWAY CO.

1935	Section Laborer and Brakeman, Northern Pacific Railway, Dilworth, Minnesota (during college years)
1941–45	U.S. Naval Air Corps, Patrol Bomber Pilot
1947	Resumed work as Brakeman; promoted to Conductor and Assistant Trainmaster at Duluth
1949	Trainmaster at Duluth
1953	Assistant to General Manager, Lines East
1954	Division Superintendent, Rocky Mountain Division, Missoula
1957	Division Superintendent, Idaho Division, Spokane
1964	General Manager, Lines West
1968	Vice President–Operations

BURLINGTON NORTHERN, INC.

1970	Vice President–Operations (Burlington Northern)
1971	Executive Vice President
1973	President, Transportation Division
1978	President, Burlington Northern, Inc.
1978	President and Chief Executive Officer, Burlington Northern, Inc.
1980	Chairman, Executive Committee, Board of Directors
1981	Retired

between Norman Lorentzsen, who headed operations on the NP, and Ivan Ethington of the Burlington. John Robson, VPO of the GN, was nearing retirement at the time. Norman had been VPO for two years, and Ethington for five years. With his background on the NP, Norman had the advantage of familiarity with the common points for the GN and NP, where the impact of the merger would be the greatest. Norman's appointment to head the operations of the merged company was strong recognition of his capabilities and potential for advancement, possibly to head the company. Two years later, when Norman moved up to executive vice president, Ethington was appointed to head the Operating Department.

In those early years, Norman and the Human Resources Department did a great deal to insure that the heads of the regions and the headquarters staff departments under operations did not limit their consideration of candidates for promotion to only those with whom they were familiar from affiliations made in their predecessor company. This kind of oversight in the process of filling vacancies insured that a broad range of candidates would be considered, to avoid in-breeding and to speed up integration in the company.

Norman spent his early years in Horace, North Dakota, a small town southwest of Fargo. His father had the job of section foreman for the NP Railway at Horace. In 1932, during the Depression years, the NP abolished every other section crew, which forced Norman's father to transfer to the foreman's position at Dilworth, Minnesota, the NP's terminal a few miles east of Fargo. In 1936 Norman enrolled at Concordia College in Moorhead, Minnesota. His earnings from summer employment as a track laborer were sufficient to pay his tuition and other expenses for three years, but he then had to stay out of school and work at the NP for a full year to make enough money to finish his education. Upon graduation from Concordia in 1941, Norman enlisted in the U.S. Navy. He was placed in the navy's flight-training program and was qualified to fly multi-engine planes. Norman served as a flight instructor for a year and also in a squadron in the Southwest Pacific. He encountered anti-aircraft fire several times, but he and his crew suffered no injuries. Following his release from the service, Norman resumed work with the NP as a brakeman. Norman was soon elected local chairman as the union representative of the Brakemen based at Dilworth.

In September 1947 Norman received a message to report to the division superintendent. He assumed he was being called in to discuss a particular claim, grievance or discipline case. Instead, the superintendent asked Norman if he would accept promotion to assistant trainmaster at Duluth. Norman was enthusiastic over the opportunity—except that he would earn less in this junior officer position than he would as a conductor. Norman and his wife, Helen, decided to take the long view of the potential that accepting a promotion should give them, and he accepted the company's offer. From that point, Norman worked up through the "chairs" of the Operating Department. Only seven years later, he was named division superintendent at Missoula, Montana, followed by superintendent of the Idaho Division at Spokane. He became general manager of Lines West in 1964 and VPO in 1969.

Norman was fortunate to gain the favor of L. W. Menk when he accompanied Menk on the trip he made by business car over the NP in 1966 while Menk was president of the Burlington. A year later, when Menk came to the NP as president, he soon promoted Norman to VPO of the full organization. The strong affiliation between Menk and Norman continued through the years on BN, with Norman succeeding Menk as president and CEO in 1977.

True to his Norwegian heritage, Norman showed a calm, polite, and respectful demeanor in all situations. He was a "classic" in adhering to the attribute of "polite but firm." This was in contrast to some operating executives of the old school who developed a style of managing by fear and intimidation. Norman would listen to the concerns and suggestions of people he met on the frequent inspection trips he made out on the railroad. He was very familiar with the work that was done at the division level and gave strong support to the needs and responsibilities of those who did the "real work" out on the railroad. Norman never forgot his roots or lost his sensitivity and concern for the needs of the farmers, grain-elevator operators, and citizens of small, rural communities along the lines of BN. The same was true for the people out on the railroad who operated the trains and yards and maintained the track, bridges, signals,

and rolling stock. He prided himself on being a "doer," rather than a bystander or manager who sat in a comfortable office in St. Paul or in a region headquarters office and took shots at the work being done by those on the ground. We were fortunate to have a person of Norman's character, values, and beliefs as the head of the Operating Department, and later as the CEO.

Norman practiced what he preached. He lived and managed by these beliefs he had developed over the years and carried into the time he held senior leadership positions. He was highly respected for the values he instilled in the company. Norman was consistent and predictable and could always be trusted to stay the course.

Norman communicated his views on leadership, the importance of attitudes, and his expectations of others in an interview conducted by BN News in 1978. He was asked, "What qualities do you look for in your fellow workers?" Norman responded: "Integrity, high personal motivation, determination, good judgment, loyalty, and the willingness to do the job immediately at hand to the best of one's ability, even though it may not seem exciting or important at the time. People who have those qualities generally also have the respect of others, because a reputation for getting the job done is what opens doors and causes your fellow workers to look for (to you) leadership."[6]

In the same interview, Norman stated the obligations of the company to provide an affirmative work environment with the opportunity for personal growth and development: "And for our fellow workers we want environments that are safe and satisfactory within a company that meets needs for personal growth, economic opportunity, reward and recognition—and we want those offered fairly and equally. Simple human justice requires that, in return, each of us gives a day's work done efficiently and well."

Norman's leadership style, and that of his next two successors in the position of VPO, Ivan Ethington and John Hertog, built a strong culture of respect and set the standards for safety, job knowledge, technical proficiency, quality of work done, and the development of

---

6   "Lorentzsen reflects on new CEO role," BN News, October 1978, p. 4.

officers who would be needed to build a successful company. It was a disappointment when some aspects of this very affirmative and successful culture were set aside in later years.

Starting on day one, Norman was expected to do everything necessary to put the new operating plan for the merged company in place. He had to lead a team of operating and maintenance officers and over 50,000 employees, with most of them in the Operating Department. There were new standards, policies, and priorities that had to be put into place without delay. More than anyone else, Norman was looked to by the company and its customers to maintain reliability and consistency in service as the merged company put its new operating plan into effect. Also, there were numerous commitments made to communities, unions, customers, and competing or neighboring railways that were conditions imposed by the ICC in approving the merger. Delivery on those commitments was mandatory, right from the start.

Overall, implementation of the merger went well. Certainly, it was done far better on BN than had been experienced in the merger of the Pennsylvania and New York Central railroads a few years earlier. This was a real credit to the planning effort for operations on BN, but also, to the leadership Norman, and the operating and maintenance teams he and the six regional vice presidents had built and let in those early days of the merger.

As a senior executive of BN, Norman had a major role in overseeing the expenditure of hundreds of millions of dollars within a very few years, in order to expand the capacity and upgrade the railroad to handle the demands of the electric-power companies for low-sulfur coal mined in the Powder River Basin. Together with Lou Menk and Bob Downing, Norman Lorentzsen worked hard with Operating and Maintenance managers at all levels to see that work was completed to high standards. As a team, they had to deal with pressure from some members of the investment community—and even from some of the company's financial officers—to cut back on the extent of work being done, in order to increase earnings in the short term. I recall being in a meeting in the summer of 1975 when Norman was challenged by a finance officer: "Norman, let's cut back on all this maintenance work and start making some money." Had he yielded to that kind of pressure, work that was badly needed would not have been completed, and the large, unprecedented opportunity BN had for growth would have been set back.

Norman took great pride with the success of Burlington Northern Air Freight, a new enterprise started under his direction. When BNAFI was sold in 1982, the company recovered many times the investment it had made to start the company only a few years earlier. It became a very successful company in a highly competitive industry.

By early 1980, when Norman advised Menk of his desire to retire, BN had improved rapidly in its ranking among the Fortune 500 companies. The heaviest work in upgrading the railroad had been done, and in the Resources Division, expert executives had been brought in to accelerate the growth in earnings in forest products, land development, and oil and gas production.

Norman left a legacy of a railroad company that was strong, well maintained, and able to efficiently handle a high level of density traffic, greater than that of any other railroad in the world. This growth had provided good jobs for thousands of people who otherwise would never have worked in the rail industry. BN had evolved from a newly formed "giant" in need of rationalization on lines and facilities to a transportation provider that ranked among the best, and was well positioned for the future. Just as important, a culture and a system of values had been established in BN that made it a safe, ethical, non-political, quality company for its employees, customers, and the general public. We took great pride in the good tools, equipment, and technology we had acquired, as well as in the excellent roadway and mechanical facilities we had built. The stage had been set for us to get even better in the years ahead.

· T. J. LAMPHIER ·

Unit train coal traffic requires a heavy duty rail system in order to withstand the continuous impacts of this heavy tonnage on the rail and roadbed.... These requirements involve enormous amounts of capital.... Unfortunately, recent ICC and court decisions have produced an uncertain pricing atmosphere to the point where it is doubtful that the revenues permit the recovery of full costs involved in the traffic.

CORRESPONDENCE BETWEEN THOMAS J. LAMPHIER AND
ARTHUR INGBERMAN OF THE U.S. DEPARTMENT OF ENERGY, AUGUST 26, 1980,
INCLUDED IN DOE PUBLICATION "AN ASSESSMENT OF DEVELOPMENT AND
PRODUCTION POTENTIAL OF FEDERAL COAL LEASES," PAGE 212

THOMAS J. (TOM) LAMPHIER WAS NAMED president of BN's Transportation Division in 1976. At the time of Lamphier's appointment, Norman Lorentzsen moved up to president and CEO of BNI, which put him over both the Transportation and Resources divisions. Early in his career, Lamphier had held a junior position in line management as roadmaster on the South Dakota branch lines of GN's Willmar Division. Several years later, he was appointed superintendent of GN's Klamath Division. The rest of his experience on GN was in managing the early computer installations and in overseeing some of the planning functions undertaken in the Operating and Executive departments.

Tom Lamphier was a graduate of the Massachusetts Institute of Technology with a degree in civil engineering. As one would expect of a graduate of MIT, Lamphier was highly intelligent and very competent on technical manners. Even though most of his experience was in the technical and staff functions based at headquarters, he kept up on advances made in the Operating and Maintenance functions over the years by spending a good deal of time out on the railroad.

From the exposure most of us had to Lamphier, mainly on hi-rail and business car inspection trips, we viewed him mainly as an expert on maintenance standards and practices. He insisted that quality work be performed in all activities and would expound in great detail on anything he saw that was below standard, nothing what was required on our part to better meet the standards.

Lamphier began his career with GN as chairman on an engineering party based in Superior, Wisconsin. A few months later he was appointed assistant to the division

roadmaster at Grand Forks. Following this orientation to maintenance work, Lamphier was named district roadmaster on the Willmar Division, and headquartered at Watertown, South Dakota, where he had responsibility for the maintenance of branch lines in that area. About three years later, Lamphier was given a new and very different assignment: He was directed to report to the vice president and comptroller in St. Paul and serve as co-chairman of a computer research committee. The committee's charge was to develop the justification for acquisition of a Univac computer, which would be among the very first such acquisitions made in the rail industry of that time.

Lamphier credited the number two man in the Accounting Department, George Norris, for "discovering" him as the person who could begin to lead GN into this early stage in the use of computers. Lamphier often recalled the first computer acquired, a Univac I, that required thirty-seven tons of water-cooled air-conditioning capacity to keep it cool enough to operate, a challenge inherent in the soon-to-become-obsolete vacuum-tube technology.

In 1967 Lamphier was placed in GN's Executive Department, where he became deeply involved in the planning for the upcoming merger. In his oral history, Lamphier states,

most of the time I served in this position was consumed with the merger studies. Bob Downing was the GN's representative on the merger committee and Doug Shoemaker was a representative for the NP and they worked very closely with Wyer, Dick and Company who were the

T.J. LAMPHIER WORK HISTORY

GREAT NORTHERN RAILWAY CO.

1949	Chainman, Engineering Department, Superior, Wisconsin
1950	Assistant to Division Roadmaster, Grand Forks, North Dakota
1952–1955	District Roadmaster, Watertown, South Dakota, Transportation Inspector, Assistant Trainmaster–Twin Cities Terminals
1955	Co-Chairman, Computer Research Committee
1958	Director, Division of Economic Research
1962	Division Superintendent, Klamath Falls, Oregon
1963	Asst. to Vice President, Operations, St. Paul, Minnesota
1967	Vice President–Administration

BURLINGTON NORTHERN, INC.

1970	Vice President–Planning
1971	Vice President–Management Services and Planning
1972	Vice President–Executive Department
1973	Executive Vice President
1977	President–Transportation Division
1980	Senior Vice President–Transportation Policy and Analysis
1981	Retired

consultants who did the economic and planning studies for the merger.... It was a tremendously large job.... We developed some pretty good systems with the help of our new computers and came up with some fairly good studies.[7]

Lamphier recalled how his work on merger planning led to an opportunity to move into line management:

At the last evening at Seattle [for a series of hearings conducted by the ICC on the proposed merger], John Budd asked if I would have dinner with him. During the course of the dinner he suggested that I should get more operating experience and that he would like me to go to Klamath Falls later in the year as Division Superintendent as there were some other changes being proposed. ...I went to Klamath Falls in June of 1962.... I got called back to St. Paul to work as Assistant to the Vice President of Operations and was principally involved in the Operating Department's use of computers.

Lamphier went on to describe his next position, vice president–Administration:

I looked after all the data processing on the GN, including the Material and Purchasing Department. I was responsible for corporate planning and we had employed McKinsey and Company to assist us with carrying out some of the things that would be desirable if we had to go it alone if the merger never did occur.... We continued [merger] planning in somewhat more detail than we had done in the past.

On M-Day, Lamphier was made vice president–Management Services and Planning. In 1973, when Lorentzsen moved from executive vice president to president, he was replaced by Lamphier. In 1977 Lamphier took over as president of the Transportation Division. Right from BN's start-up in 1970, Lamphier spent a great

---

7   The merger planning team also had participation from representatives from the Burlington and the SP&S who were experts primarily in transportation, accounting and locomotive maintenance functions in the early stages of planning. An interview with Thomas J. Lamphier by Don Hofsommer, , August 18, 1993, Railroad Executive Oral History Program, John W. Barriger III National Railroad Library, St. Louis Mercantile Library, University of Missouri–St. Louis.

deal of time out on the railroad. In this activity he put his "Roadmaster's hat" back on. It was evident that he had kept abreast of new technology and work methods that had been developed in the fifteen-plus years since he served as district roadmaster.

When Richard Bressler came to BN, he soon hired Richard Grayson of the Frisco to head the railroad company, even though the BN's application to merge with the Frisco Railroad was tied up in court, and there was a possibility it would not be approved. The merger became effective November 21, 1980. Bressler claimed he looked at other candidates from the rail industry, but he concluded, "I recognized fairly quickly that he [Grayson] was probably the best talent to run the railroad that there was around."[8] According to Lorentzsen, Menk had wanted to bring Grayson into the company for a long time, so we can assume he advised Bressler to hire Grayson.

When Grayson was put in charge, Bressler named Lamphier senior vice president–Transportation Policy and Analysis. Lamphier was directed to recruit someone qualified to head a new Research and Development (R&D) Department. BN hired Steven Ditmeyer, another MIT graduate, who had gained experience in many technical areas while working at the Federal Railroad

Administration (FRA) and Association of American Railroads (AAR). It was an appointment that worked well. The new R&D group developed within a few years the technology for train control that became the forerunner of the system named Positive Train Control now being applied on an expedited basis on the entire U.S. rail industry. Other successes in BN's program for R&D are discussed in chapter 21 of my second book, *Transformation of a Railroad Company.*

The reassignment of Lamphier was one of the first of many high-level changes in personnel and responsibilities that were to follow in a very short time, as Bressler and Grayson appointed new leaders and restructured both the Resource and Transportation divisions. Lamphier's legacy included the successful effort he led in pulling the GN into the computer age to a level ahead of other railroad companies of that day. Also, in working with Bob Downing, Lamphier deserves credit for his role in preparing the excellent operating plan for the merger that created BN. In Lamphier's years as president, we stayed the course that had been set for the railway, which is to say, to complete the program underway for expanding its capacity to handle the coal and for maintaining the track on all BN lines at the standard needed for safe and efficient service. His close interest in the progress we were making on those projects helped insure they were completed on schedule and up to standard.

---

8   Kaufman, *"Leaders Count,"* 219.

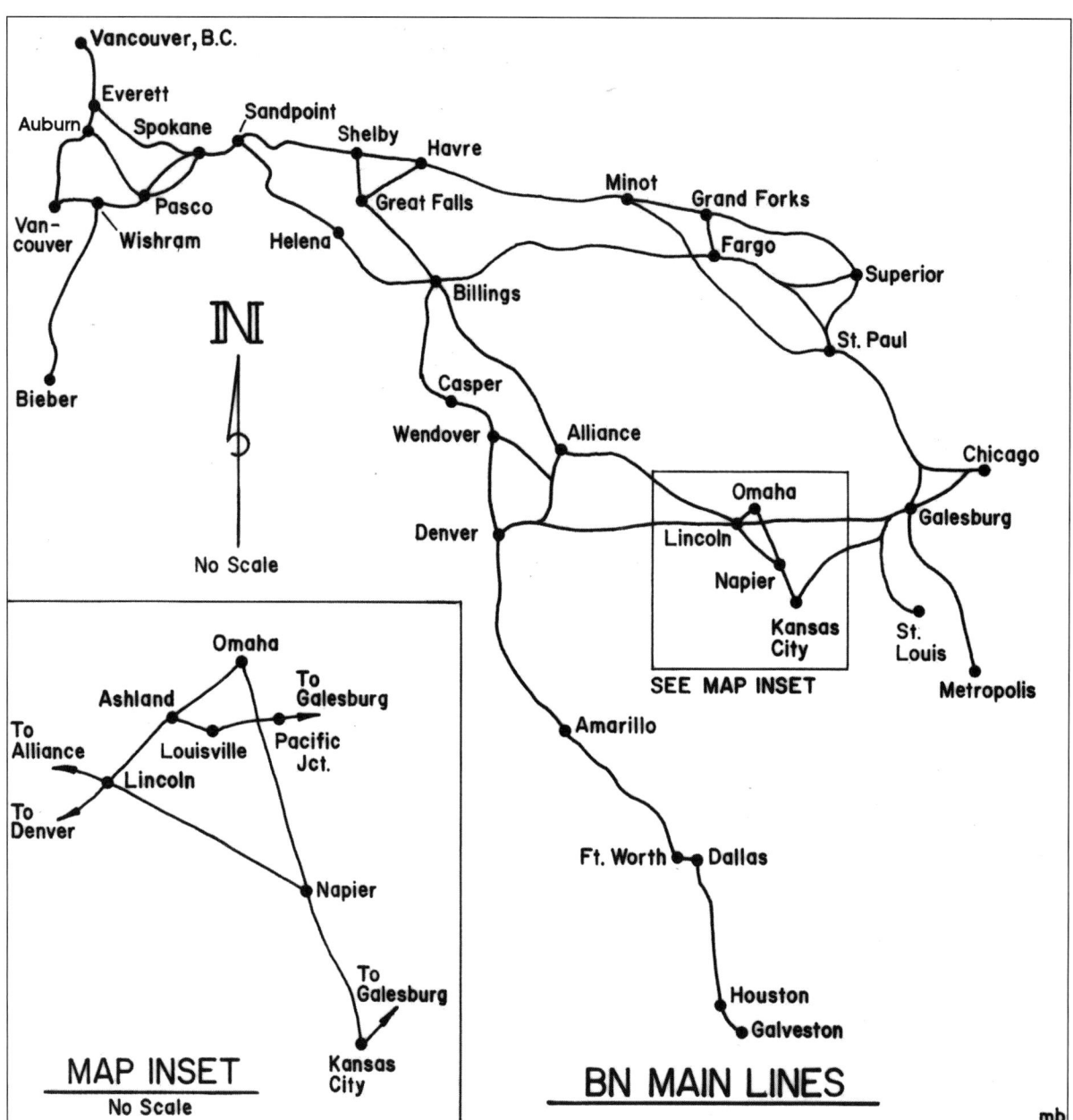

Figure 8.1.

CHAPTER 8

THE CONSOLIDATED OPERATING PLAN

*Combining the best routes and terminal facilities of each railroad into a
single unified main line between the Pacific Coast and the Middle West…*
"CONSOLIDATION: KEY TO TRANSPORTATION PROGRESS,"
A COMPANY PUBLICATION, FEBRUARY 20, 1961

The selection of the transcontinental route was the most important single decision…
BURLINGTON–GREAT NORTHERN–NORTHERN PACIFIC–
SP&S CONSOLIDATION STUDY, PAGE 129

A MAJOR PART OF THE PLANNING EFFORT
had to be devoted to determining the preferred
route for moving transcontinental business
across the northern tier of states, i.e., between the Twin
Cities and terminals in the Pacific Northwest. It was also
necessary to designate the preferred route between the
Pacific Northwest and the Kansas City gateway. On the
northern corridor, the GN and NP had competed fiercely.
For business moving to or from Kansas City, GN and
NP had had competitive routes between the Pacific
Northwest and Laurel, Montana, where both lines interchanged with the Burlington. A primary goal of merging
the GN and NP was to consolidate their business into
the most efficient combination of routes between major
terminals and to gain the economic benefits by having
a higher density of traffic on fewer lines.

The achievement of financial success in the rail industry is dependent to a great extent on its ability to build
density. It is important to apply operating practices
based on the inherent advantages that railroads have
to build on the economies of scale. The northern lines,

the Burlington, and the SP&S faced the disadvantages
of light density on hundreds of miles of secondary main
lines, where it was uneconomical to have schedules for
"through" trains that would not have local or intermediate points between terminals to set out or pick up cars.
There simply was not enough through or end-to-end
business available to operate trains with tonnage high
enough to make the operation efficient and profitable. It
was costly to operate enough local train service to enable
a light-tonnage through freight train with a competitive
schedule to be run over the same territory.

Among the strongest reasons for proposing the merger
was the opportunity and need to consolidate traffic of
the GN and NP on a single line, so faster schedules could
be established and marketed. With more competitive
schedules, it was expected, the new company would
have a better change of competing with truck service.
The dilemma the applicants faced before the merger was
brought out well in a statement given by E. L. Potarf,
vice president–Operations of the Burlington, in his testimony in merger hearings conducted by the ICC:

*Burlington Northern's mission is "to fulfill the
transportation and distribution needs of the shipping public is serves and
to develop a sustained, profitable expansion from all of the
company's physical and financial resources."*
BURLINGTON NORTHERN ANNUAL REPORT FOR 1974, PAGE 9

*Our basic goal is to provide better service to our shippers from the first day of the merger.
Changes in operations will be made as rapidly as possible consistent with this goal.*
BURLINGTON NORTHERN ANNUAL REPORT FOR 1969, PAGE 2

*Employee concern is our concern, for we must rely on loyal and devoted employees
to keep the wheels of Burlington Northern turningsmoothly and efficiently.*
BURLINGTON NORTHERN ANNUAL REPORT FOR 1969, PAGE 2

*It is gratifying to report that target dates set in pre-merger planning
have been attained and progress is on schedule.*
BURLINGTON NORTHERN ANNUAL REPORT FOR 1970, PAGE 2

*We attribute our ability to move this record volume of freight with greater efficiency and safety
to the combination of able workers, good management programs and the ongoing
modernization programs initiated by the company at merger in 1970.*
BURLINGTON NORTHERN ANNUAL REPORT FOR 1974, PAGE9

*Seldom are there presented to a company as many opportunities in such essential enterprises—
transportation and the development of natural resources—as
there are before Burlington Northern.*
BURLINGTON NORTHERN ANNUAL REPORT FOR 1972, PAGE 3

Our Laurel–Kansas City traffic is pretty thin west of Lincoln, Nebraska, and particularly so west of Alliance, Nebraska. It is so thin that now we cannot possibly justify the supporting service we would like to run, and which in turn, would permit us to operate our through trains non-stop except for operating requirements, instead of having to pick up at numerous locations enroute as we much do now.[1]

The GN and NP had the same disadvantages in the service as they ran on somewhat parallel lines in several areas, among them the routes between the Twin Cities and Duluth–Superior and the Twin Cities and Winnipeg. Also, the GN had very light density on its secondary line through central Montana (Shelby–Laurel), where it was in competition with the NP in moving traffic to and from Kansas City, Denver, and Texas.

To put the question of determining the preferred routes in perspective, it is worth making a quick comparison of the characteristics of all the northern transcontinental rail lines in the United States, as well as those in Canada. In looking first at the GN, it had by far the most favorable grades between the Twin Cities and Spokane. In determining the route on which the GN would be built, James J. Hill sought the lowest possible grades across mountain passes, and even across the prairies of North Dakota and eastern Montana. Having much lower grades made much of the GN's route more favorable than that of the NP. Even the Milwaukee, having the advantage of steam-powered earth-moving equipment by the time it built its West Coast extension (several years after the GN was completed), had more severe grades than the GN on its lines in North Dakota, Montana, and Idaho.

The GN and NP lines had equally severe grades on the lines they built between Spokane and Seattle. The grades on the Milwaukee's line over the Cascade Mountains were far more favorable than those of the GN and NP. However, the Milwaukee had a much heavier grade of 2.2 percent in eastern Washington at its crossing of the Columbia River. Both the GN and NP had maximum grades of only one percent in Washington, east of the Cascades.

While not direct competitors of the GN and NP, it is interesting to note how the routes of the two Canadian transcontinental lines compared. Canadian National's route through the mountains has much more favorable grades than even the GN. The Canadian Pacific Railway was built with a series of heavy grades west of Calgary, although all of its steep ascending grades for westbound trains were reduced to a maximum of one percent from Golden, Alberta, west to Vancouver in the 1980s. Several major projects, including construction of a tunnel of nine miles, made this possible. Upon completion of this project, the CPR had a route with more favorable grades for westbound trains than BN, given the 2.2 percent grade BN encounters in crossing the Cascade Mountains. However, the CPR still has the disadvantage of an ascending grade of 2.4 percent for its eastbound trains.

The magnitude of grades on the CN and CPR has become more significant to the U.S. rail companies with the increase in competition to move coal and grain westbound to Pacific Coast ports, and with the opening of large container ports at Prince Rupert and Vancouver, British Columbia. It is especially significant that the CPR now has a grade of only one percent for the heavy-unit trains it runs to the west coast. As a result, its operating costs and the investment in helper locomotives have been greatly reduced. The low grade on CN's route gives it a real competitive advantage in moving containers from Prince Rupert to points in the United States.

Since the Union Pacific was a major competitor of the GN and NP for long-haul business to and from the Pacific Northwest, it is important to recognize its grades as well. After building a third main track over Sherman Pass in the early 1950s, with a grade of only 0.82 percent, the UP had no severe grades for its westbound trains until it reached the mountainous areas of eastern Oregon, where it has grades in excess of 2 percent. See Tables 8.1 and 8.2.

A comparison of grades in North Dakota (in South Dakota for the Milwaukee) also shows significant differences:

GN   no grades over 0.4 % between Wahpeton Junction and Minot; no grades over 0.65% west of Minot, to the North Dakota–Montana state line.

NP   grades of 1.0% at 15 locations.

MILW   grade of 1.0% at the Minnesota–South Dakota state line; grade of 1.0% west of Aberdeen and between Marmain and Bowman.

---

1   Statement by E. L. Potarf in support of the GNP&B merger, October 27, 1960. Testimony at hearings conducted by the ICC for Finance Dockets Nos. 21478, 21479, and 21480, Merger application filed by Great Northern Pacific and Burlington Lines, Inc.

Table 8.1. Maximum Grades—
Northern Transcontinental Lines
between the Twin Cities and Spokane

GN	Westbound: 1% over Marias Pass
	Eastbound: 1.8% over Marias Pass
NP	Westbound: 1.8% over Bozeman Pass
	Eastbound: 1.9% over Bozeman Pass
	Westbound: 2.2% over Mullan Pass
	Eastbound: 1.4% over Mullan Pass
	Westbound: 2.2% over Evaro Hill*
	Eastbound: 2.2% over Evaro Hill*

*The line over Evaro Hill was used by passenger trains and light-tonnage freight trains. Regular freight trains were run on the alternate route via St. Regis, on a river grade.

MILW	Westbound: 1.4% over Belt Mountains
	Eastbound: 1.0% over Belt Mountains
	Westbound: 2.0% over Pipestone Pass
	Eastbound: 1.7% over Pipestone Pass
	Westbound: 1.7% over St. Paul Pass
	Eastbound: 1.7% over St. Paul Pass
CN	Westbound: 0.4% over Yellowhead Pass
	Eastbound: 0.7% over Yellowhead Pass
CPR	Westbound: 1.0% over four locations
	Eastbound: 2.4% over Selkirk Mountains
UP	Westbound: 2.2% over Blue Mountains
	Eastbound: 2.0% over Blue Mountains

Table 8.2. Maximum Grades over the
Cascade Mountains

GN	2.2% in both directions over Stevens Pass
NP	2.2% in both directions over Stampede Pass
MILW	0.7% westbound and 1.7% eastbound over Snoqualmie Pass

Table 8.3. Comparison of Mileages

Between Minneapolis and Spokane:

| GN | 1,432 mi. (on freight line via Breckenridge, Casselton, and New Rockford) |
| NP | 1,508 mi. (via Helena and St. Regis) |
Difference of 76 miles

Between Spokane and Seattle:

| GN | 330 mi. |
| NP | 396 mi. |
Difference of 66 miles

Table 8.4. Comparison of Routes,
Spokane–Laurel

| GN | grades: 1.1% westbound, 1.8% eastbound over Marias Pass distance: 66.2 miles |
| NP* | grades: 2.2% westbound, 1.4% eastbound over Mullan Pass distance: 561 miles |

*Assumes operation on line with river grade via St. Regis, between Desmet (Missoula) and Paradise.

Another important criterion in evaluating alternate routes was the distance between major terminals. See Table 8.3.

Grades and distance were two of the most important factors to be considered in determining the preferred route for long haul business. Other factors evaluated were: the capacity and condition of the track on alternate routes, the weight of rail, the number of miles of welded rail in service, the type of train control (signal) system in effect, allowable train speed, the length of sidings, the intervals between sidings, the number of miles of second main (double) track in service, the design and capability of intermediate terminals, the amount and type of local business, the number and length of crew districts, and the running time for priority freight trains in both directions. See Table 8.4.

The GN's route was on a secondary line of 321 miles between Laurel and Shelby with no signal system in service, considerable mileage of 90-pound rail, and maximum authorized speed of 49 miles per hour. The entire line of the NP was maintained to main line standards, with 60 miles per hour authorized for priority freight trains. In making these comparisons, the NP line was determined to be more favorable, and so was designated the preferred route.

The severity of grades is an important factor in comparing the operating characteristics of two lines of railroad. In this case, the GN line with its favorable grades was so much longer than the NP line that it had to be ruled out. Also, a sizable investment would have to be made in the GN line to increase its capacity enough to handle the combined number of trains, and to be able to reduce the scheduled running time to at least equal the time that could be made on the NP route where no upgrading was required. Twenty-five years later, the question of which of the two routes should be used was revisited. The conclusions made from the new study were the same as at the time the merger plans were prepared.

Shortly after the presidents of the GN and NP, J. M. Budd and R. S. Macfarlane, decided it was time to move in the direction of the proposed merger. William J. Wyer Company, a noted transportation-consulting firm, was engaged to make a study of the economics of the proposed consolidation. A major part of that study was preparing the plan for conducting operations on the merged company. The consultants worked together with the railroad companies' planning team on plans for combining operations at the locations where two or more of the component roads had facilities and operations in place. The major common points were the Twin Cities, the Head of the Lakes (Duluth–Superior), Fargo–Dilworth, Laurel, Spokane, Seattle, and Portland–Vancouver.

In addition to determining the preferred routes and preparing plans for rationalization of major car and locomotive shops, the team had to prepare schedules and overall operating plans for train service for the entire system. A working committee also was set up to determine changes in crew districts and labor agreements needed to facilitate implementation of the plans for the consolidated operation. The planning group had to determine which line segments of each of the component roads should form the preferred routes. It was important to fulfill commitments made to improve service by reducing transit times on major corridors, while also reducing

costs by using the line segments with the most favorable grades, the shortest distances, and the highest capacity. All of this was necessary to take full advantage of merging the two somewhat parallel, competing systems.

Some arguments were made during the planning process that operating on a line with heavier grades had not been a disadvantage since dieselization. "All you have to do is add a [locomotive] unit," was the argument made. This kind of thinking failed to recognize the cost of ownership of the additional units required, the additional fuel consumed, the time required to add and cut out the extra units at one or more intermediate points, and the additional payment to be made to locomotive engineers due to operating a locomotive consist that had a higher weight on driving axles.

At locations where adding a unit still would not provide enough power to get over a grade, or where train tonnage had to be limited to avoid exceeding drawbar strength, a manned helper locomotive would have to be used. Helper service is very expensive due to the low utilization rate of both labor and the locomotives employed. Even with the advantages of distributed power in today's operations, operating on heavy grades still raises the cost of operation. For all of these reasons, the low-grade alternatives will most often be designated the preferred route.

In designating the GN line between Casselton and Sandpoint as the preferred route for northern transcontinental service, the arguments favoring a shorter, low-grade route prevailed. In deciding whether to designate the GN or NP route between Sandpoint and Laurel as the preferred route for traffic moving between the Pacific Northwest and Kansas City, the arguments for the NP's line prevailed, even though it had heavier grades. In that case, being 100 miles shorter, having CTC in service, and heavy welded rail on much of the line gave the NP line a decided advantage (refer to table 8.1 for a detailed comparison of line characteristics).

In evaluating all characteristics of each of the transcontinental routes, the planning team proposed the double-track line of the NP be designated the preferred route between Minneapolis and Casselton, North Dakota, where the NP line crossed the GN's main route for freight trains. The corresponding GN line via Willmar was a single-track route with CTC, and although it had favorable grades, it had a lower capacity than the NP's

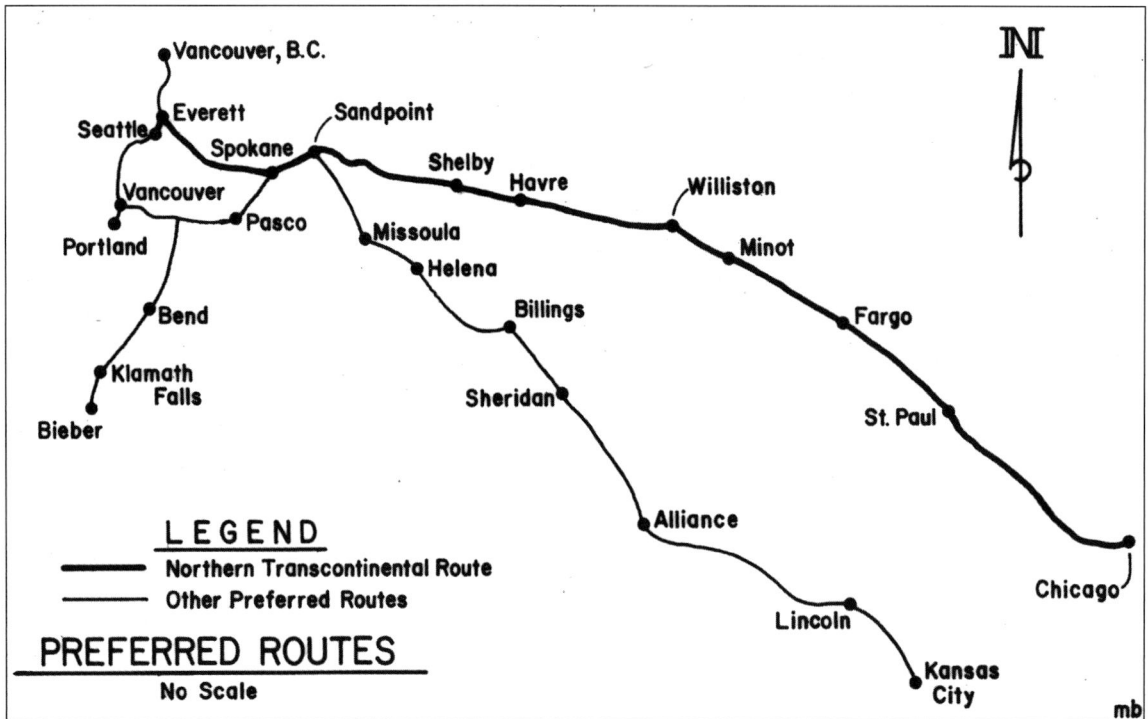

Figure 8.2.

double-track line. From Casselton to Sandpoint, the GN line was designated, due to its more favorable grades.

For the link between Sandpoint and Spokane, the NP's line would be used. It was a high-capacity line with CTC and had the advantage of being built on a viaduct through downtown Spokane, thereby avoiding conflict with motor vehicle and pedestrian traffic. Also, the NP line was adjacent to a large piece of property the company planned to acquire for the construction of a new yard after merger. Between Spokane and Seattle, the GN line had the advantage of a shorter distance, with CTC in service on all but 120 miles, and at Everett, it connected with the line to Bellingham and Vancouver, British Columbia.

The NP line between Sandpoint and Laurel would become the preferred route for business moving to and from Kansas City, Denver, and Texas points. East of Laurel, the NP main line was to be relegated largely to local service, to the junction with the preferred route at Casselton. However, starting in the early 1970s, the need to move large tonnages of low-sulfur coal out of mines in southeastern Montana and the Powder River Basin in Wyoming changed all of that. The NP line soon became

a major coal corridor carrying heavy tonnage and a high density of traffic.

The SP&S main line between Spokane and Vancouver may have been the best engineered railroad in North America, with its very favorable grades and minimum curvature. It provided an ideal operation for trains of all classes, whether high-speed merchandise trains or heavy trains of bulk commodities. There was no need to make a study to determine the most favorable line between Spokane, Pasco and Vancouver. Since the line of the SP&S was the most direct route and did not cross the Cascade Mountains, it was designated the preferred route for service to Portland and Vancouver.

In the Twin Cities, plans were made to acquire additional property adjacent to the NP's Northtown yard for construction of a new hump classification yard. Having this new capability would allow closure of nearly all tracks in the eleven smaller yards in the terminal. At Hauser, Idaho, twenty miles east of Spokane, a new hump yard was planned to replace the flat switching yard of the NP at Yardley and the GN's yard at Hillyard. A 4,260-foot-long new bridge would be built across Latah Creek at the west end of Spokane to connect the main line of the NP with the SP&S and the GN's line to Seattle. Construction of a new connection of 1.2 miles, including a bridge across Sand Creek, was planned at Sandpoint to connect the GN and NP lines. A number of less impressive but equally important new connections were planned at Carlton, Minnesota; Hinckley, Minnesota; Fargo; Everett; and Helena to enable trains and yard switching assignments to use the most efficient combination of line segments. The volume of business and the projected savings were lower at these common points than from the consolidations made at major terminals or in building connections at strategic locations on the main line.

To the extent possible, the parallel or overlapping lines used to reach these common points were rationalized through abandonment of some or all of one line. In some cases the savings in maintenance and operating costs were substantial, such as in providing service to Butte, Montana. Butte was served by secondary main lines of both the NP and GN. The NP's line through Butte also was the route of one of its transcontinental passenger trains, the *North Coast Limited*. The GN reached Butte on a line built on the "spine" of the Continental Divide between Helena and Butte with 2.2 percent grades, heavy curvature, and a tunnel of 6,000 feet. It handled only two light tonnage trains per day. Under the operating plan for the merged company, the GN's train between Great Falls and Butte was rerouted on the NP main line over Mullan Pass between Helena and Garrison, and on a new connection built at Garrison between the main line and the NP's line to Butte. Most of the business handled on this train was "bridged" between the UP at Silver Bow (connection near Butte) and the CPR at Sweet Grass, about forty to sixty cars daily in each direction. Only four miles of the GN's line between Helena and Butte had to be retained, for service to a cement plant at Montana City.

Both GN and NP provided service to and from Winnipeg. The GN's trains operated on the St. Cloud line to Barnesville, and then north on the line via Ada to Crookston, and from there, to Noyes at the border. GN and NP reached Winnipeg by trackage rights on the CN, using CN crews and power. The NP trains operated on its main line to Manitoba Junction, twenty-eight miles east of Fargo, then to Grand Forks, and to the border at Pembina, North Dakota. After the merger, BN moved its

Winnipeg business on the new preferred route between Minneapolis and Manitoba Junction. This allowed abandonment of forty-six miles of the Ada line and ten miles of the segment between Barnesville and Glyndon. Later, the trains to and from Winnipeg were moved to the main line via Fargo, Grand Forks, and Crookston, which made it possible to abandon forty-two miles of the NP line between Manitoba Junction and Crookston. By being able to combine the Noyes–Winnipeg traffic with other business moving to or from Fargo and Grand Forks, more of the economies of scale were achieved.

Together with routing of the "Fargo Fast" train from the St. Cloud line to the preferred route, the St. Cloud line was used only for local service soon after the merger. While the tonnage handled on the St. Cloud line was not heavy, it was in excellent condition, rated for 79 miles per hour for passenger trains and 50 for freight trains, but it was needed only for local service after the merger. The entire line of 156 miles between Moorhead and St. Joseph (six miles from St. Cloud) was soon sold to a short-line operator. These changes are representative of the benefits of a consolidated operation, which provided the opportunity to concentrate traffic on fewer lines and thereby realize gains through the economies of scale that are inherent in the railroad business.

The report prepared by the planning team and the William J. Wyer Company identified one hundred common points at which two or more of the component roads provided service. At thirty-nine of the common points, the report showed that consolidation would produce efficiencies in operation. Annual savings of $15.1 million were projected, based on the elimination of fifteen hundred jobs and reducing the number of yard engine shifts by 751 per week, on the basis of a seven-day per week operation. Capital investment of $43.2 million would be required to achieve these savings in the large terminals that were common points: the Twin Cities, Spokane, Seattle, Duluth–Superior, Portland, and Fargo–Moorhead.[2]

By making the capital investment recommended in the report, several parts of the railroad network could be

2   *Burlington–Great Northern–Northern Pacific–SP&S Consolidation Study* prepared for the applicants by the William J. Wyer Company.

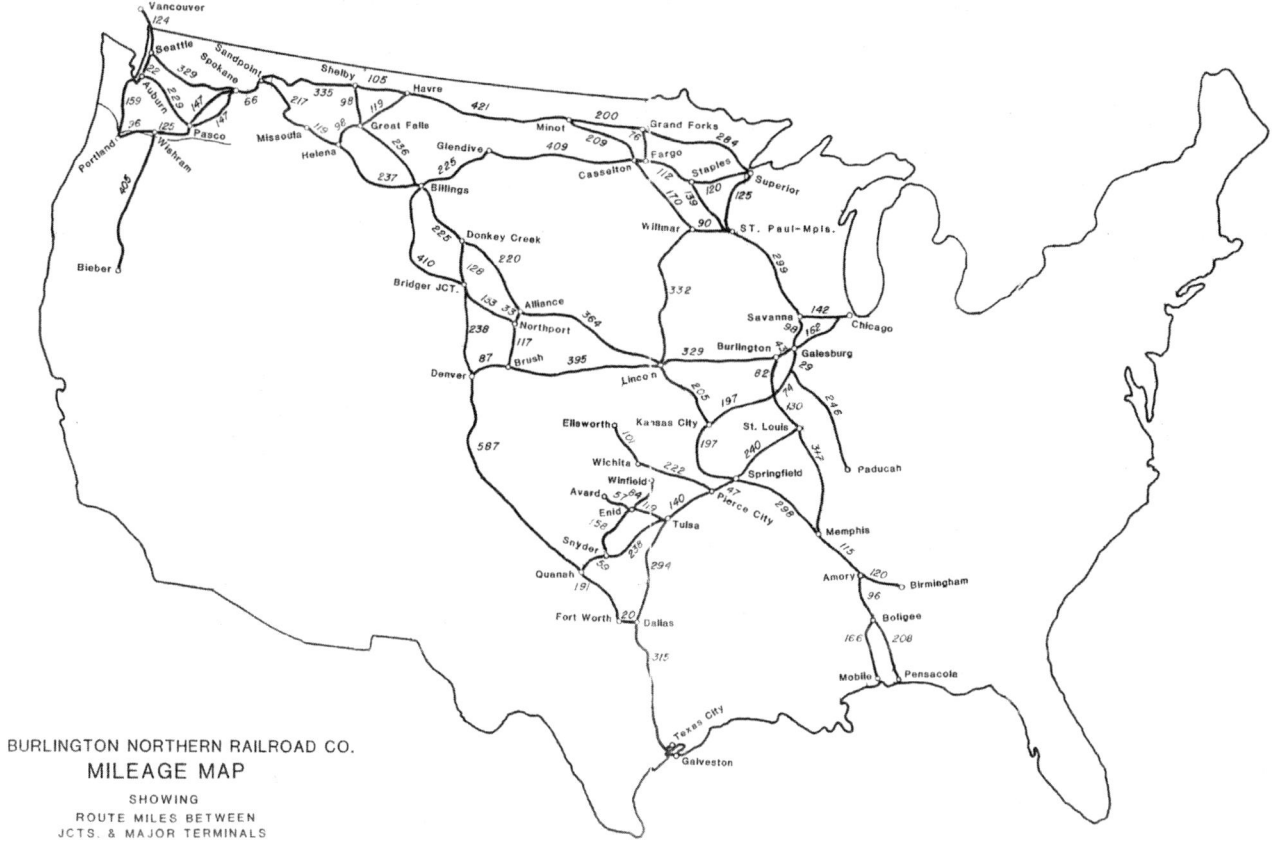

Figure 8.3.

rationalized. Following are the major reductions made on routes and in terminals after the merger took place:

BETWEEN THE TWIN CITIES AND DULUTH–SUPE-RIOR   The NP line of 65 miles between Hinckley and Carlton was abandoned. Parts of the line between St. Paul and Hinckley (76 miles) were either sold or abandoned. With a new connection at Carlton, the NP line of 19 miles between Carlton and Superior was abandoned, as well as the 16-mile line between Carlton and West Duluth.

CONSTRUCTION OF NEW CONNECTION AT SAND-POINT   This allowed sale or abandonment of 64 miles of the GN line between Sandpoint and Hillyard. Also, the GN double-track line of 4.5 miles through downtown Spokane, including the large bridge over Latah Creek were abandoned.

NP "A" LINE   These three miles in Minneapolis,

between Park Junction and the industrial area at 20th Avenue South, were abandoned, including a bridge over the Mississippi River.

NP LINE BETWEEN BLACK RIVER AND SUMAS, WASHINGTON   Use of GN line between Seattle and Sedro Woolley allowed abandonment of most of the NP line between Black River and Sedro Woolley.

A FEW RELATIVELY SMALL, INEXPENSIVE CON-NECTIONS   These were built at other common points as part of the agenda to reduce costs by eliminating duplicate lines to simplify the operation.

In concert with the program for line and terminal ratio-nalization, a number of mechanical facilities were closed or integrated. A large new locomotive-maintenance facil-ity and car shop were built at the Northtown complex in Minneapolis. Consolidation of numerous support func-tions also was part of the planning process, including

Accounting, Purchasing, Material Management, Law, Sales and Marketing, Public Relations, Property Management, Data Processing, and various administrative activities. The Accounting Department consolidated the activities of the Burlington's office in Chicago into the General Office building in St. Paul immediately upon merger. The clerks who chose not to follow their work and relocate to St. Paul were put in "surplus" status in Chicago and left to the newly formed Chicago Region staff to handle. The employees in surplus status would have to be worked off through normal attrition, and by forcing them into vacancies in the clerical staff of the Operating Department at Cicero as positions opened.

The Accounting Department carried no responsibility for this problem. Some relief was obtained when the region's labor-relations officers began to offer individual settlements to employees who were not yet placed in other vacancies.

Overall, the consolidation of functions went well, without interruptions to the flow of traffic due to poor planning, delays in construction, or inability to implement the new operating plan due to restrictions in labor agreements or in the flexibility needed to take full advantage of the opportunities the consolidation provided. The quality of the planning process, together with a general resolve to make the merger work, got BN off to a good start.

Figure 9.1. Burlington Northern built a bridge of 3,952 feet over Latah Creek to connect the new preferred route through Spokane with the SP&S line to Portland and the GN line to Seattle. The switch that the westbound train is going over is the junction switch to those two lines. In the background is the connection to the NP line to Pasco and Seattle. Construction of this new bridge was a major part of the $18.8 million project for line improvement at Spokane. The complete project included 10 bridges and a line change of 6.4 miles. Completed in 1972, the new bridge over Latah Creek was named a "prize bridge" by the American Institute of Steel Construction, describing it as "simple yet majestic." BURLINGTON NORTHERN.

CHAPTER 9

BUILDING THE CONNECTIONS
PROJECTS VITAL TO CONSOLIDATION

*• construct new connections between segments of former GN, NP, and SP&S
lines needed to form the "preferred route" for service between
Chicago, the Twin Cities and the Pacific Northwest •
• connections to be built at other "common points." • consolidation and closure
of duplicate yards, lines, and equipment-maintenance facilities • construct
new yard in the Twin Cities to speed up the handling of shipments and
to allow eleven old yards to be closed or reduced in activity •*

AN IMPORTANT PART OF THE OPERATING plan for the merged company was the planning and design of new connections that would link the line segments designated for the preferred route for northern transcontinental trains. The Engineering and Operating departments of the GN and NP worked closely together for several years before M-Day to finalize issues involving property acquisition, engineering standards, agreements with contractors, and the timing of delivery of track and signal material so the connections could be built as soon as possible after final approval of the merger in the courts.

Once these connections were in place, trains could start to run over all segments of the preferred route, and a large part of the projected savings in operating costs and reductions in transit time could be realized. The operating plan also provided for new connections at a number of locations other than those needed to set up the preferred route. Connections to be built on several secondary routes were to enable the elimination of some parallel, duplicating routes. Once those connections were in place, significant savings in ongoing track and bridge maintenance would be realized, as well as reduced train operating costs.

A few large projects were vital in getting the operations of the component roads integrated in order to achieve the cost savings projected in the merger plan. The most urgently needed projects were the new hump classification yard to be built at Northtown (Minneapolis), the bridges at Spokane that would connect the main lines of the NP, GN, and SP&S, and the connection at Sandpoint between the GN and NP main lines.

SPOKANE, WASHINGTON

At Spokane, it was necessary to build a large, impressive bridge over Latah Creek to connect the lines of the three roads serving Spokane. At the new junction at the west end of the new bridge, the line splits, for trains going to or from Portland on the former SP&S, or to Seattle on the former GN route. In addition to this large bridge, fourteen smaller bridges had to be built or modified. The project included a line change of 5.5 miles on the west side of Spokane. The entire project was completed in December 1972 at a cost of $18.8 million. Upon completion, it was possible to take full advantage of the NP's elevated low-grade double-track line through downtown Spokane and to abandon operations on the GN's line that

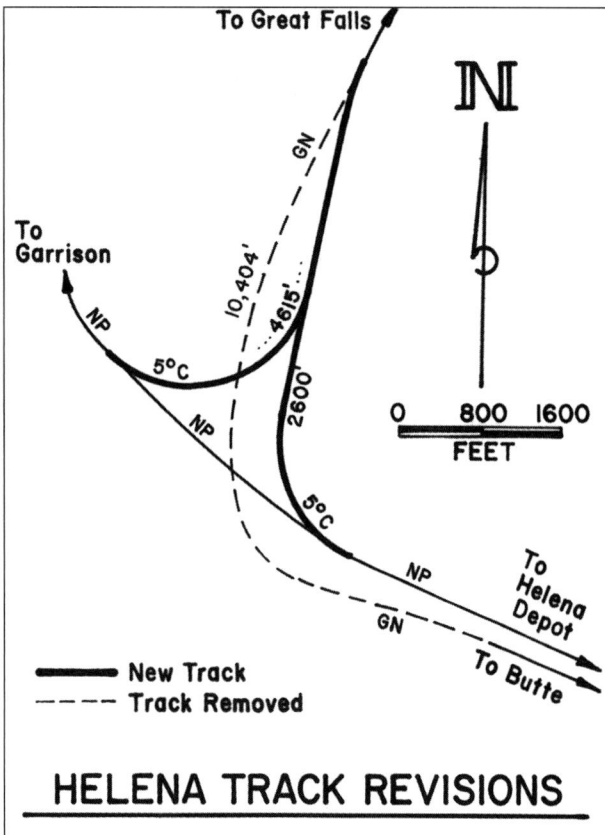

Figure 9.2.

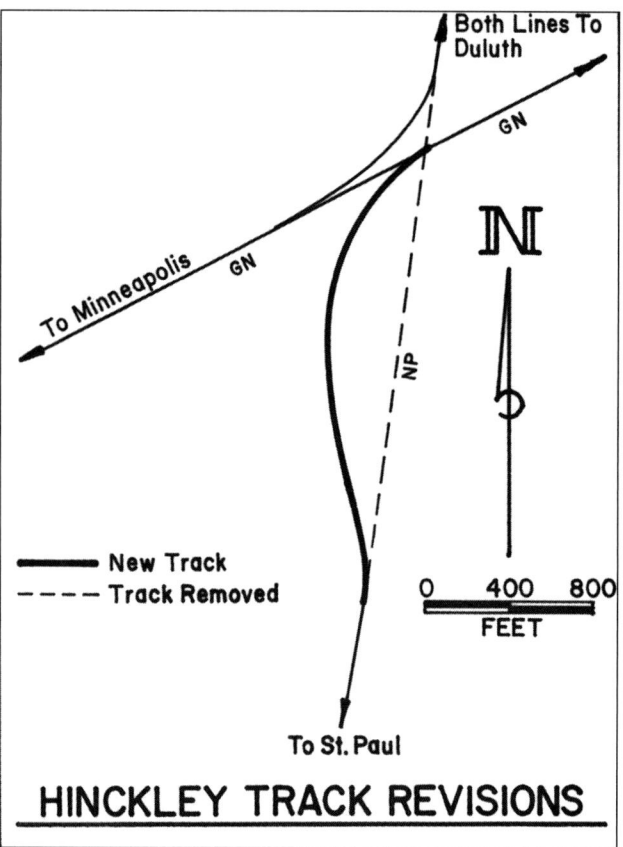

Figure 9.3.

had a grade of one percent and numerous road crossings. Listed below are some details covering this project:

- Bridge over Latah Creek: 3,968 feet in length, height of 212 feet
- Bridge over Indian Canyon: 730 feet long and 140 feet high
- Line change of 5.5 miles: reduced the length of the line toward Wenatchee by four miles, eliminated sharp curvature and a tunnel
- Consolidation within the Spokane terminal also required construction of a new bridge over the Spokane River for access between the NP yard and industries near Hillyard (GN yard) and the branch line to Kettle Falls and Nelson, British Columbia
- Elimination of GN track and freeing property in downtown Spokane enabled the city to fête its Expo '74 celebration
- Many awards were received from engineering societies for the aesthetic and design features of the bridge over Latah Creek.

SANDPOINT, IDAHO

At Sandpoint, a connection of two miles had to be built, including a bridge of 997 feet over Sand Creek, to facilitate movement between the main lines of the GN and NP. The GN's line was designated the preferred route between Sandpoint and Casselton. Between Sandpoint and Spokane, the NP's line would serve as the preferred route. The trackwork and bridge were completed in 1973 at a cost of $2.7 million. With the new high-speed, direct connections built at Spokane, Sandpoint, and Casselton, the new preferred route was in service as envisioned in the operating plan for the merged company.

CASSELTON, NORTH DAKOTA

At Casselton (twenty miles west of Fargo), where the GN's main freight line between Minneapolis and the West Coast crossed the NP's double-track main line at grade, a new 2,100-foot-long connection was built two miles to the west of the old "diamond" crossing. Between Minneapolis and

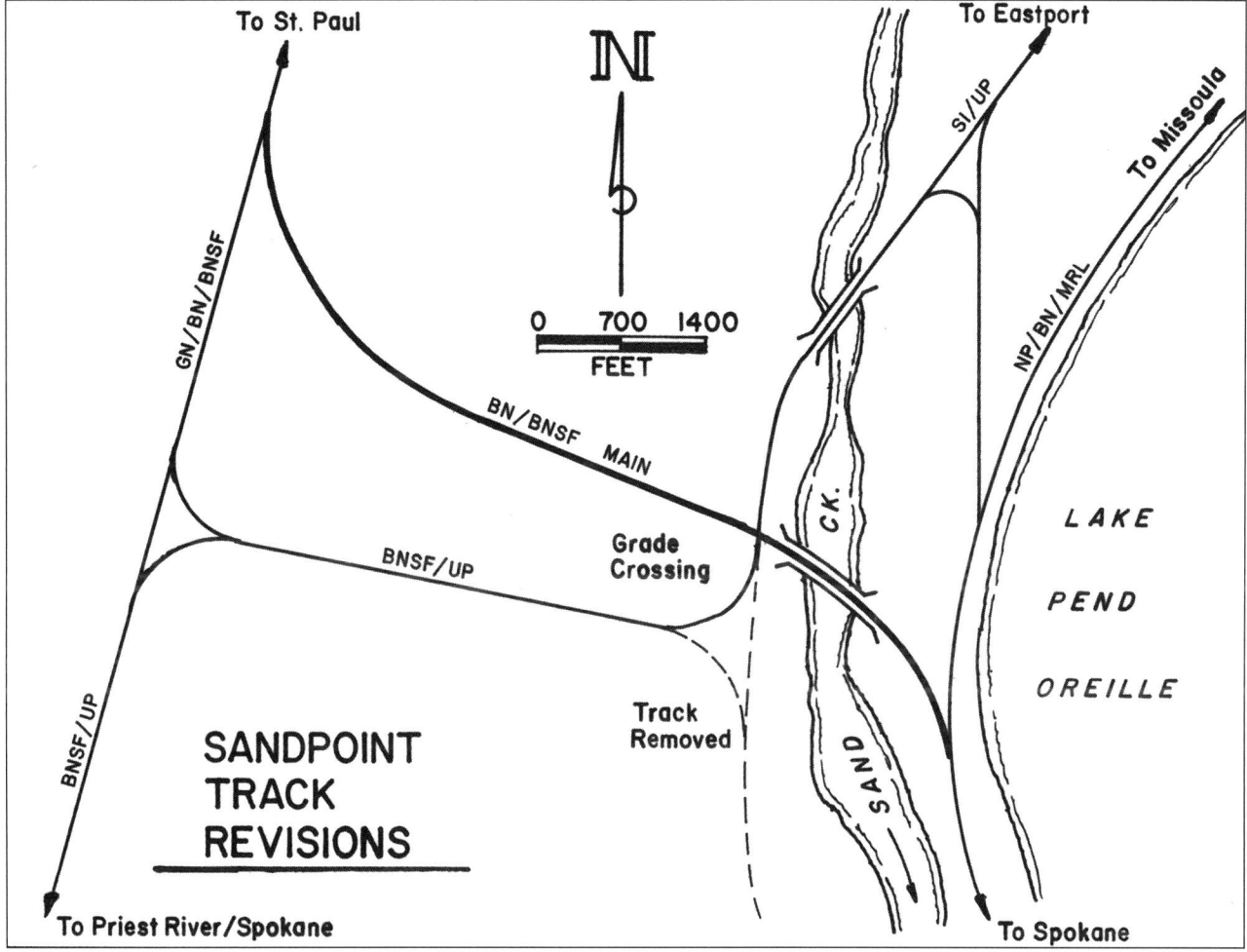

Figure 9.4.

Casselton, the NP's line was named the preferred route. West from Casselton and on to Sandpoint, the GN's line was designated. A power crossover and junction switch were installed on the NP's track to facilitate movement between the two main lines. With heavy growth in the number of trains operated on both lines, CTC was later installed between Casselton and West Fargo. At the time of this writing (2012), between sixty and seventy trains per day operate through Casselton.

New connections were built at a total of nineteen common points to facilitate the movement of trains and switching crews on the tracks of both the former GN and NP. While these were small projects in comparison with those completed at Spokane and Sandpoint, they were necessary to achieve the large savings projected in the operating plan by combining operations over one line

instead of two somewhat parallel or duplicating lines, and by being able to handle switching of some local industries with one crew instead of requiring two crews if these operations had remained separate and independent.

HELENA, MONTANA

A particularly good example of the benefit of building small, relatively inexpensive connections was at Helena. By being able to route trains operating on the former GN line between Great Falls and Butte to and from the NP main line at Helena, it was possible to abandon most of the GN's seventy-three-mile branch line between Helena and Butte. This was a high-maintenance line, with grades of 2.2 percent, heavy curvature, five tunnels (including one of 6,145 feet), with only two trains per day.

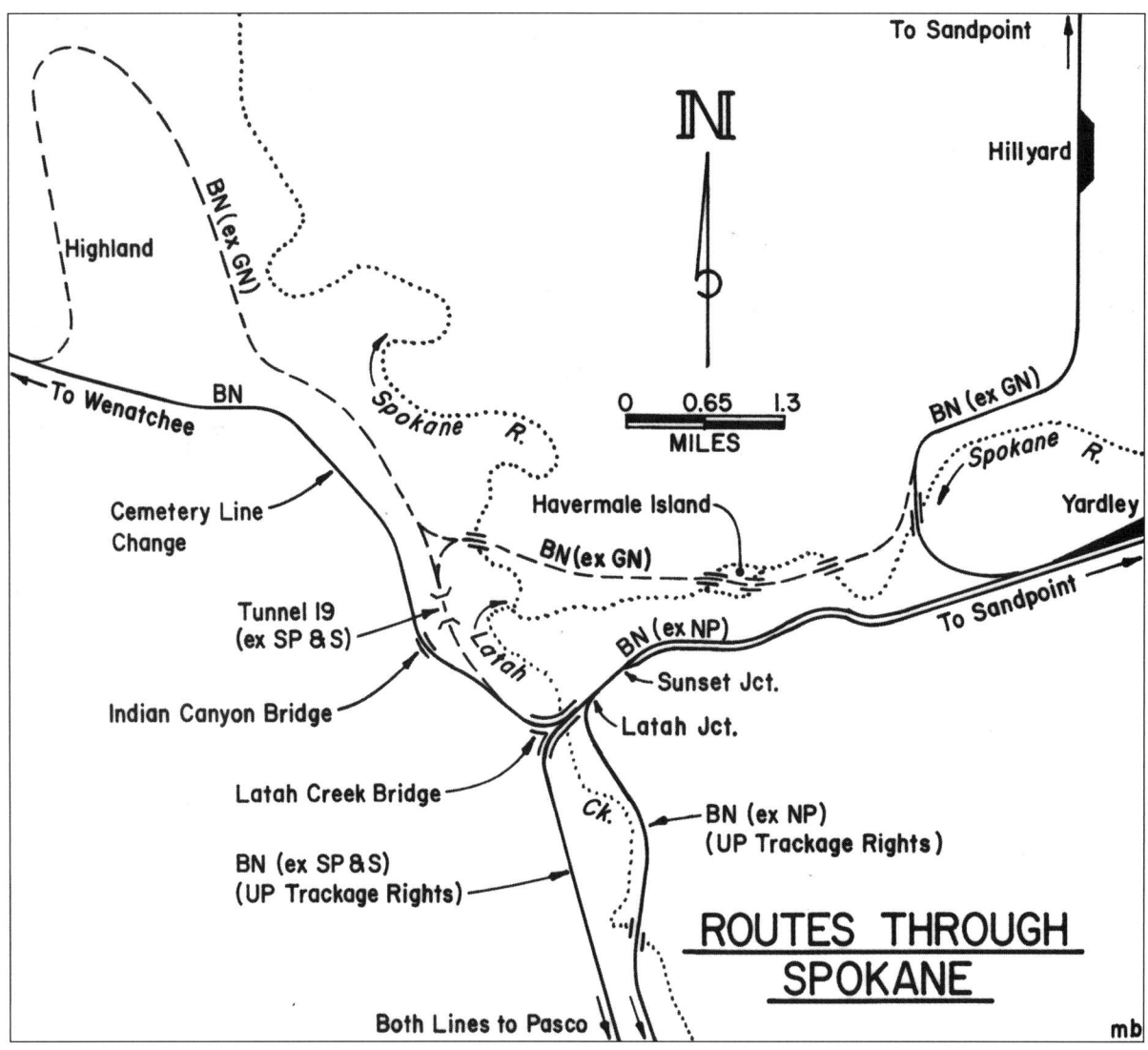

Figure 9.5.

The new connection at Helena allowed trains running between Great Falls and Butte to use the NP's main line west from Helena over Mullan Pass, and to the junction at Garrison. At Garrison, a connection was built to allow a through movement between the NP's line over Mullan Pass and its line that served Butte from the west. Even though these trains still had to operate on a grade of 2.2 percent moving west from Helena, they were operating on a main line instead of on the GN line that has been built and maintained to a lower standard as a secondary line.

The predominant flow of tonnage on the new consolidated route was to and from interchange with the UP at Silver Bow, seven miles west of Butte. The train moving northward in the direction of Great Falls usually carried between 5,000 and 6,000 tons, consisting mainly of diatomaceous earth destined for the oil fields in Alberta, to be interchanged with the Canadian Pacific at Sweetgrass, and copper anodes for the Anaconda Company's smelter in Black Eagle, near Great Falls. Most of the business moving southward was liquid petroleum gas and various other refinery products from Canada for interchange to the UP.

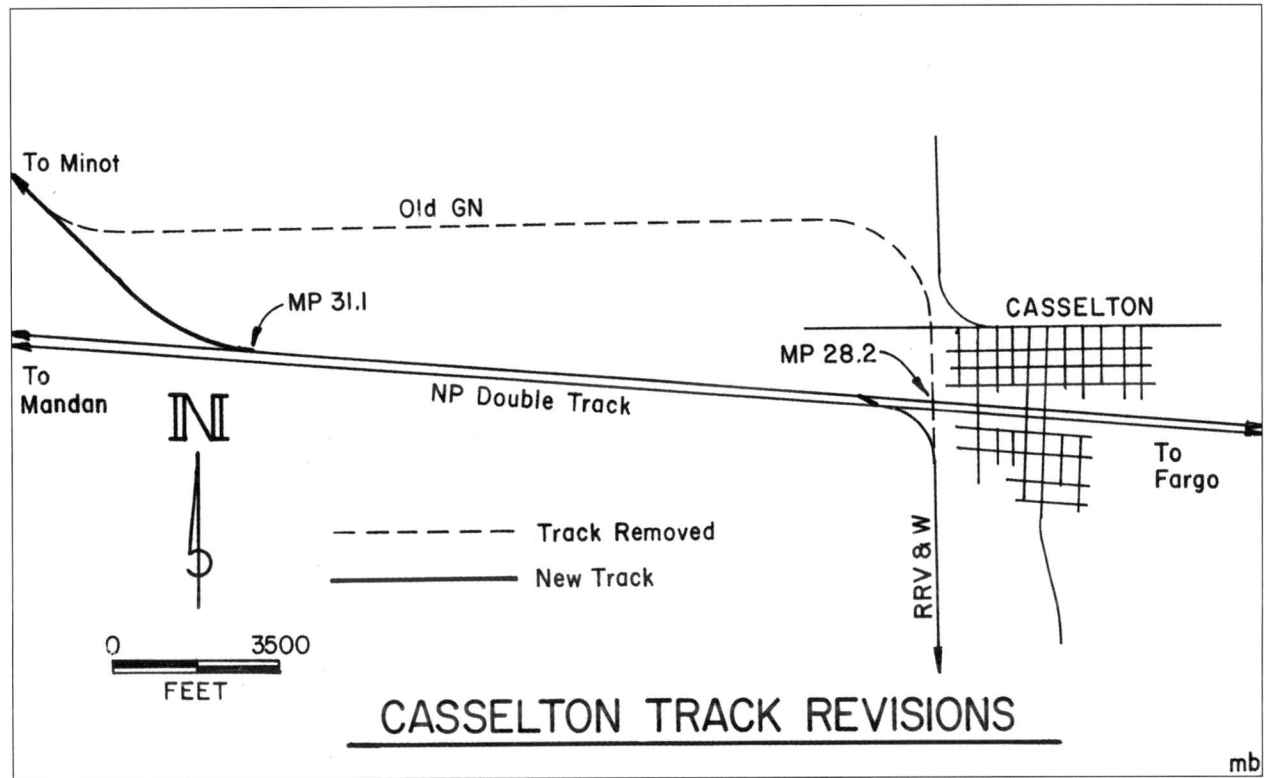

Figure 9.6.

CARLTON, MINNESOTA

Similar benefits were gained between Carlton, Superior, and Duluth through a connection built at Carlton (twenty-four miles west of Superior) to connect the NP's Staples–Duluth line with GN's line between Superior and the Iron Range and Grand Rapids. It allowed the abandonment of thirty-one miles of NP track east of Carlton.

HINCKLEY, MINNESOTA

A good example of the benefits of coordination of parallel, duplicate lines was between the Twin Cities area and Duluth–Superior. By constructing a connection between the GN and NP lines at Hinckley, the GN line could be used exclusively between Hinckley and Duluth–Superior, making it possible to abandon the somewhat parallel lines of the NP north of Hinckley. Later, parts of the NP line between St. Paul and Hinckley were abandoned, and the viable segments were turned over to a short-line operator.

This line rationalization allowed BN to save maintenance and ownership costs on 163 miles of former NP track and concentrate all business on the GN line between the Twin Cities and Superior. The GN line had a maximum grade of only 0.6 percent, compared to the grade of 1.92 percent on the NP line in St. Paul. Another advantage for using the GN line was the direct access trains would have to the new hump yard built at Northtown, whereas the NP line to Duluth came into St. Paul fourteen miles east of Northtown.

In this process of rationalization, an interesting challenge developed with regard to the Milwaukee Road, which had trackage rights on the NP over the entire distance between St. Paul and Duluth. Once BN had completed the connection at Hinckley and was able to consolidate its operation on the GN line, only the Milwaukee's trains used the NP line for through-train service. To head off the need for a heavy tie-renewal project on the NP line to keep train speeds above 10 miles per hour in some areas, it was necessary to negotiate with the

Milwaukee and its successor, the Soo Line, for operating rights on the GN line.

If the two daily scheduled trains operated by the Milwaukee had been small, local-type trains, having to operate at a slow speed line might not have mattered too much. However, since these were heavy-tonnage trains, a full-scale maintenance program was necessary. Further, neither the Milwaukee nor the Soo Line had the funds needed to acquire and maintain the NP line. The C&NW also became a tenant on the GN line when it decided to abandon part of its line north of Eau Claire, Wisconsin, which was part of its route between Chicago and Superior. By granting trackage rights on the former GN line, BN made it possible for other railroads to continue providing competitive rail service to the twin ports.

WEST FARGO, NORTH DAKOTA, BY-PASS

A new by-pass connection of 4.3 miles was built to connect the GN line that ran between Fargo and Grand Forks with the NP main line running through Fargo. The new route was intended to be used mainly for coal trains destined for an electric-power plant at Cohasset, Minnesota. Before the new connection was built, eastbound loaded trains had to be run into Dilworth, through the oft-congested area in Fargo and Moorhead. At Dilworth, the locomotive and caboose had to "change ends," a move that soaked up capacity of both yard and main tracks. Then, the train had to "retrace" its route by running west for 2.4 miles back to Moorhead Junction, then on to the GN line running through Fargo, and finally, onto the line running north to Grand Forks. All of this could take two hours or more until the new direct connection was built.

CROOKSTON, MINNESOTA

GN and NP each had a route between the Twin Cities and Canadian border crossings in northern Minnesota (Noyes for the GN) and in extreme northeastern North Dakota (Pembina, for the NP). The operating plan for the merger provided for concentration of traffic on the NP's largely double-track line between Minneapolis and Manitoba Junction, and then on NP's secondary line to Crookston. A connection with the GN's line to Noyes

was built at Crookston. Between the border and Winnipeg, traffic was moved on a line owned and operated by the Canadian National. Diversion of traffic from the GN lines via St. Cloud and Ada was a major factor in the decision to abandon 105 miles of the GN's main line between St. Joseph (just west of St. Cloud) and Fergus Falls, Minnesota. Parts of the Ada line (between Barnesville and Crookston) also were abandoned.

In the early 1980s, a plan for further line rationalization was carried out. Instead of diverging from the main line at Manitoba Junction en route to or from Crookston and Noyes, these trains were routed on main-line trackage via Fargo and Grand Forks to Crookston, thereby concentrating more business on main line segments. This change allowed the abandonment of forty-two miles of the NP line between Manitoba Junction and Crookston.

SEDRO-WOOLLEY, WASHINGTON

The NP operated a line between Auburn, Washington, and the Canadian border at Sumas, primarily for interchange with the Canadian Pacific. The volume handled through that interchange was large, in the range of sixty to one hundred cars daily in each direction. At the time of the merger, it was not possible to make an agreement with the CPR to combine this interchange at Vancouver or New Westminster, British Columbia, on BN's main line into Canada. The GN had a branch line running east from Burlington that crossed NP's line to Sumas at Sedro-Woolley at grade. A connection with a curve of 13 degrees was built to allow BN to run trains to and from Sumas on its main line through Interbay (Seattle) and Everett. That made it possible to abandon 48 miles of the NP line of 124 miles between Renton and Sumas. This change, along with diverting traffic from the NP line between Auburn and Pasco, helped make it possible to close the yard at Auburn.

CONNECTIONS ALREADY IN PLACE AT SIOUX CITY, IOWA

The connection between the Burlington and GN at Sioux City formed a through route for business moving between points in Minnesota and Omaha and Kansas

City. This connection was, of course, established long before the merger that created BN, but the corridor of 332 miles between Willmar and Ashland, Nebraska, needed to be upgraded to develop its potential for heavy unit trains of moving agri-business and coal for electric-power plants on a direct route. Under pre-merger labor agreements, a separate group of transfer crews known as "Pingers" had to be used to handle the Burlington's trains between Ferry (on the south bank of the Missouri River) and the GN yard in Sioux City, a distance of only 4.8 miles. Under the implementing agreement for the merger, road crews originating at Lincoln and Pacific Junction were allowed to operate into the GN yard, thereby eliminating the costly Pinger operation.

Development of business on the corridor through Sioux City was hampered by the weight restrictions on the C&NW's bridge over the Missouri River on which BN had operating rights. The rating for the bridge did not allow the movement of loaded 100-ton cars or loco-motives heavier than the smaller, lightweight GP classes of locomotives. By the mid-1970s, full trains of loaded 100-ton cars had to be run on a circuitous route via the Twin Cities and Galesburg to reach Kansas City and other points south of the crossing of the Missouri River. This movement was 312 miles out of route. The prospect of moving coal mined in the Powder River Basin to des-tinations in the Twin Cities area and to the rail-to-water transfer facility in Superior brought the decision to a head as to when to build a new, higher capacity bridge. A new bridge was completed at BN expense in 1982. The C&NW was given the right to use the new bridge.

LAUREL, MONTANA

There were only two points at which the three large rail-roads connected: Laurel and the Twin Cities. The NP operated a large yard at Laurel to which the Burlington and GN had access on a joint-facility basis. Burlington trains operating to and from its line to Sheridan, Alli-ance, and Lincoln reached Laurel on trackage rights on twenty-six miles of the main line of the NP between Huntley and Laurel. Trains running on the Burlington's line to Casper, Denver, and Texas arrived and departed from the west end of Laurel yard, through trackage rights

on twenty-three miles of the NP from Fromberg. The GN's line into Laurel was a secondary main line built south from Shelby and Great Falls to connect with the Burlington.

The GN's line from Great Falls and the Burlington's line running north from Casper were not completed until the early 1900s. Those lines and the creation of a joint terminal at Laurel were part of James J. Hill's vision to have only one terminal facility to connect the three railroads and, ultimately, to merge them into one large system. Under the operating plan for BN, the NP's main line via Missoula was designated the primary route for business moving between the Pacific Northwest, Kansas City, and Texas. The GN line through Great Falls would be retained for local service only. No physical changes had to be made at Laurel for it to fit into the operating plan for the merged company. The configuration of routes and all necessary connections for a consolidated operation had been determined by Mr. Hill many years prior to the 1970 merger.

CONSTRUCTION OF A NEW YARD AND UPGRADING OF THE PREFERRED ROUTE IN THE TWIN CITIES TERMINAL

Construction of a new hump classification yard at North-town (Minneapolis) was vital in getting the operations of the component roads integrated in order to achieve a large part of the cost savings projected in the merger plan. Construction of the new yard was started immedi-ately following M-Day. It allowed the closure of eleven yards of various sizes and capabilities throughout the Twin Cities area. The project also included construction of large maintenance shops for locomotives and cars, thereby replacing facilities at locations scattered in the terminal. The yard was built with sixty-three classifica-tion tracks, twelve receiving tracks, and nine departure tracks. The locomotive shop was built to handle the maintenance of three hundred units. Construction of the Northtown complex was completed in 1976 at a cost of $43 million. The new locomotive shop was built to handle the maintenance of 300 locomotive units.

Of all the capital expenditures BN made in its early years, there can be little doubt that construction of

Figure 9.7. An aerial view of the new hump classification yard built in the north side of Minneapolis and officially dedicated in January 1976. Construction of a new yard on property of the former Northern Pacific Railway allowed consolidation of eleven smaller yards at scattered locations in the Twin Cities, with greatly reduced operating costs and a substantial reduction in the time for moving shipments through the terminal. The new yard contained sixty-three classification tracks and opened with capacity for switching two thousand cars per day by gravity and electronic controls. Note the new locomotive-maintenance facility in the lower left area and the shop for car repair toward the upper left corner of the photo. With the economies gained with these new facilities, BN's investment of $43 million was recouped by 1978. BURLINGTON NORTHERN PHOTO.

the Northtown yard produced the highest "hard dollar" return on invested capital. By 1977 the project had already paid for itself. It enabled full integration of operations in the eastern part of the railroad.

In the Twin Cities terminal, all necessary connections between routes were already in place at the time of the merger. However, the line of the former NP that was used as the preferred route between Northtown and the Burlington's line to the east had to be upgraded considerably to facilitate movement of a much greater volume of business than apparently was anticipated when the consolidated operating plan was prepared. CTC had to be installed and two short segments of single track restored to form a continuous two-main-track operation. CTC also was installed on the 17.7 miles of paired track owned by BN and the Milwaukee Road between St. Paul and

St. Croix. At that time, BN took over the dispatching of the joint trackage. Upgrading of the primary route in the Twin Cities area made it possible to eliminate freight operations over the famed Stone Arch Bridge in Minneapolis, along with removal of two of the four main tracks on the former GN main line between St. Paul and Minneapolis.[1] The operating plan for the Twin Cities terminal is covered in detail in chapters 8 and 10.

---

1   Due to restricted clearance through the Minneapolis passenger station, freight cars higher than a standard box car could not be moved on the route over the Stone Arch Bridge and through the depot. Other than coal trains and "box car extras," freight trains could be moved only on the more congested route around the south side of Union Yard and through Minneapolis Junction (refer to figure 18.1).

Figure 10.1. Taconite pellets handled through BN's new dockside facility in Allouez, Wisconsin, being loaded in the *Mesabi Miner*, a new ore carrier owned by Interlake Steamship Company. BN's new facility contains thirty-six concrete silos on the dock, a car unloading facility, and a long conveyer belt from there to the dock. The new facility is located adjacent to the ore docks that had been used for loading "red ore." BURLINGTON NORTHERN PHOTO.

CHAPTER 10

MAJOR CAPITAL PROJECTS

*• covers projects not directly related to upgrading the roadway
or expanding its capacity for the movement of coal from the Powder River Basin •*

IN THE 1970S, A NUMBER OF LARGE CAPITAL improvement or expansion projects were carried out, in addition to those required for handling the massive increase in coal tonnage to be moved on BN and those required for implementation of the consolidated operating plan.[1] This chapter contains brief summaries of those projects not related to coal, ranging from a new rail-to-water transfer facility for taconite produced on Minnesota's Mesabi Range to replacement of several large bridges and expansion of the classification yard at Denver. The major projects directly related to coal are covered in chapter 31.

A list of the major investments BN was able to make in its early years is impressive and worth listing in this book. This chapter contains a list and some explanation of the capital investments made in those years. The funds for those investments came from capital generated or raised by BN through its earnings or from funds it raised from private institutions. Any projects for which funding was obtained from a government entity for bridges over navigable waterways or where a shipper that invested its capital to assist BN are identified with a note. Other than those few cases, all projects completed in those years were financed with private capital raised by BN or with cash generated from earnings.

THE TACONITE FACILITY AT ALLOUEZ, WISCONSIN

In 1977 a second facility was opened at Allouez (near Superior) for the transfer of taconite pellets from rail to lake freighters. The GN had built such a facility in 1967, but with the opening of additional taconite plants on the Iron Range in Minnesota, a second facility was needed. The new facility replaced the three large ore docks that had been used for unloading red (natural) ore, which was nearly mined out by the late 1950s. Together, the two loading facilities were projected to handle 12.5 million tons per year. The new facility was set up with an 18,000-foot-long conveyor belt to move taconite pellets from a rotary car dumper to concrete pellet-storage silos built on the dock. Four 16,000-ton unit trains of taconite can be unloaded in a twenty-eight-hour period.

Financing for the $67.4 million project was negotiated by Ray Burton, BN's assistant vice president–Financial Planning, in 1975 under a leveraged lease agreement in which BN sold the entire property then under construction to a trustee for the owner trust. Upon completion, BN leased the facility from the trust with a guarantee to meet the obligations under a $30 million credit agreement. The steel companies shipping taconite from the facility were required to ship minimum tonnages or prepay rentals to the lessor. The leveraged lease agreement was for twenty-five years and renewable for an additional fifteen years. BN had the option to repurchase the facility at the expiration of the lease.[2] This financing arrange-

---

1   Refer to chapter 9 for discussion of projects undertaken at Spokane, Sandpoint, and Casselton to connect segments of GN and NP lines to form the new preferred route between Minneapolis and the west coast.

---

2   Burlington Northern Annual Report for 1975, page 29, and for 1977, page 38.

ment has been evaluated by some financial experts as among the most complex and brilliantly negotiated agreements that had been completed up to that time.

REPLACEMENT OF MAJOR BRIDGES

Railroad bridges seem to have a character, strength, and durability that rival the Great Pyramids in Egypt. They were built long ago and have endured one hundred years or more of extremes in weather, floods, and impact from barges and ships that veer off course and strike them. Many railroad bridges are strong enough to handle cars and locomotives vastly heavier than those in service at the time the bridges were built. However, some bridges on BN had design load limits too low to carry 100-ton capacity cars and modern locomotives weighing over two hundred tons, and they had to be replaced in order for the most direct routes to be used for unit grain and coal trains. BN was fortunate to have the financial resources available to handle such large capital expenditures as replacing major bridges. Following is a list of bridges replaced with higher capacity structures in the 1970s and early '80s.

RULO, NEBRASKA   The bridge over the Missouri River replaced in 1977, due to the high tonnage projected to move through Kansas City. With the new bridge in service coal trains could be operated to Kansas City on the shortest, most direct route (16 miles shorter). Following are details of the project:

- retained original piers when new truss sections were moved in to replace the old trusses
- cost: $6.7 million
- train speed increased from 10 miles per hour to 40
- three 375-foot truss spans were replaced by two truss spans on the west end of the bridge and three 125-foot deck-type girder spans on the east end
- eliminated weight restriction of 220,000 pounds gross weight on cars and on 6-axle locomotives operated in multiple

SIOUX CITY, IOWA   BN was using a bridge owned by the C&NW Railway with load limits too low to allow 100-ton capacity cars and locomotives heavier than the older four-axle GP series. A new high-capacity bridge was needed to handle unit trains of coal and grain. Together with upgrading the track between Sioux City and Willmar, this corridor became a main line for coal moving to power plants in the Twin Cities area and the rail-to-water transfer facility in Superior. BN completed a new bridge in 1981 and gave the C&NW operating rights on it. Following are details of the project:

- length: 1,465 feet
- 425 foot truss span over a navigable river, 81 feet above water
- cost: $14 million (including cost of dismantling the old bridge)
- eliminated circuitous routing of 308 miles (via Galesburg) for loaded 100-ton cars

BEARDSTOWN, ILLINOIS   The 280-foot swing span over the Illinois River was replaced in 1974 with a 325-foot-long lift span. The new bridge was named the Prize Bridge in the movable-span category by the American Institute of Steel Construction for being "well proportioned" and for having "a clean look." Its cost of $4.5 million was borne in large part by the U.S. Army Corps of Engineers as a result of the old bridge being declared a hazard to navigation.

PASCO, WASHINGTON   A 392-foot-long lift span was placed over the Snake River in 1972 for $6 million. The old bridge had weight restrictions and was evaluated as a hazard to navigation; hence, the project was federally funded.

PORTLAND, OREGON   The bridge over the Willamette River was damaged by a ship in October 1978. A 521-foot swing span (the longest in the world) was replaced with a lift span of the same length for $34 million, with U.S. Coast Guard funding in 1989.

WILSONVILLE, OREGON   The bridge over the Willamette River was replaced in 1975 for $5.2 million. The new bridge is 1,224 feet long and was replaced due to restrictions on weight and train speed.

PRESCOTT, WISCONSIN   A single-track 369-foot swing span bridge over the St. Croix River was replaced with a 279-foot lift span in November 1984, with federal funding. A line change was made on the east side of the bridge to reduce curvature and allow an increase in speed from 10 to 25 miles per hour.

Figure 10.2. Steel truss sections for the new bridge at Rulo were assembled on each side of the river and then lifted onto falsework, the temporary structure built to support new truss spans until the old trusses would be removed for replacement with the new spans. Once the old bridge spans were taken off the piers, barge-mounted cranes moved the new spans into place. The original 1887 piers were still in a condition and at the strength suitable for the new bridge. An interruption to rail traffic of only forty-eight hours was required to get the new spans in place and open the bridge for service. With the new bridge, the haul of 100-ton loaded cars between Lincoln and Kansas City was reduced by twenty-eight miles. The bridge at the right side of the photo carried a state highway. BURLINGTON NORTHERN.

Figure 10.3. New bridge under construction over the Missouri River at Sioux City, completed in 1982. At this time the old bridge to the left was still in service, with BN holding trackage rights on it. Weight restrictions on the old bridge prevented unit trains of 100-ton cars and modern, six-axle locomotives from using the most direct route to reach Kansas City and Omaha from the north and west parts of the BN system. With the new bridge in service, 312 miles were saved on 100-ton loads that previously had to move on a circuitous route via Galesburg. BN NEWS.

PASCO, WASHINGTON   Three of the 250-foot-long truss spans over the Columbia River were replaced for $3.1 million in 1983.

CORAM, MONTANA   In 1984 BN replaced its steel trestle bridge at Coram, fifteen miles east of Whitefish, on its northern corridor main line. The new bridge was constructed as a deck plate girder bridge of 929 feet over the Flathead River.

BRAINERD, MINNESOTA   A new 632-foot deck plate girder bridge was constructed over the Mississippi River on the coal route between Staples, Minnesota, and Superior, Wisconsin, in 1983. This line carries the trains destined to Midwest Energy's facility in Superior, used for transfer of coal to lake freighters for movement primarily to power plants operated by Detroit Edison.

EXPANSION OF YARD AT DENVER   During 1975 and 1976, two long tracks were built next to the existing main track from the 38th Street yard office to the 31st Street yard office to hold coal trains. Having these additional tracks in service made it possible to keep coal trains from using track capacity, thereby taking away some of the operating capability needed for "general switching" in the yard. Coal trains had to be held at times for crew availability, to swap out locomotives due for maintenance at the Denver shop, or to add the locomotive units needed on the heavy grade south of Denver.

Also at this time, fifteen additional tracks were built at 38th Street, to provide enough capacity to bring in the switching done by the Colorado and Southern Railway (the C&S, a subsidiary of BN) at its Rice yard. Getting the yard operations combined reduced operating and maintenance expenses and eliminated the delays and expense inherent in transferring cars between yards. In the labor agreement negotiated for consolidation of the two yards, C&S employees were allocated the jobs on two-yard switching crews and retained the right to continue switching the industries that the C&S had served. The 38th Street yard also took over the switching of Santa Fe trains, which the C&S had performed under contract in the Rice Street yard. Part of the Rice yard property was then turned over to the local government authority that was in charge of building the new stadium for the Denver Broncos professional football team. The Elitch Gardens amusement park occupies much of the former Rice yard property.

PROJECTS CRITICAL TO START-UP IN 1970S[3]
A few large projects were vital in getting the operations of the component roads integrated in order to achieve the cost savings projected in the merger plan. Three of the

---

3   For additional detail, refer to chapter 9, "Building the Connections."

most urgently needed projects were the new hump classification yard to be built at Northtown (Minneapolis), the new bridges at Spokane that would connect the main lines of the NP, GN, and SP&S, and the connections at Sandpoint and Casselton between the GN and NP main lines. Construction of the new yard at Norththown was started almost immediately following M-Day. It allowed the closure of eleven yards of various sizes and capabilities throughout the Twin Cities area. The project also included the construction of large maintenance shops for locomotives and cars, thereby replacing several small facilities at scattered locations in the terminal. The Northtown complex was completed in 1976 at a cost of $43 million. A locomotive shop was built to handle the maintenance of three hundred units.

At Spokane, it was necessary to build a large, impressive new bridge over Latah Creek to connect the lines of the three component roads. In addition to the large bridge, fourteen smaller bridges had to be built or modified. A line change of 5.5 miles was made on the west side of Spokane. The entire project was completed in December 1972, at a cost of $18.8 million. Upon completion, it was possible to take full advantage of the NP's elevated low-grade double-track line through downtown Spokane and to abandon the operations on the GN's line that had a grade of one percent and numerous road crossings in the city.

At Sandpoint, a new connection of two miles had to be built, including a bridge of 997 feet over Sand Creek. The new line was needed to allow trains to move without delay between the main lines of the GN and NP. The trackwork and bridge were completed in 1973 at a cost of $2.7 million. With the new high-speed, direct connections built at Spokane, Sandpoint, and Casselton, the new preferred route was in service as envisioned in the operating plan for the merged company.

A new connection had to be built at Casselton, North Dakota, to enable trains to move to and from the GN portion of the preferred route (between Casselton and Sandpoint) and the former NP's line that was designated the preferred route between Minneapolis and Casselton.

North Kansas City

There were a few major projects undertaken by BN's predecessor roads in the late 1960s that were completed shortly after the merger. Construction of a large computerized hump classification yard at North Kansas City was started by the Burlington (CB&Q) in 1967 and completed in 1970 at a cost of $10 million. It replaced a large flat switching yard that had become inadequate to handle the increasing business from the large new industrial complex located on the Burlington, as well as the large and frequent surges of grain moving to and from several large grain elevators in Kansas City or to ports on the Gulf of Mexico.

Pasco

A program to modernize and extend automation in the Northern Pacific's hump yard at Pasco was started shortly before the merger. Tracks in parts of the yard were re-engineered to improve the rollability of cars. A new system for computer control of the hump process was installed. Two receiving tracks were added, and the pull-out end of the bowl was reconfigured to improve productivity and speed up the flow of cars from the class yard and into the departure yard. The hump yard at Pasco was opened as a new facility in 1955, but by the mid 1960s, technology had improved to the extent that this investment was needed. Also, it was expected that the volume of cars moving through Pasco would increase under the operating plan for the merged company.

Libby line change

Another large "carryover" project was a line change necessitated by construction of a dam on the Kootenai River on the Great Northern's main line in western Montana. The project cost, of $110 million, was financed by the federal government through the U.S. Army Corps of Engineers. Having the new line in service reduced the amount of track BN had to own and maintain by 14.5 miles and led to seventy-one fewer curves to maintain. However, the new line had the disadvantage of grades of one percent in both directions, compared with lower grades of only 0.3 percent and 0.7 percent on the old river-grade line along the Kootenai River.

Having to operate trains through a tunnel seven miles long and maintain a high-capacity ventilation system also carried some disadvantages. In being limited "forever" to a single track through a seven-mile tunnel, with grades of one percent on each side of it (and with a much greater volume of trains after the merger), a bottleneck was created, causing train delays to build up rather often. The options for increasing line capacity through this area are limited by the high level of expenditure that would be needed to overcome these limitations. Operating on the new line was projected to cost an additional $306,400 per year. To offset that cost, BN was awarded a one-time payment of $6.3 million in the agreement for construction of the line change.

Granite Lake line change

A few large roadway-improvement projects completed shortly before the merger are worthy of note as well. It was anticipated that the operating capability provided by these projects would fit in well with the plans of the new company. One such project was the major line change the NP made at Granite Lake, Idaho, on its line between Spokane and Sandpoint. Curvature was reduced from 8 degrees (restricted to 30 miles per hour) to 1 degree, 20 minutes. This line segment was to become part of the new preferred route for transcontinental traffic.

Line change at Marias Pass

Another was the line change on the GN's line a short distance east of the Continental Divide. It was one of a series of line improvements the GN had carried out in almost every year of its history. Curvature was reduced, but in so doing, the grade was increased slightly from 1.0 percent to between 1.1 and 1.2 percent. This line also was to become the preferred route after the merger.

New locomotives maintenance shop at Lincoln

On the Burlington, a large new locomotive-maintenance shop was built at Lincoln in the mid-1960s with capacity to maintain 410 locomotive units. Because Lincoln would continue to be a crossroads location after merger, BN was fortunate the new shop was already built and in service before M-Day.

A FEW LARGE PROJECTS COMPLETED SHORTLY before the merger are worthy of note as well. It was anticipated that the operating capability provided by these projects would fit in well with the plans of the new company. One such project was the major line change the NP made at Granite Lake, Idaho, on its line between Spokane and Sandpoint. Curvature was reduced from 8 degrees (restricted to 30 miles per hour) to 1 degree, 20 minutes. This line segment was to become part of the new preferred route for transcontinental traffic.

Of all the capital expenditures BN made in its early years, there can be little doubt that construction of the Northtown yard produced the highest "hard dollar" return on invested capital. It was reported that by 1977, the project had already paid for itself. More than any other single project, it got the operation fully integrated at the east end of the railroad.

Soon after completion of the new yard at Northtown, a major investment was made on the thirty-three-mile line between Northtown and St. Croix, at the Minnesota-Wisconsin border. CTC was installed on both main tracks, including the 17.7 miles of Milwaukee-owned track that were used as a paired track arrangement between St. Paul and St. Croix. BN took over the dispatching of the joint trackage. This project also made it possible to eliminate operations over the famed Stone Arch Bridge in Minneapolis, and two of the four main tracks on the former GN main line between St. Paul and Minneapolis were removed.

CHAPTER 11

MODERNIZING AND EXPANDING MECHANICAL MAINTENANCE FACILITIES

*• construction of new locomotive-maintenance and -servicing facilities to handle
a growing fleet of locomotives • hiring and training of hundreds of
skilled employees to staff the new maintenance facilities •
• setting precedence in technology, work practices, and productivity •*

IN ADDITION TO THE MAJOR PROJECTS CARried out to improve BN's roadway and train-control systems, a number of large, impressive projects were undertaken by the Mechanical Department to improve its capability for the maintenance of cars and locomotives.[1] Following is a brief listing of the most significant mechanical projects completed in BN's first fifteen years.

NEW CAR AND LOCOMOTIVE SHOPS IN THE TWIN CITIES TERMINAL

As part of the consolidation and modernization of yards and support facilities in the Twin Cities terminal undertaken upon the merger in 1970, new car- and locomotive-maintenance shops were built at Northtown. The locomotive shop was built to handle running maintenance for 350 units. Two tracks inside the shop were set up for fueling and servicing. Later, when these functions were moved outside to a new facility, enough capacity was freed up so an additional two hundred units could be brought into Northtown for running maintenance. The

car shop contained four tracks with capability to repair an average of about twenty-five cars per day. The cost of the new shops was included in the amount of $44 million invested in the new hump yard complex at Northtown. The new shops permitted old repair facilities to be closed at Union Yard, Minneapolis Junction, Daytons Bluff, Mississippi Street (in St. Paul), and Northtown.

TECHNICAL TRAINING

John Blaha, manager of Technical Training for the Mechanical Department in the 1970s, recalls visiting many technical-training schools to recruit graduates of their programs. To illustrate the all-out effort made by those schools to produce candidates for employment by BN, among them the Western Nebraska Technical College in Chadron and Scottsbluff: Michael Holley, an employee in BN's Havelock Shop, was loaned to the school as an instructor. Mike later became the general foreman at Alliance. BN supplied the training aids and cars for the students to work on. Later that year, the school added a course to train new machinists. Together, three hundred students were graduated from the two courses. John also recalls BN moving old passenger cars into Alliance to be used as classrooms for training new car-repair personnel.

---

1   For major mechanical projects undertaken on the Alliance Division to support the increasing tonnage of coal being moved, refer to chapter 42.

BN employees were offered the chance to improve their skills through the "Keep Pace" program in which BN would finance their enrollment in a correspondence course offered by the Railroad Educational Bureau. Many participants advanced to supervisory and managerial positions. Through BN's "This Way Up" program, shop craft employees could gain additional experience by working as relief supervisors. In a short time, many were sent to a two-week supervisory school at the headquarters building in St. Paul. Graduates of the program continued to work as relief supervisors until they were selected to fill a full-time supervisory position.

SHOP CONSOLIDATION AT THE TIME OF THE MERGER IN 1970 AND LATER

Soon after the 1970 merger, heavy-locomotive repairs were consolidated in the West Burlington and Livingston shops by closing the shops at Dale Street in St. Paul and South Tacoma. In the early 1980s, heavy-repair shops at Livingston, Hillyard, and Springfield also were closed, leaving the large facility at West Burlington as the only shop designated for heavy repairs. Limited capacity for heavy repairs was built into the new shop at Alliance. Heavy repair of freight cars was consolidated in the Havelock shop at Lincoln. The car shops at St. Cloud, Brainerd, and Springfield were closed by the mid-1980s.

CONSTRUCTION OF A NEW WHEEL SHOP IN HAVELOCK IN 1977

A fully automated wheel shop was equipped with the state-of-the-art, best-available technology developed and built by the Hegenscheidt Company in Germany and the Simmons Niles Company, major suppliers of heavy industrial machinery. The new shop contained a level of technology in robotics that put it ahead of its time. It had the capacity to supply all of the 97,500 wheel sets used across the entire BN system, and replaced the five old wheel shops in service at various locations. The new shop cost $8.6 million.

CHAPTER 12

A NEW ORGANIZATION STRUCTURE

THE REGION CONCEPT

• to place responsibility for tactical decision-making close to the local level •
to keep the new company from becoming top-heavy • to be a merger of equals •

FROM THE EARLY STAGES OF MERGER planning, there was great concern over the size of the company and what it would take to manage it effectively. It would be about three times the size of any of the three component roads and much larger than any other railroad company in existence at that time. The Penn Central was an exception, but it could not be used as a model either for organization structure or on how to run a large railroad. None of the senior executives at any of the companies (or any other railroads) had experience in managing a company as far flung or with as much activity as BN would have.

Each of BN's component lines had been organized with centralized administrative and support departments, among them Personnel, Law, Finance, and Accounting. The Sales and Marketing function (historically known as the Traffic Department) had maintained small offices at key locations, both on and off the lines of the railroad, in addition to a large staff at headquarters. For Operations, the head of the department (vice president–Operations) was based at the headquarters office. The Engineering, Mechanical, and Transportation functions were set up with staffs at headquarters and reported to the vice president–Operations. At the field level, each railroad was organized with two general managers, for "Lines East" and "Lines West." Each general manager had three or four operating divisions reporting to him.

The operating divisions were headed by a superintendent, who held responsibility for both line of road and terminal operations. In addition to having responsibility for the transportation function, the division superintendents on the GN and NP were responsible for the basic activities in roadway maintenance and the maintenance of equipment. The Burlington had been similarly organized for many years, but responsibility for roadway maintenance and the mechanical function had been moved from the division superintendent to the headquarters staff in 1966. In the merged company, the basic engineering and mechanical functions on the former Burlington territory were put back under the division superintendent.

Initially, six regions were set up, each headed by a vice president.[1] The regions had responsibility for both the operating and marketing functions. The region vice presidents (RVP) reported directly to the executive vice president. They were seasoned, well-established executives, although not all of them had held operating positions with line authority. In the first ten years, one of the RVPs had headed the Labor Relations Department, and another, the Human Resources (the Personnel Department, at that time) Department, and two had backgrounds in Sales and Marketing. The regions were given a great deal of authority to deal with tactical and

---

1   Six regions were established at the time of merger.

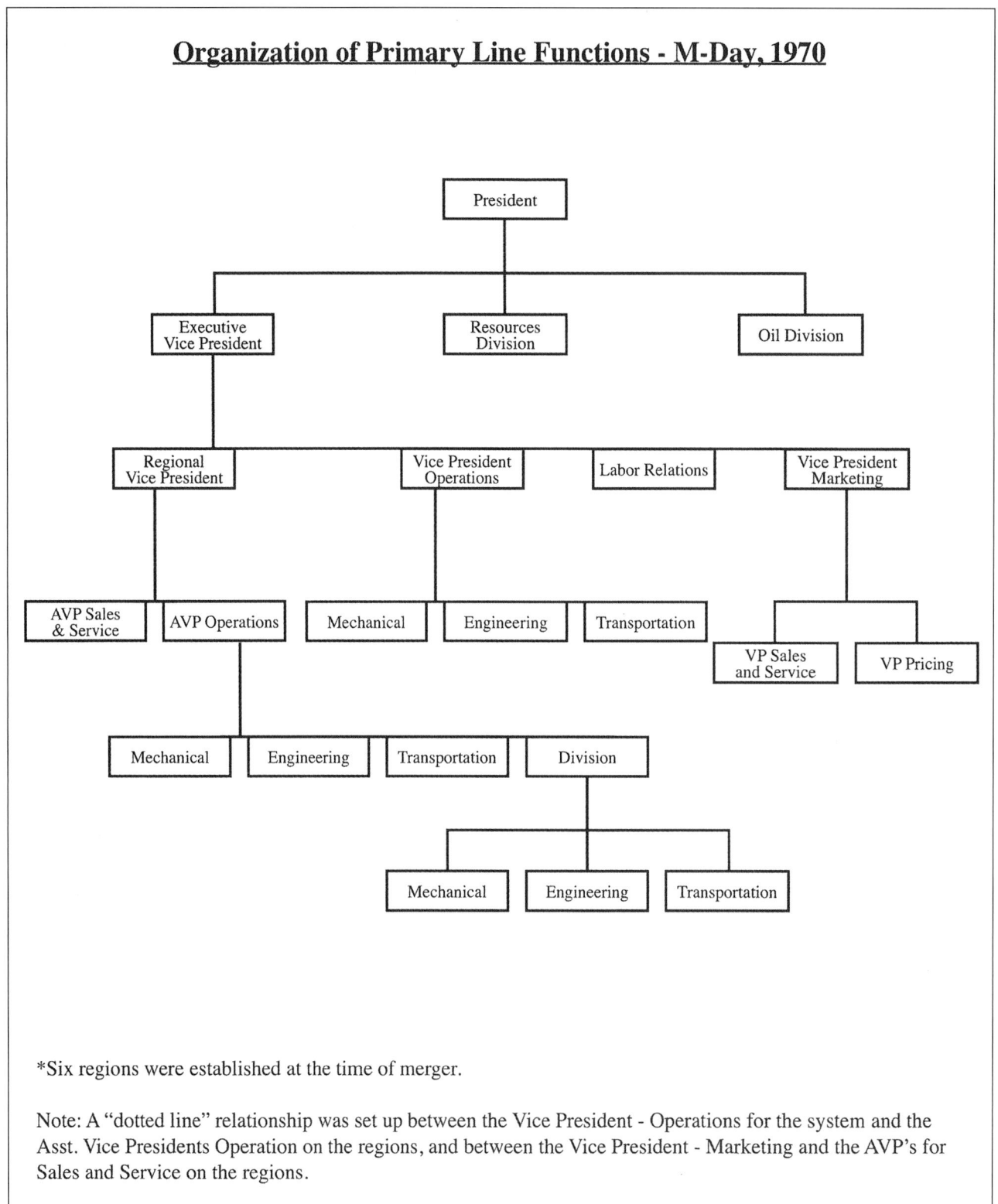

Organization of Primary Line Functions - M-Day, 1970

*Six regions were established at the time of merger.

Note: A "dotted line" relationship was set up between the Vice President - Operations for the system and the Asst. Vice Presidents Operation on the regions, and between the Vice President - Marketing and the AVP's for Sales and Service on the regions.

Figure 12.1.

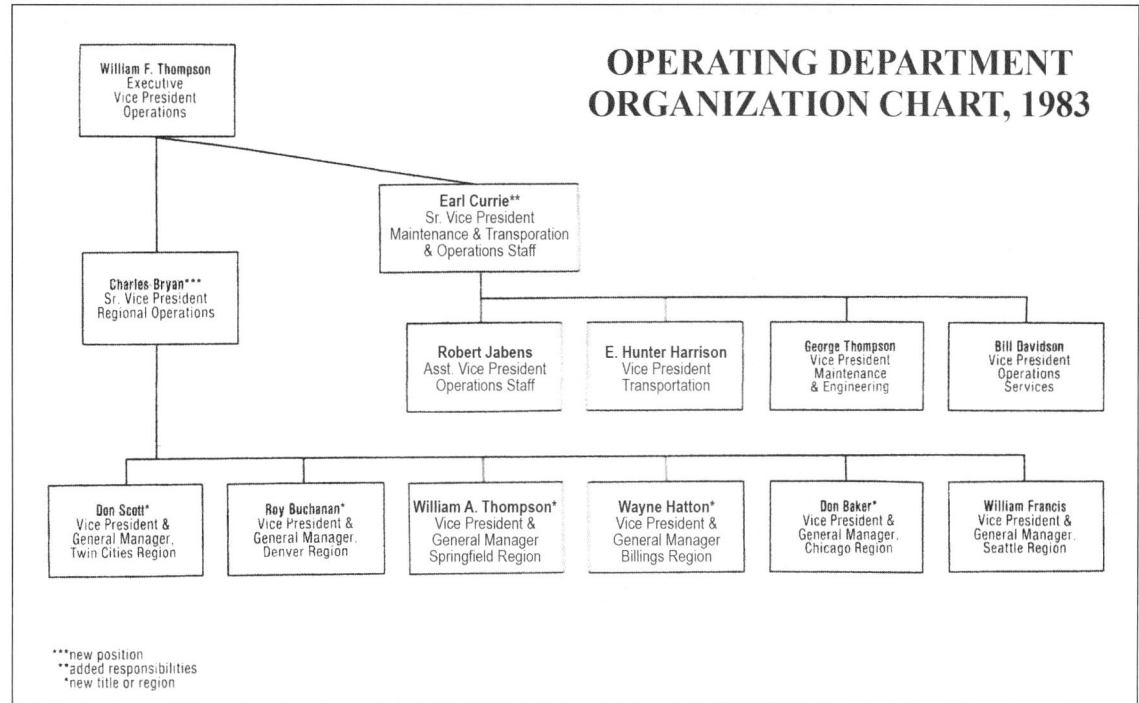

Figure 12.2.
BN News,
December
1983.

procedural matters, to help insure there would be a quick resolution of problems affecting service. It was felt that managers at the region and division level were in better position than those at the headquarters level to gather the facts needed for satisfactory resolutions or answers, and to give the direction needed to correct problems arising from customers, employees, union officials, and community leaders.[2]

By having joint responsibility for Operations and Marketing, the RVP had the authority and stature needed to handle day-to-day tactical decision-making. For cases of major importance requiring additional resources or changes in policy, the RVP had direct access to the Executive Department. Officers at the headquarters level were expected to set policy and evaluate proposals made by the regions for additional resources and performance of the regions.

Having the regional vice presidents reporting directly to the Executive Department instead of to either the

headquarters Operations or Marketing departments created some tension with the system officers who headed those functions. This was a major change from the traditional "straight up" organization structure that all of BN's line officers had worked under for their entire careers.

In the first two or three years after the merger, having the regional concept in place helped establish BN as a company that was responsive, staffed with people having the ability to act and who had direct access to the senior level of management for assistance in resolving problems. Gradually, over the first ten to twelve years after the merger, the region concept was modified. In the late 1970s, the region Marketing function was taken from the RVP and placed under the "traditional" Marketing organization. The reporting relationship of the RVPs was then changed from the Executive Department to the senior vice president in charge of Operations on a system basis.

Shortly after the merger with the Frisco Railroad in 1981, it was decided that there should only be "operating people on operating jobs." With that decision, the regions would no longer be headed by executives who had not advanced through the operating chain of command in their careers. That meant that by 1981, we had

---

2   A "dotted line" relationship was set up between the vice president–Operations for the system and the assistant vice presidents–Operation on the regions, and between the vice president–Marketing and the AVPs for Sales and Service on the regions.

come the full circle and were back to working in a very conventional railroad organization.

When the regions were set up in 1970, they had large staffs for managing the operating function. There was considerable overlap of responsibility and in the division of work among the three levels of organization in place to manage operations, i.e., at the system, region, and division level. Nearly all of the people working in the regional offices had line-operating or maintenance experience on the division or terminal level. Because there was not always a great deal of work for some of them to do on their new region-level jobs, there was a tendency for some of them to second-guess, overlap, or otherwise involve themselves unnecessarily in activities traditionally managed at the division level. Since all exempt (management) employees were given lifetime job protection under the merger conditions, it had been necessary to establish more positions than required to manage operations activities at the region level. This problem was not fully resolved until the early 1980s when major reductions in staffing were made at the region and system level.

In addition to Operations and Marketing, each region was also provided staff members to handle the Human Resources, Community Relations, Public Relations and Labor Relations functions, reporting to the RVP. These resources gave the regions even more capability to keep day-to-day, tactical matters at the region level, rather than having them handled by the staffs at headquarters. In time, the reporting of these specialties reverted from the regions to their respective departments at headquarters.

At the start of the merged operation in 1970, regional headquarters were established at Chicago, Omaha, Minneapolis, Billings, Seattle, and Portland. Two years later, the region headquarters office in Omaha was moved to Denver. In the late 1970s, the territory and the operating responsibility of the Portland Region was combined with the Seattle Region, although a regional office was maintained at Portland for four more years. Even though there was only one division in the Portland Region, a complete staff of officers was maintained in Portland from M-Day until 1978. Since Portland had been the headquarters of the SP&S Railway, it was important for the merged company to show recognition of the importance of the customer base we had in Oregon and along the Columbia River in Washington, and the

highly competitive north-south route for traffic moving between the Pacific Northwest and California, by having a senior-level executive based in Portland. Also, since the Union Pacific had a general manager headquartered in Portland, it was important for BN to have a strong presence there as well.

In the original organization plan, regions were to be headquartered at only four locations, Chicago, Omaha, Minneapolis, and Seattle. In addition to adding a region at Portland, it was decided shortly before M-Day that because of the distance between headquarters of the Twin Cities and Seattle regions, a regional office should also be established at Billings. There also were some substantial political considerations that made it desirable to have an executive-level officer and staff located in Montana. As mentioned earlier, there was also a need to have enough jobs for all of the people who would still have to be employed in those early years. The regional office in Billings was maintained until 1987.

When the Frisco Railroad was merged into BN, Springfield, Missouri, was set up as headquarters for the additional territory brought into BN. Shortly after that merger, another region was set up at Fort Worth, with only one division reporting to it. That region remained in place for a very short time. Its territory was soon split between the Denver and Springfield regions.

Overall, the regional concept worked out quite well, and in time, the region staffs were adjusted to the level commensurate with the responsibilities that a regional office should have. It took some time to sort out the roles and accountabilities for each of the three levels in the operating organization. Having a high-ranking executive level position at each of the region headquarters locations on the system was of great value in resolving local customer-service issues in the early days under the consolidated operation. It was possible to handle questions, communicate with employees, and settle their concerns about the new implementing agreements much more quickly than if it had been necessary to elevate all or most of them to the system level. RVPs also were able to give BN a strong presence in communities of all sizes and at the state level.

In 1981 the title of the heads of the regions was changed to vice president–general manager, once the regions had fully evolved back to traditional operating organizations.

PART II

THE PRICE PAID FOR THE MERGER

CHAPTER 13

LIFETIME JOB PROTECTION AND GUARANTEED WAGES IN EXCHANGE FOR MANAGERIAL FLEXIBILITY

*...the best merger agreement in history.... While some western Congressman may question this merger,
our members realize we are much better off after the merger than we are now.*
C. L. DENNIS, PRESIDENT, BROTHERHOOD OF RAILROAD AND
AIRLINE CLERKS, "BEST EVER JOB PROTECTION AGREEMENTS SIGNED,"
BURLINGTON BULLETIN, JANUARY–FEBRUARY–MARCH 1968

*If everybody lives up to commitments, there is
the potential here to make this one of the best railroad mergers.*
UNNAMED GENERAL CHAIRMAN, QUOTED IN *RAILWAY AGE*, JULY 27, 1970, PAGE 22

*...the protection thereby afforded, providing at it does job security
as well as monetary benefits, could hardly have been achieved except for the merger.*
ICC DECISION APPROVING MERGER ISSUED NOVEMBER 30, 1967,
QUOTED IN LETTER TO EMPLOYEES, DECEMBER 29, 1967

*Our employees will enjoy a level of employment security which is
unprecedented in American industry. The merged company, in turn, will function with
a freedom from union restraint which is of a degree unknown in railroading for almost a century.*
L. W. MENK, PRESIDENT, BURLINGTON NORTHERN,
BN NEWS, SEPTEMBER 1970, PAGE 8

WITH LABOR AND BENEFITS COSTS RUNning close to 50 percent of total operating expenses in each of the four roads involved in the plans for merger, it was obvious that a large part of the projected savings were to be achieved through reductions in employment. It had to be expected that the labor unions would strongly oppose the merger in hopes of stopping it or, at least, to obtain protection or severance payments for employees who stood to lose their jobs. In denying the application for merger, the ICC clearly stated its concern that appropriate labor-protection conditions be worked out with the unions.

The ICC stated that "negotiated agreements of a tentative nature ... reported in the press since closing of the formal record, indicate that the parties themselves have been voluntarily resolving these issues." The ICC expressed its belief that "attrition conditions would ... be desirable and without undue harm to the interests of either management or labor."[1]

In March 1966 the ICC handed down its decision to deny the application to merge, but stated the applicants could apply for reconsideration if they would negotiate agreements with the railroads opposing the merger, so as to preserve and enhance rail competition. Also, the applicants would have to negotiate agreements with the unions for protection of employees who stood to lose their jobs, or whose compensation would be adversely affected. The ICC recognized that progress had been made in resolving those issues, but that the unions remained opposed to the proposed merger.

Getting approval for the merger was seen as vital to the long-term health and survival of the northern lines and the Burlington. The applicants negotiated agreements with the C&NW, Milwaukee, and Soo Line and applied to the ICC for reconsideration in July 1966. By the end of 1967, the labor-relations officers and representatives of employees in all crafts successfully negotiated agreements for wage guarantees and job assurance. The ICC handed down a favorable decision sixteen months later, on November 30, 1967.

To get the unions to agree to withdraw their opposition to the merger, the applicants agreed to provide lifetime job protection for all contract employees who would be on the payroll as of the effective date of the merger. It was also agreed that the level of wages earned by employees in a pre-merger "test period" would be protected, provided they exercised their seniority on the highest-paying position they could hold. Any shortfall in earnings through no fault of the employee would be made up by the new company.

However, before management agreed to the provisions of the employee-protection agreement, it was vital that implementing agreements would be agreed to and signed. Those agreements would give the company the flexibility needed to establish new, combined routes for train service, and to determine the level of activity to be established at the various yard and shop facilities across the system. Once the implementing agreements were in place, the employee-protection agreement was finalized. Listed below are several key elements covered in the implementing agreements:

- Consolidation of seniority districts, e.g., from thirty-eight to five for train-service employees on the northern lines, and from over a thousand districts for clerical employees to six new districts (four on-line and two in the General Office). For the shop crafts, change from a point-seniority basis to seniority by one of the fourteen districts that were established.
- For employees to be transferred, the option of accepting a payment of 25 percent of the fair market value of their house and retaining ownership, or accepting payment of 100 percent of fair market value and having BN take ownership of the house.
- Allowing BN to designate whether the NP or GN schedule of agreements would apply in each consolidated terminal.
- No provision for severance payments except at the company's option, and no severance to be paid to employees who refused to accept new assignments or transfers.
- Agreement to apply the contract (work rules) of the former road having the "predominant miles" on the runs for through trains that would be operated on portions of lines owned by both the GN and NP.

SENIORITY: ISSUES AND TENSION

The consolidation of seniority rosters was left to the unions. Determining the fairest, most equitable way that it should be done proved to be a difficult task in some areas. Of all the issues that arose in the first few years after the merger took place, there probably was more tension on the preparation of new seniority rosters than any other matter that directly affected operating employees. An example of the complexity and stress that employees felt at some locations has been revealed by John Langlot, who was a Great Northern employee in train service at

---

1　　Great Northern Pacific & Burlington Lines, Inc.—Merger, Etc.—Great Northern Railway Company, et al., Interstate Commerce Commission, Finance Docket No. 21478.

Spokane. John was promoted to conductor in 1966. At the time of the merger, he did not yet have enough seniority to work regularly as a conductor or to hold a regular job as a brakeman. However, he was old enough to be the senior man on the Brakeman's extra board. On the day the merger took effect, John had caught a ten-day vacancy on an "outside" job, on a local assignment based at Oroville, about 310 rail miles from Spokane, or 180 miles by highway. A few days after M-Day, at the close of his ten-day stint at Oroville, John drove to the yard office at Spokane, intending to visit the crew office to mark back up on the Brakeman's extra board.

He found he had to park two blocks away from the crew office because of the large number of employees who were there dealing with seniority issues. It took John thirty minutes just to get inside the door. He recalled that the crew clerks looked like they were about to have nervous breakdowns. With the consolidation of seniority that had been made in those first few days after the merger, John found that instead of being at the top of the Brakeman's extra board, he was now seventy-five times out.

John went on to say:

As unhappy as my experience was, it was far from the worst. I know that some men went five or six days without working because they would bump onto a job [assigned or regular job, i.e., not on the extra board] and then get bumped just before the job was called to go to work. Then, they had to go through the same thing on some other job. For me, it was back to the Brakemen's extra board for a long time.

In the first effort to set up a consolidated seniority roster, the separate rosters for the GN, NP and SP&S were set up by dovetailing, based on date of hiring without regard to an employee's prior employer. On this basis, John had dropped behind seventy-five former NP and SP&S employees. However, there was no lack of work, since business was good at that time. The company needed "pilot" conductors and engineers for qualifying trips with employees who were to start working on lines they had never worked before. As a result, the Brakeman's extra board was able to absorb most employees, even though they had been set back many positions on the dovetailed roster.

Six months after the merger, the union developed a new approach to the problems of consolidating seniority,

through what was called the "ton-mile roster." It was based on the relative amount of business each of the component roads had been handling before the merger. On this new roster John recalls he moved up about twenty-five positions. However, by that time, employees had become qualified on all lines they could work on, so there was less need for pilot conductors to assist in the familiarization process.

About nine months or a year later, the ton-mile roster was replaced by a new version of a dovetailed roster, but this time, it was based on prior rights on employees' former railroad. Even though this change was found to be more reasonable, fair, and equitable, the seniority issue continued to cause hard feelings and dissension among employees of the three railroads for years to come.

In recalling those difficult times, John wrote,

I blame the United Transportation Union, the UTU, for the seniority mess. Not long before the BN merger, the ORC&B [Order of Railroad Conductors and Brakemen], and the BRT [Brotherhood of Railroad Trainmen] had merged to create the UTU. As the BN merger approached, there was still a power struggle going on among the union top management about who would control the new union. The UTU spent all their energy fighting the merger right up to the time the merger was approved. They made no effort to look ahead to settle the issue of seniority consolidation if the merger was approved. If only they had agreed to the merger, the union could have written their own ticket in terms of dealing with seniority issues.

In John's opinion, the company also should have wanted to resolve seniority issues before the merger actually took place.

John has also commented on the seniority issues involving the Locomotive Engineers. The Brotherhood of Locomotive Engineers (BLE) used the Fireman's date of hiring in preparing the consolidated seniority roster. As a result there were NP firemen who were not yet promoted to engineer who were put ahead of GN engineers who had been working as engineers for between ten and thirteen years. John recalls,

This caused real hatred between the GN and NP Engine-men that has lasted to the present day, even with the

THE EVOLUTION OF SENIORITY ISSUES
FOR OPERATING EMPLOYEES AT SPOKANE

1. Before the date of the merger, the ton-mile (equity roster) arrangement as stated in Implementing Agreement No. 2 was to govern. However, it had not been ratified by the effective date of the merger. Initially, the dovetail arrangement was applied in order to "have something" in place at start-up.

2. About a month after merger, Implementing Agreement 2 was finalized. It contained the work equity (ton-mile) formula that was of relative advantage for the former GN employees.

3. That caused representatives of former NP employees to appeal the matter to the UTU's Internal Board of Appeals. The ruling of that board was favorable to NP employees. In a meeting with BN Labor Relations officers, UTU leadership asserted that Implementing Agreement No. 2 should not apply. Instead, the ruling of UTU's Board of Appeals should apply, even though BN was not a party to nor bound by the UTU board's decision. BN and UTU agreed to form a new agreement, Implementing Agreement No. 5, dated October 15, 1971, Mel Winter, general chairman representing Conductors and Brakemen working under the GN contract, was angered over the action that highly favored former NP employees, but he was without recourse.

retirees. When the subject comes up, you can still feel the strong negative feelings on this issue. The BLE would not back down on the way the enginemen's seniority roster was made up using the fireman's date of hiring. The BLE general chairman was a former NP engineer and he was the one who set up the way the seniority roster was arranged and took care of his former NP enginemen. To be fair, this roster should have been either a ton-mile or prior rights roster.

DIFFICULT SITUATIONS ENCOUNTERED

There probably was more tension and frustration at Spokane than any of the common points where employees of the component roads had to be integrated. However, the work force had to be combined in order for the new operating plan to be implemented. John Langlot's stories reveal what life was like at BN in those early days for some of the people who did the "real work" of the railroad. While many examples can be cited to show how well the merger worked, it is important to recognize that it did not come without considerable pain and turmoil in the lives of some of the people of the railroad.

The implementing agreements did not require that a new consolidated schedule or compendium of the agreements built up over the years on each of the four railroads be prepared for employees working in train and yard service. It was recognized by all concerned that getting that done in time for the start-up of the new company would not have been possible, due to the complexity and the magnitude of differences in the contracts that applied on the four roads. However, new consolidated schedules were negotiated for the non-operating (clerical and shop craft employees) in time for the start-up on M-Day.

Shortly after M-Day, Labor Relations encouraged the general chairmen of the operating crafts to go though the schedule of agreements for each of the former roads and designate which rules they thought should be placed in a new, consolidated schedule for BN employees. By "cherry-picking" the most favorable rules, operating costs would have increased about 40 percent. Obviously, that approach had to be rejected by Labor Relations. Today, forty-seven years after M-Day, the old schedules remain in effect. Another interesting sidelight through the process was how the general chairmen came to realize how their jobs and status depended on their having to administer a unique schedule and represent a unique body of members. That recognition caused any remaining enthusiasm for a consolidated schedule to wane.

Over the next several months and into the early years after M-Day, the difficulties and differences that arose in the early days of the merger were ironed out. More and more employees began to think of themselves as being part of BN, a company with a good future, with less and less on "how things used to be" on their prior road. As generally happens with all of us, as time goes on, we have better memories of the good times rather than the difficult times.

Although other examples could be cited where employees felt the merger disadvantaged them, the transition to the consolidated operation was accepted in good spirit by most of the employees affected by changes in the location of their jobs, having to learn new work procedures and work under supervisors from one of the other roads.

THE HANLON BOARD

Transition to the new operating plan generally went well over the BN system, although there were some locations where employees running over a new territory or working in a different yard raised concerns that they were not being treated in the spirit of fairness and evenness they had been promised by the leadership in both the unions and the company. Strong, sincere efforts were made at both the local and headquarters levels to ameliorate concerns when the problems rose above and beyond the kind of moderate complaining that is sure to happen any time that comfortable, long-established routines are changed. In the first few years, a number of situations arose in field operations where expert, unbiased intervention proved to be helpful in determining how a new implementing or

protective agreement should be applied or interpreted, and in harmony with the historic prior road agreements that still applied.

The company and the general chairmen of the operating crafts and the clerks agreed that Paul Hanlon, a well-known, respected arbitrator and referee, should be enlisted to handle such questions. Hanlon's decisions were to be accepted as final and binding. The "Hanlon Board" handled literally hundreds of such cases in the span of six and a half years, on a timely basis. Having the services of Hanlon was of great benefit in working through local situations that could not have been anticipated at the time the new agreements and new operating plan was developed. It is interesting to note that the Hanlon Board was convened to handle questions and anticipate problem situations even before M-Day. His work continues to receive high praise from union and management personnel who worked closely with him in those early years.

THE PRICE PAID FOR MERGER

A great deal of credit should be given the general chairmen and the companies' Labor Relations officers who worked hand in hand to produce the agreements the applicants needed to present to the ICC during its reconsideration of the proposal for merger. Until those agreements were finalized, detailed operating plans for all departments could not be finalized and made ready for M-Day. The last agreement was finalized on February 5, 1968, with the Switchmens Union of North America (SUNA), a division of the UTU. Altogether, five different organizations representing employees had to be dealt with in the nine months between the date the ICC first denied the merger and when the ICC began its reconsideration of the proposal for the merger. A good spirit of cooperation, mutual respect and commitment to make the merger work developed, and that laid the foundation needed for dealing with new issues that were sure to lay ahead.

The agreements for lifetime job protection and the guarantee for wages were sure to be very expensive. However, including those provisions in the revised application for merger was essential in gaining approval for merger from the ICC and in the challenges that were certain to be made through the courts. Through attrition, the cost of these provisions would go down, but for many years, the new company's earning power would be impaired by the cost of having more employees on the payroll than were needed at many locations, as well as in the departments where savings in personnel had been made through consolidation at the outset of the consolidated operation.

For some time, it would be difficult for BN to produce financial returns at the level of the best performers among the large railroad companies of that time. Strong leadership at the senior level of management was needed to move the company ahead in implementing the plans for the consolidation of functions and facilities, and in revamping the operations so the preferred routes and designated equipment repair facilities would be used intensely.

A cloud hung over BN when the newly formed Penn Central had to declare bankruptcy on June 21, 1970. Financial institutions had become reluctant to issue financing to even the strong railroad companies for some time. On BN, the inability to finance the $60 million of CB&Q bonds that matured early in 1970 caused a major concern that had never been anticipated in planning for the merger. Until an alternate plan for financing was set up with a group of lending institutions, that issue was a major distraction for BN's senior management, away from the attention they should have been giving to managing a very large, new organization that was going through a time of unprecedented challenge. The failure of the Penn Central and the impact it had on the reputation of the rail industry was a sign of general concern over its future, and whether BN could develop to the level of success that so many constituents had anticipated. Employees as well became concerned about BN's ability to avoid the kinds of problems the Penn Central and several Midwestern railroads—among them the Milwaukee, Rock Island, Illinois Central, and C&NW—were having, which appeared to be making those railroads weaker and perhaps even moving toward bankruptcy.

Strong leadership at all levels of management was needed to convince the entire BN organization that our plans were set up to prevent the mistakes and shortcomings that had caused the Penn Central to fail and were causing some of the weaker roads to struggle to survive. By merging, we had established a new company strong enough to head off the problems that were affecting a large part of the rail industry. We had the opportunity to survive and prosper by running a better railroad. It was up to all of us to move ahead and take full advantage of the opportunities we had created. We had to have faith and resolve to do our part: to do a good job of railroading in the new environment we found ourselves. By and large, there was a spirit out there among BN people to do that and more—to make it a successful company. The leaders of the unions, Labor Relations officers, and those who developed the operating plan had done their job. Now, it was up to the rest of us to put it in place and make it work.

In the next several years, countless details had to be worked through regarding the interpretation and application of the new implementing agreements, transferring of work, and consolidating work centers. For the most part, there was a spirit of willingness to clear up any disputes or uncertainties as to what should and could be done to move the company forward by making such changes. We were fortunate that in the 1970s the economy was moving ahead (except for a recession in 1974 and 1975), international business was increasing, and that as the operation improved, customers were shifting more business to BN. Most of all, the unexpected shift made by electric-power companies to low-sulfur coal mined in Wyoming and Montana was beginning to increase the amount of traffic on several corridors in the early 1970s. As a result, there were fewer surplus or unneeded train-service employees on most of the former CB&Q territory and on the NP lines east of Laurel. Wage-guarantee claims began to level off as well.

TO IMPROVE COMMUNICATIONS

During those years, BN made a strong effort to improve communications with employees on issues of concern to them, and to update them on the progress and direction the company was taking. The basis for such attempts to improve the dialogue with employees was to overcome the perception of many employees (and some officers) that bureaucracy was building up within BN and making it difficult to get information out to the field and make routine decisions promptly at some level. A program was set up in which the regional vice presidents (RVP) conducted meetings twice each year with the local chairmen of each craft. Division superintendents met monthly with the local chairmen. It wasn't long before these meetings became known as "love-ins," to pick up on parlance of the day. In part, this program was an outgrowth of an initiative of Al Chesser, president of the UTU, called "Project 70s." In announcing this new program, Chesser wrote to UTU members, "When I became President of the UTU a few months ago, I said that I wanted to create the best possible line of communication between the members and the officers because I feel that we cannot serve well unless that line of communication is open, both ways, at all times."[2] In a brochure Chesser and UTU sent to its members, railroad presidents, and many entities who had an interest in the well being of railroad employees, Chesser wrote, "They [the railroad presidents] have indicated a desire to change course, to end long years of controversy and to enter into an era of mutual respect and cooperation." Chesser suggested that top-level officers conduct meetings periodically with both general chairmen and local chairmen to hear local complaints and try to resolve those problems by upgrading conditions in the workplace and through a systematic means of solving problems of any nature.

This new approach to improving labor-management relations was highly touted by Menk and Downing. In addition to building on the working relationships that had been established in preparing the agreements necessary for merger, it was hoped that it would improve the possibilities for management and labor to work together on legislative issues of mutual interest and concern. At first, this program was not well received by some line officers, since they thought it gave local chairmen the unprecedented opportunity to make inaccurate, exaggerated, spiteful statements to a RVP who might overreact to accusations made against local supervisors. It was feared that since many of the RVPs of that day had no operating experience, they would not know how to filter out or evaluate vociferous charges that might be made against managers, nor would they have a good understanding of the situations the local chairmen might complain about.

At that time, there were many operational changes underway—specifically authorized in the new implementing agreements all of us were working under—that some employees, union officers, and occasionally, some

managers, found unfavorable. Some local chairmen either did not know or refused to recognize that the general chairmen they had elected had signed those agreements only a short time before the merger took place. The "love-ins" provided an opportunity for line-operating officers and labor-relations officers to answer questions and provide information to employee representatives on how the new implementing agreements were being applied to the operation. Also, discussions in those meetings were an eye-opener for the RVPs and the senior officers they reported to, hearing about the intransigence that was developing in the big company that was making it difficult to get small matters progressed to the point where a decision could be made—whether pro or con—on things that employees and many managers thought should be done to improve the operation.

Having these regularly scheduled meetings throughout the system turned out to be helpful in clearing the air. We were better able to communicate current and correct information to employees, thereby overcoming unbased rumors and perceptions about what was going on at both the local and system levels. Local union officers had an opportunity to voice concerns, raise questions, and provide information to officers who, in their minds, were not always aware of problems in the operation due to blockages in communication within the many levels of management.

Another program that was effective in breaking down barriers, getting problems resolved, and providing employee input was called "DIAL BOSS." Employees could dial an 800 number and have questions or complaints recorded. The company promised to respond within three or four days. In this way, an individual employee had the opportunity to be heard and to be given a prompt answer to a question or situation that had become perplexing to him or her. Together with the employee newsletter, BN News, the response from employees to these new initiatives to improve communication was quite favorable.

VALUES, ETHICS, AND STANDARDS OF CONDUCT FOR THE NEW COMPANY

During the process of building BN into a strong, competitive, and respected corporation, Menk issued clear

---

2   Al Chesser, President, United Transportation Union, "Project Seventies," 1970, pages 1 and 4.

statements as to the style of leadership and the values under which the company would be managed, and on how he expected employees would be treated:

> It will be the policy of Burlington Northern to maintain its relations with the labor organizations in a manner which is constructive, effective, and enlightened and which is conducted on the highest ethical level. There is no place in our organization for an officer who is high-handed, unfair or sharp in dealing with the representatives of our employees.... I am convinced that poor labor relations inevitably grows out of distrust. No matter how difficult a problem may be or how sharply divided people are over an issue, the officers of Burlington Northern should always conduct themselves in such a way that they will be trusted.... The railroad employee must be a participant in our growth, not a victim of our prosperity. Because of this, collective bargaining is becoming a process of mutual problem solving.... This will be the cornerstone of our labor relations policy.[3]

Employees learned in those years that Menk and other senior executives insisted that such values be in place and applied throughout the organization. They also expected employees to put forth their best efforts in performing their jobs, to work according to established rules and safe working practices, and to not be allowed to use their time or the company's tools, equipment, or supplies in any way that was dishonest. An example of Menk's expectations came through very clearly in a letter he wrote for *BN News*. Menk clearly expresses his expectation that employees work productively, report to work on time, and not quit working until the end of their assigned hours. It was discussed widely throughout the system as to how it should or should not apply in various work situations. As would be expected there was a wide range of reactions to the implications of this letter.

While reasonable people would not take issue with the simple, basic expectations Menk expressed, there were some employees who saw them as a threat to a way of work life with bad habits that had been baked into the culture at some locations throughout the years. Indeed, it was a wake-up call to the fact that we were fortunate to

have good jobs with a railroad company that had a future. It was up to all of us, in any craft or level, to be sure that we were putting in "eight hours' work for eight hours' pay" every day. Those who did not think the old adage applied to them would have to be given some coaching and an opportunity to change their ways.

At the end of BN's first decade of operation, employees hired before the merger had benefited from having lifetime job protection and guaranteed wage, provided they worked the highest-rated job their seniority allowed. Employees found themselves fortunate to be working for a railroad company in which business was growing, and not in a state of decline as was the case in major parts of the industry at that time. BN had a bright future compared to the unfortunate but inevitable struggles underway at the Milwaukee, Rock Island, and Penn Central.

The heads of the Labor Relations Departments[4] of each of the four roads that formed BN, along with the general chairmen representing all of the crafts, deserve a great deal of credit for the success they had in negotiating the complex set of agreements for job protection and the guarantee of test-period earnings for each employee. The implementing agreements they negotiated gave management the flexibility needed to implement the consolidated operating plan.

Overall, there was "peace" in the ranks of labor. Employees had more job security than was available in perhaps any other industry. Working conditions were improving through employee participation in programs for safety and training, as well as new programs that provided access to senior management. A serious effort had been started to change some aspects of the historic railroad culture, to move away from practices that sometimes were seen as militaristic, arbitrary, and at times demeaning for professional people. Employees were being recognized and shown more respect for outstanding records of success and demonstrating professionalism in their trades.

A. E. (Al) Egbers returned to BN in 1977 to head its Labor Relations Department after holding the same position at Conrail for two years. Al was highly respected

---

3   Gus Welty, "BN's Employees Help Make the Merger Work," *Railway Age*, July 27, 1970.

4   T. C. de Butts and Clyde M. Illg for the GN, A. E. Egbers for the CB&Q, G.M. de Lambert for the NP, and Hugh Tierney for the SP&S.

BURLINGTON NORTHERN

LOUIS W. MENK
Chairman

176 East Fifth Street
St. Paul, Minnesota 55101

Recently, I learned that at one of our smaller terminals a yard crew going to work between 7 and 8:00 a.m., promptly at 8:15 a.m. stops its engine and the crew walks about a half block for a coffee break. I wonder how many times similar practices involving many crafts are repeated every day over the system.

The word "featherbedding" has always to me been a distasteful and undeserved term. I have often said that switching cars, doing maintenance work, and checking yards in below zero weather has no similarity to a feather bed.

On the other hand I am devoted to the achievement of full productivity, which I define as every employee, including myself, doing his or her job to the full extent of his or her ability within the confines of union rules and safe practices.

Studies have shown that happy, contented people are those people who are productive to the extent of their abilities.

Burlington Northern has had a tough first half resulting in a net loss for the period. The third quarter dividend was omitted, and 79 senior officers received salary cuts.

We are hopeful that all employees recognize that their best interests are closely allied to the best interests of this fine Company and that to do less than their best is, in reality, a disservice to themselves.

We are only asking that you report on time, quit on time, give the Company eight hours productive work, and join with all of your fellow workers in making this the finest railroad in the United States.

Sincerely,

Figure 13.1. BN NEWS,
OCTOBER 1976.

by the general chairmen of the unions representing BN employees, due to the reputation he had built at the former CB&Q and at BN for his honesty and straightforwardness in his dealings on issues involving all crafts of railroad workers. Al took reasonable approaches to working through complex situations that required new interpretations in applying the historic labor agreements in combination with the new implementing agreements that became effective at the time of the merger.

Under Al's direction, the company was generally seen as reasonable in its response to appeals filed by the unions on discipline cases or alleged carrier violations of agreements that caused employees to think they were entitled to penalty payments. While Labor Relations did not have authority to make decisions on management promotions or transfer of supervisors, its input to senior management of the Operating Department as to

evaluations made by union representatives and Labor Relations officers carried some weight. As a result, Al was looked upon as a carrier representative whom the unions could work with.

However, several major issues lay ahead for BN's second decade that would require some serious transformation in the manning of trains and yard-switching assignments, employee productivity in all crafts, and the level of wages earned by railroad workers in comparison with employees in the trucking industry. Deregulation was on the horizon and with it, railroads would have to make major reductions in their costs. The prospect of having to maintain jobs for hundreds of protected employees who might become surplus through shop closings and the application of technology to clerical and other administrative functions in the field was becoming an increasingly serious threat to BN's profitability.

CHAPTER 14

MERGER CONDITIONS TO STRENGTHEN THE MILWAUKEE ROAD

11 new ways—gateways—your shipments can now travel farther on the Fast Track
MILWAUKEE ROAD ADVERTISEMENT IN *RAILWAY AGE*, MARCH 30, 1970, PAGE 21

◦ *request in 1973 for inclusion in BN; rejected by BN in 1974 and by ICC in 1976* ◦
1977: declared bankruptcy ◦ *1980: operations discontinued*
on all lines west of Miles City, Montana ◦

MILWAUKEE ROAD'S LAST FIGHT FOR SURvival began in 1954, when it filed a formal complaint with the ICC against the SP&S and its subsidiaries (the Oregon Electric and Oregon Trunk railways) and the GN and NP, as connections with the SP&S. Milwaukee charged that willful failure on the part of those competitors to agree to make joint rates to and from Spokane was in violation of the Interstate Commerce Act. It was an effort to persuade the ICC that it would be in the public interest to require the northern lines (the GN and NP), through the SP&S, to allow the Milwaukee access to customers located on the northern lines. Milwaukee enlisted a large number of rail shippers to testify on its behalf before the ICC.

Soon after the ICC hearings ended, the SP&S issued public tariffs providing for joint rates on eastbound business moving via Spokane and Milwaukee Road. However, the ICC ruled against Milwaukee. The SP&S cancelled the new joint rates as soon as Milwaukee's request for such rates was denied. Milwaukee appealed the case to the U.S. Supreme Court, which also ruled

against it on June 5, 1961. It is interesting to note that the northern lines, Burlington, and SP&S had applied for authority to merge on February 17, 1961, less than four months earlier.

The case presented by Milwaukee became known as the "Spokane Gateway Case." In the challenge Milwaukee made against the proposed merger during the next several years, it continued to argue for the opening of several gateways, to enable it to compete for additional business. The basis for this part of Milwaukee's opposition to the proposed northern lines merger was the longstanding principle under which NP and GN had protected their long haul on business moving to and from customers located on their lines. GN and NP were not required to "short haul themselves" by allowing a competitor to haul business for those customers over a portion of the route to maximize revenue.

Under that principle, Milwaukee was required to turn westbound business over to GN or NP at the Twin Cities rather than haul it further on its line to a junction or "gateway" located much closer to the customer's facility.

Only the higher "combination" rates rather than the lower through rate would apply, if the shipment was to be interchanged with Milwaukee short of the maximum distance GN or NP could haul it.

REQUEST TO OPEN ELEVEN GATEWAYS[1]

In its objections to the merger, Milwaukee claimed the long-haul provision unjustly precluded it from serving customers located at intermediate points on lines owned by GN and NP. As a result, the only customers Milwaukee could serve along its line of 1,765 miles between the Twin Cities and the West Coast were those who were located directly on its own line, but none of those located on GN or NP. Milwaukee requested the ICC to open eleven gateways at common points (see figure 18.1) that allegedly would provide longer hauls for Milwaukee by requiring the northern lines to set joint rates upon request. Milwaukee believed it would then be able to compete for traffic moving to or from those intermediate points. Under the conditions granted in the merger, Milwaukee could receive shipments at a point close to its origin and get the long haul. The lower through rate would apply.

Milwaukee was especially anxious to serve Billings, the largest city in Montana, a local point on NP main line. Milwaukee could not reach Billings on its own rails. In addition, Milwaukee wanted to be granted trackage rights on the former NP line for entry into Portland and Vancouver, including the right to interchange with the Southern Pacific in Portland. In the absence of conditions that would protect and enhance Milwaukee's competitive capability, the ICC estimated Milwaukee's gross revenues. With the conditions favoring Milwaukee in place, its gross revenue was projected to increase

$19.964 million, and its net railway operating income would increase about $9.8 million.

On March 3, 1966, the ICC rejected the merger application of the northern lines, Burlington, and SP&S. The commission found that approval of the proposed merger "would not be consistent with the public interest." Concern for the loss of effective rail competition was among the reasons cited in the decision. However, the ICC stated the proposal for merger "should be approved" if the applicants would accept conditions that the ICC would specify to protect the competing roads, as well as the employees of the railroads involved in the transaction and the general public. In July of the same year, the applicants filed for reconsideration. They agreed to accept all conditions requested by Milwaukee and C&NW. Together with conditions established to protect the earnings and jobs of employees working for the applicants, the ICC approved the merger on November 30, 1967. However, due to court challenges, the merger did not become effective until March 2, 1970.

Milwaukee management asserted with confidence that, with access to the new markets it could reach with the eleven gateways to be opened, it would be able to compete effectively with the merged company. As Bob Downing often stated, "At least at that time, Milwaukee didn't consider itself as failing." Consummation of the BN merger set the stage for Milwaukee to aggressively move ahead to build up its business. At the time of the 1970 merger, Milwaukee offered a schedule for westbound merchandise traffic that was competitive with GN and NP.

Milwaukee made an all-out sales effort to generate new business for the improved service. Before long, Milwaukee was running three pairs of scheduled freight trains between the Twin Cities and its terminals at Tacoma and Seattle. It was the highest volume of business Milwaukee had ever handled on its line serving the West Coast. Jointly with Union Pacific, Milwaukee built a large facility at Kent, Washington, for the transfer of new automobiles from the inbound rail mode to trucks for delivery to dealers in the Puget Sound area. Milwaukee invested over $800,000 to improve tunnel clearances so it could handle tri-level loads of automobiles and the new trailer-on-flat-car business.

---

1   A gateway refers to a junction between two railroads where traffic moves on through rates. Cars can be interchanged between railroads at any junction, but if that junction is not a gateway, traffic will move through it at a combination of local rates that will be higher than a through rate. Any junction can be designated a gateway upon agreement by both railroads. Without such agreement, combination rates cannot be offered to shippers. Richard Saunders Jr., *Main Lines: Rebirth of the North American Railroads, 1970–2002*, 392.

ACCESS TO NEW MARKETS

Perhaps Milwaukee's biggest success under the merger conditions was the rapid buildup of its volume of interchange with Southern Pacific (SP) at Portland. SP was glad to have a partner in addition to UP, to reach destinations in the Midwest and Canada. The volume through Portland soon taxed Milwaukee's ability to furnish enough crews and locomotives to handle it without delay. Milwaukee had to move all of its new business on BN's already congested line of two main tracks between Longview Junction and Portland, where it was given new operating rights.

To provide Milwaukee with better-quality routes, the ICC ordered BN to grant trackage rights on the northern lines route between Renton and Snohomish, Washington, and between Everett and Bellingham, Washington. These conditions allowed Milwaukee to operate on much better track, compared to the lines it owned, enhancing its capability to compete for business. Other merger conditions provided for elimination of dual switching charges at some locations and the restoration of rate relationships that had been in dispute with the northern lines.

Unfortunately for Milwaukee, the glitter from the new business opportunities afforded by the merger conditions soon began to wear off. It still did not have enough earning power to maintain its track to a standard high enough to prevent track-caused derailments and consequent disruptions to service. At one time in the late 1970s, there were reports that Milwaukee's Montana-based wrecker was out on derailments for nearly one hundred consecutive days. Having to set up frequent detours on neighboring BN lines was expensive and delayed the competitive traffic it had just taken from the northern lines. Milwaukee's net income did not improve appreciably in the early to mid-1970s, even though it was able to build up its business with the access it had to the new gateways and by being able to operate its own trains into Portland.

REQUEST FOR INCLUSION

In assessing its situation for both the long and short term, Milwaukee had concluded for several years that it would have to merge with another railroad. Milwaukee and C&NW had discussed that possibility many times, but by April 1973, Milwaukee concluded its best strategy was to petition the ICC for inclusion in BN. In approving the northern-lines merger, the ICC provided that for the next five years, it reserved the jurisdiction to consider petitions for inclusion by any other railroad in the territory involved. BN argued that Milwaukee did not "qualify" for inclusion since unique conditions for the benefit of Milwaukee had been imposed. It took the ICC almost four years (until March 2, 1977) to render a decision on Milwaukee's request. The commission denied Milwaukee's application on the basis that its "illness" was not caused by the northern-lines merger itself. In other words, the creation of BN was not the cause of the financial, operating, and maintenance problems Milwaukee was facing.

During the time the ICC had Milwaukee's application under consideration, BN put considerable effort into joint studies with Milwaukee to determine where their operations could be coordinated by extending trackage rights to allow Milwaukee to use BN track between a number of intersecting points or where the lines of BN and Milwaukee had been built in close proximity and could be connected at a fairly low cost. Likewise, BN and Milwaukee looked at opportunities for BN to use tracks owned by Milwaukee. Through coordination, both companies would be able to reduce some excessive and overlapping track mileage. The lines involved had relatively low traffic density and were located mainly in eastern and central Montana, where lines of the former NP closely paralleled Milwaukee. Short connections would have to be built at designated common points to a standard that would allow the trains of one company to move "seamlessly" and without delay to or from the track of the other.

Following is a list of the line segments where BN and Milwaukee closely paralleled each other and where all or a large part of the track of one company or the other might have been removed:

Terry–Miles City	39 miles
Miles City–Forsyth	45 miles
Three Forks–Butte	64 miles
Butte–Garrison	54 miles
Garrison–Missoula	66 miles
Missoula–St. Regis	78 miles
TOTAL	346 miles[2]

---

2   Based on mileage of the lines owned by BN.

Most of Milwaukee's track that would be shared with BN would have to be upgraded with the replacement of between 1,000 and 1,500 ties per mile and upgrading the ballast section with higher-grade material. BN insisted that the segments of Milwaukee-owned track it would operate on would have to be at the standard of maintenance allowing its trains to operate at a speed of no less than on the track it was using and that would be abandoned. To prevent either the BN or Milwaukee from incurring a new and heavy burden of joint-facility expense from using the track of the other company, BN attempted to keep in balance the number of miles of track each company would "contribute" to the new joint operation. By 1976 Milwaukee found that even with that approach, the burden it would have to bear in additional joint-facility costs and its costs for upgrading track and building new connections would be too much for it to assume. As a result, the joint effort to identify and agree to specific line and yard coordination had to be abandoned.

The Milwaukee's petition for inclusion was denied on February 16, 1977. Seven months later, its petition for reconsideration was denied. The ICC concluded that merging the Milwaukee into BN would be totally inconsistent with the decision it made in 1970 to impose conditions that would strengthen the Milwaukee enough so rail competition could be maintained in the northwestern states. If the Milwaukee were placed in BN rather than remaining independent, rail competition in much of that territory would be eliminated. On December 19, 1977, only three months after the ICC ruled against the Milwaukee's request for reconsideration and inclusion in BN, the Milwaukee filed for bankruptcy.

Over the years, the Milwaukee was involved in several attempts to arrive at terms for merger with the C&NW. Both companies had the problem of light density over nearly all of their systems. Because they were competing in the same markets, their routes overlapped, causing them to have to maintain duplicate facilities and trackage. With a combined system, the Milwaukee and the C&NW would have been able to reduce operating costs and consolidated their operating and maintenance facilities. Attempts were made to arrive at the terms for such a merger as far back as 1946. Again in 1955, 1961, and 1964, attempts toward merger were undertaken, but in the

lengthy process of negotiating the terms for a merger, and preparing an application to submit to the ICC, a merger never came even close to being consummated.

The Milwaukee petitioned the ICC to require the Union Pacific to merge with the Milwaukee as a condition for the UP's proposed acquisition of the Rock Island. After trying for ten years to get approval for acquiring the Rock Island without success, the UP withdrew its application, thus ending yet another attempt by the Milwaukee to merge with or be acquired by another railroad.

DETERIORATION OF TRACK CONDITION

The conditions that the ICC granted the Milwaukee in the 1970 merger that created BN still did not generate enough additional revenue and net income to enable the Milwaukee to adequately maintain its track to the standard needed for it to provide consistent, reliable service to its customers. To reduce its costs, the Milwaukee took the bold, controversial step of shutting down its electrified operation on 656 miles in Montana, Idaho, and Washington on June 16, 1974. The electrified operation had become increasingly uneconomic; hence, the decision was made to use only diesel-electric locomotives on the Pacific Coast Extension.

Having to increase the number of miles of track under speed restriction (down to 10 miles per hour at many locations) due to poor track conditions significantly increased the cost of train operations. Relief crews often had to be called to take over trains on which the first crew had to be relieved en route under the federal law that limited the maximum time on duty to twelve hours. The expense of holding trains or detouring on BN lines due to derailments was growing, in addition to increased costs for car-hire (a time-based payment for the use of cars owned by other railroads or other parties), repair of damaged track, and motive power getting out of cycle.

By December 19, 1977, the financial situation became desperate, and the Milwaukee had to declare bankruptcy for the third time in the twentieth century. Under bankruptcy, the trustee faced the immediate challenge of conserving enough cash to keep the company operating until a plan for reorganization could be developed and approved. An emergency loan of $5.1 million was obtained from the federal government in 1978. Later, additional loans were obtained under the Emergency Rail Service Act. At the end of 1982, the Milwaukee owed $182.3 million to federal financing agencies. Toward the end of 1978, deferred roadway maintenance for the entire Milwaukee system was estimated at $587 million, including between $120 million and $140 million for the Pacific extension alone. Even in the face of its severe financial problems, the Milwaukee was able to invest $100 million from early 1978 to the spring of 1979 in track and equipment maintenance. The money for this heavy maintenance work became available as a result of having no tax liability due to bankruptcy, from dividends and loans from a subsidiary (the Milwaukee Land Company), and $56.2 million of federal aid.

The Milwaukee and the FRA estimated that to continue operating all of its 9,800 mile system for another eight months would require $80 million to $90 million more cash than it would be able to generate. With that crisis at hand, the trustee requested the bankruptcy court in April 1979 to authorize the Milwaukee to embargo business on all but 2,500 route miles and to reduce employment accordingly. In August 1979 the trustee applied to the ICC for authority to abandon all of the Milwaukee's lines west of Miles City, Montana.

As stated earlier, the major conditions imposed by the ICC to enhance the Milwaukee's competitive position was the requirement on BN to establish reciprocal through rates and joint competition through rates via the eleven gateways that were to be opened. In its ruling, the ICC used the routing of grain from origins on BN through one or more of the new gateways as an example of how it intended to impose this condition on BN. The Milwaukee was to do likewise for BN: in other words, the Milwaukee would be required to open these gateways to allow business to flow from it to BN, causing the Milwaukee to have a shorter haul and less revenue on such business.

ADVANTAGE TO BN
FROM OPENING GATEWAYS

In his book, *The Nation Pays Again: The Demise of the Milwaukee Road, 1928–1986,* Tom Ploss wrote,

> Never was it suggested that, following its merger, Burlington Northern would, by the reciprocal provisions of Condition 23, raid the traffic indigenous to Milwaukee Road so as to prove false the increased revenue projections stated above. Indeed, when the ICC had cause to construe that condition formally, it held in a supplemental opinion that "…we look upon Condition 23 as requiring [BN] and Milwaukee to establish reciprocal through routes and joint competitive rates via the 11 gateways, on grain … [which] would permit routing of grain … from an origin on the lines of [BN] to one of the 11 new gateways and Milwaukee beyond." No possibility was seen in this official ICC view of BN taking Milwaukee road's grain traffic under the same condition. The possibility was clearly antithetical—it was not to happen. After its merger, conditioned as it was to assist the Milwaukee Road to compete with itself, BN was expected to refrain from using its enormous corporate power and four-times-larger sales force to raid Milwaukee Road's business.

There were some who argued that BN's success in taking advantage of this new "opportunity" was in violation of the spirit of a merger condition intended to help only the Milwaukee. In other words, it was not intended to provide a new competitive opening for BN. The BN Sales and Marketing Department certainly did not see it that way. As far as I can determine, BN was not faulted by the ICC or other legal authority for failure to conduct its operations and marketing efforts in line with the merger conditions or the public interest by pursuing the new business opportunities it recognized at that time.

Table 14.1.

MILWAUKEE ROAD OPERATING INCOME, 1970–1977

Year	Operating income (millions)
1970	d $7.9
1971	2.6
1972	d 6.8
1973	12.6
1974	8.8
1975	d 23.6
1976	d 8.8
1977*	d 29.4

*Declared bankruptcy on December 19, 1977

d = deficit

Source: Moody's Transportation Manual

Table 14.2.

DENSITY—RAILWAYS OF THE MIDWEST AND NORTHWEST
(AVERAGE TONNAGE PER MILE OF ROAD, 1969)

Milwaukee	1.59 million tons
C&NW	1.64
Rock Island	2.52
Soo Line	1.74
NP	2.12
GN	2.34
CB&Q	2.41
UP	4.91
Southern	4.22
Santa Fe	3.17

THE BASIC PROBLEM: LOW DENSITY

Table 14.1 shows the magnitude of the Milwaukee's losses in operating income from 1970 through 1977, the year in which the Milwaukee declared bankruptcy. Note that the Milwaukee earned a positive level of operating income in only three of those eight years. Financial support from various government agencies shown as "transfers" helped offset this deficit in operating income starting in 1976. But by 1982, even with that assistance, the Milwaukee owed $182.5 million to all of the federal agencies that had provided some amount of financial support. It is interesting to note the substantial gain in revenue in 1983 and 1984, and that a modest amount of operating income was earned just before the Milwaukee was sold to the Soo Line.

In the end, it has to be recognized that the basic problem the Milwaukee had was the very light density of traffic on its network (see table 14.2). This primary deficiency put the Rock Island into bankruptcy and dissolution. Low density also caused the C&NW to struggle throughout its existence until 1984, when the C&NW finally was able to raise the money to pay $76 million to BN for 50-50 ownership in the new line BN had built in the Powder River Basin to serve several high-production coal mines. The C&NW also was able to increase its density in the 1970s when it was established firmly as the Union Pacific's preferred connection for UP business moving to and from Chicago.

CHAPTER 15

CONDITIONS AFFORDED TO THE C&NW AND SOO LINE

• attempts to enhance the competitive capability of railroads to be in competition with the consolidated "northern lines" • to help ameliorate the concerns of shippers for competitive rail service • a case of competitors using the northern lines merger and an opportunity to persuade the ICC to enhance their market position •

THE CHICAGO AND NORTHWESTERN WAS A tough, aggressive organization, tough enough to avoid bankruptcy and survive, unlike some of its competitors, Rock Island and Milwaukee. During the northern lines' and Burlington merger proceedings before the ICC, C&NW sought conditions it thought would help it compete and stay in business. Because C&NW was seeking a merger with Milwaukee at the time, it was not in position to oppose the northern lines' merger. Instead, it requested that the ICC require the northern lines to open a gateway at Oakes, North Dakota, a remote interchange point on branch lines owned by NP and C&NW, where "slow route lumber" could move in large volumes.

C&NW also requested a gateway be opened at Crawford, Nebraska, to enable it to get a longer haul, instead of having to interchange with BN at Omaha or Sioux City, giving C&NW a much shorter haul. As with opening a gateway at Oakes, having a gateway at Crawford was intended to get more revenue and establish a stronger marketing position with lumber producers. Westbound business originating on C&NW for destinations on BN also could be interchanged at Crawford instead of at a connection on or near the Missouri River. Also, C&NW wanted the ICC to set a merger condition in which the

merged company would be required to lease it to long tracks in the GN's Union Yard in Minneapolis.

C&NW stood to lose some of the large volume of business it received from the northern lines at Minneapolis. Before the merger, roughly 30 percent of the eastbound business GN and NP hauled into Minneapolis was interchanged to Burlington, and most of the balance of 70 percent was split among C&NW, Milwaukee, Rock Island, and Soo Line. By far the greatest amount of that business was delivered to C&NW at its East Minneapolis yard. After merging, the northern lines and Burlington would of course extend their haul east of Minneapolis, thereby causing a large reduction in business delivered to the C&NW. Opening of the Oakes gateway for the slow route lumber was successful for C&NW, with large numbers of cars being routed to Oakes for interchange to C&NW soon after the merger.

In recalling the aftermath of this merger condition, T. J. Lamphier, president of BN's Transportation Division in the 1970s, stated,

> delivery could be delayed until the broker found a customer for a carload of lumber.... They [the North Western] provided tri-weekly service to Oakes from Huron

[SD], but we were handling train loads of this lumber from New Rockford and Jamestown, North Dakota, and we would just shove it down the C&NW branch line. . . . This [practice] died a natural death, as the lumber companies started selling directly to lumber retailers, and in a few years [after the merger] the Oakes gateway dried up.[1]

Allowing cars to be used as warehouses for some commodities was a wasteful commercial practice of long standing in the rail industry. It was a practice that ran contrary to all principles of efficiency in car utilization. Shipments that should have reached their destinations within about seven days could use three weeks or more to finally reach their destination if they were interchanged at points on branch lines where service would be slow due to the slow train speed being allowed on the line, having to move those cars through several yards, and because the railroad company had no reason to make any effort to move such shipments efficiently. This counterproductive practice contributed to the widespread problem of the shortage of 50-foot-wide-door box cars built mainly for lumber loading in those days. In time, both the lumber producers and the railroad companies agreed it should be stopped. It was a big relief when this wasteful practice was ended.

Over the years, GN and Burlington were quite generous in accommodating C&NW on trackage rights so it could continue serving locations that it might have had to give up because of poor track conditions or because the expense of maintaining its own line had become too much to bear. One such case was when Burlington gave the C&NW 87.2 miles of operating rights so it could continue to reach Lander, Wyoming. Another was the right to use BN's "preferred route" (the former GN line) between Minneapolis and Superior when C&NW decided to abandon part of its line to Superior. BN did not force C&NW to operate instead on NP's line between St. Paul and Duluth–Superior, which by that time had deteriorated due to a delay of several years in starting a tie-replacement program, in anticipation of sale to Milwaukee, which had rights on that line, or abandonment of parts of that line.

Granting operating rights to C&NW on the former GN route allowed the it to maintain service on a high-class line of railroad to markets it might otherwise have had to give up. In the Duluth–Superior terminal, Duluth, Winnipeg and Pacific (DW&P Railway, a Canadian National subsidiary) and C&NW were granted the use of small pieces of northern lines' trackage to facilitate their interchange of traffic.

This kind of friendly and cooperative relationship came to an abrupt end in the mid-1970s when C&NW decided to either build a new line into the Powder River Basin or somehow force its way onto the new line BN planned to build to serve several new mines.

A U.S.–BASED SUBSIDIARY OF THE CANADIAN Pacific Railway, the Soo Line, operated several lines serving markets in North Dakota and Minnesota in direct competition with the GN and NP. The Soo Line was granted conditions that would give it "full and complete" use of three GN branch lines in North Dakota and Minnesota on which it had held trackage rights since 1960. These rights were expanded to include access to industries on each line. It was also granted the use of GN track in Minneapolis that would provide a better route for interchange with Minnesota Transfer Railway and the Rock Island Railroad.

During the early years of negotiations on merger conditions, BN also agreed to consider Soo Line's request to have its cars switched in the new hump yard that the northern lines planned to build on the north side of Minneapolis. Soo Line had an old, outmoded flat switching yard called Shoreham, located just south of NP's yard at Northtown. The northern lines' merger-planning team agreed to consider taking on that additional work, once the new yard was operational and a better assessment of its capability could be made. After the merger, Soo Line did not progress its request, nor did BN attempt to convince Soo Line to move its terminal work into the new yard. By the time the new yard was completed, business on both roads had grown enough that there was not enough capacity to take on the volume of work the Soo Line had suggested in the negotiations several years earlier.

Some of the conditions requested by C&NW, Soo Line, and Milwaukee are typical of the concerns and requests

---

1   Conversation between Roger Grant and T. J. Lamphier, in H. Roger Grant, *The NorthWestern: A History of the Chicago and North Western Railway System*, July 31, 1995, 205–206.

that some neighboring railroads may advance when a competitor is seeking regulatory authority to merge with another railroad. Those conditions may be to strengthen a smaller or weaker railroad so it will be able to maintain competitive service, or to give it access to the business of customers who have concerns they will have only one railroad available instead of two or more. There may be cases in which a railroad will use the prospect of a merger involving a competitor to obtain new or more favorable joint-facility agreements that will reduce their costs or gain access to new business opportunities they might not have been able to get in the normal course of negotiations with the owner of the track they wish to use. They may then look to the regulatory agency to force the owner who is attempting a merger to grant them those concessions.

PART III

POST-MERGER OPERATIONS

CHAPTER 16

DISPARATE CULTURES

· interesting differences and common bonds within the Hill family of railroads ·
to reform old cultures into a new culture for Burlington Northern ·

I N MANY WAYS, THE CULTURES OF THE FOUR component roads were more similar than disparate. Before the merger, there were no cultures at the opposite ends of the spectrum, as had been the case in the merger of the Pennsylvania and New York Central (or, in later years, the BN and the Frisco). Each of the three large railroad companies that made up BN, and the SP&S, had been under the influence and direction of James J. Hill in the years in which they were molded into large, strong railroad companies.

RELATIONSHIPS OF LONG STANDING

Through joint or common ownership and usage of numerous lines, facilities, and subsidiaries, as set up under Hill, managers in all departments and at all levels of these companies knew each other well. The NP and GN were both competitors and partners. Officers at the senior level had worked together for a long time while serving on the boards of the Burlington and the SP&S, as well as while preparing the plans and strategy for the merger. The Burlington interchanged large volumes of business with the NP and GN through the Twin Cities and Laurel, and also operated passenger trains for both of the "northerns" between Chicago and St. Paul. Ralph Budd, who had roots on the GN and SP&S, served as president of the Burlington from 1931 to 1949—a long tenure that helped build even stronger ties to the NP and GN. The SP&S was jointly owned in equal shares by NP and GN, and its presidency rotated between the two "northerns" every two years.

All of these factors had caused all of the companies involved in the 1970 merger to work closely together and cooperatively for a long time. Managers at all levels were well familiar with the strengths and weaknesses of each property, the leadership styles of principle officers, and where each company held a competitive advantage.

Each of the railroads had given high priority to property maintenance. Their middle- and upper-level operating officers all had some experience and skill in managing the engineering and maintenance function (although under Menk in 1965, the Burlington switched to the departmental rather than divisional organization structure, which is to say that the responsibility for basic track and bridge maintenance was taken from the general managers and division superintendents and put directly under the Engineering Department). On each of the four lines that formed BN, the work of the maintenance-of-way and engineering forces was held in high respect throughout the company. This probably was due to so many of their top-level officers having held responsible positions in that kind of work and having mastered it. I have noted on a number of other railroads that the people in the maintenance and engineering functions are held in less regard. In those organizations, there are many who look upon that work as dirty, not glamorous or as important as running trains or managing terminals. That was not the case, especially on the northern lines, where great interest was shown by other departments in line improvement projects and the annual programs for upgrading track and bridges.

The GN, NP, and SP&S maintained "clean" railroads: their right of way was not cluttered with scrap track material that was not picked up. After work programs and general maintenance work was completed, and after track damaged in derailments was rebuilt, it was all picked up. Buildings at field locations were well maintained and kept clean and painted so they looked respectable. The Burlington had been deficient in this respect, and this was an area where attitudes, priorities, and expectations on the part of managers and employees had to be upgraded, once we merged. Good "housekeeping" was important in maintaining a safe work environment, making the railroad a credit to the community, and giving employees a sense of pride in the company.

With most of its lines in the "lower" midwestern states, the Burlington had faced much more highway and rail competition than the NP and GN. It had to compete head to head with the six other railroads that fed California traffic to the Union Pacific. The Burlington also had to work with the Rio Grande–Southern Pacific or UP–SP combination, in addition to several railroads that provided service to Texas over the Kansas City gateway. As a result, its marketing and operating people had worked under more intense pressure and scrutiny than their counterparts on the NP and GN. By having to run a money-losing, but very high-profile, commuter service in Chicago, Burlington people were more accustomed to working in a mode or environment of zero tolerance for errors or lack of quality in the of service they performed.

DIFFERENCE IN QUALITY OF MAINTENANCE

The only other difference of great significance the component roads brought to the merger was in the quality of track maintenance performed. The main tracks on the lines of the NP, GN, and SP&S had been in excellent condition throughout their history. That had also been the case on the Burlington until the 1960s, when not enough rail and ties were being replaced to keep the track in good shape and avoid getting into a situation of deferred maintenance. In addition, not enough had been done to mechanize the basic maintenance work required between cycles of renewal. Only a small amount of continuous welded rail had been laid, which caused a large opportunity to reduce maintenance costs to be missed. These deficiencies are covered more thoroughly in chapters 32

through 40, which discuss the heavy work that was done in the 1970s to put the lines of the Burlington in shape to handle the heavy tonnages of new unit coal traffic.

LEADERSHIP STYLES

There were some subtle differences in the leadership styles of managers at all levels on the GN compared to those on the NP and the Burlington. However, those differences were not so extreme as to cause BN to have to deal with "disparate cultures" on M-Day in 1970. Most GN people had developed traits and a style of leadership in line with the character exemplified by John Budd, who served as GN's president for twenty years before the merger. John Budd's long tenure had given him many years to select people who could be trained and developed in line with his norms and standards for conduct and leadership. A fairly large number of GN officers were graduates of Yale, as was Budd. Although he was clear and firm in his expectations, Budd did not use outbursts of temper or foul language or dress people down in front of others when he was out on the railroad. Budd ran a tight ship, as the GN's financial and operating performance numbers show, but those good results were not achieved through management by fear and intimidation. These norms of conduct ran through the GN organization with remarkable consistency.

On the NP and Burlington, many officers were "out of the same mold," in terms of leadership style. More so than on the GN, there were a few notable exceptions among operating and maintenance officers at all levels who would sometimes resort to less than affirmative conduct when they faced (or thought they might face) even small instances of substandard performance of a subordinate. Within a few years after the merger, most of the excessively tough managers had adjusted their conduct and persona to the mode of leadership that was clearly manifested by such leaders as Budd, Downing, Lorentzsen, Ethington, and Hertog. In the first decade of BN's history, there was little evidence of politics or gamesmanship, and no "in groups" or "out groups" among the officers, as sometimes will develop in organizations of any size. However, the disparity in cultures became a major challenge when BN merged with the Frisco in 1981.

CHAPTER 17

TENSION AT MERGER

• merging at the workforce level •
while implementation of the merger went very well overall,
there were some employees who experienced tension and disruption in their work lives •

THE MERGER THAT CREATED BURLINGTON NORTHERN in 1970 has been widely acclaimed as a merger that worked, particularly in contrast with the unfortunate outcome when the Pennsylvania Railroad and New York Central were merged a few years before. Overall, the BN merger worked quite well, from the time of start up and going forward. The shareholders, freight shippers, and general public all benefited. With lifetime job protection and wage guarantees, the employees also benefited, if one looks at the big picture and over the first ten years following the merger.

However, the merger did not come without stress, disruptions, and some setbacks for employees at some locations, mainly at the common points. Tension developed when employees had to follow their work to a crew district, yard, shop, or office on "the other" railroad, if they found they were in a worse position because of how their seniority was handled. Some were no longer able to hold jobs with weekends off or on the shift they preferred. Some conductors, engineers, and brakemen were set back from assigned jobs or from the "chain gang" pool to the extra board when employees from "the other road" moved in. Perhaps more than any other aspect of their work, railroad employees value their seniority and the options it gives them in placing on the kinds of runs or assignments they prefer, as they move up the ladder on the seniority roster.

Employees did not always take kindly to suddenly have to work with those from a "foreign" railroad, even though the agreement for consolidation had been negotiated

on their behalf by the general chairmen they elected to represent them. It was difficult for many of them to drop the belief that they were supposed to be in competition with the other railroad, that the railroad they had worked for was superior, and that somehow, they were better at performing their jobs than their peers who had worked for a rival company.

SENIORITY AS A MAJOR ISSUE

When seniority was affected in the integration of people from one of the other roads, or by changes in the operation, some employees got angry, upset, and bitter at whatever and whoever they thought had caused them to no longer have the same working conditions they had before the merger. At times, these strong feelings caused otherwise good citizens in the work place to take their frustration and anger out on the "foreign" employees who had suddenly come onto their turf. Having the unprecedented benefits of lifetime job protection and a guaranteed wage were not always sufficient to offset the feelings of resentment or general discomfort from changes in routines or other working conditions.

The contrary behavior of some employees most often amounted to refusal to welcome the new people into their new place of work, and instead, giving them the cold shoulder or outright ridicule. On a few occasions, it resulted in threatening notes or messages, slashing tires, harassment in the presence of others, or deliberately causing the newcomers difficulty in being able to get their work done satisfactorily. Fortunately, these kinds of behavior were isolated and the work of only a few people. But even a few such incidents resulted in a hostile workplace for employees who had to start working in a different facility or location to protect their earnings and to retain railroad employment.

GARY NELSON, BRAKEMAN AND CONDUCTOR IN THE MINNEAPOLIS–STAPLES CHAIN GANG

An example of what some employees had to endure and overcome was the experience Gary Nelson had when forced to move from Willmar to Minneapolis to maintain his employment in train service. Gary and ten other GN brakemen, who were the most junior employees at Willmar, were directed to report to the crew-management office in Minneapolis and assume the allocations given to Willmar employees in the Northtown (Minneapolis)–Staples freight pool, commonly referred to as the "chain gang." This was NP "territory."

Since four regular trains (two per day in each direction) were to be rerouted from the GN's Willmar line to the former NP line as provided in the new operating plan, an agreement between the company and the general chairmen representing employees of both the GN and NP provided that Willmar men would have to follow their work by moving to Minneapolis and taking the two allocations in the Staples pool that had been established for crews that had GN seniority at Willmar. At the time of the merger, there were eight chain-gang crews working east and eight working west from Willmar. The pool was reduced to five crews in each direction when the four trains were rerouted. The GN crews based in St. Cloud were given one allocation in the Staples pool, due to rerouting of their trains to the former NP line. In November 1971, when the allocations were set up, there were to be eighteen crews in the Staples chain gang, with the crews from Willmar holding five turns, and the St. Cloud crews, two turns. The agreement provided that BN would pay those employees the appraised value of their homes, assume the cost of moving their household goods, and reimburse them for temporary living expenses in Minneapolis.

Gary and the other former GN employees from Willmar

and St. Cloud walked into a hostile environment when they assumed their positions in the Staples pool and began to make their qualifying trips. Some of the NP crew members were not cooperative in acquainting the GN employees with the characteristics of the railroad, the operating practices and standards in effect, and basically, how to get a train over the road safely and on schedule. There were cases of NP crews trying to set the GN crews up for of some kind of failure that would get them in trouble with the management. Some GN crews would be told by a crew of former NP employees on a passing train that there was a car in the train with a safety defect, which would require the crew with former GN employees to stop the train to make a walking inspection of as much as a mile or so, only to find no defect. Such episodes could easily delay a train an hour or more, and required the crew to unnecessarily walk along the entire train to try to locate a defective car, at times in locations with poor footing, in darkness, in the rain, or in severe winter-weather conditions.

It was difficult for field operating supervisors to deal with these kinds of behavior. Train crews usually work without much direct supervision, and often it was hard to reconstruct a particular incident to prove it was a case of harassment. Causing these types of incidents was not unique to crews of the former NP. Similar incidents occurred on former GN territory, in the Spokane, Whitefish, and Seattle areas where NP employees had to take allocations in chain-gang pools that ran predominately on former GN territory. Also, such "fraternity antics," harassment, or hazing was not always limited to operating crews. A few such incidents also occurred when clerical forces were consolidated in yard offices and when mechanical forces were combined in car- and locomotive-maintenance facilities.

Gary Nelson's challenge in working on "foreign" territory became even more tense and difficult when he was suddenly notified to take the class for promotion to Conductor. He was given only ten days' advance notice of this examination. The ten GN employees who would soon be working the two GN allocations were junior employees, and not yet qualified to work as a conductor. Yet some of them had to become qualified to work as conductors in a matter of days, and without having the benefit of several years of experience from working under experienced and well qualified conductors. Even with only having seventeen days to take the examination for promotion, Gary passed the test, and on the same day, made his first trip as a BN conductor on an empty coal train. At the time of his promotion, he was only twenty-five years old, far younger than any of the NP conductors working in the same pool—most of them were in their fifties and had at least twenty-five years of service. This quick promotion of much younger former GN employees caused an additional stirring of resentment that took a few years and the retirement of some former NP people to overcome. Gary continued to make this run between Minneapolis and Staples for the next twenty-three years. In 1994 an interdivisional run with chain-gang crews was set up between Minneapolis and Dilworth, which Gary then worked until he had enough seniority to work as conductor on the local train that served the Monticello branch line out of Minneapolis.

It was disturbing to hear about some employees reacting so defensively and exhibiting mean, threatening behavior toward fellow railroaders. These were employees who by and large had carried out their responsibilities professionally and in compliance with the company's rules for conduct for many years. The basis for resorting to such behavior seems to be in the resentment they had for the company deciding to merge with a competitor (and an adversary, in the view of some) and for the union, who did not stop the merger or at least prevent changes from being made they thought would adversely affect their work lives.

Most of all, they resented the company giving "their work" to a group of employees who were not part of their group. On top of that, the foreigners were being promoted ahead of older employees in their group. The new employees coming into their work environment were an easy and readily available target for the resentment they had for their work lives being disrupted. Of course, such feelings and behavior were unreasonable, unfair, and threatening to those who were required to move in and disrupt the "straight line" that some employees had thought would always be there and take them through to retirement.

Over the years, the intensity of these negative feelings dissipated. Employees came to evaluate each other more on the basis of their capability as railroaders, and how well they looked out for each other, than on their heritage.

Gary Nelson closed out his career after forty-two years of service, and as the number one employee on the combined seniority roster. He served on the division safety committee for many years, assisted in making safety audits throughout the BN and BNSF systems, chaired the annual United Way fund drive for the Twin Cities, and was selected for the Chairman's Award in 1994 for his service. Although he retired after a good career, Gary's experiences in the early years of the BN merger are an example of what some employees had to endure in getting the work done at the level needed to make BN successful. These stories about difficult times are not widely known, but they are an important part of BN history.

JOHN LANGLOT, BRAKEMAN AND CONDUCTOR AT SPOKANE

John Langlot has been writing extensively on his experiences while working in the first few years following the 1970 merger.[1] With Spokane being a major terminal for the GN, NP, and the SP&S, consolidating the terminal operations and changing the routing of most trains to the preferred route through Spokane was a complex process. These changes had a major impact on the work force, affecting employees in train and engine service the most.

There probably was more tension and frustration at Spokane than any of the common points where employees of the component roads had to be integrated. The work force had to be combined in order for the new operating plan to be implemented. John Langlot has revealed what life was like at BN in those early days where people were doing the "real work" of the railroad. While many examples can be cited to show how well the merger worked, it is important to recognize that this success did not come without considerable pain and turmoil in the lives of some people of the railroad.

---

1   John hired out as a brakeman for the Great Northern in Spokane in 1960, and was promoted to conductor in 1965. He retired from the BNSF Railway in 2004. The quotes from his writings are from the draft of a book that will contain experiences John had while working as a conductor in the chain-gang pool and on the branch line between Wenatchee and Oroville, Washington. His writings also give a good review of the maintenance and operating practices used on lines in the Spokane area.

John wrote that the biggest problem for employees at the time of the merger was what he considered poor handling of seniority on the part of the unions. As mentioned elsewhere in comments about railroad life and culture, "seniority is everything" to employees in any craft on the railroad.

As a general principle, the Labor Relations Department abdicated from any involvement in how seniority of former GN, NP, and SP&S employees at Spokane (or at any other location) would be combined. For operating employees, decisions on seniority were left to the leaders of the United Transportation Union (UTU), which represented the Conductors and Brakeman, and to the Brotherhood of Locomotive Engineers (BLE) for the Locomotive Engineers.

Determining the fairest, most equitable way that this should be done proved to be a difficult task at many locations. Of all the issues that arose in the early stages of the merger, there probably was more tension on the preparation of new seniority rosters than any other matter that directly affected operating employees. At the time of the merger, John Langlot did not yet have enough seniority to work regularly as a conductor or to hold a regular job as a brakeman. However, he was old enough to be the senior man on the Brakemen's extra board.

On M-Day, John was temporarily working on an "outside" job based some distance from Spokane. A few days later, when he completed his assignment on that ten-day vacancy, John returned to his home terminal at Spokane, intending to visit the crew office to mark back up on the Brakemen's extra board to make himself available for his next assignment. He found he had to park two blocks away from the crew office because of the large number of employees who were there dealing with seniority issues. It took him thirty minutes just to get inside the door. He recalled that the crew clerks looked like they were about to have a nervous breakdown. With the consolidation of seniority that had been made in those first few days after the merger, John found that instead of being at the top of the Brakemen's extra board, he was now seventy-fifth on the list.

John went on to say,

As unhappy as my experience was, it was far from the worst. I know that some men went five or six days without

working because they would bump onto a job and then get bumped just before the job was called to go to work. Then, they had to go through the same thing on some other job. For me, it was back to the Brakemen's extra board for a long time.

The seniority issues with the BN merger were horrible. I went from first out from being able to hold [a job on] a local out of Hillyard [the former GN yard in Spokane] to about 75 times out to hold any local out of Hillyard. It took almost a year and a half to straighten out the seniority mess. I blame the UTU for the seniority mess. . . . The UTU spent all their energy fighting the merger right up to the time the merger was approved. They made no effort to look ahead to settle the issue of seniority consolidation if the merger was approved. If only they had agreed to the merger, the union could have written their own ticket in terms of dealing with seniority issues. . . .

On merger day the consolidated roster was created by dovetailing the separate rosters based on hire out date with no regard to a man's prior employer [i.e., whether GN, NP, or SP&S]. . . . Fortunately business was good, and the company needed pilot conductors and engineers to train everybody on the new territories now available to them. The Conductors' and Engineers' extra boards were increased dramatically to accomplish the training. As a result, the Brakeman's board was able to absorb most of us who had lost so many positions because of the dovetailed roster.

John has also commented on the seniority issues involving the locomotive engineers. The Brotherhood of Locomotive Engineers (BLE) used the fireman's date of hiring in preparing the consolidated seniority roster. As a result, there were NP firemen who were not yet promoted to engineer who were put ahead of GN engineers who had been working as engineers for between ten and thirteen years. John recalls,

This caused real hatred between the GN and NP enginemen that has lasted to the present day, even with the retirees. When the subject comes up, you can still feel the strong negative feelings on this issue. The BLE would not back down on the way the enginemen's seniority roster was made up using the fireman's date of hiring. The BLE general chairman was a former NP engineer, and he was the one who set up the way the seniority roster was arranged and took care of his former NP enginemen. To be fair this roster should have been either a ton-mile or prior rights roster.

John recalls how the inefficiencies in the early stages of implementing the plan to route trains on the new preferred route generated a heavy load of work for operating personnel:

Before the merger we didn't do much dog catching [a railroad term for having to send a relief crew out to replace a crew that has had to stop work due to being on duty for the twelve-hour maximum allowed under the federal Hours of Service Law] on the GN except in a bad winter or if we had a derailment. After the merger, it seemed like we were dog catching all the time, which kept the extra boards turning, which supported more men. The major factor contributing to the increase in dog catching was that the density of trains on some routes, such as the NP's line between Spokane and Sandpoint and the GN's line east of Sandpoint, increased as traffic was rerouted. In addition, trains were getting longer, which made the short sidings useless, effectively reducing the number of available meeting points. The excessive dog catching kept everybody working.

Both the GN and the NP had postponed hiring trainmen and switchmen for some time before the merger so we really didn't have a surplus of men. . . . Between the short staffing before the merger, the training, the dog catching, and good business volumes, I can't think of one time that any pre-merger trainmen were furloughed at Spokane.

About nine months or a year after the ton-mile roster came out, we went back to the dovetailed roster, but this time everybody had prior rights on their former railroad. This let me hold a brakeman's job in the Troy, Montana, chain-gang freight pool. The seniority issue caused hard feelings and dissention between the GN men and the NP men, with the SP&S men caught in the middle for many years to come. In the Spokane area, the GN and NP trainmen still had separate local union lodges when I retired at the end of 2004 more than thirty-four years after the merger. . . .

We trainmen had an income guarantee due to the merger, but few Spokane trainmen ever collected on it.

You had to be a mathematician to figure it out, and there were many ways the company could refuse to pay the guarantee by holding some junior trainman who was making more than you were against you. To get the guarantee, you had to be working the highest-paying job available to you. If some junior guy happened to have a good month, your claim would be refused because you did not work the job he was on. The NP trainmen had their choice of one of three years to use for their guarantee year. The GN trainmen had only one year and it wasn't our choice, another blunder by our union. Some guarantee claims took three to five years before they were settled through the union local chairman and the trainman received the money.

Over the years John worked up as a conductor in the Spokane–Wenatchee chain-gang pool and the interdivisional runs between Spokane and Whitefish. He often placed on work train assignments that were set up to run for several days at a time to move and unload track material such as rail and ballast, handle snow removal, and assist on ditching and bank-widening projects. He was recognized for his commitment to safety and the "can do" attitude he had toward getting a train over the road on a safe and timely basis in the face of bad weather or mechanical problems experienced during a trip. He was a good, professional operating man by any measure. When out on the job, he knew how to convince employees to put aside such distractions as the issues they had on seniority, and to keep their attention focused on the work at hand. Students of railroad operations will find John's writings very revealing of the conditions that railroad people sometimes had to work under, while they were earning the high level of wages they are more often known for.

One could fault BN management for not taking a leadership role or at least working jointly with the unions on the very important issue of seniority, rather than leaving it entirely in the hands of the unions alone. Seniority had a major impact on the morale of the employees involved. Having this issue not resolved satisfactorily for so long made it difficult to get some employees to become more focused on understanding and carrying out the changes in operations that would help make the BN merger successful.

OTHER INSTANCES OF HEAVY STRESS
ON EMPLOYEES AND SUPERVISORS

At the time of the merger, the Accounting Department quickly centralized all of its work in the General Office in St. Paul. No provision was made for the responsibility of handling the large number of employees who were made surplus in Chicago, as a result of their decision to not follow their work to St. Paul. The problem fell to the Chicago Region staff, with no plan from Labor Relations or the Accounting Department on dealing with the approximately one hundred employees who had no work but were still under pay, per the agreements made for job protection and wage continuation. This situation soon developed into such a state of frenzy that television stations in Chicago stormed right into the rooms in the Burlington's former General Office building where these employees were required to spend their time. These "surplus" employees were required to report to that room each day to remain qualified for compensation. While "on duty," they were allowed and encouraged to use the telephone to seek other jobs or to leave the building for job interviews. Even though they still were being paid, having nothing to do and the general boredom that resulted sometimes turned into emotional outbursts and expressions of strong resentment for the situation they were in. Considerable tension built up until these employees were able to find work elsewhere, or until they became ready to request and negotiate the terms for individual severance payments.

Several months into the problem, Louis Johnson, the region's director of Labor Relations, took it upon himself to begin negotiating individual settlements or "buy-outs" with employees who were not content to sit any longer in what amounted to a detention hall, even if it meant giving up their lifetime job protection. That effort was successful in reducing the number of surplus employees. When clerical vacancies opened in the yard or freight office in Cicero, about seven miles west of the offices in downtown Chicago, employees were required to place on such vacancies in order to retain their qualification for job protection. However, many were reluctant to place on these assignments, due to the difference in work environment and because most of the jobs that opened up required work at least some nights, weekends, and holidays.

Instead of commending BN for giving its employees unprecedented benefit of lifetime job protection, even for employees for whom it had no jobs, the news media and much public sentiment turned against the company. They would have been far better off if the agreement had provided for them to take a lump-sum severance payment at the time they lost their jobs and either declined to follow their work to St. Paul or if they were unwilling to exercise their seniority on jobs they could hold at Cicero. In that way, they would not have been subjected to the tension, ridicule, and indignity of having to sit in the same room, day after day, with nothing to do in order to keep collecting on their job protection.

Adding to the already tense and stressful situation at Spokane (caused by consolidation of the operations and integration of the operating employees of the GN, NP and SP&S) was the decision to establish a single-ended chain-gang pool for operations on the two somewhat parallel lines between Spokane and Pasco. Having a combined pool was thought to be better than the double-ended pool set up on M-Day with SP&S crews based at Spokane and NP crews at Pasco. In setting up bi-directional running on the two lines, it was necessary to have all crews based at Spokane. That required the SP&S crews to relocate to Pasco. This proved to be a difficult transition for the employees and their families who had to relocate and for the supervisors who had to administer the change, even though the agreement provided for the company to take over real estate owned by the employees and compensate them for expenses related to their change of residence.

Although other examples could be cited where some employees felt the merger disadvantaged them, the transition to the consolidated operation was accepted in good spirit by most of the employees affected by changes in the location of their jobs, by having to learn new work procedures, and having to work under supervisors whose roots were with one of the other roads.

CHAPTER 18

TRANSFORMATION OF THE TWIN CITIES TERMINAL

*• to build a large hump yard to replace the network of
eleven smaller flat switching yards • the single most critical project
in the early years • a return on invested capital that exceeded all expectations •*

To A GREAT EXTENT, THE SUCCESS OF THE 1970 merger depended more on consolidating and modernizing the Twin Cities terminal than on any other single project. All three of the railroads that would form Burlington Northern had large-scale operations in the Twin Cities. A large volume of business was interchanged between the two northern lines and the Burlington. The terminal consisted of eleven yards that had been built over the years on a rather disjointed, uncoordinated manner. None of these yards contained the capacity or kind of modern equipment, technology, or track layout needed to efficiently handle increased volume under a consolidated operation.

The "bulge" or greater length of the line between St. Paul and Minneapolis marked BN (NP) compared to the more direct BN (GN) line was caused by NP's decision to build its line on a lower grade of one percent. To accomplish that much of a reduction in grade, compared to the grade of 1.65 percent on the GN line, required the NP's line to be built one mile longer.

A large, new automated hump classification yard had to be built to get the consolidated Twin Cities terminal performing at the level that would deliver good service at a low cost. A large part of the savings in operating costs projected for the merged company was to come through improvements in operations in the Twin Cities. In addition to having yard operations scattered throughout the terminal, both the GN and NP had large mechanical facilities that needed to be consolidated and replaced with modern, efficient facilities.

In planning for the merger, efforts were concentrated almost entirely on the planning and design of a new hump yard to be built on NP property at Northtown. Acquisition of the additional property needed for a yard as large as specified in the plan was started shortly before the merger became effective. The yard had to be designed with enough capacity to handle all of the traffic being handled through the large yards of the component roads, i.e., the Union, Daytons Bluff, Lyndale, and Northtown yards. A small part of Union Yard and a few tracks in some of the smaller yards would be retained to serve as industry yards needed for cars moving to and from nearby industries. Unneeded yard trackage would be removed to free up the property for industrial development or other uses.

CONSOLIDATED TERMINAL AGREEMENTS

It was vital to have consolidated labor agreements for the terminal in place on the date of the merger. To the credit of the Labor Relations departments of the three roads and the general chairmen representing the Switchmen, Locomotive Engineers, Yardmasters, Clerks, mechanical forces, and track-maintenance employees, that ambitious challenge was met. The agreement allowed road crews from La Crosse to yard their trains at any point in the

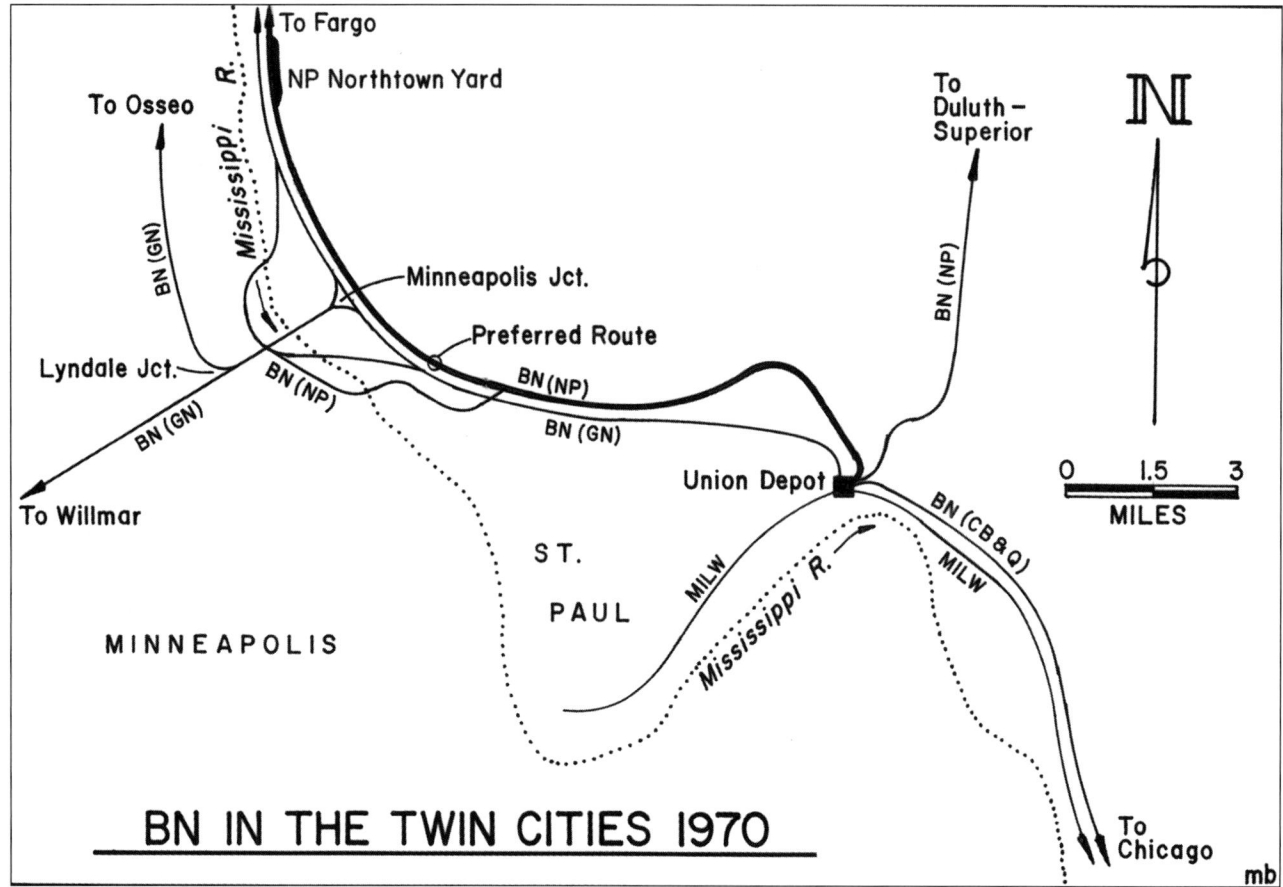

Figure 18.1.

consolidated terminal—they would no longer be limited to Daytons Bluff on the east side of St. Paul as their destination. These crews also would be able to handle an eastbound train out of any yard. They would be allowed to make setouts and pickups at intermediate yards. A separate group of yard transfer crews would no longer have to be used to move cars between Daytons Bluff and other yards instead of using road crews.

Train crews working on former GN or NP lines likewise would be allowed to deliver trains to any yard in the terminal, and take outbound trains from any yard. Road crews on the Burlington and GN would continue to go on final terminal delay payments (FTD)—a penalty payment made to crews when kept on duty in excess of thirty minutes from their time of arrival at the switch or other location designated at the entrance to the yard until relieved from duty—from the historic "boundary" point where they entered the terminal. This was a

very expensive concession made by Labor Relations in negotiating the new agreement. However, it would not apply to employees hired after M-Day. Most employees considered final terminal delay payments as "blood money," as they would rather get their train to its final yard and go off duty soon after entering the terminal than spend a lot of time in working their way through the often congested routes. Also, the FTD payments were an incentive for supervisors to move the inbound trains to their destination yard as quickly as possible.

THE CONTRIBUTION
OF ROBERT (BOB) HANSON

Another important aspect of the implementing agreement for the Twin Cities terminal was having the flexibility needed to transfer yard switching crews to the yards where the most work was to be concentrated. This advantage was available both at the outset of the merger and when the new yard at Northtown was opened for service. This agreement removed barriers that otherwise would have made it difficult to close yards and shift traffic to other yards during the construction at Northtown or—in the interim—to the yards best positioned to handle business as required in the new operating plan for train service. Together with the Labor Relations departments and the general chairman, Bob Hanson, who supervised the consolidated crew office, deserves a great deal of credit for getting this agreement written and implemented. Bob led the effort to formulate an agreement that would be essential in meeting the needs of the consolidated terminal operation. Bob often said that Jim Horne, the general chairman who represented the Switchmen working under the GN contract, also deserved a great deal of credit for formulating the set of implementing agreements for consolidation of the yard operations in the Twin Cities.

Bob had served as general chairman for the Switchmen of the NP and accepted the company's offer to head the crew-management function for the entire terminal after merger. BN was fortunate that Bob agreed to accept this offer, as that job was one of the most challenging and stressful jobs of any on BN at that time. He managed a large staff of crew callers and clerical employees who also handled the complex task of verifying monthly guarantee claims of employees who failed to make their test-period earnings, even though they had worked the highest-paying job their seniority allowed them to hold. In addition to managing the placement of yard crews, Bob's office called the road crews on duty for the five main-line subdivisions that radiated out from the Twin Cities, each of them with their own set of agreements governing their calling and placement. At that time and to this day, nearly forty-five years after the merger, there has been no success in negotiating an agreement to consolidate and simplify the complex schedules of agreements for road crews that are still in place from before the merger.

Because of his knowledge, fairness, and general demeanor, employees had great respect for Bob Hanson. He was able to work through disagreements or questions about the application of any part of the consolidated terminal agreement or the old agreements still in effect for road crews, with results satisfactory to everyone involved, including the supervisors and employees, Labor Relations officers, and the general chairmen. His capability to work through problems made it possible to settle things locally rather than letting them fester unsolved until Labor Relations and the general chairmen could schedule time to work them out. There were many individuals who stand out yet today for their excellent work in making things go reasonably well in the Twin Cities terminal until the new yard at Northtown was put in service, but Bob Hanson stands out among "the best of the best" who were working in those times.

In addition to doing excellent work in administering the agreements that governed the operation of the consolidated terminal, Bob helped individual employees deal with the stress and uncertainty of having to go to work at a new location, sometimes with a different set of co-workers who were not friendly toward a new face suddenly coming into their midst from a rival company, and perhaps displacing one of their friends. Bob understood the differences in the culture, norms, and work practices of each yard or facility. He was able to get the rival factions calmed down and accept the new order. There were times when the challenge of adapting to changes in the workplace applied as much to supervisors as to employees who worked under a contract. Bob coached them along, as well.

SETTING UP ROUTINES AND CONSISTENCY

As the superintendent of the Twin Cities terminal in the early 1970s, while the new yard at Northtown was under construction, I was told by the three heads of the Twin Cities Region—W. R. Allen, vice president of the region, Bob Shober, AVP–Operations, and Chuck Moehring, AVP–Sales and Marketing—of the importance of getting the operation set up to overcome service deficiencies. Many shipments were getting delayed because of the additional time required to transfer cars from one yard to another of the eleven yards we were operating. We tried to run inbound trains to the yard located closest to the industries to which the majority of the cars in a train were destined, or the yard where the outbound train would originate to which most of the cars would connect. But even that practice still left many cars that had to be transferred from one yard to another and then switched again in making up the outbound train on which they would finally depart the Twin Cities.

The transfer of cars from one yard to another could add up to forty-eight more hours to get a car handled through the terminal. Before the merger, many cars required fewer handlings in getting them switched to a customer's facility or placed in the outbound train to which they were to connect. The new large, consolidated yard under construction at Northtown was needed to overcome the disadvantages we faced in cost and transit time through the terminal due to having so many yards at locations scattered throughout the Twin Cities.

The situation we faced in running the terminal was made worse by the amount of capacity that had to be taken out of the yard at Northtown as construction of the new yard progressed. Having to build a new yard under traffic was difficult, but having to also minimize the delay or interruption to shipments during heavy yard construction is an even greater challenge for a railroad company to deal with. Most customers understood and accepted the need for delays to their shipments and were willing to stick with BN until the new yard was opened and the time required to handle their shipments would be reduced.

While construction was underway and we continued to have delays from having to transfer cars from yard to yard, customers demanded that we set up scheduled routines for making such transfers. Having consistency in the time it took us to handle their business through the terminal satisfied most customers. Once we got those routines and schedules established, the complaints were reduced and there were fewer delayed cars for the "hot sheet" we prepared. We also found ways to reduce the time for handling many regular movements. At the same time, we managed to reduce the number of daily yard-engine shifts from 126 to 100. The three heads of the region were very supportive of our efforts to improve the operation and did well in convincing staff personnel in both the system and region offices not to deluge us with second-guessing and demands for special moves.

We were fortunate the company had placed a team of outstanding officers at the region level. They were our superiors, and also served as effective coaches and mentors for our young group of field officers. Eastbound trains were blocked at Gavin Yard in Minot for the large volume connections in the Twin Cities and for cars going beyond, such as to Chicago and Galesburg. Cars on the "through" trains that originated in Cicero were blocked so they could be run through the Twin Cities without any switching being required. Those trains required only a crew change and the time required to set-out blocks of cars with a Twin Cities destination or to add cars for intermediate points short of Minot. However, large numbers of trains from points other than Minot, Chicago, and Superior arrived unblocked, which required us to switch them, car-by-car, and the need to transfer many of them from one yard to another in the terminal.

SUPPORT FROM TOP REGION OFFICERS

Company leadership continued to make positive contributions. W. R. Allen, the regional vice president at that time, had been brought from the Frisco to the Burlington and BN by L. W. Menk. Allen had served as vice president–Operations of the Frisco for many years and is among the most capable operating managers I have worked under. Bob Shober had been general manager–Lines East of the Great Northern before the merger and had particular skill in the development of younger officers in matters of any kind. Shober was tolerant of minor errors in judgment on the part of field people as long as he could see they learned from their mistakes and moved ahead. He was not a nit-picker, nor did he second guess us. When he found weaknesses and the need to improve our operation, we would often get a two-sentence advisory from him, usually ending with the same challenge: "Let's get the show on the road." Chuck Moehring was a well-established career marketing and sales officer of the Northern Pacific and never overreacted to service failures. Together, Allen, Shober, and Moehring made a good team. They had left their past, pre-merger affiliations behind and demonstrated a strong commitment to the goals and values established for BN. Overall, the working environment they established for subordinates in charge of running the Twin Cities terminal was most affirmative.

At the same time that we were running the terminal, we were expected to prepare the plans for running it after the new Northtown hump yard was put in service. This was an opportunity that not too many line operating officers ever had. More often, a group of experts from such headquarters departments as System Transportation, Industrial Engineering, and the office of the operating vice president would be set up to prepare the operating plan for a major new facility such as BN was building at Northtown. Terminal officers would participate in the planning process, but for the most part, the plan for running the operation would not be prepared by the local officers who would be running the new yard.

Our terminal officers enthusiastically took on this project, and as we made progress, we made presentations to the region transportation staff and System Transportation to determine if there were any considerations for the flow of traffic, service commitments, and volume forecasts we had overlooked or were not aware of. The process worked well and built a strong sense of ownership into the terminal operating officers and such key personnel who participated in it, among them yardmasters, car foremen, chief clerks, and our administrative staff. Allen, Shober, and Walter Crum (region director–Transportation) presented good insight and advice along the way. All of this helped insure the terminal would be ready for the cut-over to the new operation and to achieve the cost savings projected when the AFE (Authority for Expenditure, a formal process for approving expenditures for track bridges, signals, cars, and locomotives) for the new yard was approved, as well as reduce the activity and resources employed in all of the old yards in the terminal.

In our planning, we projected a reduction in the number of yard crews worked daily from one hundred to fifty. Two years after the new yard became operational, the actual number of crews worked was down to twenty-five. Of all the major capital expenditures made by BN in its first decade, the new consolidated yard in the Twin Cities must have had by far the highest return on investment. The flow of cars over the entire northern corridor between Chicago and the Pacific Northwest improved dramatically, making BN's service more competitive. This project was the fulfillment of perhaps the most important link and component of the total operating plan for the merged company.

After the new yard was open for service, attention had to be directed toward improvement of the main route

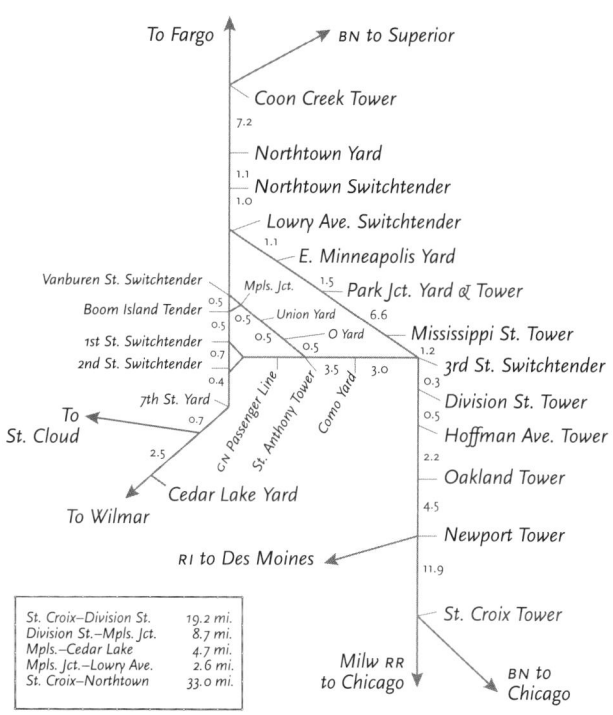

Figure 18.2. Twin Cities Terminal, 1972, Operators and Switchtenders. BASED ON GRAPHIC BY W. H. FERRYMAN.

through the terminal for the entire 44.0 miles between the crossing of the St. Croix River at Prescott to Coon Creek Junction, 7.5 miles west of Northtown Yard (see map in figure 18.1). If there were any deficiencies in the plan for operations for the merged company, it was the failure to prepare a plan for modernizing the routes through and approaching the Twin Cities terminal from both the east and west. However, having the financial resources available to take on yet another major project in the early years of the merger very likely was a limiting factor.

Raising the money needed for the new yard in the Twin Cities, plus the expensive new connections between lines at Spokane and Sandpoint, consolidating functions in the General Office, and establishing the six new regional offices right at the time of the merger was very challenging. Labor protection and wage-guarantee payments also would be very expensive for several years. However, by the late 1970s, BN was ready to appropriate the funds needed to upgrade, expand capacity, and improve operating flexibility on the route used to get the increasing number of trains running through the Twin Cities terminal.

Starting from the east end of the terminal, a new drawbridge at the crossing of the St. Croix River at Prescott was needed. The speed over the old bridge was limited to 10 miles per hour. Since this was a single-track bridge in the midst of a double-track railroad, having a restriction of 10 miles per hour was a limitation on capacity and a cause of delay to many trains. The old bridge was replaced in 1984 with a new structure. A line change to reduce curvature on the Wisconsin side of the river allowed train speed to be increased to 35 miles per hour.

The most restrictive area began at St. Croix Tower, 2.9 miles west of the bridge, the beginning of a paired track operation for 19.2 miles. One of the two main tracks, owned and maintained by the Milwaukee Road, signaled for operation in one direction only. The Milwaukee's track was in fair to poor condition, a reflection of the financial problems the Milwaukee had at that time. The other main track was owned by BN and operated under CTC rules. Because the Milwaukee's track had a grade of 0.7 percent ascending westbound, and the BN's track was built on a river grade, eastbound trains were usually routed on the Milwaukee's track. This segment of joint track was controlled by dispatchers located in the tower of a "strong arm" interlocking plant at Newport. This was the only territory handled by these dispatchers. A few years after the merger, BN was able to work things out with the Milwaukee to get its track upgraded. Also, CTC was put into service on the Milwaukee's track, which gave BN and Milwaukee a modern, high-capacity joint route. At that time, BN took over the dispatching of the joint track and put it under the dispatching position

BN had established to cover the entire Twin Cities terminal.

From the end of the paired track at Division Street in St. Paul to Northtown, the preferred route was on a main-track segment built by the NP. All but 1.2 miles of this 13.0-mile segment had double track. There was an automatic block-signal system in service on only 8.4 miles of that segment. The balance of 4.4 miles (between Park Junction and Northtown) was not signaled. On top of that, the flow of traffic on this line was not controlled by a dispatcher. Instead, trains moved under informal instructions from a series of switchtenders, yardmasters, and tower operators, from the "territory" of one of them to the next point where movement "authority" resided for movement on the next short segment of the line. It is easy to envision the difficulty we often had in moving a priority freight train over this line without delay from yard crews already occupying a main track for switching at an industry, or in making a delivery from one yard to another (refer to figure 18.2 for a drawing of the locations of the "points of control" on this line). To modernize this line required that the two short single-track segments created sometime in the past be restored to service as double track, and that CTC then be installed to give us a continuous two main-track operation signaled for movement in both directions.

The line of the NP rather than the GN's line was designated the preferred route through the Twin Cities because the NP's line would give the most direct access to the new yard at Northtown. It also had a much lower grade (one percent ascending westbound) compared to the GN's line, which had a grade of 1.65 percent ascending westbound. A difference in grade of that magnitude requires significantly more motive power to move a heavy tonnage westbound train, and avoid having to use a helper locomotive. While most of the NP's line had heavy rail and crushed-rock ballast, the condition of the line was further upgraded to allow a speed of as much as 50 miles per hour. With these improvements, trains could get over the road without delay and the capacity of the line was increased. By getting control of the traffic under a dispatcher, all of the low-productivity "controller-type" positions along this line were eliminated, giving this project a high return on investment. A new single main track was built around the new yard. Between the west end of the yard and the 5.7 miles to Cook Creek Junction (the junction of the preferred route with the former GN's line to Superior), two-main-track CTC was installed to increase capacity and provide flexibility in moving trains in and out of the yard.

While the preferred route through the terminal was being upgraded, action was underway to rationalize the high-capacity line the GN had built between St. Paul and Minneapolis. That line had four main tracks, two for passenger trains and two for freight trains, including trains and transfers operated by the C&NW, Chicago Great Western, and Rock Island to and from interchange with the GN or for access to their own yards. The two passenger main tracks were maintained for 55 miles per hour. Trains moving on the freight mains were restricted to 30 miles per hour. Due to clearance restrictions through the Minneapolis depot, freight trains could be operated on the passenger main tracks only if they contained no cars higher than a "standard" box car of that era. Because operating on the passenger "mains" was completely independent from the freight operation, trains using those tracks did not encounter interference from movements in or out of yards, nor were there any junction points to contend with. Except for the ascending grade of 1.65 percent for westbound trains in St. Paul, the route via the Stone Arch Bridge was ideal for the movement of through trains. Since there were no restrictions in running unit coal trains on these tracks, we took advantage of the passenger route until the decision was made to remove the tracks through the Minneapolis depot and to turn the Stone Arch Bridge over to a government agency for conversion to a pedestrian walkway.

The two freight main tracks were signaled for only the 7.0 miles between Third Street in St. Paul and the interlocking at the St. Anthony crossing. In rationalizing trackage in the terminal, two of the four main tracks were removed between Third Street and St. Anthony. West of St. Anthony, the two freight main tracks were retained for the 2.5 miles to Minneapolis Junction. The passenger main tracks west of St. Anthony, going over the Stone Arch Bridge and through the GN's passenger station in Minneapolis, were removed. By this time, Amtrak trains had been rerouted on the Milwaukee's line between St. Paul and the Minnesota Transfer Railway tracks to reach the new Amtrak station in the Midway District.

The double-track line between Minneapolis and Wayzata (15.0 miles) was reduced to single track. With the diversion of through trains from the Willmar line to the preferred route, the extra capacity of a double-track railroad was no longer needed. The NP's "A-line" between Park Junction and the flour-milling district of Minneapolis was removed shortly after the 1970 merger. This line included a large steel trestle bridge over the Mississippi River, commonly referred to as "Bridge 9." As of this writing, Bridge 9 still stands but is not in service.

The GN had major shops in St. Paul for the overhaul and heavy maintenance of locomotives. That work was moved from the GN's Dale Street Shop to heavy-repair shops at Livingston, Montana, and West Burlington, Iowa. A large shop for running maintenance and servicing of locomotives was built at Northtown to replace the NP's facilities for running maintenance at Northtown and Mississippi Street in St. Paul and the GN's facility at Minneapolis Junction. Other shops that were shut down in those years were the passenger-car-maintenance shops of the GN and NP in St. Paul, and the GN's roundhouse at Jackson Street.

Together, the changes described in this section greatly reduced the asset base employed by the merged company in the Twin Cities. The new hump yard at Northtown replaced the eleven flat switching yards operated by BN's predecessor companies. A few tracks in some of the old yards were retained as staging areas for cars moving to or from local industries. The old routes and support facilities were replaced with the major investment made in new shops and in improving the preferred route through the terminal. The new and fully upgraded asset base provided BN a more functional and efficient foundation for providing the quality of service and the capacity needed to handle its growing volume of business.

CHAPTER 19

TRANSFORMATION OF THE SPOKANE TERMINAL

• to be able to concentrate the movement of all trains on the best-constructed line through Spokane • construction of a large bridge to connect the lines of the three component roads •

IN ADDITION TO THE TWIN CITIES, THE 1970 merger required a major transformation of the network of yards and rail lines in Spokane. The changes that had to be made were centered on the decision to make the Northern Pacific's 67-mile line between Spokane and Sandpoint the preferred route for transcontinental business. That designation provided the opportunity to rationalize the infrastructure through Spokane that the Great Northern had built and maintained. It also required the construction of a major bridge and connecting links needed for trains to have smooth, efficient access between the various lines in and radiating out from Spokane. The map in figure 19.1 shows the configuration of lines in place at the time of the 1970 merger.

Following is a brief description of the network of lines owned by the GN, NP, and SP&S in the Spokane area:

NP   The NP had by far the best route through the city and in approaching Spokane from both the east and west. In addition to its more favorable grades, the NP's line had the advantage of an elevated structure through downtown Spokane that had been built in 1909–11 to eliminate all grade crossings. The line had 9.6 miles of either two-main-track CTC or double track with ABS (Automatic Block System) through the city. Twenty-six miles east of Spokane, at Hauser, Idaho, property was available for the construction of a new hump yard, an additional feature that made the NP's route even more favorable.

To be able to use the NP's line through Spokane, a new two-mile connection had to be built at Sandpoint, Idaho, to connect the lines of the NP and GN. Construction of this new line also required construction of a bridge of 997 feet over Sand Creek.

GN   The GN had a less desirable route through Spokane. The NP built its line through Spokane several years before the GN had to decide whether to serve Spokane or to build a line that would bypass the city some distance to the north. After convincing the city to provide a right-of-way, the GN decided to build through Spokane, even with the disadvantage of a one percent grade and having to cross major streets at grade. After crossing Latah Creek on a high single-track bridge at the west end of Spokane, another grade of one percent was encountered. While a grade of one percent is not unduly severe for a line of railroad, it was in excess of the maximum of 0.6 percent that James J. Hill wanted on all lines, except when crossing mountainous areas. The GN built a large yard and shop complex at Hillyard, just east of Spokane. Because the yard was built on a grade, extra switchmen had to be assigned to assist some crews in operating hand brakes to control the speed of cars rolling free during switching operations. This requirement added to the cost of operating the yard, but the yard the NP built in Spokane did not have this kind of restriction. Overall, the NP had a far superior route through Spokane.

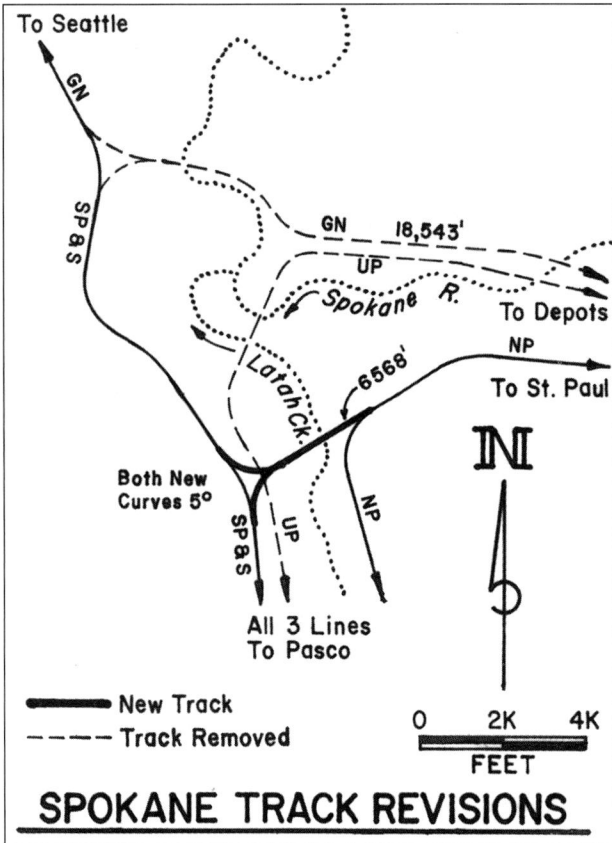

Figure 19.1.

SP&S   Trains of the SP&S operated in and out of the GN yard at Hillyard under trackage rights.

RATIONALIZING ROUTES

Use of the NP line as the preferred route through Spokane required the construction of a large bridge over Latah Creek on the west side of the city to connect the NP line with the GN line to Wenatchee and Seattle, as well as the SP&S line to Portland. The new bridge, 4,260 feet long and 215 feet high, was completed in 1972. BN received many awards for the design of this bridge. The new connection to the GN line on the west side of the bridge also required construction of a bridge 730 feet long and 140 feet high over Indian Creek. A line change of 5.5 miles had to be built, although it reduced the length of the line between Spokane and Wenatchee by four miles. The new line was built on a grade of one percent.

Once the new bridges and the line change were completed, BN was able to rationalize the track, yards, and mechanical facilities located on the GN route. The line between Hillyard and downtown Spokane was reduced to a single-track operation. The large steel trestle bridge that GN used to cross Latah Creek was removed. The large locomotive shop and the facilities of the Western Fruit Express Company facilities at Hillyard were shut down, and the switching and train-makeup operations were transferred to the NP's facility at Yardley, over a period of twelve years.

Removal of the GN's depot and its main line in downtown Spokane freed up a large amount of property for Expo '74, an international exposition put on by the city of Spokane. For the city to have all of the land it needed for the exposition also required removal of tracks and facilities owned by the Union Pacific. To make that possible, BN agreed to give trackage rights to the UP on its route through downtown Spokane (on the NP line), on the new bridge over Latah Creek, and to a new connection with the UP's line at Fish Lake, twelve miles west of Spokane. All of this was a huge undertaking that greatly benefited BN, UP, and the city of Spokane.

Bob Downing deserves credit for having the vision and skills for planning this major transformation of railway facilities and routes in Spokane. The plan he developed was vital in moving BN ahead in its objectives to quickly get the operations of the GN, NP, and SP&S integrated. Having an efficient network with adequate capacity through Spokane enabled BN to greatly improve its operating capability, and also to no longer have to maintain the less-efficient route the GN had built in Spokane. Bob had remarkable success in negotiating the terms for the major changes involving the UP, the city, and numerous other constituencies who had an interest in having the railroad network and facilities rationalized in ways that would meet the needs of all concerned.

While the transition to a rationalized terminal in Spokane went very well, there was some tension among employees in train and engine service as they made their transition to a combined work force, working under new sets of agreements governing their seniority placement and other types of new or modified work rules. These issues are covered in some detail in chapter 17.

DECISION NOT TO BUILD
NEW HUMP YARD AT HAUSER

The new hump yard proposed at Hauser, Idaho, was not built. With the switch to intermodal equipment and unit train service for many commodities in the 1970s, the need for additional large-scale capability to switch cars declined. Also, since BN had a large, modern hump yard at Pasco, only 145 miles from Spokane, there was no need to have another large hump classification yard so nearby. That decision saved BN upwards of $50 million at a time when major capital investments had to be made to handle the rapidly growing volume of coal being mined in the Powder River Basin.

In the late 1990s, BNSF built the additional tracks needed to move the crew-change point from Spokane to Hauser, and locomotives on through trains could be fueled there at the same time. This capability was needed to reduce the congestion at Yardley from the increasing number of trains being run on the northern corridor.

In more recent times, a planning agency called the Spokane Regional Transportation Council has been promoting a large project (estimated at $270 million in 2002) for construction of grade-separation structures to eliminate most of the seventy-five grade crossings on BNSF and UP tracks in the 42-mile corridor between Spokane and Athol, Idaho. In the interest of improving public safety, the council has worked with BNSF and UP on a plan for construction of an additional main track adjacent to the existing BNSF main tracks, to provide the additional capacity needed to handle the UP trains on what would become a jointly used corridor. With this new route, the UP line between Spokane and Athol would be removed, making it possible to eliminate all of the crossings on its existing line. The UP's yard in Spokane would be relocated a few miles to the east. This ambitious project could be of great benefit to the public and provide enough capacity to handle the approximately sixty trains per day that UP and BNSF were running in 2005 when the council completed its study.

It has been reported that this project has been delayed over the issue of whether the UP should be given access to industries in the corridor that are presently served exclusively by BNSF. At the time of this writing, it is uncertain if or when financing will become available to move this project forward without undue delay.

CHAPTER 20

SHORTFALLS IN THE 1970 MERGER OPERATING PLAN

> • *the consolidated operating plan worked very well but*
> *left some issues for the post-merger team to handle* •
> *the initial capital program contained several*
> *large, expensive projects. Good financial results would make*
> *it possible to invest in projects that would further improve the property,*
> *reduce operating costs, and take advantage of new market opportunities* •

OVERALL, THE PLAN PREPARED FOR THE consolidated operation proved to be well thought out and realistic. The preparations undertaken for implementation of the plan were at the level of detail necessary for validating it. Managers at all levels and from all functional areas participated in the planning sessions. The transition in the early days of the merger generally was smooth, with no major disruptions to service. No serious deficiencies or shortfalls in the basic elements of the plan came to light in the first critical weeks after M-Day. The plan for setting up a network control center (commonly referred to as "Diesel Control") and for connecting the line segments that would make up the preferred route turned out to be a splendid piece of work. Local operating officers saw to it that local service, including industry switching, was carried on without disruption. It was of great advantage to all concerned that the new long-distance trains in the plan would be established over a period of several weeks and not all at once, as had been done in some other mergers in recent times.

A RESTRICTIVE ROUTE THROUGH THE TWIN CITIES

There were a few areas in the operation that in my opinion were either not recognized or given as much attention in the pre-merger planning as they should have been given at that time. The one that stood out the most clearly was a failure to plan for upgrading the fourteen-mile segment of the preferred route through the Twin Cities terminal between the junction at Third Street in St. Paul and the yard at Northtown. With the large volume of priority freight trains to be run on that line (sometimes called the NP freight line, lines A and B), it should have been put under the control of a dispatcher right from the start. Even though the money needed to install CTC and eliminate the short segments of single track in St. Paul might not have been immediately available, plans for those improvements should have been prepared before the merger so the work might have been started as soon as one or two years after M-Day, and not seven years later, as happened.

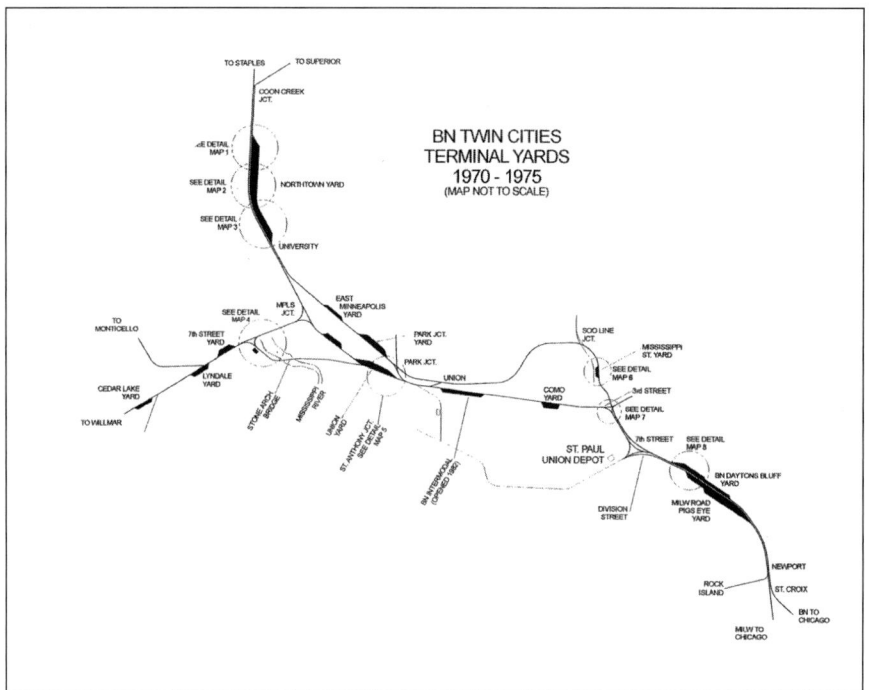

Figure 20.1. By KRISTOPHER JOHNSON.

OVERLOAD AT YARDLEY (SPOKANE)

The second factor was the number of additional trains and switching put into Yardley, the main classification yard in Spokane. Too many westbound trains carried blocks of cars for the lines both to Pasco and to Everett and Seattle, requiring block-swapping and a considerable amount of car-by-car switching in Yardley. Eastbound, too many trains from Pasco and the line from Everett and Seattle arrived in Yardley with cars for both the line to Missoula, Laurel, and Kansas City, and the other line, leading to Minot, the Twin Cities, and Chicago. The plan for train operation should have had eastbound trains made up in yards west of Spokane with cars for only one line or the other, to reduce the need to bring so many trains into Yardley for switching.

The merger plan called for a large new hump yard to be built at Hauser, twenty-one miles east of Spokane. However, with so much demand for heavy capital expenditures in BN's early years, it was known that it would be five years or more before construction could be started. Until the new yard could be built, the plan for the first several years after the merger should have avoided putting such a work load that would exceed the capability of Yardley.

THE "FUNNEL": NOT ENOUGH CAPACITY

Another issue either not recognized while the planning was underway, or perhaps set aside from consideration due to not having enough capital to do everything at once, was the capacity of the former NP line between Sandpoint and Yardley. This was a single-track line operated under CTC rules with only minimum grades. The lack of operating capability on this 63-mile line segment of the preferred north transcontinental route became an issue as soon as the trains operating on both the former GN and NP lines east of Sandpoint were concentrated on it (i.e., trains from the preferred route via Whitefish, Minot, and Minneapolis, plus those moving on the corridor leading to and from Missoula, Laurel, and Kansas City had to run on it to reach Spokane). This chapter contains some detail on the characteristics of that line segment and what has been done in the nearly forty-five years since the merger to meet the need for additional capacity.

In planning for the consolidated operation, the GN's somewhat parallel line was to be shut down as a through route, leaving only the line of the NP in service for all of the through trains. The NP's line was designated for the preferred route for many reasons, among them its lower grades[1] and because the NP operated on a double-track viaduct through downtown Spokane. The GN's line was plagued with a number of grade crossings and had a less-favorable grade. Another advantage of the NP line was having CTC in service, while the GN's line was still operated under the timetable and train-order system of dispatching.

A major line change completed at Granite Lake on the NP line shortly before the merger eliminated a 25-mile-per-hour speed restriction by reducing curvature from 8 degrees to 1 degree, 20 minutes, which allowed freight train speed to be increased to 60 miles per hour. With completion of that line change, the only speed restriction due to curvature that remained on this segment was a 6-degree, 40-miles-per-hour curve at Cocalalla.

Even with the superior characteristics of the NP line, it still had limitations for handling the large number of trains that had to be moved on it after the merger. One limitation was in not having segments of two main tracks in place for some distance on at least one side of the junction at Sandpoint. Since trains will often become "bunched" at such a junction, a common practice is to build a second main track for eight to ten miles on one or both sides of the junction. That was not done in preparation for the merger, nor was it included in plans for improvements to be made following the merger.

About one mile west of the junction, there was a 4,750-foot-long bridge over Lake Pend Oreille, which would limit the combined line or "funnel" to a single-track line over the lake for a very long time, if not forever, due to the expense of building a second bridge of that length. However, it was very encouraging to note the announcement BNSF made late in 2014 that a second bridge will be built over the lake, with a goal of having it in service by 2018. Having this new segment of second main track will be of great benefit in speeding up the flow of traffic through this congested area.

A large hump yard was to be built along the NP line at Hauser, seventeen miles east of Spokane. The property to

be acquired for that yard would not have been accessible from the GN's line unless a long, expensive connecting track of ten miles or more were built between Hauser and the vicinity of Milan or Dean on the GN line. There was no practical way to build a connection between the GN and NP lines through commercial property close to downtown Spokane, for "directional running" between Spokane and Sandpoint. Using the existing interchange route between Hillyard and Yardley for through trains would require the trains run into a yard track at Yardley, where they would have to change ends, i.e., move the locomotive and caboose from one end of the train to the other end. This awkward and time-consuming move would require the use of a long and badly needed track in a yard that already was severely taxed.

A new single-track connection one mile in length was built at Sandpoint between the NP and GN lines, including a bridge of just under 1,000 feet over Sand Creek, but no double track was built either east or west of the junction. The controlled siding at Boyer (one mile east of the junction) was extended to 10,363 feet. As already mentioned, very little would have been gained by building a mile of second main track between the junction and the east end of the long, 4,750-foot single-track bridge that crosses Lake Pend Oreille. With about forty trains operating daily over all or part of the NP line, congestion developed right from the start of the merged operation.

The operating plan called for the GN's line to be kept in service only for local business. Part of the line would be abandoned in time, which further eliminated any possibility of having its capacity available for relief of the congestion that developed on the single-track line of the former NP.

By the mid-1970s, the limitations of having only single track on all three sides of the junction made it necessary to prepare plans for increasing line capacity west of the bridge by building a second main track over the entire fifty-eight miles into Irvin, where the ten miles of two main tracks through Spokane began. Having so many trains delayed on the single track between Sandpoint Junction and Irvin (five miles east of Yardley) added greatly to the problem of having to bring so many of them into Yardley for switching and block-swapping. Also, some of the interdivisional crews set up in 1972 to operate trains to and from Whitefish and Missoula

---

1   The GN's line of five miles between downtown Spokane and its yard (at Hillyard) was built on a grade of one percent, while the maximum grade on the NP line between Spokane and Sandpoint was only 0.8 percent.

were having to tie up on line under the federal Hours of Service Act due to delays in meeting so many trains on the single-track railroad. Having these trains tied up on a siding until a replacement crew arrived took a meeting point away from the dispatcher, thereby adding further to the congestion on the line. By this time, the number of unit grain trains and intermodal trains was increasing, a nice problem to have, but remedy of the line congestion would require that large capital expenditures be approved for construction of additional second main track.

Had enough money been available at that time, the entire line between Yardley and the west end of the long bridge over the lake at Sandpoint might have been upgraded to a two-main-track operation within three or four years of the 1970 merger. However, it was not possible to move ahead on such expensive projects due to the expensive projects of an even higher priority that were required in those early years. A large hump classification yard had to be built in Minneapolis, plus the new connections at Sandpoint and the large new bridge and connecting tracks on the west side of Spokane that were needed to integrate the three GN, NP, and SP&S lines. Also at that time, BN was beginning to face heavy expenditures to expand capacity and upgrade the railroad to handle coal originating in the Powder River Basin.

If the need to expand line capacity on this corridor had been recognized in the merger plan, it might have been given more attention by senior management at that time and ranked higher on the priority list for future capital expenditures. However, in the years the planning was done, expectations for growth in business were quite low. A short time before the merger, John Budd, president of the Great Northern, stated, "the Great Northern requires a strong increase in traffic. Yet, on its own, prospects for such increase appear elusive in the extreme."[2] In working under a viewpoint this restrictive, the planning team might have assumed the funnel between Sandpoint and Spokane would have sufficient capacity to handle the combined traffic of both main lines without difficulty.

Over the years, BN and BNSF have built second main

trackage on 46.2 miles of the 69.7 miles and 3.3 miles of five miles of five main tracks (at the Hauser yard and fueling facility) of the NP line between Sandpoint and Sunset Junction (Spokane) to increase its capacity. The amount of single-track mileage remaining in 2016 had been reduced to 20.2 miles of the 69.7-mile segment.[3] This includes the long single-track bridge over the lake just west of Sandpoint. The hump yard proposed at Hauser was not been built. With the shift to intermodal equipment and unit trains that happened by the early 1980s, it was concluded in 1982 there no longer was a need for it. Instead, a "flat" switching and staging yard of twelve long tracks and a high capacity fueling facility and for staging through trains have been built at Hauser.

While this change reduced congestion at the old yard (Yardley) in Spokane, having a large number of trains reducing speed to enter and depart from Hauser might have introduced a new point of congestion on the funnel. To prevent this from occurring, BNSF installed high-speed turnouts and signaled five tracks as main tracks for 3.9 miles for movement through the yard and leading to and from the new fueling facility. Trains are authorized to run up to 40 miles per hour on the three main tracks that run through the yard. Without block-signal protection, trains would be limited to restricted speed not exceeding 20 miles per hour while moving through the yard. It would take much longer for trains to clear the main tracks while entering or leaving the facility at Hauser. With crews from Whitefish having seventeen fewer miles to run, the number of crews "dying on the road" has decreased. The locomotives on most trains are fueled in only forty-five minutes, much less time than was required when fueling had to be done at the older servicing facilities at various points west of Spokane. The crew change point for through trains has been moved from Yardley to Hauser. It took about thirty-five years from the 1970 merger until 2005 to complete the revised plans for expanding capacity at Hauser, as well as the additional miles of second main track that were needed to overcome the congestion at Yardley and in the funnel that had been troublesome problems from the time of the 1970 merger.

---

2   John F. Strauss Jr., *The Burlington Northern: An Operational Chronology, 1970–1995*, 3, quoted from letter from Budd and Robert S. Macfarlane, president of the NP, to employees in 1955, exact date not specified.

3   BNSF Railway Northwest Division Timetable No. 17, Spokane Subdivision, effective November 5, 2014, with update through May 3, 2016.

OVERLOAD IN EASTBOUND YARD
AT CICERO

Another issue not foreseen in the merger plan was the large number of unblocked eastbound trains that would have to be run from the Twin Cities and into the yard at Cicero. In the operating plan, only one additional train was to be operated from the Twin Cities to Cicero. However, within only a few weeks, the actual volume increased by far more than one additional train per day, with three or four unblocked trains being run into Cicero each day. The number of crews in the Cicero–Savanna and La Crosse pools was nearly doubled in a short time.

At Cicero, the hump had the capacity to handle only the westbound interchange deliveries received from the eastern lines. Eastbound trains had to be brought into a flat yard called the "eastbound" in which they were broken up for direct deliveries to the eastern lines and the Belt Railway Company (BRC) yard at Clearing. An unplanned increase in volume of the magnitude experienced after the merger put a real burden on the eastbound yard. With the lack of capability in the Twin Cities to block eastbound trains by connecting line for delivery in Chicago before completion of the new hump yard at Northtown, eastbound trains arriving at Cicero were not blocked except for the one or two daily trains that were humped and blocked by connection at Minot. Handling the unblocked trains required time-consuming car-by-car switching in a yard with very limited capacity. The intermediate yards at North La Crosse, Savanna, and Eola did not have the capability to switch cars into blocks and then hold the blocks long enough to be able to swap blocks between trains, or to hold the blocks long enough to make up solid trains for delivery to the eastern lines in Chicago.

Trains made up in the hump yard at Galesburg were blocked by connection, but those coming from the Twin Cities were unblocked, causing the eastbound yard to back up. At times, trains had to be held out of the yard for several hours. The only outlet for relief was to send some of the unblocked trains to the BRC for switching. However, it took longer to move cars to the eastern lines if they had to be switched at the BRC instead of BN delivering them directly from Cicero to the connecting line. Also, the per-car charge for cars switched at the BRC was quite steep, making the cost far greater than if those cars could be moved in direct interchange as planned. In summary, the operating plan did not provide for the large increase in eastbound business coming out of the Twin Cities terminal.

Of course, having this large increase in business was another nice problem to have, but it did present a major challenge to the company until 1974 when the new yard was opened at Northtown and eastbound cars could be blocked for connections. The transportation plan for the consolidated operation to take effect in 1970 should have provided for blocking eastbound business destined to Chicago at a combination of yards as far west as Staples, Willmar, and Superior, as well as the La Crosse, Savanna, and Eola yards on the Chicago Region, so some solid, run-through trains could have been made up for large volume connections at Chicago. The blocks made up at each of these yards might then have been assembled at Eola with blocks coming from Galesburg, into solid consist trains for delivery to the eastern connections.

Had such a plan been developed, the worst of the congestion and car delay that occurred in the eastbound yard at Cicero might have been avoided. The Chicago Region could not have simply decided on its own to set up such additional switching capability at its intermediate yards since the flow of transcontinental business was involved. Because there was not enough capacity in the region's yards to do all of the blocking required, a central planning group would have to determine if yards west of the Twin Cities would be able to do enough blocking to give some relief to the eastbound yard at Cicero. A change this radical in the operation of eastbound trains into Cicero would also require changes in the scheduling of motive power to insure power was available for on-time departure of westbound trains. The underlying problem was failure either to acknowledge, anticipate or reveal that such a large increase in eastbound business from the Twin Cities would occur, presumably from diversion of business to a single-line haul for BN that had been interchanged to other railroads before the merger.

UNANTICIPATED LARGE-SCALE MOVEMENT OF COAL

During the years when the consolidated operating plan was developed, there was no way to anticipate or plan for the movement of the heavy tonnages of coal that started to move in 1972. The impact of the passage of the Clean Air Act of 1970 hit BN by surprise. If the planning team and the executives who were to head the new company had known they would be facing that challenge, they might have had to decide to set money aside for coal-related projects at that time, and construction of some of the merger-related projects might have been set back.

It may be fortunate they were not aware that the impending legislation on emissions would force the electric power companies to begin to burn coal from the Powder River Basin in large quantities starting in the early 1970s. In its early years, BN might have had difficulty in raising enough money to start work on the challenge to move the coal on top of the expense of $43 million for construction of the new yard at Northtown,

$13.5 million for the new connections at Spokane, and numerous other start-up costs. It might have been necessary to set back those costly but vitally needed projects, even though they had a high return on investment and were so fundamental to getting the operation integrated.

Overall, as stated at the beginning of this chapter, the operating plan was realistic and well thought out. The purpose in writing this chapter is to point out some of the problems in the early stages of the merger that were not recognized or did not rate serious consideration at the time. Also, with the company's financial resources already strained from the demand for capital to finance the new yard at Northtown and the bridges and new connections at Spokane and Sandpoint, it is likely the planning team recognized that the new company would not be able to handle additional high cost projects at the outset. They would have to be set back for review in later years as events unfolded (i.e., success or failure in the marketplace, and in the company's financial performance).

CHAPTER 21

OVERLOAD AT CICERO

• handling more eastbound business through Chicago than anticipated in the plan •
a fast-paced yard operation, perhaps the fastest on BN •
a yard designed and built for rapid throughput, not
for holding or staging cars for later movement •

IF YOU HAVE READ CHAPTER 22, ON COMPASS (Complete Operating Movement Processing and Service System), you already have a good idea of the character of operations in the BN's yard in Cicero, Illinois. Managing that yard provided numerous day-to-day challenges for operating managers, both at the terminal level and for the staff in the system Transportation Department. Some of these challenges were unique to Cicero, or at least, much more complex and difficult to manage than at other terminals. No other yard on BN had the kind of fast-paced operation, culture, demands, and complexity that was faced in Cicero. The only other yards on BN that had even a hint of the intensity of Cicero were the Union, Northtown, and the Daytons Bluff yards in the Twin Cities (before the new hump yard at Northtown was completed), Kansas City, and Spokane. There may be some who worked on BN who would challenge that claim, but anyone who has ever actually held a line-management position in Cicero, and "learned the railroad" by having other assignments as well, would agree. As a result, I believe it is safe to say that the Trainmaster and Assistant Superintendent positions in Cicero were the toughest operating jobs on the BN.

The factors that made Cicero tough to manage were the backgrounds and general attitudes toward work of many of the switchmen and clerks; the difficulty in filling yardmaster positions; the fast pace in the operation; and the complexity of interchange, train makeup, and service commitments that had to be filled (with no excuses accepted by senior operating officers).

A FAST-PACED OPERATION

The highlight of each day's operation was the makeup of the morning fleet of outbound trains. Cars for these trains were received on interchange deliveries from several eastern railroads between midnight and 8:00 A.M. All of these cars had to be switched in time to be placed on the fleet of trains due out before noon on the same day. On most days, the total volume of cars received for the BN's outbound fleet ranged between four hundred and five hundred cars. All of the cars received in interchange had to be inspected before humping (switching) could begin. Cars with defects that could not be repaired in place within a few minutes had to be switched out and moved to a repair facility. In addition to connections from the east, a BN transfer job brought in a "hot" block of between fifty and sixty cars of merchandise that had been loaded by forwarders at freight houses located on BN on the west side of Chicago. Every car loaded by the forwarders had to make its intended outbound train without exception. The forwarders were a very demanding group. They had no tolerance for delays of any kind to their shipments and were quick to divert business to our competitors if they thought they had to teach us a lesson.

Cars had to be humped into several classifications for delivery to connecting lines: three blocks for the Union Pacific connection at Grand Island, three for connections in Kansas City, two for No. 97 (BN train for Seattle), one for Galesburg, and three for the Rio Grande connection at Denver. The first of the outbound trains had to depart by 6:00 A.M. to beat the morning "dinky rush" (commuter trains operated between Chicago and Aurora). This train handled connections for both the Union Pacific and the Rio Grande and sometimes operated in two sections. Additional trains for the UP at Grand Island and for the Rio Grande at Denver were scheduled to leave at 10:00 A.M., followed by a train for Kansas City at 11:00 A.M. and No. 97 for Seattle at 11:30 A.M. In the evening, another fleet of westbound trains was run for the same connections.

The pressure was just as intense on eastbound (incoming) business. Trains arrived early in the afternoon with blocks of perishable freight (fresh meat, fruit, and vegetables) that had to be delivered to the yards of the eastern roads by cutoffs they set for their evening departures. Due to limitations in capacity, eastbound business could not be brought into the receiving yard and humped. Instead, eastbound trains were yarded in a large flat switching yard called "the eastbound." The blocks brought in by these trains were set over for yard transfer assignments that would make direct deliveries to the yards of the eastern lines. We made such direct deliveries to the yards of the former Pennsylvania, New York Central, Grand Trunk Western, C&O, B&O, Nickel Plate, and Erie, as well as for the Belt Railway Company of Chicago, a large terminal switching company. It was expensive to make these direct deliveries, but a day's delay was avoided by not using a terminal company for intermediate handling.

Cars that moved through Galesburg were blocked by connecting line. Due to the limitations in the capability of the Twin Cities terminal before the new hump yard at Northtown was completed, trains from "the north" arrived unblocked, except for a block of perishable freight to be unloaded at facilities in Chicago. With the large increase in eastbound business from the north that started to move soon after the 1970 merger, there was difficulty at times in handling all of the car-by-car switching required at Cicero. Once the new yard at Northtown was opened, blocks could be made for each of the eastern roads. That capability made it possible to run solid trains to some of the connections, with no switching required in Cicero. Also, it freed up some of the capacity of the eastbound yard at Cicero for conversion to intermodal service.

To be able to run the morning fleet of westbound trains on schedule required close attention to having enough motive power, cabooses, and crews in place at the right time. At times, this could be challenging, since the volume of westbound business was much heavier than what would arrive from the west on some days, generally on weekends, in addition to some imbalance in the number of trains operated in each direction. Meeting the requirement for having cabooses available became a major problem at the time of the merger, when responsibility for the management of motive power and cabooses was placed in the consolidated operation center in St. Paul. Despite appeals from Chicago Region officers for more attention to getting enough cabooses coming into Cicero in time for each day's fleet, some trains were being held due to no caboose being available.

When no relief came, Ivan Ethington, the regional vice president in Chicago, told the terminal officers in Chicago one day to put a passenger car on No. 83 (the evening merchandise train for the Twin Cities) to serve as a caboose. It happened that the passenger car that was available to use was a surplus vista-dome, round-end observation car, a very classy piece of equipment. This move really got the attention of even the senior operating officers in the General Office, when they learned that Ethington had made the decision to use the passenger car rather than delay or annul the train for want of a caboose. This gutsy move on Ethington's part was seen by some as an in-your-face cheap shot at the people in the Operations Center who said they were doing their best to meet the needs for locomotives and cabooses all over the system. However, it was the end of the problem of not having cabooses at Cicero at the right time.

THE NEW ENTRANTS INTO THE LABOR FORCE IN THE 1970S

Much of the labor force at Cicero and most of the maintenance and operating facilities in the Chicago area was made up of nationalities whose parents or grandparents had settled in the Cicero and Berwyn area two or three generations earlier. Generally, they were a tough, independent lot, but hard working and dependable. In the mid- to late 1960s, hiring a large number of people as switchmen and clerks became necessary due to the draft for military service, and because wages at some local area industries were higher than what we paid at that time.

Also, many younger people did not want jobs on the railroad that would require them to work on weekends, at nights, and outside in all kinds of weather. Indoor, daylight, Monday-through-Friday jobs in factories and warehouses were readily available, making it challenging to hire enough people for railroad work.

Also at the time, there was a well-intentioned effort to bring disadvantaged people into the labor force. The former Burlington was among the companies in the Chicago area that agreed to hire large numbers of people who'd had little or no work experience of any kind. Getting them acclimated to railroad work life was quite a challenge. With large numbers being hired within a few months, it was unreasonable on the part of the company to expect the terminal operating officers to be able to train these new hires on the elements of a work ethic, to say nothing of how to do the job of a switchman or yard clerk. The work of a switchman was somewhat complex, and in an environment that was unforgiving if mistakes were made. Clerical work, though much of it was repetitive, required knowledge of the details of the operating data-information systems and the flow of data that at the base of the system for processing cars through the yard.

The rate of turnover among the new hires added to the problem. Classes should have been set up and conducted by supervisors specifically assigned to teach the rudiments of life in the railroad workplace, on what constitutes a work ethic (our expectations of employees), and then on the details involved in switching cars and working in the yard office. Instead, the entire burden was thrown on the assistant superintendent, trainmasters, and yard office supervisors, on top of the challenge they already had of running a yard under the "normal" conditions of pressure and tension that existed at Cicero, even with an experienced work force. The historic method of on-the-job training and learning by osmosis were inadequate in this kind of situation.

Wayne Hatton was one of several operating officers who worked as a trainmaster at Cicero in those difficult years of trying to get the work done, plus shaping these new hires into responsible railroad workers. Wayne recounted examples of the challenge he and other supervisors had to work under:

There were two trainmasters assigned to the terminal. The daylight trainmaster spent the morning hours making the connections and making sure the priority trains left on schedule. The afternoons were spent holding formal investigations for missed calls, no shows, etc. I recall holding as many as five investigations [disciplinary hearings] in one afternoon.

The night trainmaster spent an inordinate amount of time making sure there were adequate crew members on every crew and that yardmaster positions were filled, especially on weekends. Many crew assignments were abolished. I recall weekends and particularly holidays, having to work the third trick hump Yardmaster job because there was no one qualified. . . . During that period we were instructed by the Executive Department to provide transportation [as requested by newly hired crew members] from the CTA "L" [the rapid transit system] station on Cicero Avenue to the yard office for employees who were afraid to walk to the east end tower in that this was the border between Cicero and Chicago. . . . And with all due respect, I think that many of the decision makers in St. Paul in 1970–71–72 had little if any sensitivity of the environment that existed in Chicago.

Bill Greenwood recalls in the early 1970s how the personnel situation at Cicero made it difficult to get the work done and to install a major project such as COMPASS:

There was a sizeable segment of the Cicero work force that would not work weekends, were habitually late, didn't perform on the job, and had a very high rate of laying off. Part of the solution to bringing Cicero back was a concerted effort in 1973 and '74 to "cleanse" the work force of people who didn't want to work. That was an ugly, difficult and time consuming task, but it was achieved. . . . My recollection is that we averaged 15 investigations a day for several months, and ultimately fired for cause [after three offenses in the year] probably 100 people.[1]

Bill states that even with the payouts that had to be made in settling legal challenges from those terminated, BN was ahead with the efficiencies we gained. It was necessary to go back to hiring residents of the neighborhoods in or near Cicero who had made up the bulk of the workforce since the yard was built three generations earlier.

In the 1980s, more and more of the merchandise and

---

1 W. E. Greenwood, email and telephone communication, various times between 2000 and 2005.

perishable business was being moved in intermodal equipment (containers and trailers) rather than box cars or refrigerator cars. That allowed us to convert more of the yard to an intermodal terminal over the next several years. In time, the hump was shut down and leveled, and nearly all of the yard switching tracks were removed. The new yards built at Galesburg and Northtown were set up to do more blocking for connections to the eastern lines and to the Belt Railway Company of Chicago. By then, we were able to consolidate the volumes enough to run solid trains directly to some eastern roads, using road crews coming from Galesburg and North La Crosse. This greatly reduced the need for the eastbound yard. Without that capability for handling eastbound business, together with the shift most merchandise customers made to intermodal service for their westbound shipments, Cicero was no longer needed as a classification yard. It was the end of an era.

CHAPTER 22

CUTOVER TO COMPASS

• moving BN into the Information Age • to provide up-to-the-minute knowledge
of the location, status, and movement information for the 120,000 cars on BN lines •
a complex, expensive undertaking, too cumbersome for a fast-paced operation, as in Cicero •

I N 1968 THE NORTHERN PACIFIC HIRED Frank H. Coyne from the Southern Pacific (SP) where he held the positions of manager–Systems Research and general auditor. At the SP, Coyne was in charge of developing what was to be "the railroad industry's first complete 'real time' data processing system" that would underlie business processes and information used throughout the company. The SP named the new system TOPS (Total Operations Processing System). With the capability of the TOPS system, it was expected that the utilization of freight cars would improve. By having real-time information available to operations managers at every point of action on the SP system, it was believed they would be able to do a superior job of modern-day railroading.[1]

To say the least, it was a very ambitious project, unprecedented in most industries at that time, but certainly in railroading, where the operations are so far flung, and with hundreds of data-entry points. To his credit, L. W. Menk saw the need for similar capability on the NP and, of course in the operation of the new consolidated railroad about to be created. He was able to recruit Coyne, a Harvard MBA with strong skills in finance and planning. Coyne advanced to the position of senior vice president–Finance and Administration.

In planning for the merger, a team was set up to consider what kinds of operating and data systems should be adopted and developed. It did not take long for this team, together with those who were to be assigned to principle executive positions, to decide in favor of the NP's COMPASS (Complete Operating Movement Processing and Service System), an adaptation of TOPS. That decision came about after debate and some tension over the merits of the systems being developed at that time by the Burlington and the GN. Some of those involved in the decision process argued that a system of the design and scope of COMPASS would be far too expensive to implement and operate, relative to what seemed to be soft, hazy, unmeasurable, and uncertain results.

It was argued that it would take a huge staff to design and implement COMPASS, plus additional clerical forces to input the massive amount of data at hundreds of locations in the field. It was even argued by some that it would cost less to hire an "army" of clerks at the system level with each clerk being responsible to watch over a "handful" of cars and enter data in the old-fashioned way to record the movement of each car. The arguments against having a massive new computerized system on BN did not prevail. And, it was decided the system would be COMPASS.

It was anticipated that the implementation of COMPASS would cost $10 million. As in the case of SP's decision to develop and implement TOPS, the return on investment was based on the projected improvement in freight-car utilization. BN determined that this improvement under COMPASS would be so great that we would not have to invest at least $150 million to acquire new freight cars.

---

1 "The Computer Age Comes to Southern Pacific," *Southern Pacific Bulletin* (December 1967): 15.

A well trained, highly motivated team of about seventy managers, trainers, and technicians was assembled to begin implementation on the West Coast in 1971. Getting the necessary input/output devices installed, with the enhanced communications capability needed for transmission of massive amounts of data, was a very ambitious project. Even more challenging was the training of hundreds of clerical employees; data supervisors at operating terminals, junction points, and mechanical facilities; all operating officers; and dispatching offices for the new responsibilities and workload they would have under COMPASS.

Getting the entire management team at the division and regional levels to learn how their jobs would change was a vital and ambitious undertaking. Those managers would also have to accept responsibility for operating and managing the system once implementation was complete in their territory. All pockets of resistance to the new system, whether among employees or managers at any level, had to be dealt with. Implementation could not be allowed to falter or fail anywhere along the way from the West Coast to Cicero. A smooth flow of traffic would have to be maintained to avoid interruption to service or a chain reaction of yard congestion develop during the implementation process.

It was disheartening to learn that the some additional clerical positions had to be established at several locations to handle what was a very labor-intensive system. To some executives, that was not to be considered a major problem, since we had surplus or underworked people at several of the common points where yards had been consolidated after the merger. With their lifetime job protection, these employees had to be paid, so bringing in additional work would cost us no more in the short term. Generally, the clerical employees and the union representing them accepted COMPASS very well, since they could clearly see that the additional workload throughout the system would require us to establish more clerical jobs.

SHOULD WE SKIP THE TWIN CITIES?
Seventeen months after getting started in June 1971, the time for cutting over to COMPASS was approaching the Twin Cities terminal. Numerous senior managers in all departments thought the COMPASS team should skip over the Twin Cities, out of fear that the operation would become gridlocked upon implementation, and that in a few days, the cascading effect would put the entire northern corridor in a state of congestion and unable to function. Before long, they believed the entire BN system would be bogged down, and it would take many weeks to recover. Customers would pull their business away from BN, revenues would be lost, and our operating costs would skyrocket. In the face of these concerns, however, the decision was made to proceed to implement COMPASS in the Twin Cities as scheduled.

A SUCCESSFUL EFFORT
Extra careful planning and training was undertaken, and we proceeded with confidence that we could handle the task. The cutover in the Twin Cities terminal started on a Sunday morning in October 1972. In the next several weeks, some problems came up that caused operating difficulties for a few hours in one or more of the eleven yards we had open at that time, but overall, the cutover went far better than anticipated. The problems we had in the terminal were largely overcome before they affected other terminals and before the worst fears of anyone were realized.

The implementation of a massive new information system was a very large project, unlike the large operations-based projects that senior operating and executive officers at the headquarters and regional levels had experienced earlier in their careers. To their credit, none of them yielded to the tendency we had all seen at times in the past—for them to want to call the shots and critique what was going on during the cutover. COMPASS was a different ballgame, that none of them really knew anything about. Terminal officers did not have to host tours for headquarters and regional people to show them what we were doing, as was customary when other large projects were underway. They did not show up to be in photos taken on the job, as usually happened when a new facility was opened or a new type of locomotive was put in service. None of them tried to take credit or ownership of the success in which COMPASS was implemented. Instead, credit was given to the COMPASS team and the fine corps of yard office supervisors, trainmasters, chief clerks, and yardmasters whose dedication and job knowledge made it work.

Of course, part of the reason for this aloofness toward COMPASS was that it was something totally new, which they had no familiarity with from their years in field operations. Also, because they did not understand it, they would not want to be identified with it in case trouble developed during implementation, or if it overran its estimated costs of implementation, or there was a failure to produce the targeted rate of return on the capital invested in it. In short, the value of COMPASS and the uncertainty of the effect it might have on service and operating performance, on top of the possibility of colossal failure, caused many to distance themselves from it. To those of us who held the primary accountability for any and all aspects of COMPASS (the COMPASS team and operating officers), we had to make it work, no matter how labor intensive and cumbersome we found the processes imbedded in it.

In planning for the cutover, we knew that our most vulnerable area for problems would be the Northtown yard, due to so much of it being out of service for construction of the new hump yard. Our second-greatest concern was the Union Yard complex, which consisted of several subyards, all of which were linked in the processes we had in place for the classification of cars in switching. Also, Union Yard had the lowest capability for productivity, due to being built on a grade, its overall track layout, and poor track conditions that prevailed, pending transfer of switching to the new yard at Northtown.

The Daytons Bluff yard in St. Paul was well laid out and had an experienced and productive work force, both in the office and on the switching lead. We had already moved all of the switching into Daytons Bluff that the yard could handle without having to hold trains out, or to cause double handling of cars to get them lined up for departure from the Twin Cities terminal. The workforce at Daytons Bluff took on about double the workload that was anticipated in the operating plan for the consolidated terminal. This was done with almost no increase in force in the yard office. It was a real credit to Rube Roelofs, the terminal manager in the St. Paul zone, and his staff of operating officers, together with their fine staff of clerks, switchmen, yard transfer crews, yardmasters, and car inspectors. They did a fine job to keep the Twin Cities terminal fluid in the years before the new yard at Northtown opened, and especially in those early weeks of implementing COMPASS.

In addition to the team at Daytons Bluff, we had an outstanding corps of managers at the consolidated yard office and agency at Northtown and at the Union and Northtown yards. Their collective knowledge of the intricacies of operations in the Twin Cities made it possible to plan the cutover to a level of greatdetail, and to know how to overcome problems or questions that might arise in those critical weeks of the cutover. Again, it is difficult to name all who deserve credit for this success, but here is a list of those I have been able to recall, together with input from managers I have been able to reach this many years later: Ron Sherve, Don Erickson, Carl Russert, Sam Lasser, Jim Lasser, Pat Peyton, and Arnie Santa.

The system COMPASS-implementation team also deserves a great deal of credit for what they accomplished in getting such a complex system installed across BN in only about two years. Forty-plus years later, it would be difficult to name all of the principle managers who led the effort, but at the risk of leaving some out, I think it is important to name as many as we can in this writing. Implementing COMPASS was one of the biggest undertakings in the company in those early years, so even though there is no edifice such as a bridge or shop building we can point to as a symbol of their success, we still need to recognize the leaders of the COMPASS team: Hank Gagnier, Red Tilsworth, Dave Harlan, and Ed Kelly.

APPLICATIONS AND USES OF COMPASS

Among the thousands of applications and uses made possible by the vast amount of data generated by COMPASS:

- status of every car and locomotive on line
- car movement updated when passing each of 175 field reporting locations
- information accessible to customers on status of shipments or cars they own or lease
- information available on demand to sales offices, operating officers, car management personnel, car-repair facilities, and the Finance and Accounting departments

- car inventory in yards by destination or commodity on hand and en route to terminals
- availability and status of locomotives
- to manage the flow of empty cars to locations of demand
- cars ready for movement at customer facilities or at any location on BN
- provide data to the Accounting Department for determining anticipated revenue and operating expense
- to provide real-time, immediate updates on car movement status or changes on locomotive availability
- a complete overhaul of the practices used to generate operating data and make it available to users in all functions of a railroad company
- a response to demands of customers for better information on the status of their shipments, and to serve as a base of data to be analyzed for improving service and the utilization of equipment.

Having a system such as COMPASS installed in the early 1970s was necessary to bring a railroad into the modern age of computerization of basic operating data, as a platform for several aspects of service delivery and to better manage the flow of cars. Having COMPASS fundamentally transformed the way BN was monitored and managed at all levels of the Operating and Accounting departments, in particular.

The COMPASS team had good reason to celebrate its success. The next (and final) terminal to be cut over was Cicero, the BN's primary yard in Chicago. The leaders of the COMPASS team thought that after beating the odds for success in the Twin Cities, Cicero would be a snap. Those who had visited Cicero—planning the resources they would need for implementation and how COMPASS would fit in with the processes used in running the yard—noted "how small" the yard was, how few industries we had in Chicago compared to the Twin Cities, and how simple the process of interchange with foreign lines was. From that evaluation, they anticipated no difficulty—a rapid and easy cutover. When challenged because of the required pace of the operation in Cicero in contrast with the much slower pace of operations in the Twin Cities, Spokane, or Seattle, they countered that they had noticed nothing major or otherwise "impressive" that

would cause them to have any concern or anticipate any need to adjust COMPASS procedures to better fit in with the reality of "life in Chicago."

Some of us suggested to our counterparts in Chicago (in both the Cicero terminal and the regional office) that they should send their data-management specialists to the Twin Cities to have a look at the methods under which data had to be entered into the system, and how information had to be processed and formatted before switching of inbound interchange deliveries could begin. A "rapid fire" operation, Cicero needed to keep cars flowing through the yard and to be able to get many "hot" cars in the early morning deliveries from connections switched within two to six hours to make the five high-priority trains due out between 6:00 A.M. and 11:00 A.M. There was no place for a slow, unduly complex data-entry procedures, and a labor-intensive system that would be much less efficient and much slower than the system in place.

The reality was that in Cicero, between 1,000 and 1,200 cars were moved westbound every twenty-four hours and about the same number were received on inbound trains from the North and the West. This was far in excess of the volume handled by Union and Northtown yards combined, and with much less time allowed for handling these cars. In addition, the clerical force in Cicero was far less experienced and qualified to take on a new, complex system than the "army" of seasoned, capable yard clerks we had in the Twin Cities. To our surprise (and great concern), no favorable response or even curiosity resulted from the yellow flag of caution we raised. The response I received was along the lines of "If you can handle it [COMPASS], so can we." That reaction set the stage for what was supposed to be the final, and likely the easiest, cutover the COMPASS team would have to make.

UNDERESTIMATING THE SPEED OF THE OPERATION IN CICERO

By the second day of the cutover in Cicero, the yard was already in big trouble. We could not get cars coming in from the eastern lines switched fast enough to make the outbound trains as specified in the service plan because of the complexity and volume of data that had to be entered into COMPASS before switching could begin. Trains arriving from Galesburg and the Twin Cities could not be broken up quickly enough to keep the east-bound operation fluid and deliveries made in time to make the cut-off times of the eastern lines.

In recent years, there has been much talk in the rail industry about operating a "scheduled railroad," as if that were a totally new concept and it had never been in place on any railway until it was touted as something just invented. In the yards of at least the C&NW, Sante Fe, Indiana Harbor Belt Railroad, Erie, Nichol Plate, and even the Penn Central, despite all of its problems, the priority trains with merchandise and perishable shipments departed on schedule and with the "right" cars in the train: with the cars scheduled for today's train (not yesterday's train or an unscheduled extra one), in the right block, and on schedule at the train's destination. The procedures that had to be followed under COMPASS were not conducive to running an operation of that pace and under that form of discipline in the operation.

Furthermore, with little to no room to hold any amount of cars delayed for COMPASS data input in Cicero, cars had to be moved through the yard as scheduled in the service plan. That was necessary to keep the receiving yard from getting congested, and then, the bowl (the main classification yard) from filling up. The fallout from the problems of a congested yard was the need to hold inbound deliveries out of the yard, and to delay receiving our inbound trains from the west and north. Immediately, the problems at Cicero cascaded into Galesburg and the Twin Cities, and soon, to all of the connecting lines and terminal companies we had to work with in Chicago. The situation was alarming, and none of the experts on the COMPASS team had a solution that could be implemented to immediately correct this situation. These problems soon got the attention of senior management in the Operations, Marketing and Executive departments. Complaints were rampant, and customers were soon diverting their business to our competitors.

L. W. Menk, then chairman and CEO, directed Ivan Ethington, vice president of Operations, to go to Cicero and not return to headquarters until the problem was solved. Ethington had a business car placed right outside of the yard office and stayed in Cicero for about three weeks.

Bill Greenwood was part of a team of officers who had prior experience in running Cicero and were sent there

to help get the terminal (and by that time, all of BN) through a very difficult time. Bill recounted what had to be done at Cicero to overcome the difficulties in the COMPASS system:

> I was part of the beefed up team sent to Cicero to right the ship after the COMPASS installation brought the place to its knees.... The way Cicero was brought back was to get the COMPASS leaders to significantly modify COMPASS process requirements for Cicero only.... As I recall there were dozens of shortcuts created, but the most critical one was [that] we were relieved of the absolute restriction that no movement activity could be initiated on a car without the physical waybill being present and entered into the system.
>
> Cicero practice was to receive, inspect, switch, and classify cars from eastern connections, and make up outbound trains, all of this in as little as a few hours to a maximum of about six hours. That activity was started primarily with advance teletype train list information, because waybills frequently followed the actual deliveries. The new shortcut allowed us to restore the old practices, and enter the data from Xerox copies of the waybills, sometimes even after the outbound train departed. The tragedy is that it took four months for the COMPASS leaders to relent and to allow these changes ... permanent losses of business resulted from months of 5,000 plus cars being held out of Cicero on connecting railroads unable to deliver to the BN in Chicago.
>
> They [the COMPASS team] failed to grasp the enormity of the importance and system impact of the place [Cicero]. The recovery at Cicero only occurred when the power of Menk descended on the place and dramatic shortcuts and compromises to the COMPASS systems "integrity" were sacrificed instead of the operation.

Wayne Hatton, another seasoned officer who knew from the years he spent there what it took to run Cicero, recalls the directions Ivan Ethington issued upon his arrival in Cicero. The "ICE man" told Wayne to get trains made up and out of the yard, and to not delay the switching of inbound deliveries until all of the detailed data input was complete. Giving the conductor a manually prepared list of cars in his train would have to be sufficient until the COMPASS procedures could be modified enough to overcome the major weaknesses in that and other parts of the COMPASS system. In looking back on that time, it may have been the worst episode in BN's first ten years, other than serious accidents that took lives and destroyed millions of dollars of cars, locomotives, and infrastructure.

In the 1971 annual report, the implementation of COMPASS was given credit for the decision to cancel plans for an investment of $150 million in new freight cars. This projection was made even with COMPASS having been installed only in the Portland and Seattle regions by the end of 1971. The cost of getting COMPASS installed across the system still was estimated at $10 million. In the annual report for 1972, it was asserted that with COMPASS installed on four of the six regions at the end of 1972, freight-car utilization had improved 10 percent. By 1975, it was claimed that revenue ton-miles per car-day had increased by 20 percent since 1970 (pre-COMPASS). Again, this improvement in car utilization was said to be the result of investing in and installing COMPASS.

Even with the passing of the stressful and contentious time of implementation, the desire to scrap COMPASS and acquire or develop a better system remained with many. Even with the modifications made to COMPASS during and following its implementation, it still was very labor intensive, requiring us to have more clerks working at data-entry points and to continue programs of instruction needed to improve their performance. The system seemed unduly complex: difficult to train new hires and to modify to correct an endless stream of challenges that needed to be overcome. We knew from visits to facilities on some other railroads that we were overstaffed with clerks and junior-level clerical supervisors needed to help produce data integrity. We also had a large team of experts set up to handle inquiries on a 24/7 basis from employees all over the company on how to overcome immediate problems. A large staff also was at work year after year in the Management Information Services (MIS) Department on modifications requested from a host of users of operating data.

When senior managers of the Frisco and other managers hired from outside the company first faced the COMPASS system, the reaction of many was that it should be scrapped and that we start over with a better system. However, after evaluating the cost and disruption in undertaking such a major effort, it was judged better to undertake revisions to COMPASS one module at a time, over a period of two to three years. By making one big jump at a time in an orderly fashion, the job eventually got done. There are some who say that merger of the BN and Santa Fe systems was necessary to finally bring about the extent of change needed in the COMPASS system.

CHAPTER 23

AMTRAK

RELIEF FROM LOSSES

In 1970 BN's deficit on passenger train operations was $40.8 million on a direct cost basis.
BURLINGTON NORTHERN ANNUAL REPORT
FOR THE YEAR ENDED DECEMBER 31, 1971, PAGE 12

As the cost of joining Amtrak, Burlington Northern is paying $23,447,000 over a three-year period.
BURLINGTON NORTHERN ANNUAL REPORT
FOR THE YEAR ENDED DECEMBER 31, 1973, PAGE 10

We feel a commitment to operate these [Amtrak's] trains to the best of our ability.
BURLINGTON NORTHERN ANNUAL REPORT
FOR THE YEAR ENDED DECEMBER 31, 1973, PAGE 10

THE DECISION OF THE U.S. POSTAL SERVICE in the mid-1960s to no longer use railroad passenger trains to haul mail could be considered the crowning blow to any chance these trains could continue to run. For a long time, the losses on many passenger trains had been offset to a considerable extent by the income received from having Railway Post Office (RPO) cars in the train, together with anywhere from one to fifteen or more cars of "storage mail" that was not to be sorted en route by Postal Service employees. But once the mail business was removed from those trains, railroad companies moved quickly to consolidate the schedules of the passenger trains that remained and, soon after that, to request authority from the ICC to discontinue those trains.

At the time of the merger in 1970, BN's predecessors had been aggressively reducing their passenger service. Even with the large losses they were incurring, those railroads continued to run their remaining trains with a high quality of service, right to the last day of their operation. It would not be fair to say that any of them took deliberate measures to discourage patronage.

In 1970 the U.S. Congress passed legislation to create a quasi-nationalized company called the National Railroad Passenger Corporation (NRPC), soon to be identified as Amtrak, to set up a skeletal network of intercity passenger trains, with start-up on May 1, 1971. The NRPC would make decisions on the routes to be retained (or reconfigured) for passenger service, on pricing, and on the kinds of accommodations to be offered. Some of the constituencies with a stake in this transition thought it would be much simpler to contract with the railroad companies to continue running a few of the trains that remained in service, and with a subsidy to cover their operating (i.e., the "above the rail") costs.

However, there were more voices who thought that having a fresh look taken by a new government agency would breathe new life into rail passenger service. In their opinion, subsidizing the operation as it was being run at the time would not provide an opportunity to reverse the trend of declining patronage. The views of those favoring the setting up of a new government-sponsored agency to manage the operation and market the service clearly prevailed. It was figured that without such action, rail passenger service would soon come to an end, and that there still was a need to have a minimal level of service retained on a few routes. Hopefully, it could be restructured to make it more appealing to the public.

The railroad companies were not required to put their passenger service under Amtrak. Three of them, the Southern, Rio Grande, and the Rock Island, decided not to join, and instead, to continue to run their trains and suffer the operating losses. Within a few years, the Southern and Rio Grande changed their position and jointed Amtrak. The Rock Island soon received authority to discontinue its remaining service.

The opportunity to join Amtrak and get relief from losses did not come without a price. There were options in the terms of "admission" to the Amtrak program, and each railroad had to evaluate those terms in light of its financial condition, the magnitude of its losses in running passenger service, and the number and value of the types of cars and locomotives Amtrak wished to acquire from it for the new service. Participating railroads were required to pay the NRPC over a three-year period, a total amount equal to the lesser of:

1   50 percent of the fully distributed passenger deficit for 1969, which was $66,893,000, or

2   100 percent of the avoidable loss of intercity passenger service in 1969, as determined by the ICC, or

3   200 percent of the avoidable loss for 1969 of intercity passenger service operated over routes between points within the basic system designated by the NRPC.

Such amount to be payable in cash at the option of the NRPC or by the transfer of rail passenger equipment or by providing future services to the NRPC, in exchange for which participants would receive stock of NRPC or a federal income-tax deduction of the amount contributed. The NRPC would be able to contract with participating railroads to operate passenger trains over the basic system it had designated. The railroad companies would be required to provide job, wage, pension, and seniority guarantees to employees affected by the discontinu.... of trains.[1]

BN decided to transfer its intercity passenger service responsibility to the NRPC. At that time the maximum estimated cost to BN for that transfer was $33,450,000. On the basis of fully distributed costs, BN reported a passenger-train deficit of $66,894,000 in 1969, with $38,320,000 of that amount being solely related to passenger service. BN established a program to retrain and place 812 employees whose jobs were made surplus. Within a few months, 696 of those employees were placed on other jobs.

In 1971 an extraordinary charge of $52.6 million (net of provision for federal income taxes) was set up to cover estimated future costs of severance and protection payments, losses for the retirement of passenger train equipment, and payments made to acquire stock in Amtrak. BN was required to pay $33.4 million to the NRPC in equal installments over thirty-six months. Amtrak purchased passenger locomotives and cars from BN for $2.6 million. In 1971 BN wrote down the value of its stock in Amtrak to one dollar. In its annual report for 1977, BN reported that the Amtrak service it was providing was operating at a "slight" loss.

Thus ended the Hill roads' obligation (and desire) to provide intercity passenger service. Although there was a big relief in knowing that the ever-increasing deficit from providing passenger service was finally at an end, there was at least a little regret in the BN team that these "flagship" trains and symbols of pride had become history. It was disappointing to see how the quality of service and equipment provided by Amtrak on the trains it decided to run on BN lines deteriorated. Rather than taking away most of the high-quality and uniquely decorated and furnished cars that BN and its predecessors had used on those trains and substituting a hodgepodge of cars from other roads, Amtrak should have decided to keep the cars used for the *Empire Builder, North Coast Limited* and *Denver Zephyr* intact and paid BN a subsidy to run those trains.

In its early years of operation, Amtrak made some decisions on the assignment of equipment that needlessly

---

1   *Burlington Northern Annual Report for the Year Ended December 31, 1970,* page 14.

caused some service and operating difficulties on the *Empire Builder*. For example, Amtrak began to use dining cars that had been in service on lines in the South. Those cars did not have the kind of insulation needed to protect them from freezing up in sub-zero weather. They did not have the amount of food-storage space needed for service on a trip of about forty hours. For a while, Amtrak insisted on using the E-type locomotive units on the mountain grades. The GN's experience with that type of power on passenger trains in the 1940s had shown it was not suitable for service on heavy grades due to the wheel slip caused by the low weight on driving axles. Having a policy to retain the type of equipment and operating practices the railroad companies had been using for many years would have saved some grief for Amtrak and its patrons, as well as for BN as its contract operator. It is disappointing to see that the kinds of cars used on today's trains, after over forty years of operation under Amtrak, are less appealing and without the classy décor of those used in pre-Amtrak service.

The one area in which Amtrak bettered the old equipment was in the conversion to head-end power to eliminate the use of steam to heat the cars. Also, the air-conditioning system on today's cars is more reliable and functional than the technology for temperature control that was used before the new equipment was developed for rail passenger service.

In addition to getting relief from the losses that had become too much for BN or any other railroad to bear, other advantages were gained from the elimination of passenger service on some routes or, at least, in reducing the number of passenger trains operated on the routes selected by Amtrak:

1 Capacity was freed up to handle increasing volumes of freight traffic.
2 Delays to freight trains that had been caused by having to give priority to passenger trains were eliminated.
3 Management was freed up from the burden of having to supervise employees involved in the maintenance of passenger equipment and the operation of the trains, and from the time being consumed in dealing with "challenges" from various constituencies that used or had an interest in passenger train service.

At the time Amtrak was created, some industry observers expected that it would be an unsuccessful venture (other than in the Northeast Corridor), and that in about two years it would be allowed to quietly fade from the scene. Clearly, that has not happened. Despite all the problems Amtrak has had in financing and maintaining some quality of service, it has developed enough support from our elected representatives and the lobby for maintaining rail passenger service to keep operating. In the corridors between Chicago, Galesburg, and Quincy, and between Portland, Seattle, and Vancouver, British Columbia, Amtrak service has been expanded. The popularity of the *Empire Builder* and the *San Francisco Zephyr* appears to have been built up enough to insure they will continue to operate for the foreseeable future.

FINANCING IN THE 1970S: ISSUES, CHALLENGES, AND INNOVATIONS

CHAPTER 24

FINANCING FOR THE NATION'S LARGEST AND NEWEST RAILROAD

BY ROBERT F. GARLAND

• pre-merger planning of finance functions •
• March 3, 1970: "M-Day" •
• unanticipated troubles: The Penn-Central bankruptcy •
• financing growth •
• budgeting and other on-going functions •
• financial communication (including betterment accounting) •
• the 1980s: BN finance in transition •

PART IV CONTAINS TWO CHAPTERS ABOUT the challenges BN faced in managing its finances in its first decade. The first chapter was written by Robert F. (Bob) Garland. Bob covers the major financial challenges BN faced shortly after the 1970 merger was consummated, as well as a number of issues involving the merger plan, budgeting, merging with the Frisco Railroad several years later, and the changes in senior management in 1980.

The second chapter was written by the author of this book and covers work done by Raymond C. (Ray) Burton to obtain the financial resources BN needed to carry out the many large capital improvement projects required to handle its many growth opportunities.

Before BN established its holding company structure in 1980 and 1981, Bob Garland served as vice president and controller, and Ray Burton as vice president and treasurer. Bob and Ray reported to Frank H. Coyne, executive vice president–Finance and Administration. Upon formation of the holding company, Frank Coyne and Ray Burton assumed positions in BNI's new headquarters in Seattle. At that time, Bob Garland was named senior vice president–Administration and Planning for Burlington Northern Railroad headquartered in St. Paul, and reported to Richard Grayson, newly named president and CEO of BN's transportation holdings.

PRE-MERGER PLANNING[1]

As this is the story of the overall Burlington Northern Finance Department, it is ironic that none of predecessor companies (Chicago, Burlington and Quincy, Northern Pacific, and Great Northern) had modern comprehensive finance departments headed by a chief financial officer. Of the three classic commercial enterprise functions—production, sales, and finance—the first two were clearly fixed in the Operating and Traffic or Marketing departments of the CB&Q, NP, and GN. However, the finance function, as later developed by the BN, was divided primarily between the Accounting, Treasury, and Executive departments.

The function of planning the funding of capital needs seemed particularly diffused, ordinarily being led as needed by a senior officer of the Executive Department, with assistance from Accounting, Treasury, Law, and others as required. These financings were largely responsive to the routine maturities of mortgage-bond issues and the placement of equipment trust certificates to finance the acquisition of new locomotives and freight cars. There was little or no recent history of financing the expansion of railroad services.

Guided by the three railroad presidents, Harry Murphy, Robert Macfarlane, and John Budd, beginning in the early 1960s, the post-merger functions of the future BN Finance Department were subject to comprehensive planning by numerous "merger committees." There were steering committees of, for example, the treasurers, Leo Assell (CB&Q), Bill Montgomery (NP), and Dick O'Kelly (GN). There was a corresponding committee of the Accounting Departmentsconsisting of Bill Ernzen (CB&Q), Earl Ordell (NP), and John Tauer (GN) that guided sub-committees representing each of the various divisions of the department. For example, there were committees for freight-revenue accounting, disbursements accounting, general accounting, tax, and so forth. All these committees met frequently and took a conscientious approach to their assignments, reporting back to their steering committees.

There were many, many merger meetings over the decade during which the three railroads planned the merger and sought necessary approvals from the Interstate Commerce Commission and the courts. My own first experience was as a relatively newly hired student officer assigned to the GN's computer systems and programming office. I was working on Saturday morning and was unexpected called by the GN's general auditor (second in the Accounting Department to the comptroller) to accompany him to an important merger meeting on the second floor of St. Paul's Minnesota Club. I don't recall the exact purpose of the meeting, but it was attended by a number of senior executives and our various outside advisors, including lawyers from the Chicago law firm of Sidley, Austin, Burgess, and Smith. Interestingly, two of these lawyers, Mr. Burgess and Mr. Garrett, at the date of the meeting rather elderly men, had previous assisted the railroads in their failed merger attempt in the late 1920s! As the meeting ended, I found myself standing near Mr. Burgess just as he stumbled, fell, and rolled all the way down the club's long stairway, ending on the first floor. I ran down to assist Mr. Burgess, but he assured me that he was all right and, as he suffered from some condition causing such falls, he had learned how to fall and avoid injury!

The fact that the approval of the merger took a long time probably benefited the future BN in more ways than one. In addition to allowing thorough planning and the development of needed procedures, those who were to be the executives and managers of the future finance department had many opportunities to get to know and generally like one another, share experiences, and enjoy meals and travel together.

Just as in the Operating Departments, where the merging railroads made various trackage changes and

1   The author of this chapter, Robert F. (Bob) Garland is an experienced writer on historical topics, having published a number of articles and a book on topics of interest in the St. Paul area. Bob was directly involved with the financial and accounting affairs of Burlington Northern both before and long after the March 3, 1970, merger. He served successively as a student officer, special accountant, auditor, and assistant comptroller of the Great Northern Railway Company. On Burlington Northern, Bob held the positions of Assistant Vice President and Vice President and Controller of Burlington Northern, Inc. Upon formation of the BNI holding company and the activation of Burlington Northern Railroad Company, given the choice, he elected to remain with the railroad, serving as senior cice president–Finance until his resignation in May 1984. This chapter was written from his personal knowledge and experience.

improvements in anticipation of the merger, systems work that was done in the various departments that would comprise the BN Finance Department paid dividends after the merger. For example, the GN's very active Systems and Computer departments provided much of the hardware and software that was able to expand quite smoothly to handle most of the basic accounting functions of the merged company, including freight revenue, interline settlements, payroll, disbursements, equipment service, and general accounting.

External consulting services also served to assist the components of the future BN finance department in preparing for the merger. For example, the GN had a lengthy experience with McKinsey and Company. Among other things, this produced an up-to-date budgetary-reporting and responsibility-accounting computer system that, with improvements, served the merged company, having the capability to assemble and report against the comprehensive operating budgets prepared by the operating and other departments.

The McKinsey engagement also taught us procedures for cash-flow management and administrative-cost reduction, and was the origin of BN's long term variable-cost-accounting procedures. This was a frequently used industrial-costing method that recognized that certain costs, such as those of administration and some long-lived facilities, do not increase with additional volume. Thus, additional revenues that do more than cover long-term variable costs add to the bottom line. Use of this method requires an accurate assessment of the additional long-term volume of business that will actually result from pricing strategies based only on long-term variable costs and not on a full apportionment of total costs.

The long-term variable-costing process supported decisions to offer lower freight rates that would take advantage of what was then perceived as excess capacity, and it was seen by marketing officials as improved support for their efforts. Later, with the unexpectedly large increase in BN coal traffic, some of these costs did not adequately reflect the true cost of handling the traffic, as they assumed much of the cost of the track structure was fixed and "sunk" and didn't need to be recovered from the newly added traffic volume. This was not the case.

M-DAY, MARCH 3, 1970

Among the preparations the CBQ, NP, and GN made for the merger was a comprehensive management organization study with the consulting firm of Booz, Allen, and Hamilton. The "deliverable" from this was a very detailed organization manual (about three inches in thickness) that was widely distributed and used (at least initially) by those planning our merger and later by the BN. As outlined in the manual and by the personnel-assignment decisions made by top management, the new Finance Department was to be headed by Bill Ernzen, vice president–Finance. Directly reporting to him were John Tauer, vice president and controller, and Leo Assell, treasurer. Dean Wigstrom (who had replaced the recently retired Earl Ordell) and Bill Montgomery were the most senior of the executives reporting to John Tauer and Leo Assell, respectively. Dick O'Kelly, the GN's secretary and treasurer, became corporate secretary of the merged company. As with all other departments of the new BN, appropriate positions were found for all managers and executives of the merging railroads. This doubtless made us a little "top heavy" at first, with rosters of senior assistant vice presidents, assistant vice presidents, directors, assistant directors, and managers. (We even joked that some of these positions might fall within the "scope" of our labor agreements.) However, the vast majority of us felt this was part of treating everyone fairly, a very basic principle underlying the BN merger. Just as with the agreements negotiated with the railroad labor unions, we tried to leave no one behind!

Also reporting directly to the vice president–Finance were the executives and managers responsible for financial planning, property taxes, internal audit, and costs and statistics. The reporting relationship of the assistant vice president–Costs and Statistics directly to the head of the Finance Department was a significant change, as this function had always been part of the Accounting Departments of the merging railroads. We thought at the time that it was possible that the Marketing Departments had advocated this because the Accounting Departments were seen as not flexible enough in their approach to cost accounting. The change may well have deprived those determining costs for rate making of valuable support and oversight and may have contributed to some of the subsequent difficulties with our long term variable costs, including those used for pricing early unit train coal traffic.

Also, again on the basis of treating all interests fairly, key decisions were made as to the various outside advisors to be used by the merged company. We understood that because we wanted to retain the GN's law firm of Sidley and Austin as outside counsel, Haskins and Sells, the NP's CPA firm, was selected as our external auditors. While this may have initially disappointed those from the CBQ and GN, the decision quickly proved to be a good one. While the firm's name was later changed to Deloitte, Haskins, and Sells, and then to Deloitte and Touche ("Deloitte" being of significance to their international practice), we soon developed an excellent relationship with the CPAs, the only problem supposedly being the difficulty the CPA partners and clients had in learning to spell "Deloitte"!

The decision was made to locate BN's headquarters at St. Paul, Minnesota, in the so-called railroad building, completed in 1916 by Mr. Hill. The building occupied most of the block demarked by Fifth (NP), Jackson, and Fourth (GN) Streets. For whatever reason, most of the headquarters departments expanded north from the floor on which the particular GN department was located, into space formerly occupied by the NP. For example, the executive department was located on the 10th floor, operations on the 9th, and marketing on the 7th. Accounting occupied most of the 3rd, 4th, 6th, and 8th floors, with the 5th housing the computers. The newly formed office of the vice president–Finance and the offices of the treasurer were at the southwest (Fourth and Jackson) corner of the 10th floor.

The fact that the headquarters was in St. Paul made the merger relatively easy for the Finance Department executives, managers, and other staff from the NP and GN. They did not have to acquire new homes or move their families, and they were generally "on familiar ground." However, their opposite numbers from the CBQ did have to move. Most of us felt that these fine people deserved extra credit for the success of the new Finance Department. They made more sacrifices for our future. Further, it always seemed that one thing that made the BN merger successful where the Penn Central was to fail not long after our March 3, 1970, M-Day, was that it is much harder to have a fight between three people than

between two! In any event, CBQ accounting executives including Assistant Vice Presidents Leo Meyer and Don Lamb, and Assistant Controller Erv Numrich made fine contributions as they joined Assistant Vice Presidents Ken Woodruff (NP) and Bob Garland (GN) and Assistant Controllers Glenn Pederson (GN) and Dick Molitor (GN) to complete the Accounting Department's management team.

After so many delays, court challenges, and the like, the top management of what was about to become the BN naturally wanted to execute the merger the very instant it was legally possible. As already indicated, this proved to be on March 3, 1970, a date that one might have thought would be greeted by the accountants with horror, as it was not even the end of a calendar month, much less the end of a year. However, Ervin Peterson (NP), who was designated to be our director–General Accounting, ingeniously managed this with little trouble. Perhaps this was a good omen!

UNANTICIPATED TROUBLES:
THE PENN CENTRAL BANKRUPTCY

The June 21, 1970, bankruptcy of the Penn Central (formed in the 1960s by merger of the Pennsylvania and New York Central railroads) caused very serious problems for the U.S. railroad industry, including the newly merged BN. Casting doubt on railroads in general and newly merged railroads in particular, the bankruptcy forced the BN to grapple with one financial problem unique to it, and to take the lead in solving another very serious industry-wide problem.

With BN's new Finance Department less than four months into the merger, a significant pending matter was the August maturity of over $40 million of mortgage bonds of the former CBQ. As far as I know, the intention was to refinance them with a new issue of BN mortgage bonds, an approach traditionally followed by major railroads. Very unfortunately, the sudden bankruptcy of the Penn-Central destroyed the market for railroad securities of any kind, including the planned BN bond issue. This left the BN with the immediately pending need for over $40 million and, it was revealed, no so-called "Plan B" to provide it. As we were told, it was necessary for John Budd, chairman and CEO, and Lou Menk, president and

COO, to travel to New York and persuade the banks there to lend the BN the needed funds. No doubt, certain board members helped with this behind the scenes. The result was a "revolving credit agreement" with, in view of the Penn-Central bankruptcy, severe terms and covenants concerning income, cash, other new debt, and so forth. Our lack of such a backup arrangement already in place was seen as a shortcoming of the BN's new finance function.

The second Penn Central–related problem faced the railroad industry as a whole, but had particularly serious potential impacts on the BN, which therefore led the effort to solve it. This had to do with so-called "interline revenues" and requires a brief detour into one of the recesses of 1970s railroad accounting. (There will be a few more!)

Many railroad freight shipments required the services of two or more railroads to complete the haul. For example, a carload shipment of lumber originating on the BN in the state of Washington and destined for, say, a lumberyard in Pittsburgh, Pennsylvania, required the services of the BN and the Penn Central, with the freight car of lumber "interchanged" to the Penn Central perhaps at Chicago, Illinois. However, a single freight rate applied, and the shipping lumber mill or the receiving lumberyard naturally wished to make only a single payment. By tradition, this payment was usually (but not always) made to the delivering railroad, in this case the Penn Central. After claiming any funds collected by the BN on "prepaid" shipments, it was the Penn Central's responsibility to divide the payments between itself and the BN, in accordance with so-called "interline divisions" agreed to by the Traffic or Marketing departments. The net results of all such transactions in a particular month between Penn Central and railroads such as the BN, Union Pacific, Southern Pacific, and so forth, were payments of millions of dollars by Penn Central to other railroads.

With the Penn Central in bankruptcy, it was the initial reaction of the bankruptcy trustee and his staff that all such monies belonged to the bankrupt's estate. If so, they should not be paid to the BN and other railroads, but instead held to settle accounts with Penn Central's bondholders and other creditors. In addition to depriving railroads such as the BN of large amounts of money,

this had the potential to very seriously disrupt railroad freight movements, with the originating western and other railroads insisting on receiving "prepayment" for all shipments terminating on the Penn Central or routing such shipments via other railroads that could also serve the customer.

BN immediately recognized that firm action had to be taken. As a result, Frank Farrell, vice president and general counsel; Lou Harris, attorney; and I were soon on our way to Philadelphia. There, after a laborious and frustrating process of finding local co-counsel that was not already involved in the bankruptcy, we appeared before the trustee. It was our position that these so-called "interline funds" were not monies belonging to the bankrupt Penn Central, but rather were in the nature of "trust funds" held by the Penn Central in trust pending the routine settlement of interline accounts of all kinds (passenger revenues, loss-and-damage freight claims, and freight-car-usage payments were also involved). Guided by the BN attorneys, I had the opportunity to explain all this to the trustee, who ultimately ruled in BN's favor.

This restored normalcy to the interline movement of freight and recovered millions of dollars due the BN from the bankruptcy. It also established a good precedent for the other bankruptcies that plagued the railroad industry in the 1970s. Perhaps the reader who is appraising the BN Finance Department's success in dealing with the results of the Penn Central bankruptcy may well rate us "one for two"!

After a year of major challenges and continuing work to fully merge the various finance functions and reduce our clerical costs, there were further changes at year-end, with new leadership put in place. John Tauer, vice president and controller, retired. Bill Ernzen, vice president–Finance, moved down to become vice president and controller, and Frank Coyne became head of the Finance Department as vice president–Finance. Frank Coyne had joined the NP from the Southern Pacific to lead the NP's COMPASS program for advanced computerized train- and car-movement information. A graduate of the Harvard Business School, he bought many valuable attributes to the BN Finance Department and led it well through out the rest of the 1970s.

FINANCING GROWTH

Although under new leadership, BN's Finance Department organization structure continued intact. Along with the day-to-day functions of accounting, treasury, property tax, costs and statistics, and internal audit, a small but important staff section, headed by an assistant vice president reporting directly to Frank Coyne, guided the financial planning function. This key position was successively held by Richard Stumbo and Robert Ortlip, and, beginning in 1974 or 1975, by Raymond Burton, who added the position of treasurer in 1978. All were very able, and, among other duties, coordinated the Finance Department's work, and that of other departments, in supporting our external financing efforts. These were clearly going to be necessary, in part because as of the end of 1970 the BN had already borrowed a total of $60 million under the new revolving credit agreement.

As the BN moved past its 1970 merger year, the attention of all departments, including Finance, was gradually and increasingly drawn to the exciting opportunity presented by the enormous reserves of low-sulfur western coal and the demand for them by electric-power companies in the Midwest and Southwest. This demand was driven by clean-energy concerns and the practicality of moving the coal in low-cost unit trains.

Emphasis is placed on the word "gradually," as BN's coal traffic grew incrementally and was financed incrementally also. As far as I ever knew, no one in the first few years of the 1970s recognized the ultimate need to plan to finance an investment of $2 billion in the coal-transportation business. The opportunity was gradually recognized, along with other financing needs of the transportation and resources businesses, also including maturities of existing debt.

Those planning to meet BN's financing needs also understood that both the railroad and the resources businesses had the capability, year in and year out, to internally generate very substantial flows of funds for dividends, reinvestment, and so forth. A few net total figures in table 24.1 covering BN's first ten years (1970–1979) tell the story.

Money is fungible and the reader may best decide whether it is logical that internally generated funds or externally generated funds or some proportion of the two are seen as applied to repayment of the maturing debt.

Table 24.1. SOURCES OF FUNDS

Funds generated from operations	$1,600 million
Mortgage and debenture bonds	$245 million
Equipment and other debt obligations	$794 million
Revolving credit agreements—net	$10 million
Preferred stock	$151 million
One use of funds:	
Repayment of long-term debt	$786 million

Whatever the view (and no doubt railroad finance or accounting officials of the period may be found to argue to the contrary!), it is clear that from somewhat less than half to somewhat more than half of the funds that BN had to invest in coal and other railroad traffic and in its resources properties were earned from internal sources.

As to the external sources, through the 1970s, in addition to equipment trusts, and capital and operating leases, BN's Finance Department planned and issued a variety of debt and equity securities. These were:

1971	$60 million	Consolidated mortgage bonds These funds were seen as for repayment of borrowings under the Revolving Credit Agreement.
1972	$65 million	Convertible debenture bonds These were recognized as the first equity-oriented railroad security since the 1930s.
1974	$60 million	Consolidated mortgage bonds
1977	$100 million	Convertible preferred stock
1978	$60 million	Consolidated mortgage bonds
1979	$51 million	Preferred stock

These securities issues were all guided by Frank Coyne and the officer serving as assistant vice president–Financial Planning at the time. They were always a team effort, including representatives from the Accounting and Law departments, Deloitte and Touche, Sidley and Austin, and our investment bankers and other advisors. For bonds and preferred stock, our investment bankers were from Morgan Stanley. One of them, Parker Gilbert, later became chief executive officer of that firm. For equipment trust certificates, financing the acquisition of locomotives and freight car, the BN often used Salomon Brothers.

However, as years passed, such equipment was frequently leased. At times, leasing also financed other railroad facilities. For example, Ray Burton devised an ingenious structure of "sales and lease backs" and leases to fund a major taconite transfer facility near Superior, Wisconsin.

It was always necessary to balance the capital needs of the expansion of BN's coal and other rail traffic against the need to maintain BN's credit worthiness and financial stability, maintaining the correct balance between equity and debt, and a strong assurance of our ability to continue to cover our interest payments, debt maturities, and dividends.

Frank Coyne maintained a valuable continuity among those working on BN's securities offerings. For example, I was always the Accounting Department's team member. Those involved learned to work well together and were confident in their ability to assemble the offering documents and issue the securities smoothly and on a timely basis.

As the Finance Department staff, CPAs, and investment bankers accumulated shared experiences and got to know one another, the work was not without its dry humor. For example, when a prominent New York lawyer found convoluted language too late to change it, he exclaimed, "Every finance document must have its element of mystery." Or, perhaps as a supreme example of exactitude or "nit picking," when it was necessary to show the printers that they had inserted one extra dot on one of the dotted lines of one of the financial statements, one team member resorted to using a straightedge and never heard the end of it.

If one studied the succeeding BN securities-offerings documents over the years, one noted a shift in emphasis to equity offerings as BN's financial reputation improved. The purpose of the issue was always cloaked in such general language as "invested in working capital, railroad property and other corporate purposes," but there was increasing reference to expanded capacity for coal traffic.

As the 1970s ended, the BN Finance Department had reason to look back with some pride at what had been accomplished in financing BN's expansion.

BUDGETING AND OTHER ONGOING FUNCTIONS

All through the decade of the 1970s, in addition to the more attention-getting external financings and other more publicized events and activities, the BN Finance Department also continued its daily functions of facilitating and assembling budgets, billing and collecting revenues, payrolling and other disbursement processes, interline settlements, managing cash, accounting and controlling, internal auditing, cost finding, and management reporting. There was a constant effort to reduce clerical costs and realize the planned merger savings, however, in the context of the promises made in the late 1960s agreements with railroad labor. New and improved computer systems played an important role in all this.

The preparation of BN's multi-year strategic plans and its annual capital and operating budgets were another example of BN's multi-department teamwork. Each year, the assistant vice president–Strategic Planning, first worked with the Executive, Marketing, Operating, Finance, and other departments to formulate and update BN's multi-year plans and traffic-volume and financial projections, with numerous alternative scenarios (perhaps a few too many of these, some of us thought). Once discussed and approved, these plans and projections formed the basis for the annual capital and operating budgetary processes.

Some recent writings concerning the 1970s BN reveal misunderstanding or lack of knowledge of BN's budgeting. Just as the best writings on the early British Admiralty rely on the diary of Samuel Pepys, who was actually there, it is best to rely on the knowledge of those who were actually involved. Each of the predecessor companies (CB&Q, NP, and GN) had comprehensive operating and capital budget processes, and the BN continued the best experience of each.

Annually, for the upcoming year, all departments received appropriate budgeting goals, directives, and assumptions. Operating budgets were prepared in detail, submitted for higher-level approval, and then assembled by the Finance Department. After executive and board approval, the annual budget details became a part of the BN's computer-based responsibility accounting system, installed just before the 1970 merger with the help of

McKinsey and Company. With detailed reports for each budgetary unit and summaries for higher-level managers, this system was fully comparable to those used in industry generally.

Annual capital budgets were similarly developed. In his June 16, 2011, memorandum, John H. Hertog, who was for a number of years BN's operating vice president, describes specific procedures for budgeting capital expenditures and the related operating expenses associated with them. Proposals were developed by the railroad operating divisions, regions, and headquarters departments. All were carefully reviewed. Those that survived the review process were then taken to an annual three-day meeting for further review and approval.

Capital dollars were always limited. After including those projects required for safety and legal reasons, and those that were a part of approved multi-year plans, where appropriate, we evaluated new projects financially using the net present value of cash flows and the BN's cost of capital, a process also developed with the help of McKinsey. Final annual capital budgets were approved by the board and then closely monitored by the Operating and Finance departments to be sure that authorized amounts were not exceeded.

There were also numerous special projects and new assignments requiring Finance Department support. One of the latter was our opportunity to provide finance and accounting support to the startup and early months' operations of BN's airfreight forwarder, Burlington Northern Air Freight, Inc. This highly successful new venture was led by Norman Lorentzsen, then president of BN's Transportation Division. It took a maximum working capital investment of about $2 million to support the new company's rapid growth. Thereafter, we received over $6 million in dividends and later sold the company for over $80 million. We were proud to have had a part in this success.

We were by no means perfect. An expensive GN passenger revenue-accounting system from the late 1960s had to be "junked" when Amtrak took over all intercity passenger service on the BN. Another unexpected consequence of this was that the BN found itself arguably the sole and final user of the St. Paul Union Depot Company, owned jointly with several other railroads. Under the outdated, but seemingly applicable joint-facility contract, it first appeared that we would be obligated to pay all the property retirement costs, mostly to the benefit of other railroads who were owners but no longer users of the depot. Only after a prolonged accounting and legal wrangle did the BN largely extricate itself from this archaic entanglement.

One mistake received more public attention. Once, in the early 1970s, well-meaning tax accountants gave premature affect to recently enacted federal tax law changes. This mistake required restatement and republication of quarterly earnings results, an embarrassing matter for any corporation. In some ways, however, this taught us a good lesson. Led by Assistant Controller Dick Molitor and our CPAs, comprehensive federal-income-tax-computation software was developed, making assumptions and processes crystal clear for those reviewing our monthly tax computations.

FINANCIAL COMMUNICATION
(INCLUDING BETTERMENT ACCOUNTING)

Perhaps because of the traditions of a regulated industry, BN top management and Finance Department of the 1970s had an admirable tradition of extensive external financial reporting and disclosure. Quarterly and annual reports were detailed and comprehensive. Board members received all necessary information, and board meetings included complete financial presentations. Every effort was made to appropriately inform investors, security analysts, trustees, and other "stakeholders" about BN's financial results and future plans. BN held meetings with security analysts at financial centers, and almost annually they conducted inspection trips, using a special train of so-called private or business cars and a specially constructed auditorium car whose large seating capacity facilitated simultaneous communication with two- or three-dozen analysts or other interested persons. In addition to the information exchanged, these shared experiences were thought to build trust and mutual understanding.

One reason for this effort was that a number of railroad accounting and reporting procedures were unfamiliar to certain investors, including, for instance, the interline settlement processes that had potentially jeopardized BN's cash flows after the Penn-Central bankruptcy. Another example was the railroad industry's traditional betterment accounting for track structure.

Virtually all businesses follow the practice of writing off as expense or "depreciating" the cost of productive assets (e.g., factory buildings and machinery) over their estimated productive lives, thus in each year "matching" a portion of the original cost with the revenues produced by use of the assets. Railroads had long followed the same practice for locomotives, freight and passenger cars, bridges, buildings, shop machinery, signal and communications equipment, and so forth. However, for the track structure (such as rails, ties, and track fastenings), through the early 1980s railroads followed the so-called "retirement, replacement, betterment" accounting, which then constituted "generally accepted accounting principles" for railroad track structure.

Perhaps the theory was that, for safety reasons, the basic track structure could never be allowed to physically "depreciate" to the point at which it was worn out. Thus, for example, when replacing 90-pound rail with 115-pound rail, railroads recorded the current cost of a 90-pound rail as a current expense, and capitalized as an asset only the cost of the additional 30 pounds' weight of rail placed in the track. In theory, with replacements using the same weight of rail being made as needed to maintain the original safe productive capacity and with no price inflation, the annual track-maintenance cost under betterment accounting would have differed little in aggregate from that which would have been recorded under depreciation accounting.

In practice, however, there were several problems. First, since no annual depreciation expense was being recorded for the rail already in the track structure, a railroad in financial difficulties could reduce operating expenses by not replacing rail as needed. This was the central concern with betterment accounting in the 1960s and 1970s. Second, if inflation was continually increasing the price of rail, a railroad that was replacing worn-out rail as needed, would, because of inflation, record higher current expenses under betterment accounting than under depreciation accounting. Third, as in the case of the BN, a railroad that was replacing perfectly serviceable existing rail with heavier rail, in order to carry much more traffic, heavier loads, and so forth, would likely record even more current expense under betterment accounting than under depreciation accounting.

BN's Finance Department believed this was the case. The betterment-accounting procedure and its potential problems were always explained at great length to any interested security analysts or others, including the possibility that our recorded annual expenses might be overstated.

We were vindicated when, in the mid-1980s, railroads converted to depreciation accounting for track structure, and BN's prior years' maintenance-of-way and structures operating expenses were restated downwards.

THE 1980S: BN FINANCE IN TRANSITION

The 1980s brought significant change to BN and its Finance Department. Within a short time, Louis Menk, chairman of the board and former president and CEO, and Norman Lorentzsen, president and CEO, retired. Both were splendid executives and their support and guidance contributed greatly to whatever success BN's Finance Department may have achieved in the 1970s. Richard Bressler, a top executive of Atlantic Richfield Oil Company, was hired as chairman, president, and CEO. As BN executives, we understood the board wished to have a CEO with expertise to support the development of BN's natural-resource properties. Dick Bressler had a background in finance, and we looked forward to working with him. On certain financial matters, his knowledge did, indeed, prove helpful to us: He quickly recognized the uncertainties with the costs underlying our early coal rates. Later, as another example, he gave us firm guidance as to the appropriate cost of capital to use in analyzing very large potential capital projects.

In 1980 we merged with the St. Louis and San Francisco Railway (Frisco). This brought the BN some new board members, additional trackage in the south-central states and access to important new territory, and some capable new executives and managers. It also brought to some departments a culture that differed from that of the BN.

In 1981 BN adopted a "holding company" form of corporate organization, with its headquarters in Seattle,

Washington. The subsidiary Burlington Northern Railroad headquarters remained at first in St. Paul, Minnesota, but then moved to Fort Worth, Texas, and Overland Park, Kansas.

Viewed from the vantage point of thirty-plus years of history and of the fine success of what is now the Burlington Northern Santa Fe Railway, it is certainly possible to believe that these first steps helped pave the way to an excellent future for BN and its shareholders. It is, of course, also possible to believe that the same ultimate results could have been achieved differently.

Soon, there were further top management changes. Within a year or two, Frank Coyne, executive vice president–Finance and Administration, Tom Lamphier, president–Transportation Division, and Bob Binger, president–Resources Division all retired. They had made strong contributions to BN's success during the challenging 1970s.

With the Frisco merger, Richard Grayson, the Frisco's president and a highly respected railroad chief executive, became president–Transportation Division and later, president and CEO of Burlington Northern Railroad Company, by far the holding company's largest subsidiary.

I elected to remain with the railroad—rather than join the holding company in Seattle—as controller. As the railroad's senior vice president—Finance and Administration, I had the opportunity to guide five important railroad departments. These were Accounting, led by Charlie Roberts; Information Systems, led by George Clinkenbeard; Human Resources, led by Fran Coyne; Labor Relations, led by Al Egbers; and Strategic Planning, led by Mike Donahue. All were highly capable executives and we made a good team.

With Dick Grayson's support, in order to assure full recognition and coverage of all true costs, we refined the long-range variable-cost approach that had caused difficulty in the early 1970s. We sharpened our planning focus, moving away from multiple scenarios that tended to confuse managers and, supported by George Clinkenbeard's expertise, we increased Information System's emphasis on operating department function. Charlie Roberts and George Clinkenbeard had joined us from the Frisco. Al Egbers had been with us at the 1970 CB&Q/NP/GN merger, but then spent several years leading Labor

Relations for Conrail, the reorganized Penn Central. We were very glad when he returned to the BN in 1977.

Beyond the change in perspective resulting from my new assignment, the change to a holding company structure placed external financing, key treasury responsibilities, and external financial reporting with the holding company. Thus, as to years after 1980, I am unable able to write of such matters from direct knowledge.

The first holding-company chief financial officer was Chester Rose, who joined us from the Union Pacific Corporation, where he had been the controller. We believed he was a good choice because he was obviously familiar with the workings of a holding company with a very large railroad subsidiary. Unfortunately, he resigned in 1982 and was replaced by John Parrish, who had been hired from General Electric to lead the holding company's strategic planning. We, at the railroad, also welcomed this appointment, but it ended about a year later. Our colleague, Ray Burton, who had made a fine record as BN's treasurer, joined the holding company in that capacity, but soon left to become president of Trailer Train.

Another change reflected new doctrine concerning the published financial disclosure of railroad information, which was reduced in holding-company annual reports to shareholders. This may well have been appropriate, given the new emphasis on non-railroad assets and operations, but it also may have deprived current investors and future historians of potentially useful information.

As a whole and as viewed with the benefit of hindsight, the early 1980s were busy and successful years for Burlington Northern Railroad Company. While with a "rough edge" or two, Dick Grayson proved a likeable and aggressive CEO from whom we learned a lot. It is clear to me that he foresaw the time in which the railroad would be totally and appropriately separated from the vast resources properties of the former Northern Pacific, and would have to live on its own earnings. I think most BN Railroad officers welcomed this challenge, although a few felt the resource properties should still be dedicated to support railroad operations and capital needs.

Dick Grayson once said to me, "Everything we got on the Frisco had to be 'scratched up' from between the right-of-way fences." I always remembered this as he led us in a relentless and continuing drive to reduce railroad operating costs. Mike Donahue, in his role as vice president–Strategic Planning, educated us on the "Driving Force" strategic concept, first introduced in 1980 by Tregoe and Zimmerman. We used this effectively in focusing on cost reduction, always repeating, "Low Cost Carrier, Low Cost Carrier."

Many strategies and tactics of this continuing effort met with wide support. However, a few important ones caused serious concern. The abrupt departure of valued senior colleagues who had contributed much to the BN created an uneasy situation. This even happened with members of the board of directors, as in 1982 when two longtime BN board members, Robert Wilson and W. John Driscoll, had to resign. I was told this was due to a perceived or purported conflict of interest in their membership on the board of directors of the Weyerhauser Company. This was hard for some of us to understand, given railroad industry practice and the very long-standing relationships between the two companies.

With the Frisco merger and the formation of the holding-company structure, the question of the location of the various companies' headquarters naturally arose. While most of us appreciated the objective and subjective benefits of our location in St. Paul and Minnesota, BN studies had told us that, if starting over "from scratch" with no "sunk costs" and no existing skilled and committed work force, a number of the major cities in the central time zone would be suitable as a railroad headquarters.

Moreover, when the decision was made to locate the headquarters of the holding company in Seattle, Washington, the relatively few employees involved, the location of BN's important natural-resources operations in Montana and Washington, and the benefit of separating the holding company from the railroad made this seem a logical decision.

Next, however, we were confronted with a plan to divide the railroad headquarters between Fort Worth, Texas, and Overland Park, Kansas, with only a relatively small administrative center remaining in St. Paul. This seemed a much different matter, involving very substantial moving costs and the disruption of the lives of many hundreds of people. Initially, an attempt was made to justify the large costs by linking the planned moves to savings achieved from a concurrent administrative-cost-reduction campaign. We were able to show that there was little connection between the two, but then other justification was found.

It is clear to me that chief executives and boards of directors are fully entitled to locate company headquarters wherever they believe makes the company most effective. However, unless one further believes that it is really justifiable to make such moves so that executives may avoid state income taxes or so that corporate social responsibilities may be avoided, it would be best not to advance such arguments. I, for one, grew tired of hearing them from newly hired or transferred executives.

A final area of serious concern grew from the experiences of newly placed executives with what they appeared to see as difficulties and inefficiencies arising from the various labor agreements entered into before the 1970 CB&Q/NP/GN merger received final approval. It was soon apparent that plans were being made to evade or break these agreements that many of us had always viewed as serious commitments that we intended to keep. Indeed, without them, the 1970 merger would probably have not been approved at all.

These plans had substance. Executives were put in place with track records of confronting unions and achieving work-force reductions with or without the agreement of organized labor. I was even told that the 1983 acquisition of the El Paso natural gas company was, at least in part, intended to provide a stream of earnings to support the BN holding company should these labor initiatives provoke a lengthy railroad work stoppage.

In summary, over the first several years of the 1980s, BN's culture and leadership changed markedly from that of the company that had been guided by John Budd, Lou Menk, Bob Downing, and Norm Lorentzsen through the 1970s. While some obviously thought this new company culture was necessary or at least desirable, more than one of us concluded that it was best not to have a continuing association with it. Mine ended in May 1984.

CHAPTER 25

FINDING BILLIONS
FOR NEW VENTURES

I was on the front line in raising $2.6 billion to meet the challenge. . . . It was a hectic period involving over 150 transactions. In 1980, we were doing a financing every three weeks, and in early 1982, I had to establish bank lines of credit for $1 billion in four days for other developments the company suddenly had in mind. But we did it and by all signs the results have been good.

"RAY BURTON: PIGGYBACK LEADER/CORPORATE HEALER,"
MODERN RAILROADS, JANUARY 1986

MOST OF THE ACTION IN A RAILROAD company is centered around the work of the Operating and Marketing departments. The Operating Department is charged with delivering the company's product, i.e., the service it provides for its shippers. The Marketing Department develops service and pricing packages that will attract and hold the business of the shipping community. Other departments are expected to dedicate their efforts to support the goals and mission of the two major departments.

In BN's years of rapid growth and challenges to prepare its roadway, equipment, and capability to handle its rapidly growing coal business, its Operating Department needed vast quantities of new track, signal and bridge material, and the financial resources to pay the wages of the thousands of additional people who had to be employed. These expenditures could not be delayed or held up until the projected increases in revenue would be realized. This meant the work would be completed well before the revenue would start to flow to the company's coffers.

Since some of the rates for the large blocks of new business that were already moving by the mid-1970s had been priced too low, cash flow from operations could not generate the cash needed to finance the planned level of capital expenditures. Also, operating expenses were running at a very high level, resulting from the all-out effort underway to raise the standard of the roadway above what had been a secondary, low-density operation. There was a much larger fleet of locomotives to maintain and a rapidly increasing number of trains being run. Having such a high level of operating and maintenance expense made it impossible to generate enough cash from current operations, and it was planned that that would have to be the case through 1980 when most of the programs would be completed.

It was in the midst of these challenges that Ray Burton was hired by BN in the fall of 1974 as an assistant vice president in the Finance Department. Before coming to BN, Ray had worked for the Santa Fe Railway for eleven years. His first assignment was in the Santa Fe's Transportation Department. Later he was promoted to assistant trainmaster and assistant to the general manager. In those years he took time to become qualified as a Locomotive Engineer, a skill he maintained throughout his career. Ray's next assignment was in the Cost Research Department, and then as a staff assistant to the vice president–Finance. Ray graduated from Cornell

University with a BA degree in Liberal Arts, and after completing military service, he earned an MBA degree from the University of Pennsylvania's Wharton Graduate School of Business.

Soon after coming to BN, Ray was charged with raising the $2 billion needed to finance the plan for capital expenditures that had been approved by the board. All of us were enthusiastic, energized, and proud to know that our company would have the financial resources we needed for the projects that would enable us to take advantage of the opportunity we had to turn the fortunes of BN into a real success story. It was the creative, innovative, and brilliant work done by Ray Burton that made it possible for us to undertake the ambitious programs we had laid out.

Once financing was obtained, it was up to us to get the work done on schedule and at the standard of quality defined by the experts in the Engineering and Mechanical offices at the headquarters level. We needed to hire and train thousands of people, devote time to plan and organize the work, and see that it was done safely and up to standard. The coal-marketing section was charged with negotiating rates that would give us a fair return on the capital investments we were making, and help raise net income to a higher level so shareholders could be rewarded with increased dividends and realize a higher value in their stock.

Following is a list of some of the financing arrangements Ray negotiated in his years of service in BN's Finance Department, starting with $200 million in leveraged leases for locomotives. In the 1970s, the path for negotiating these leases was tortuous due to the documentation required and the arguments raised by lawyers for the investors. Leveraged leases were of benefit to capital intensive industries such as railroads that could not take the accelerated depreciation and tax credits. In recalling those transactions, Ray wrote,

Financial companies such as GE Credit, Ford Motor Credit and large banks had large tax bills that could be mitigated by being the "owner" of assets of capital intensive companies, e.g., BN, by owning our rail cars, etc., and leasing them back to the borrower and taking the tax deductions surrendered by the lessee. Some of these deals were getting rather cavalier in their structure. Of course, this trading of tax benefits to finance companies

allowed them to reduce their tax payments to the government. The IRS said "stop" until appropriate regulations could be instituted. . . . I was appointed the railroad representative to work with the counsel and the IRS toward a solution. Such was done and the IRS published guidelines to be followed by lessors and lessees. Problem solved. This was critical to BN as we could not use the tax benefits of ownership on any timely basis. During my time we did many leveraged leases for motive power and rolling stock, all in place and legal before deliveries. . . . I was giving speeches around the country on "how to do a lease."[1]

The second challenge Ray handled in those early days was financing of the new taconite dock to be built by BN in Superior, Wisconsin. The shipment of iron ore and taconite had been significant for the Great Northern, and Bob Downing was insistent that BN build the dock. Ray recalled,

The Board of Directors were not. The dock was needed and demanded by Bethlehem and National Steel companies. . . . Accomplishing this $70 million financing was given to me to handle. I held discussions with Morgan Stanley on how a lease financing could be structured that would not be on the balance sheet of our company [to satisfy a concern of some board members]. A model for this was the recently completed Standard Oil of Indiana building in Chicago. The BN deal was completed by the Fall of 1975. It included a commitment on the part of the steel companies to ship a specified amount of taconite through the new dock or pay a certain amount to BN if there was a shortfall. . . . Financing was issued by the City of Superior in the form of tax-exempt bonds. . . . This entire transaction made my reputation within BN.

Numerous equipment trust certificates were issued to finance the acquisition of locomotives and cars. In using this method of financing, the railroad company put up 20 percent of the purchase price and the parties in the financing syndicate, the remaining 80 percent. Because equipment trust certificates could not exceed $10 million by custom, BN had to issue a number of certificates each year to be able to acquire enough equipment to meet its needs.

Ray recalls putting together a $100 million revolving

---

1   Ray Burton, correspondence in the early 2000s.

credit agreement with Citibank. Interestingly, no ICC scrutiny was required for this type of loan.

To finance construction of the new 127-mile line between Gillette and Orin, Wyoming, and expensive upgrading required on many existing lines, BN issued consolidated bonds that were supported by additions and betterments made on lines of the Northern Pacific that were registered with the trustee for a series of NP bonds issued in 1893 containing that requirement. In this time of rapidly increasing interest rates, a member of the offering syndicate suggested to Ray that BN consider hedging the forthcoming BN bond offering, as interest rates would probably go up by the time of the offering in the next month or two. It turned out to be a good bet, since interest rates did increase as expected. According to the Chicago Board of Trade, this was the first time a corporate bond issue had been hedged. The Board of Trade was so elated over this success of this new approach to financing that they had a picture taken of Ray in front of a BN locomotive for its annual report. The picture hung on the wall of the board's visitors gallery for many years.

The new car- and locomotive-maintenance shops in Alliance were financed by an issue of a fixed facility mortgage of $35 million through Morgan Stanley. In 1980 Ray arranged financing for a secondhand 727 airplane that was acquired in a trade-in of two smaller corporate aircraft.

An offering of preferred stock made in the early 1980s was the first equity offering made by a railroad company in several generations.

Early in 1982, BN's new president, Richard Bressler, requested the Finance Department to raise $1 billion for possible acquisitions. In only four days, Ray was able to complete the deal on favorable terms, including no commitment fee. Ray recalls getting it done by early April, and on April 30, he left BN: "It was a busy time for me, but a successful one for all concerned."

In evaluating what was accomplished in those years, Ray believes the Powder River coal business made BN exceptional as a railroad company, with growth ahead of it. Other railroads benefited as well, as the growth in coal and intermodal business brought new life to the industry.

Upon leaving BN, Ray was invited by Lou Menk to join him as vice president–Corporate Planning at International Harvester, which Menk was heading on a temporary basis after retiring from BN. Menk had been on the board of IH for several years and moved in as CEO for several months at the board's request to help IH work out of its deep financial trouble. In the fall of 1982, Ray accepted an offer to head Trailer Train Company, later named TTX. He came into the company at the time it had severe financial problems with its Railbox and Railgon subsidiaries. It was a big relief to the board and owners of TTX that Ray had the ingenuity and leadership capability needed to work through those difficult situations. He prevented what would have been the largest default on railroad-equipment paper in the history of the rail industry from occurring.

Later, Ray helped set up an "education" program for the railroad companies on the opportunity for them to begin using new double-stack technology for handling containerized intermodal shipments. Rail cars for this type of service had just been introduced by American President Lines. The United States had become an importing nation on a large scale, and imported consumer goods were suitable for containerization and movement by rail as a "land bridge" or to major consumption areas in the United States. Double-stack container service soon became the fastest growing segment of rail business. In a few years, TTX owned and provided nearly 100,000 double-stack container slots for use by North American railroads.

This and other innovations at TTX turned it into a highly valuable jointly owned company for the rail industry. Success in applying these new initiatives brought recognition to Ray when he was named the *Railway Age* "Railroad Man of the Year" in 1986 and again in 1993. Ray recognized publicly that this and other honors he received in those years "were really for all the men and women at TTX and that it was a signal honor for the company itself."

Overall, BN and the entire rail industry benefited greatly from the intelligence, business skills, and leadership Ray was able to provide, and with the complete honesty and objectivity he demonstrated in all of his business dealings.

PART V

MANAGING BN'S VAST RESOURCE HOLDINGS

CHAPTER 26

VISION, VALUES, AND PHILOSOPHY IN DEVELOPMENT OF RESOURCE HOLDINGS

… timberlands are managed on a sustained-yield basis under long-range plans that require balanced planning and harvesting.
BURLINGTON NORTHERN 1974 ANNUAL REPORT, PAGE 14

It may not have been the cheapest way (managing forest resources), but it was the responsible way. I think you have to pay the cost to do what has to be done right in managing lands.
C. ROBERT BINGER IN INTERVIEW WITH
TIMOTHY MATTHEW BECHTOLD, NOVEMBER 26, 1991

You couldn't put such an array of assets together at any price.
RICHARD M. BRESSLER, QUOTED IN
PAUL GIBSON, "A RAILROAD FOR THE LONG HAUL," *FORBES*, APRIL 27, 1981

The first of these "two key" objectives is to make Burlington Northern Railroad the most efficient carrier in the industry. … The second key objective is … to achieve a better balance between the transportation and resource business.
RICHARD M. BRESSLER, LETTER TO SHAREHOLDERS,
BURLINGTON NORTHERN ANNUAL REPORT FOR 1982, PAGES 2 AND 3

NORTHERN PACIFIC LAND GRANTS

When the Northern Pacific was chartered to build a line of railroad from the head-of-the-lakes (westward from Duluth, on Lake Superior), its founders were granted thirty-nine million acres of land. This was accomplished by designating every other section (square mile) of land for ownership by the railroad company in checkerboard fashion, for twenty miles on each side of where the track was to be laid in the states of Wisconsin, Minnesota, and Oregon. In the territories (i.e., where statehood was not yet established) of Dakota, Montana, Idaho, and Washington, the grant was extended to forty miles on each side of the line. The grant to the NP did not include land with known deposits of minerals other than coal or iron ore. If there were mineral deposits within the twenty- or forty-mile band, the band was extended out far enough to get the total number of acres allowed.

The NP would sell this land as a source of funds for construction from sales to settlers, the development of townsites, and to logging and mining companies. By the

Table 26.1. BN RESOURCES HOLDINGS AND PRODUCTION (AS REPORTED IN ANNUAL REPORTS FOR 1970 AND 1971)

Sub-bituminous coal—mineable coal reserves

	10 billion tons
Coal production	4.0 million tons
Oil reserves	19.2 million barrels
Oil production	2.7 million barrels
Taconite and ore production	6.0 million tons
Timberlands	1.3 million acres
Range land under lease	1.2 million acres
Cultivated land under lease	22,500 acres

Table 26.2. GROSS REVENUES FROM RESOURCES (AS STATED IN ANNUAL REPORT FOR 1969)

Oil and gas	$8.6 million
Timber and lumber production	$31.9 million
Timber and log sales	$8.1 million
Leasing and sale of minerals	$2.1 million
Iron and taconite properties on Mesabi Range (Minn.)	$0.6 million
Grazing and cultivation	$0.5 million

1960s, all but about 1.9 million acres had been sold. The NP retained the mineral rights on 6.2 million acres it had sold. Unlike the NP, the GN received no land grants for construction of its lines of railway, with two exceptions: its predecessors, the St. Paul and Pacific Railway and the St. Paul, Minneapolis and Manitoba Railway, received a combined grant of 5.9 million acres of land in Minnesota at the time construction was started on the line built from Minneapolis to Breckenridge, and for the line built north from Breckenridge to the Canadian border. Virtually all of that land was sold for farming and a few town sites, even before the Great Northern Railway Company was formed in 1889. The former Burlington (CB&Q) received land grants of 2.8 million acres in Iowa and Nebraska, but its land likewise was sold for farming as the line was built. The SP&S received no land grants.

Much of the land the NP still owned by the time of the 1970 merger was forested. Oil, natural gas, and coal had been discovered on a large part of the land on which mineral rights had been held. Over the years, the NP swapped parcels of the "checkerboard" land with adjacent land owners to get large, continuous parcels under its control. With the checkerboard plan for allocation, it had been hoped that with a railroad line within a short distance, the land in the sections not granted to the NP would increase in value.

In its annual reports, the NP summarized the revenue produced from activities or operations on its non-rail

assets. This revenue was comprised of royalties from oil, coal, gas, and minerals extracted, and income from leases of forested land, the sale of timber, and the sale of manufactured wood product following the acquisition of Plum Creek Lumber Company. The Northern Pacific's vice president of Administration was designated to oversee these activities. From 1962 until 1967, Edward Stanton, whose background was in civil engineering and maintenance of way on the NP and SP&S, held that position. From 1943 to 1952, he served as vice president–General Manager of the SP&S Railway.

VALUE OF NP'S RESOURCE HOLDINGS IN ANTICIPATED MERGER TRANSACTION

The value of the NP's non-rail assets and their earning power had been a major issue all the way back to the 1960s, when the management and directors of the GN and NP had to negotiate the terms for exchange of their stock for shares in the merged company. For common stock, a ratio of exchange of one-to-one was agreed upon. However, because of the GN's superior earnings over the years, it was decided the holders of GN stock would be granted one-half share of preferred stock in addition to the one share of common stock in the new company. This exchange ratio was deemed unsatisfactory by a group of NP shareholders, who formed the Northern Pacific Shareholders' Protective Association.

This group protested the terms for the exchange of stock at the annual meetings of NP shareholders for several years, as well as in the hearings conducted by the ICC.

They took their case to the courts in an effort to get recognition for the higher value they believed should have been given because of the value of resources owned by the NP. In its decision to approve the merger, the majority opinion issued by the U.S. Supreme Court upheld the decision of a three-judge District Court that "the stock exchange ratio applicable to Northern Pacific stockholders and Great Northern stockholders, which was established after protracted arms-length negotiations with approval of the companies and the large majority of their stockholders, is just and reasonable, is supported by substantial evidence, and the ICC's refusal to reopen the record for evidence to update it was not an abuse of discretion."[1] This decision of the Supreme Court settled the issue of the value and potential for higher earnings of the NP's resource holdings at the outset of the merger.

However, there was a groundswell of concern on the part of some investors and board members that continued over whether the value that could be realized by shareholders from aggressive and intensive development of BN's resource holdings was truly understood and appreciated by the management. Failure to accelerate earnings from these non-rail assets to a more impressive level led to the decision of the board to bring a new management team into BN in 1980.

MANAGEMENT OF TIMBERLANDS, HIRING OF C. ROBERT BINGER, CONFRONTATION WITH NORTON SIMON

In 1968, two years after L. W. Menk was elected president of the NP, C. Robert (Bob) Binger was hired as vice president with responsibilities for the resource activities. Binger had a Masters degree in Forestry from Yale and had held senior management positions with the Minnesota-Ontario Paper Company and Boise Cascade. Also in 1968, the NP hired George Washington to head up its activities in oil and gas. From these appointments, it was apparent Menk saw the need to bring people into the company with expertise in managing its non-rail resources.

Statements made by Norton Simon, a member of the boards of both the NP and BN, clearly indicated he believed there were large, untapped opportunities to increase revenue from the company's non-rail holdings. In an article in the *Smithsonian* magazine, Simon's approach to the management of companies in which he invested was summarized: "With Hunt [referring to Hunt Brothers, a large vegetable canning company in which Simon had invested] profits, he sought to find established companies that had everything going for them except that they were 'badly run.' He would acquire big blocks of shares in such companies and then demand of the Board of Directors that changes be made. These meetings were not always pleasant. Simon's best friends will not credit him with diplomacy." A fellow member of the board of regents at the University of California said, "He has a unique faculty for making a roomful of people know he thinks they are a bunch of ignorant slobs." Neither the NP nor any company other than Hunt was mentioned by name in the article. However, the tone of the article clearly indicated the role Simon thought he should have as a board member. Indeed, NP and BN management were under pressure from Simon to expand its earning power through more aggressive development of its under-performing resources.[2]

Simon's interest in the NP originated from the investment he made in the Ohio Match Company beginning in 1944. With the large amount of timber used in the match business, Simon had pushed hard to get long-term contracts with timber suppliers. However, when the NP showed no interest in doing business with Ohio Match on that basis, Simon's researched NP stock and found it to be undervalued at $15 per share. Ohio Match then acquired 10 percent of the NP's stock, enough for Simon to gain a seat on the NP board. The NP and Ohio Match then entered into a long term contract. Only eleven months after Simon's election to the board, NP stock rose to $92 per share. By that time, Simon, together with Ohio Match, his family, and friends, held 14 percent of the NP's stock, making them its largest shareholders. Norton Simon became a member of the NP board in April 1951. Having been a director for nineteen years put Simon into

---

1   396 U.S. 491, 494, pp. 516–22.

2   Robert Wernick, "Norton Simon's Zeal for the Best and Its Results," *Smithsonian* (September 1979): 50.

a position of standing and strength by the time BN was formed, enough for him to be named chairman of the BN board's Finance Committee. Simon advocated that BN reorganize into a holding-company concept and spin off its resource holdings to shareholders. This possibility for reorganization was another of Simon's ideas for enhancing performance that BN put into place some years later.

In addition to his business ventures, Norton Simon became well known for his art collection and his marriage to the actress Jennifer Jones. Simon clearly made his mark with BN when he addressed the shareholders present at its annual meeting in 1973:

> For more than 20 years I have pushed hard to make this company more aware of the potential and responsibility it has in respect to its outside properties. There were some accomplishments. Much of the resistance I have met has been because our management team has grown up in the railroad, steeped in the traditions and practices of the railroad industry, and when it comes to important decisions, tends to put the railroad first. As a result, the priority, financial support, and manpower applied to development of our natural resources has not progressed as it should. To this day, with more than eight million acres involved, I am advised we have only about 180 people, including all levels of employees, participating in the management of our vast natural resources.
>
> For the last several years, getting the merger accomplished was considered to be a first consideration. During this period, progress on natural resources did not move as an appropriate pace. This last year, however, with merger problems getting well on their way, I expected things to accelerate. When management once again showed an pronounced and possibly an excessive tendency to "keep its eye down the track," however, I decided to strongly assert my responsibility as a director.... I urged management to use expert consultants in property development and the development of other resources.[3]

In a letter to Congressman Brock Adams the same year, Simon stated his belief that railroad executives were not sufficiently knowledgeable or competent in managing business activities outside of railroading, although he felt they had sufficient challenge in trying to lead the rail industry through all of the problems of that time. Based on that view of railroad management, Simon advocated that railroads be required to divest themselves of non-rail holdings. He made particular reference to the diversion of funds to non-rail ventures made by Penn Central's Board. About eight years after Simon gave his challenge to BN's board, management, and shareholders, BN actually fulfilled Simon's challenge to separate the rail and non-rail assets in 1988 when its Burlington Resources Company was spun-off to the shareholders. Needless to say, Simon's views were a major cause of disruption and debate within the company. Apparently out of frustration, Simon resigned from the board a few months after the speech he made at the annual meeting in 1973.

CONTINUED STRUGGLE BETWEEN ADHERING TO THE BEST PRACTICES IN FORESTRY VS. MAXIMIZING SHORT-TERM FINANCIAL GAIN

In his article "A Railroad Merger That Worked," Rush Loving termed Norton Simon and Bob Binger as "two men who see the forest as well as the trees, but differently."[4] Simon put pressure on Binger to drastically increase timber cutting, and Binger countered that "forests must be husbanded for future harvests." To accomplish this, BN planted upwards of two million trees each year.

Simon and a few other board members believed that wood products would soon be replaced by other materials, and they saw no point in cutting timber on a sustained yield basis, to pursue reforestation, or to employ selective cutting practices. BN's management countered that the U.S. demand for timber was expected to double by 1995, and that the amount of commercial forest land was decreasing by about one million acres each year.

To be in position to meet the demand for forest products, BN's schedule for planting trees insured that mature timber would be available to cut at regular intervals. In contrast, Simon and some fellow board members pushed for aggressive clear-cutting. Increasing pressure was put on Menk, Lorentzsen, and Binger to cut timber more

---

3   Norton Simon, letter to BN shareholders mailed shortly after the annual meeting of shareholders held in 1974.

4   *Fortune,* August 1972.

aggressively and to accelerate development of BN's oil and gas holdings. That view led to the board's decision to recruit an executive from outside the company who would be much more aggressive in managing the forest resources as well as the other non-rail holdings. The purpose of writing this extensively on Norton Simon is to explain how the pressure he put on NP and BN management to extract greater earnings from non-rail resources became the driving force for the new management team BN formed in the early 1980s. Simon acknowledged to Loving that Binger had accomplished a great deal, but couldn't resist adding, "but, he's no swinger." Binger held to his position and his beliefs, right up to his retirement in 1981.

ACQUISITION AND SUCCESS:
PLUM CREEK LUMBER COMPANY

Before the merger, Binger convinced the NP to get into the manufacturing side of forestry by acquiring the Plum Creek Timber Company, which owned and operated sawmills and other wood-processing operations in Montana. It was believed that synergy could be developed between the NP's vast holdings of timber and a manufacturing unit that would process the trees harvested from those lands. In the same year (1968), Plum Creek expanded its capability through acquisition of a sawmill at Fortine, Montana. Plum Creek paid the NP a dividend of $1.064 million in the first year it was under NP ownership.

According to Norman Lorentzsen, the NP paid about $6 million for Plum Creek. At that time Plum Creek produced mainly 2×4's and 2×6 pieces of lumber. Under Bob Binger's direction, Plum Creek expanded its manufacturing capability into plywood and medium-density fiberboard. Plum Creek's new plants were set up to use 100 percent of every tree that was cut, including the mulch from sawdust, the leaves and small branches, and all of the wood chips produced. There was no waste in Plum Creek's use of timber cut on its land or purchased from other owners of forested lands. The NP and BN were able to finance all of the equipment and facilities Plum Creek needed for expansion and improvement of its operations. Lorentzsen believes that the work of the forest-products unit was never held back because of unwillingness or inability of the company or its board

to provide adequate funding. Binger asserted the board backed him 100 percent "in anything he wanted to do."[5]

Following is a chronology of investments made to expand BN's capability to use and profit from the timber harvested on its lands:

1968 Acquisition by the NP of Plum Creek Lumber Company, owner of a sawmill and plywood plant in Columbia Falls, Montana, and a sawmill in Pablo, Montana

1971 Acquisition of Ksanka Lumber Company in Fortine, Montana

1973 Acquisition of the Arden Lumber Company, owner of two sawmills in Colville, Washington

1974 Completion of a new medium-density fiberboard wood plant at Columbia Falls

1977 Construction of a new sawmill at Arden, Washington, thereby increasing capacity by one-third

1979 Construction of a new sawmill at Columbia Falls

1979 Acquired C&C Plywood in Kalispell, Montana, thereby doubling Plum Creek's plywood production capacity

With these additions and acquisitions, Plum Creek became a major player in the forest products industry.

Plum Creek was set up as a profit center, independent of BN's forest-management unit. Under that policy, Plum Creek had full flexibility in purchasing its timber. In 1978 Plum Creek bought only 19 percent of its timber from BN, at full market price, and 30 percent in 1979. By functioning as a separate business unit, Plum Creek's management had to pay closer attention to profit margins than if BN had provided its timber to Plum Creek on an artificially low, subsidized cost basis.

Binger accelerated the exchange of land with the U.S. Forest Service to consolidate holdings for greater efficiency in operations and management. The Forest Service benefited as well by moving away from the checkerboard pattern that was established at the time of the land grant to the NP and consolidating the lands it controlled. In 1980 about 30 percent of BN's acreage contained old-growth timber. Since the annual growth of those trees was much lower than that of more recently

---

5   Timothy Mathew Bechtold, "Now v. Forever: The Conflict Between Business and Forestry in the Management of Plum Creek Timberlands in Montana," thesis, University of Montana, 1992.

planted trees, BN increased the rate of cutting old trees and opened up new areas for planting trees that would grow at a faster rate. In each of the previous five years, BN planted an average of three million trees, but in 1980, increased plantings to 4.6 million. Even heavier plantings were anticipated in the future.

Not unlike the problem the railroad had with rates being set much too low in the early years of moving low-sulfur coal, years ago the NP had set the prices in long-term sales contracts for timber with inadequate provision for price escalation. This was very costly to the company, but fortunately, the terms of these agreements ran out by 1980.

In BN's first ten years, Binger, George Washington, and Leighton Steward (successor to Washington) were successful in increasing the revenue brought in by the non-rail assets. The trends in revenue and other indicators or performance in the Resources Division are shown in tables 27.1–27.6.

CHAPTER 27

OPERATING AND FINANCIAL PERFORMANCE OF RESOURCE UNITS

URING THE 1970S, THE FINANCIAL community—and a number of people holding prominent positions in BNI—strongly asserted their concern that the amount of revenue and the percent of net income produced by BNI's non-rail units was outpacing that of the railroad company. Even with its much lower revenues, the earnings of the Resources Division exceeded those of the railroad in 1977. The implication was that the railroad was not performing well, in spite of the very large investment BNI was making (in both operating expense and capital investment) to increase capacity and upgrade the track on a large part of the system from 1974 through 1980.

However, with the growth in revenue from coal and the completion of much of the heavy maintenance and upgrading work, the amount of operating income produced by the railroad increased quite dramatically in 1979, as had been planned. "The tide turned," and by 1981, the railroad's contribution increased steadily from 65 percent to 87 percent. It reached 96.8 percent in 1983. After that, the percentage from Resources began to rise again, resulting from acquiring or merging with other large energy companies. These actions were part of the oft-stated goal of getting a balance between rail and non-rail income. It was believed that with such strong income from non-rail interests, BNI would be able to sustain a prolonged strike if that was what it would take to get a lower cost base for running the railroad in the future.

In 1980 BNI restructured its organization by putting both the Transportation and Resources units under a holding-company structure. The amount of capital invested in non-rail activities was increased, both for additional exploration and development in property already owned, plus that acquired through merger and acquisitions. In those years, the earnings of the railroad continued to increase as the investments made in the 1970s began to pay off, together with success from initiatives taken to improve the railway's efficiency and productivity.

The amount of capital invested in the railroad was greatly reduced in 1986 through 1989, with the implementation of programs to extend the lives of such high-value assets as rail and locomotives. Also, capital spending was reduced to make more cash available for investments in non-rail projects expected to have a high rate of return.

Operating under the holding-company concept ended in 1988 with the spin-off of Burlington Resources, leaving BNI with ownership of only the railroad company. With the split of Burlington Resources from BNI, the railroad company took on an enormous burden of debt, resulting in a debt-equity ratio of 72. BNI was successful in reaching its goal to reduce that ratio to 45 in five years. To accomplish that ambitious goal required that capital expenditures be reduced. That requirement set the company back in acquiring replacement locomotives, in further expansion of its capacity to handle ever-increasing tonnages of coal, and kept it from carrying out the programs large enough to maintain all primary, heavy-tonnage corridors at the standard needed. However, most of the railroad's people agreed that it was worth enduring this restriction for a few years to be relieved of the burdens of having to turn the cash it generated to a holding company for investment in non-rail businesses.

Table 27.1.

COMPARISON OF PERCENT CONTRIBUTED TO
NET OPERATING INCOME, TRANSPORTATION AND RESOURCES DIVISIONS, 1970–1987

Year	Transportation	Resources*	Land and Real Estate*
1970	65.2%	19.9%	14.5%
1971	61.4	20.0	18.1
1972	56.8	25.7	17.9
1973	50.9	38.1	10.8
1974	61.9	27.5	10.1
1975	60.0	26.5	13.1
1976	53.2	32.6	13.8
1977	46.2	63.8	
1978	49.5	50.5	
1979	56.1	43.9	
1980	71.3**	28.7	
1981	87.3	12.7	
1982	92.3	7.7	
1983	96.8	3.2	
1984	67.6	32.4	
1985	60.2	39.8	
1986	†	†	
1987	57.3	42.7	

* Effective with 1977 BN showed only a combined number for Resources and Land and Real Estate

**Amounts for 1980 included the merged SLSF Railroad for December only

† In 1986, a non-cash, pretax Special Charge included a writedown of oil and gas properties and a writeoff of surplus railroad assets. The Special Charge and change in railroad depreciation method of accounting resulted in an increase of 11.5 to the 1986 operating ratio. In 1996 the company adopted the successful efforts method of accounting for oil and gas properties. This resulted in unamortized oil and gas capitalized costs to exceed the present value of future net revenues. As a result the company recorded a $605 million pretax charge to reflect the substantial decline in oil and gas properties. With adoption of a unit method for depreciation for the majority of railroad transportation properties, the company reported a $36 million after tax charge in 1986. The result of the adjustments in costs for oil and gas was an operating loss of $631.659 million, and for the railroad, operating income of $152.786 million compared with $822.146 million in 1985.

Table 27.2. CAPITAL EXPENDITURES
RAILROAD, RESOURCE UNITS, AND NON-RAILROAD TRANSPORTATION UNITS, 1970–1994

Note: Prior to 1973, the amount of capital expenditures for units other than the railroad was not specified in the annual reports. Instead, capital expenditures were broken down only to show the amounts for equipment leased, equipment capitalized, road and other. There were no separate numbers shown for the Resources units.

Year	Railroad	Products*	Forest, Oil, and Gas	Coal and Minerals	Land and Real Estate	Corporate and Other
1971	$94.3M					
1972	158.4					
1973	141.4	$9.9M	$0.6M	$0.017	$7.5M	$1.2M
1974	128.2	5.9	0.8	0.236	7.6	0.2
1975	181.4	4.9	1.6	0.004	8.2	0.8
1976	158.7	9.8	5.6	0.4	8.6	0.5
1977	249.8	11.3	2.1	0.08	6.2	0.4
1978	223.1	11.9	2.6	0.018	7.6	4.9
1979	297.8	17.0	5.2	0.056	10.8	0.4
1980	335.2	18.5	10.4	0.036	11.5	0.6

Year	Trucking	Air Freight Forwarder
1971		
1972		
1973	$1.6M	$0.2M
1974	2.1	0.6
1975	0.6	0.5
1976	0.2	1.1
1977	4.1	2.6
1978	9.5	1.6
1979	5.6	2.7
1980	2.5	3.8

* Includes both Forest Management and Forest Products Manufacturing. Note: Capital expenditures for non-rail units were not stated in BN's annual reports for 1971 and 1972

Table 27.3. CAPITAL EXPENDITURES—RAILROAD AND RESOURCES, 1970–1994 (continued II)

Year	Railroad	Natural Gas Operations	Oil and Gas	Forest Products	Corporate and Other
1981	$524.1M	$ 64.5M	$13.0M	$10.3M	
1982	538.7	83.1	13.1	10.1	
1983	551.5	80.6	16.3	9.6	
1984	610.9	$57.1	232.1	12.2	19.7
1985	650.6	95.0	334.8	15.2	35.8
1986	351.2	72.0	116.1	12.2	40.0
1987	276.4	47.3	102.6	17.4	40.2
1988	474.0				
1989	465.0				
1990	567.0				
1991	584.0				
1992	487.0				
1993	676.0				
1994	753.0				
1995 (merger with Santa Fe)					

Year	Coal and Minerals	Trucking	Air Freight Forwarder
1981	$1.7M	$0.3M	8.1
1982	0.4	0.7	*
1983	1.3	0.1	

Note: capital expenditures for Coal and Minerals and Trucking were shown separately in only 1981–1983.

* BNAFI (air freight forwarder) sold in 1982.

Table 27.4. OPERATING INCOME FROM RAILROAD AND RESOURCES, 1969–1987

Year	Railroad	Timber and Lumber[1]	Oil and Gas	Minerals[2]	Land and Real Estate
1969	$56.3M	$6.4M	$4.9M	$1.7M	$8.9M
1970	40.4	5.6	4.5	2.2	9.0
1971	55 0	8.9	5.3	3.7	16.2
1972	54.5	13.6	5.4	5.7	17.2
1973	55.0	29.7	6.5	4.9	11.7
1974	92.7	23.6	14.0	3.4	15.1
1975	74.0	13.8	11.5	7.0	15.9
1976	76.5	27.6	14.2	5.8	20.7
1977	60.2	33.9	11.9	5.8	25.6
1978	83.5	47.3	17.0	4.9	27.4
1979	123.9	55.7	21.6	7.7	25.6
1980	278.1	45.3	39.5	10.2	24.6
1981	358.0	26.0	55.1	6.7	28.2
1982	245.3	27.9	37.1	.3	—[3]
1983	733.6	53.0	29.9	(9.7)	
1984	979.7	42.4	152.2	—[4]	
1985	779.9	46.0	96.9		
1986	100.8*	53.3	(585.8)[5]		
1987	602.2	70.4	89.2		

*   Includes effect of special charge of $352 million for surplus, obsolete or otherwise unproductive assets including locomotives, rolling stock and abandoned track. (Source: BNI annual report for 1987, pp. 39 and 40.)

( ) denotes deficit

1   Also designated "Forest Products" in some years.

2   Includes coal, taconite, and other minerals.

3, 4 Starting in 1982, operating income from Land and Real Estate and Minerals were no longer reported separately; instead, it was included in "Other and Eliminations."

5   Includes the effect of reclassifying amounts for El Paso Hydrocarbons Co. from natural gas to Oil and Gas and a write-down on Oil and Gas properties.

Table 27.5. REVENUES FROM RAILROAD AND RESOURCES, 1969–1987

Year	Railroad	Timber and Lumber[1]	Oil and Gas	Minerals[2]	Land and Real Estate
1969	$907.4M	$31.2M	$8.6M	$2.1M	$10.5M
1970	953.2	31.9	8.3	2.7	10.9
1971	1.029B	36.9	9.2	4.2	18.4
1972	1.098	46.5	8.9	6.1	19.6
1973	1.223	69.3	10.8	5.6	15.2
1974	1.375	68.3	18.9	4.3	19.1
1975	1.408	61.2	20.8	8.1	20.9
1976	1.642	85.5	30.4	7.4	24.9
1977	1.802	103.8	26.9	7.1	30.9
1978	2.110	135.9	35.5	6.3	34.8
1979	2.635	185.3	86.3	9.7	35.7
1980	3.254	190.7	126.5	13.0	35.6
1981	4.088	172.9	178.0	11.8	41.2
1982	3.773	183.3	107.1		
1983	4.098	261.7	108.4		
1984	4.490	273.9	406.7		
1985	4.049	258.4	889.0		
1986	3.741	285.8	721.8		
1987	4.038	299.7	636.7		

1 Also designated "Forest Products" in some years.

2 Includes coal, taconite, and other minerals.

Note: revenue not reported separately for minerals, land, and real estate after 1981.

Table 27.6. FINANCIAL PERFORMANCE OF NON-RAIL SUBSIDIARIES PRE-TAX INCOME, 1976–1980 (MILLIONS OF DOLLARS)

	1967	1977	1978	1979	1980
Oil and gas	14.6	12.7	18.4	22.4	41.3
Forest products manufacturing	8.7	13.1	17.6	16.6	13.8
Forest management	20.4	22.1	41.0	41.4	32.9
Air freight forwarder	3.5	4.6	8.8	19.4	22.9
Trucking	1.2	2.6	1.8	0	(3.6)
Coal and minerals	5.8	5.7	4.9	7.8	—¹
Land and real estate	20.2	26.0	27.5	25.9	—¹

( ): deficit

1 Nor specified in annual report for 1980

Source: Burlington Northern Annual Report for 1980

Author's note: In this first of two books on BN history, I have given no attention to BN's non-rail subsidiaries in air-freight forwarding and trucking except for providing pre-tax income as shown above. My second book on BN history, *Transformation of a Railroad Company*, contains insights on what led to the spin-off or sale of all of BN's non-rail holdings.

VALUE OF INCOME PRODUCED BY RESOURCES DIVISION

In BN's first decade, the earnings from its resource holdings increased substantially. By 1978 the net income produced from the non-rail assets was 50.5 percent of the total earnings of BNI, although the revenues from resources were only 17 percent of the total. From this comparison came outcries from some shareholders and financial analysts that the resources side was "carrying" the railroad company, and that more capital should be invested in the development of the resources than in the railroad. As has been mentioned several times in this book, railroad earnings were low in the 1970s due to the tremendous amount of track-upgrading projects we had underway in which all or a substantial part of the cost of the work had to be charged to operating expense rather than capitalized, under the ICC accounting rules in effect at that time. It was projected that once the preponderance of that type of work would be completed, anticipated to be in 1980, the railroad's earnings would increase substantially.

CHAPTER 28

ESTABLISHING SEPARATE RESOURCES AND TRANSPORTATION DIVISIONS

I N 1973 MENK AND THE BOARD DECIDED TO set up two divisions: one to manage the transportation holdings (the railroad company, truck line, and BNAFI) and the other to manage BN's non-rail holdings. A president was designated to head each division: Norman Lorentzsen for Transportation and C. Robert Binger for Resources. Lorentzsen and Binger reported to Menk, CEO of BNI. This reorganization clearly gave prominence and equal footing to each of the company's two main activities. Especially in the making of decisions for capital investment, it was felt the resource projects would be in a stronger position when "competing" for the approval in the budget for capital expenditures.

The tables on pages 170–75 summarize the progress made by the Resources Division in increasing its earnings by exploiting opportunities to grow its businesses. However, the reorganization and the improved performance that resulted was not enough to remove the tension and pressure coming from those who thought even better results should be produced.

In addition to its decision to manage its forests for the long term by sticking to its principles, BN invested in facilities for the manufacture of lumber and other wood products.

EXPANSION AND DEVELOPMENT OF OIL AND GAS RESERVES

Starting in 1979, BN took major steps to expand development in its oil and gas unit. In 1980, the first full year of this effort, revenue grew 47 percent and pre-tax income,

85 percent, compared with 1979. For the first time, BN entered into joint venture agreements in the Gulf Coast area of Texas and Louisiana. Capital expenditures for oil and gas activities were expanded from $34.4 million in 1981, compared to only $10.4 million in 1980. These were early indications of the capability H. Leighton Steward brought to BN.

BN opened a small oil refinery at Osage, Wyoming, in 1976. Osage is located sixty-two miles east of Gillette on a main coal corridor serving the Powder River Basin. The new refinery had a capacity of four thousand barrels per day, utilizing crude oil produced by BN. With the scale of operation BN had at that time, the Osage refinery produced about 10 percent of BN's need for diesel fuel. At the same time, BN established a subsidiary, Saxony Corporation, to handle the purchase of crude oil for the refinery and to market by-products. Not long after Bressler came to BN, he and the new team of experts he had brought into run the oil business announced the Osage refinery would be shut down. In their experience, they had learned that refineries with such a low capacity were not economical to operate. It was surprising that two different groups of people who were experts in the oil business at BN would have come to a complete different conclusion within a span of only a few years,

IMPROVED FINANCIAL PERFORMANCE

It was argued by some, both within and from outside the company, that the earnings of the units in the Resources Division were being held back because so much of the capital spending still was being directed to the needs of the railroad. Since the earnings and return on investment being made at that time on the railroad were not yet showing much of a return, they felt that more funds should be allocated to projects in the resource units, where they would produce quicker and higher returns. Even with the much improved financial performance of the railroad starting in 1979, the railroad's continuing need for capital and higher operating expense to maintain its roadway, together with difficulty in setting the rates on coal transportation at a satisfactory level, continued to make the railroad a poor place to invest capital. On the resources side, some felt our management still was not being aggressive enough in seizing on growth opportunities, and was much too conservative in planning developmental activity for the future.

It is interesting to note that while revenue in the Resources Division increased 143 percent in 1979, pretax income increased by only 17 percent in that year. As was the case on the railroad, operating income would not increase immediately (i.e., in the same year) as capital investments projects and activities chargeable to operating expense were undertaken. However, it was anticipated that within a few years, the increased investments would begin to pay off handsomely. Starting in 1980, the investments made in capital and hard work and the commitments made in BN's first decade began to pay off in improved financial performance in both the railroad and the Resources Division. There were many challenges yet to be overcome, but it appeared the right decisions had been made at to the course the company should take. The leadership teams in place in both the Resources and Transportation divisions appeared to have the talent and strength to move BN forward.

PART VI

GROWTH
AND DEVELOPMENT IN
THE POWDER RIVER BASIN

CHAPTER 29

LEADERSHIP IN THE YEARS OF RAPID GROWTH

THE MAGNITUDE OF THE TASK

The advance of Burlington Northern toward the longer-range goals we have set for our company has not been deterred by temporary adversities, economic or otherwise.
1977 ANNUAL REPORT, BURLINGTON NORTHERN, INC., PAGE 4

What he [Norman M. Lorentzsen] will be faced with as chief executive is a railroad whose operating revenues are the industry's second best, but whose net income ranks eighth.
BURLINGTON NORTHERN NEWS, JANUARY 1978, PAGE 11

Because of projected increases in demand for rail transportation in the late '70s capital expenditures are likely to increase, requiring large amounts of new capital.
BURLINGTON NORTHERN ANNUAL REPORT FOR 1973, PAGE 5

Burlington Northern's maintenance policy is to maintain its physical plant, including track structure, at a level which will allow it to carry out its obligations as a common carrier and to upgrade and modernize its plant and facilities to enable it to meet competition and provide for future traffic.
BURLINGTON NORTHERN ANNUAL REPORT FOR 1975, PAGE 5

STARTING ON DAY ONE, NORMAN WAS expected to do everything necessary to put the new operating plan for the merged company in place. He had to demonstrate leadership of a team of operating and maintenance officers and over 50,000 employees, with most of them in the Operating Department. There were new standards, policies, and priorities that had to be put into place without delay. More than anyone else, Norman was looked to by the company and its customers to maintain reliability and consistency in service as the merged company put its new operating

plan into effect. Also, there were numerous commitments made to communities, unions, customers, and competing or neighboring railways that were conditions imposed by the ICC in approving the merger. Delivery on those commitments was mandatory, right from the start.

Managing a railroad under these conditions of growth required an entirely new set of skills than those railroad executives had needed for a long time. Since World War I, the mandate under which railroad officers had operated was to have their companies survive by cutting expenses, shrinking the physical plant, and shedding marginal and unprofitable lines and market segments. Railroads had to deal with an ever-shrinking share in the transportation market and ever-increasing regulation of their pricing and operations. Due to heavy regulation, they were not allowed to compete in the marketplace and could not function as a business must do in a free-enterprise system in a market-driven economy. As a result, railroad managers were trained to a great extent on how to comply with regulations, rules, and procedures. For the most part, the investments they made were limited to maintenance of plant and equipment to a desired level of utility, and in labor-saving equipment and facilities. There was limited opportunity and need to invest in projects that would generate new business or for growth in volume. But starting in 1973, that was no longer to be the case at BN.

In the seven years in which Norman headed BN and its railroad company, BN became an industry leader in many respects, not only in size but because of the rapid growth of its revenue. This high rate of growth brought attention to BN from the investment community. BN was seen by investors and financial analysts to have far more potential than other railroad companies of that time. BN had opportunities and challenges unlike any railroad faced since the days of westward expansion.

We entered a new era of market opportunity that would require very extensive upgrading, modernization, and expansion of capacity on a large part of the network as well as the car and locomotive fleets and maintenance facilities. The need to finance these major projects came not long after the bankruptcy of the Penn Central, which had made the investment community skeptical of raising capital for use on any railway company. The arguments in favor of deregulation had been raised, but the possibility for such legislation to be enacted was seen as unlikely to occur for a long way into the future, at best.

Tremendous risks and challenges were undertaken by BN in those years. It is no exaggeration to say Menk, Downing, and Lorentzsen "bet the company" when the decision was made to seek financing for in investment of $2 billion to get the railway in shape to handle the amount of coal that would be mined in the Powder River Basin and to move the electric power plants at many locations in the midwestern and southwestern parts of the country.

MEETING THE CHALLENGES
OF THE MID-1970S

With their many years of experience and responsibility for maintenance and operations at all levels of the Operating Department, both Bob Downing and Norman Lorentzsen were well qualified to judge the amount and types of capital investment and maintenance work needed to handle the immense task of upgrading the railroad and undertaking the projects needed to increase the capacity to handle the anticipated growth. If all of that coal had been planned to move on a single corridor of the railroad (for example, from the Powder River Basin to Chicago), the amount of track to be upgraded would have been limited to about a thousand miles. However, that was not the case at all. Coal would be moved on virtually every main line segment of the former Burlington, plus its subsidiaries, the Colorado and Southern and the Fort Worth and Denver Railway, to electric power plants in Texas. In addition, there were some branch lines that would have to be upgraded to main line standards to handle even the first unit coal train to be run over them. The map in figure 32.1 shows how extensive the "coal network" would become.

Even the main line segments of the former Burlington had to be upgraded. By the late 1960s and into the mid-1970s, the tie and ballast condition had become marginal on every line, and continuous welded rail had been laid on only a few miles. None of the lines were ballasted with high-quality rock such as granite or trap rock. Many lines had been maintained only to a standard needed for a secondary main line, a standard that was satisfactory with the light density of traffic handled, but inadequate for large numbers of heavy unit trains of coal and grain, with 100-ton capacity cars and heavy locomotive units.

The first unit coal train movement started in the late 1960s from Colstrip, a mine in southeastern Montana, to a power plant in Cohasset, Minnesota. This train was operated jointly by the NP and GN (on the NP from the mine to Fargo and then on the GN from Fargo to Cohasset). Because it was viewed as incremental business, the rate for hauling coal over this route was not priced at a level high enough to provide for upgrading the track. Since the original rate was so low, it was used by other customers as a benchmark as new coal movements started on many other routes in the 1970s. The low rates established in the early years of the "coal boom" were often referred to as "missionary rates," introduced to draw new business to the railroad. The higher rates set by BN on new coal movements were strongly challenged before the ICC and in the courts by electric-power companies.

Unless and until rates of an adequate level could be established, BN could not make a return high enough to justify the large investment it had to make for upgrading and expansion of capacity. Unfortunately, BN found itself "at war" with many of its coal customers through the 1970s and well into the 1980s. This vital issue consumed a great amount of time for Menk, Downing, Lorentzsen, and Grayson. To improve the return on investment expected by the BN board and the investment community, and for BN to be able to obtain financing for other projects in the future, this was a problem that had to be solved. It was ironic that the electric-power companies tried so hard to keep BN from making an adequate return on its investment, since the rates the power companies charge their customers were set by government regulatory agencies at a level that would guarantee them a prescribed rate of return, and high enough to attract the outside capital needed to finance the construction of power plants.

STAYING THE COURSE

By the late 1970s, Menk and Lorentzsen were under pressure from some board members and the financial community to scale back the amount of money being put into upgrading the railroad. Some directors thought these funds should instead be invested in BN's Resources Division, in line with their belief that the rate of return would be higher than was being achieved in the railroad. Some investors wanted the period of time extended over which investment was made in track, equipment and support facilities, so earnings and dividends could be increased.

BN's senior executives and the majority of its board members stuck to the plan that had been laid out in 1974 to invest the $2 billion necessary to upgrade the railroad and expand the capacity needed to handle coal that was being mined, plus the additional coal from new mines under development. Unless and until those large projects were completed, we would not be able to meet delivery schedules, reduce congestion, and overcome the cost and disruption of derailments caused by poor

track conditions. Had they yielded to that kind of pressure, work that was badly needed would not have been completed, and the large, unprecedented opportunity BN had for growth would have been missed, or at least delayed. By 1980 the heaviest part of the upgrading was completed. Year-to-year earnings started to improve in the late 1970s. It should be noted that before 1982, ICC accounting standards prescribed that virtually all of the expense for replacing ties, rail, and ballast and heavy repairs on cars and locomotives be handled under the "betterment" system of accounting. Under that system, these costs had to be charged to operating expense, and not capitalized. As a result, the heavy upgrading undertaken in the 1970s impacted operating expense much more than if we had been under depreciation accounting, the more commonly accepted system of accounting that was mandated by the ICC starting in 1982.

It was commendable that BN's leadership was able to withstand the pressure to cut back on the amount being invested in the railroad and thereby increase earnings in the short term, and free up cash for higher dividend payments of investment in non-rail businesses. The returns made on the investments in the railroad in later years indeed validated the decisions made in the 1970s. It was rewarding to hear Bob Downing's comment in a meeting of the Lexington Group in 2010: "I haven't run into anyone in the last twenty years who even suggests that we should have done nothing."

To conserve cash during a severe recession in 1975, BN omitted its dividend in the third and fourth quarters. Dividend payments were resumed in the first quarter of 1976. The base pay of the upper tier of executives was cut 10 percent in the last half of 1975. These were strong signals to all concerned, including employees, customers, and investors that BN had given priority to continuing as much of its track upgrading programs as possible.

It is interesting to speculate on what may have happened if BN had been unwilling or unable to make the necessary investments. Without the strong commitment of the BN board and the resourcefulness and courage of its Finance, Operating, Mechanical, Engineering, and Employee Relations teams, BN may not have been able to capitalize on the opportunity it had. However, that would not have been the end of the story. Somehow, delivery of the low-sulfur coal had to be made to the electric-power industry, due to the mandates it faced under the Clean Air Act. Other organizations with the financial strength might have moved in to take ownership of the rail routes involved. That could have been a consortium of power companies or other railroad companies. The Chicago and North Western Railway (C&NW) had a branch line running close to the mines that would soon be going into production, putting it in good position to build a new line to reach those mines. However, the C&NW did not have the financial strength to attract the capital needed for an investment of the magnitude required to build such a line. Union Pacific, of course, would have had the resources needed to acquire the BN lines used to reach the area to be mined and probably would have been glad to be given that opportunity.

In the annual report for 1979, Menk and Lorentzsen provided strong indications of progress in their letter to the shareholders:

> Our policy in these ten years (1970–79) has been to build and develop our assets to improve Burlington Northern's growth in the future and to enhance the value of its owners' investment. As we enter the '80s we have every indication we are making progress.... In ten years, Burlington Northern's consolidated net income has grown at an average annual rate of 22 percent.... Of all our assets, the railroad offers the best opportunity for growth.[1]

1   Burlington Northern 1979 Annual Report, page 3.

The maintenance-of-equipment ratio also would be going down to the level of the best industry performers, with the new maintenance methods being adopted, plus our having a more modern locomotive fleet that would require less maintenance. With improvements in the roadway and mechanical functions, we could also expect less expense from derailments and cost savings through fuel economy. With a better operation, fewer cars and locomotives would have to be in service to make the agreed upon cycle times for unit train equipment. Fewer train-operating employees would be needed, thereby reducing transportation expenses as well.

Through these early years, those of us in field operations knew that we were accomplishing a great deal and the investors would be rewarded with higher earnings—that the price of the stock was likely to increase, and in time, dividends could be increased. We felt fortunate that our board and management were able to prevail over the demands of some directors, shareholders, and members of the financial community that we reduce the amount of track work and thereby realize a higher level of operating income in the short term. However, we understood from statements in the annual report for 1979 that spending on maintenance work would have to be decreased from the levels of the past several years, with so many of the major upgrading programs completed. The time to "reap the harvest" from these large investments was at hand.

CHAPTER 30

THE CLEAN AIR ACT

• its passage set the stage for BN to move vast tonnages of new business •

I T ALL BEGAN WITH THE PASSAGE OF THE AIR Pollution Control Act of 1955 to provide funds for federal research in air pollution. The next step was passage of the Clean Air Act of 1963, authorizing research into technologies for monitoring and controlling air pollution. In 1967 the Air Quality Act was passed, which provides enforcement proceedings and for stationery source inspections.

The real take-off point came with the Clean Air Act of 1970, which contained amendments authorizing the development of regulations and standards to limit emissions from industrial and mobile sources. Limits were set on the emission of six "criteria pollutants": sulfur dioxide, nitrogen oxide, carbon monoxide, lead, ground-level ozone, and particulate matter.

With the passage of the law, electric-power companies that relied heavily on coal had to switch to a type of coal or other energy sources that would reduce the level of emission of those pollutants to meet the new standards. Power companies throughout the Midwest, Texas, and Oklahoma turned to the low-sulfur coal available in the Powder River Basin. The demand to gear up to mine and transport it by rail came quickly and put both an unforeseen demand and new opportunity on BN.

The first movements were for power plants in the Chicago area operated by Commonwealth Edison. Unit trains for plants in Joliet were delivered to the Elgin, Joliet and Eastern Railroad at Eola and to the Chicago and Illinois Railroad in Peoria. The C&IM moved the trains to a rail-to-water transfer facility on the Illinois River at Havana. Barges then moved the coal to power plants located on waterways closer to Chicago. From

eight to ten trains per week moved to Eola and Peoria at the outset. To reduce the financial burden on BN to gear up, Commonwealth Edison acquired the cars through the C&IM, a wholly owned subsidiary. BN provided the locomotives.

Much of the coal moved from mines in eastern Montana to the Twin Cities and then "down the river" to Savanna. The ninety-eight-mile line between Savanna and Galesburg was not in the condition required to handle even the "regular" base of business (usually four merchandise trains per day plus local service) moving on that line. Adding several 100-car unit coal trains of about 13,000 gross tons every week took its toll on that line. We had all-too-frequent derailments, which of course affected the turnaround time on the equipment and cost BN dearly. The railroad had to be rebuilt from the ground up, with all heavier, continuous welded rail, except for twenty-one miles that had been relayed with second-hand welded 112-pound rail in the mid-1960s. The remainder of the rail was 100-pound and 110-pound jointed rail laid second-hand as far back as the 1920s. Ties had to be replaced at a rate of 700 to 1,200 per mile and crushed granite rock had to be moved in from St. Cloud to form a satisfactory ballast section.

The problem BN had on the Savanna–Galesburg line was just the start. The same conditions soon began to show up on main lines as well. The work started on that line was the beginning of a massive program to upgrade the track on nearly all of the main-line subdivisions of the former Burlington railroad.

In the next ten years, numerous power companies began to source their fuel from mines in the Powder

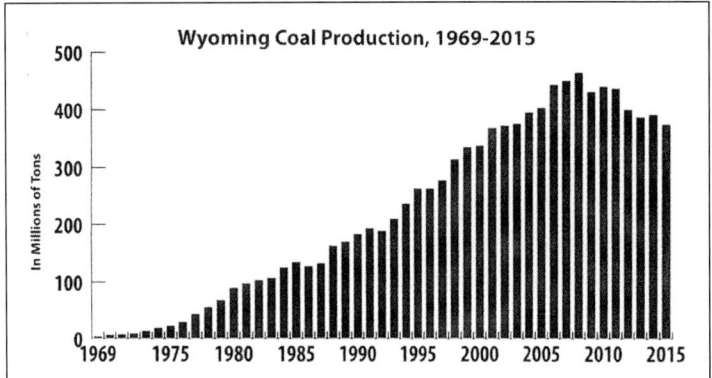

Wyoming Coal Production, 1969-2015

Figure 30.1. The tonnages shown above do not include mines of the Powder River Basin that are located in Montana. The annual tonnage from those mines ranged from about 40 million tons to 45 million tons in the peak years of production in both Wyoming and Montana. WYOMING MINING ASSOCIATION, WYOMINGMINING.ORG, UNITED STATES GEOLOGICAL SURVEY (USGS) DATA.

River Basin. The demand for service put a severe strain on BN to "catch up" on its track maintenance and at the same time build more capacity into its railroad. BN played a very significant role in support of the effort of the electric-power industry to comply with the standards of the Clean Air Act of 1970. The Institute for Energy Research reported that sulfur-dioxide emissions from coal-fired plants throughout the United States were reduced by about 40 percent between 1970 and 2006. Between 1980 and 2006, nitrous-oxide emissions were reduced about 50 percent. It was the start of a successful effort to reduce air pollution.

However, passage of the 1970 act was only the start of federal requirement for the power companies to reduce pollution of the atmosphere. Amendments were passed in 1990 to deal with the problems of acid rain, ozone depletion, and toxic air pollution. Further reductions in sulfur dioxide and nitrous oxide also were required. Achieving compliance required electric utilities to install equipment such as scrubbers to reduce emissions or to purchase permits (cap-and-trade) from other utilities that already were operating at levels below the maximum emissions allowed. These changes brought an even higher level of demand for coal mined in the Powder River Basin, and the necessity for BN and Union Pacific to further expand their capacity. The market and need for Powder River coal expanded well beyond plants in the Midwest and southwestern areas, into areas served by power companies operating in the southern and eastern states.

The favorable thirty-year trend of increases in the use of low-sulfur coal leveled off and began to decline with the new and even more stringent regulations proposed in recent years. Those changes provide new limits on the emissions of mercury, and further reductions in sulfur dioxide and nitrous oxide in twenty-eight states. On top of this came a great expansion in the availability of natural gas, at very low prices. Together, these factors have caused a reduction in the amount of coal consumed in the generation of electric power. Also, some utilities shelved plans for building additional coal-fired plants needed to meet the ever-increasing demand for power. The tonnage of coal moved from mines in Wyoming declined 14 percent from the peak reached in 2008.

However, at the time of this writing in 2014, demand has picked up to the extent that a record number of sets in coal-hauling equipment are in service on BNSF. As the price of natural gas increased, some power plants have again found it economical to burn more coal. Also, as the level of industrial activity has recovered from the recession of 2008–10, the demand for electric power has rebounded. Some of the loss of domestic consumption of coal has been offset by increased exports to China and India as well. However, the recent recovery in coal shipments may be short lived, as new restrictions on carbon emissions take effect. There is some hope that new demonstration projects to capture, store, and possibly find uses for large quantities of carbon dioxide will bring coal into compliance with the new regulations. The technology being used in a few test projects is not yet a commercially viable investment for the power companies. It may be some time yet before the pilot or demonstration projects have provided enough experience in the capture of carbon dioxide for the power companies to decide to burn coal in preference to natural gas for the long term.

CHAPTER 31

MAJOR CAPITAL PROJECTS IN THE "COAL PLAN"

• the magnitude of the task: $2 billion to upgrade hundreds of miles
of track, expand line capacity, and build additional equipment repair facilities •

THROUGH 1977, BN SPENT A COMBINED total of about $2 billion on its fixed plant. The capital budget for 1977–81 called for capital expenditures of an additional $2 billion. At the time the new plan was prepared, BN expected about half would be financed by funds generated internally and half from such outside sources as equipment trusts and leases, additional long-term debt and equity, and some through revolving-credit agreements and term loans.[1]

To say the least, these amounts for capital expenditures were ambitious. Raising this much money for investment on a railroad was made possible through the support of BN's board members and the confidence the lending institutions had in BN's ability to get a satisfactory rate of return from the investments it was making to handle the rapidly growing volumes of coal and other commodities.

As an industry, railroads must make capital expenditures at a much higher percent of revenue than other types of business. This is because of the large amount of capital that railroads require to maintain their physical plant at the level of utility that will keep them competitive, as well as for obvious reasons of safety. In BN's case, far more capital than what was needed for normalized maintenance was required in the 1970s and into the '80s to improve the quality of its plant and equipment and to expand its capacity.

It took courage on the part of senior management and the board of directors to authorize investments of this magnitude in the 1970s, when the company's return on invested capital and the record of earnings were not impressive. The railroad industry had fallen out of favor with many in the investment community because of the Penn Central bankruptcy and the lack of much promise for growth virtually anywhere in the North American rail industry. All of these factors were a constraint to BN's ability to make the level of the investment needed to take full advantage of the potential its leaders believed it had for growth and improved operating performance.

It was incumbent on us to apply our abilities and the authority we had to spend the money wisely, stick to the plan, and get the work done as scheduled to ensure the coal could be moved at the standard agreed upon with the electric-power companies. It was a unique and unprecedented opportunity for us to rebuild a large part of the railroad and build a company that would have good earning power for years to come.

It became apparent to Bob Downing in 1974 that BN had to move aggressively in preparing to handle the tonnages of coal that were forecasted at that time. He directed Ivan Ethington to undertake that effort about mid-year when the capital budget for 1975 was being prepared. This ambitious task was handed to the one of the industrial engineers, J. A. (Jerry) Pinkepank, who had as broad (and also, more detailed) knowledge and understanding of engineering, mechanical, and operation functions as anyone in BN at that time. In two weeks,

---

1   Gus Welty, "For Burlington Northern, It's Coal and a Whole Lot More," reprint from *Railway Age*, December 26, 1977, page 3.

Jerry produced a broad plan for the investment of $2 billion over the next five years that would provide the capacity required and also allowed for the massive upgrading that would be required on a large part of the BN system.[2]

That work became the "coal plan" for the company. Over the next few years, the plan was enhanced and refined by engineering and operating personnel as specific projects were undertaken, but in looking back at those times and the plan Jerry created, it is safe to say there were few, if any, cases of either gross under investment or excessive and unnecessary investment made in plant or equipment. The plan specified the resources needed to build the coal routes up to the standard needed for safety and efficiency in handling high tonnages. It also provided an estimate of the number of locomotives and cars needed both to handle the tonnage forecasted and to operate unit trains at the cycle times specified in the plan for coal service.

While considerable work was already underway to upgrade particularly troublesome areas of the railroad, the track, bridge, and signal programs were greatly expanded starting in 1975. In addition to upgrading the property and equipment, we invested more money in technology that would improve the quality of work performed and reduce operating and maintenance costs. We also invested more in the purchase and rebuilding of freight cars needed to better handle commodities other than coal.

A list of the major investments that BN was able to make in its early years is impressive and worth listing in this book. This chapter contains a list and some explanation of the capital investments made in those years. The funds for BN's capital expenditures came from capital generated through earnings or from funds raised from private institutions. Funding was obtained from a government entity only for replacement of a bridge over a navigable waterway at Beardstown, Illinois. Some explanation is provided in cases where a shipper invested its capital to assist BN. Other than those few cases, all projects completed in those years were financed with private capital raised by BN or with cash generated from earnings.

This chapter contains a list and some discussion of many of the major projects in the "coal plan." In addition to the new lines constructed, this chapter covers major line changes built to reduce curvature and thereby decrease

rail wear and other ongoing expense for maintaining track to the standards required on a high-tonnage railroad.

Toward the end of this chapter, there is some explanation of major upgrading that had to be carried out on two fairly long branch lines in Nebraska. Both lines had been maintained only to the level needed to handle the small amount of grain traffic generated by local elevators, along with a few carloads of fertilizer, farm equipment, and lumber each year. Both lines had to be rebuilt upward from the subgrade to standards necessary for handling unit trains of coal to power plants not accessible from BN's main lines.[3]

CONSTRUCTION OF NEW LINES

DECKER SPUR   Construction of a new 16.5-mile line to the Decker mine was completed in December 1971. The junction point for the new line is at Arno, twelve miles east of Sheridan. At that time, the coal mined at Decker was shipped to Commonwealth Edison though a rail to water transfer facility at Havana, Illinois. It was routed via Mandan, Minneapolis, Savanna, and Galesburg. Construction of the line was financed and owned by Decker Coal Company, a subsidiary of Pacific Power and Light Company, the shipper. In 1976 BN acquired ownership of the line.

SARPY CREEK LINE   In 1973 a thirty-six-mile line was completed from Sarpy Junction (seventy-eight miles east of Billings) to the Westmoreland Resources Company's Absalouka mine at Kuehn for $11 million. The line was built on an ascending grade of 1.0 percent in the direction of the loaded move and 1.25 percent against the empties.

BELLE AYR LINE, DONKEY CREEK–ORIN LINE   In 1972 the Amax Coal Company notified BN of its plan to open a new mine about fourteen miles south of Gillette. The new rail line to the mine became the first segment of a new 126-mile line that would serve many new mines to

---

2   Fred Frailey, "Powder River Stories," *Trains* (April 2010): 44.

3   The reader is encouraged to also note chapter 18 for description of large projects carried out to increase line capacity, through the addition of hundreds of miles of CTC and additional miles of second main trackage, new and longer sidings used for meeting trains, and new facilities for maintaining the growing fleets of locomotives and coal-carrying cars. Also, refer to chapters 32–37 for details on the magnitude of work carried out under operating expense on seven of BN's divisions to upgrade existing main tracks.

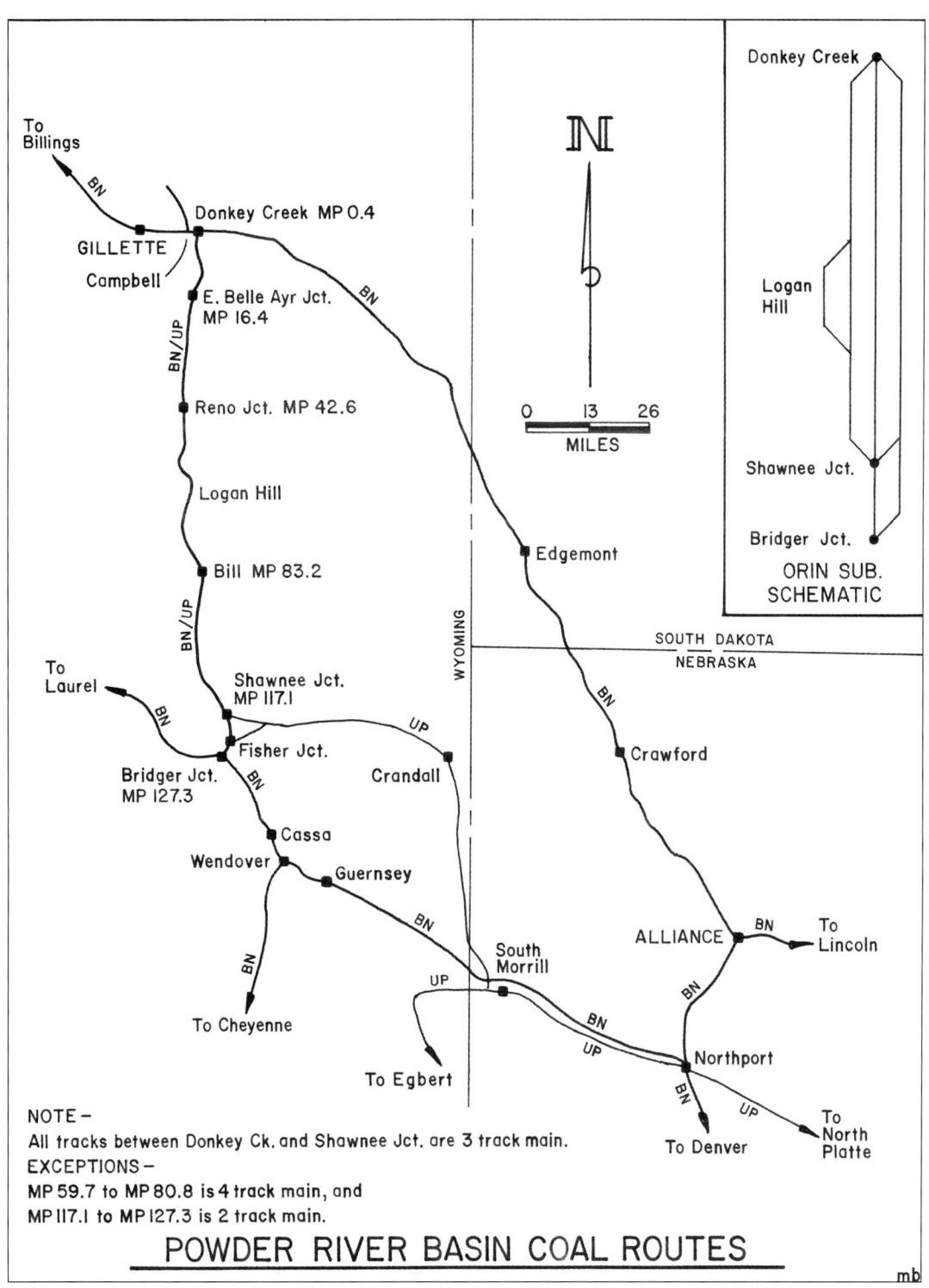

To
Billings

BN

Donkey Creek MP 0.4

GILLETTE

Campbell

BN

E. Belle Ayr Jct.
MP 16.4

BN/UP

Reno Jct. MP 42.6

Logan Hill

BN/UP

Bill MP 83.2

Edgemont

WYOMING

SOUTH DAKOTA
NEBRASKA

BN

To
Laurel

BN

Shawnee Jct.
MP 117.1

UP

Crawford

Fisher Jct.

BN

Crandall

Bridger Jct.
MP 127.3

Cassa

Wendover

Guernsey

BN

BN

BN

South
Morrill

ALLIANCE

BN

To
Lincoln

BN

UP

BN

To Cheyenne

UP

BN

UP

Northport

To Egbert

BN

UP

To Denver

To
North
Platte

N

0   13   26
MILES

Donkey Creek

Logan
Hill

Shawnee Jct.

Bridger Jct.

ORIN SUB.
SCHEMATIC

NOTE –
All tracks between Donkey Ck. and Shawnee Jct. are 3 track main.
EXCEPTIONS –
MP 59.7 to MP 80.8 is 4 track main, and
MP 117.1 to MP 127.3 is 2 track main.

POWDER RIVER BASIN COAL ROUTES

mb

Figure 31.1.

be opened between Gillette Donkey Creek and Bridger Junction (refer to figure 31.1). With the heavy start-up expenses BN had in its first years after merger, including the new yard at Northtown and restructuring its routes in Spokane, we had to turn to Amax to finance construction of the 14.5 miles of new main track between Donkey Creek, and the junction to the 3.5-mile spur track into the mine. This line was completed as far as the junction to the Belle Ayr mine and later extended the entire distance to Bridger Junction, thus forming the entire new line that became known as the Orin line or the Joint Line after the C&NW Railway was able to finance purchase of a 50 percent ownership of the 90.9 miles between Coal Creek Junction and Shawnee Junction.

The new line was built on a rather unfavorable ascending grade of 1.4 percent in the direction of loaded trains and 1.25 percent for empty trains, severe enough to require BN to use helper locomotives on the loaded trains. With the coal company financing the project, there was pressure on BN's engineers to keep the cost of construction down, and not build it to the main line standards BN would have preferred. A more favorable route with lower grades, but seven miles longer, starting four miles east of Donkey Creek that would follow Caballo Creek was rejected by Amax.

Perhaps worse than the adverse grades was the presence of large deposits of bentonite clay, a type of native soil used in the subgrade that does not drain well, causing all too frequent and costly loss of proper track surface and line. Dick Brohaugh, BN's assistant vice president–Engineering, described the problem in a presentation he made to the American Railway Engineering Association: "The soil in this area is probably about the world's worst." Much less bentonite would have been encountered if the line had been built on the alternate route, and the line could have been built on a lower grade. Many thousands of feet of wood and steel piling were driven into the cuts and fills on the new line in an attempt to stabilize the roadbed. Don Rogers, engineer–Track, Denver Region, recalls that for a long time a track patrol had to be sent ahead of every train to look for spots where the track had sunk or gotten out of line. Train speed was limited to 10 miles per hour. Even with all the precautions taken, several derailments occurred. This had an obvious impact on operating costs and at times made it difficult to provide satisfactory service to the Amax mine.

With the plans for more mines to be opened to the south of Belle Ayr, and the need to build a second main track to handle the increased number of trains, the problem of the bentonite had to be overcome. Construction standards for the subgrade material had to be modified for the 113 miles of new railroad yet to be built. Instead of using native soil, compacted granular soils were placed as a layer between the ballast and the subgrade, which would shed water rather than hold it, as happened with bentonite. Brohaugh explained:

> To meet the problem, construction procedures provided that the top three feet of the subgrade, both cuts and fills, be composed of granular material if available or non-expansive material. . . . Other areas requiring special attention were the junctions of cuts and fills where over excavation was performed to a depth of three feet and in areas where soils showed looser density than desired. These unsuitable conditions were corrected by the use of compacted granular soils. . . . As a result of the care in selection and placement of these difficult soils, the subgrade performance has been excellent. No subsidence or settlement has been observed.[4]

Bob Downing felt that at the time the location for the new line to Belle Ayr was established, the BN people who yielded to pressure from Amax probably did not know or appreciate the possibility that lay ahead for extending the line to the additional mines that would be opened in the next few years. The design and construction of the new line was managed by the Billings Region, which at that time extended southeast from Huntley and all the way to Edgemont. Bob mentioned numerous times that he wished he had intervened at the time the plans were developed for the first fourteen miles of what became the new Gillette (Donkey Creek)–Orin line. Bob had not because of the pressure of other business he had as president in those early years and because every effort was being made to keep decision making at the region and division level rather than having the headquarters executives and their staffs reviewing and approving matters that could be handled satisfactorily at a lower level.

The Belle Ayr line was extended in increments to form

---

4   R. G. Brohaugh, AVP–Engineering, Burlington-Northern, "The Gillette–Orin Line, One Year Later," *Proceedings of the American Railway Engineering Association* 82 (March 25, 1981): 440–443.

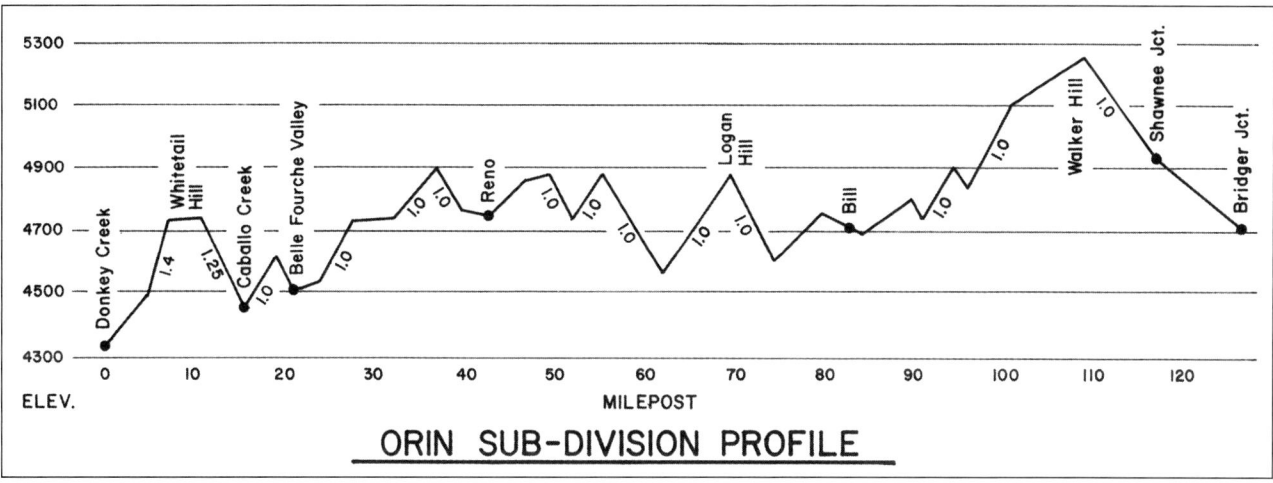

Figure 31.2. By Mike Bartenstein.

a new line of 127.3 miles between Donkey Creek and a connection at Bridger Junction with the Casper–Guernsey line. The new line had to be built generally north-south, which made it impossible to take advantage of the drainage flows, i.e., creeks and coulees, all of which flowed from west to east. As a result, the railroad had to be built with several grades. One percent was established as the maximum grade with curvature no greater than 3 degrees. Note the jagged profile in figure 31.2. With the loaded coal trains powered at a ratio of about one horsepower per ton, they would operate on those grades at a speed of between 10 and 15 miles per hour. With the large number of trains that had to be moved over the line at such slow speeds, and the need to reduce delays in their movement caused by line congestion, it soon became necessary to approve construction of a second main track over the entire distance between Donkey Creek and Shawnee Junction, where trains of the C&NW entered and left the joint line.

Following is a list of specifications and other interesting details involved in the construction and operation of the new Donkey Creek–Orin line:

- It was the longest new line of railroad built since the Great Northern completed its ninety-one-mile line between Klamath Falls and Bieber in 1931 and the Santa Fe built twelve miles of what was to become its new line between La Junta and Amarillo.
- Maximum curvature of 3 degrees.
- Having the new line in service reduced the length

of haul to power plants in Colorado, Oklahoma, and Texas by 155 miles.

- Ties were laid on 21-inch centers, requiring 3,114 ties per mile.
- A green spike, rather than the traditional gold, was driven by Harry Wilkins, director–Wyoming Construction Projects at MP 71 on October 6, 1979, to commemorate completion of the new line. That celebration was limited to representatives of the Engineering Department, the contractor who built the track, and the gang of workers who completed the trackwork. Not to be outdone, BN operating officers (headed by Jake Greeling, assistant superintendent) and employees who operated the many work trains that delivered rail, ties, ballast, and other material to the gangs, set up their own celebration, by having the first loaded train, exactly one month later, on November 6, crash through a commemorative barrier as it completed its loading at the Jacobs Ranch Mine and moved onto the new line. The intervening month from the driving of the last spike until the first revenue train operated was needed to unload all of the ballast needed to bring the track up to standard and to complete surfacing and lining of the new track.[5]

---

5   Refer to chapter 38 in *Transformation of a Railroad Company* for a summary of work done in the 1980s and later years to expand capacity to three and four main tracks between Donkey Creek and Shawnee Junction.

Figure 31.3. Harry Wilkins, director of Wyoming Construction Projects, had the honor of setting and driving the last spike to complete the construction of the new Donkey Creek–Bridger Junction line in October 1979. Harry was preceded by William H. (Bill) Ferryman, who had directed the grading, bridge construction, and signal work from 1977–78 until his promotion to chief engineer of the Denver Region. Another three weeks were required to unload and place ballast and other details to get the track ready to handle the coal trains. Note that even the spike maul and the last spike were painted in stripes of green and white for the occasion. BURLINGTON NORTHERN PHOTO.

Figure 31.4. The first train to operate on the newly completed Donkey Creek–Bridger Junction line was loaded at the Jacob's Ranch mine. The train was staffed by a five-person crew, made up of (from left bottom) Conductor Jim Rich, Brakeman Mike Tollman, Engineer Rick Stein, Fireman S. H. Aured, and Brakeman Kathy Tollman. The Tollmans are husband and wife. With the application of new technology and revisions in the new crew-consist agreement in the 1980s, crews on unit coal trains now consist of a conductor and locomotive engineer only. BN News.

LINE CHANGES

LINE CHANGES ON CRAWFORD HILL   Between Edgemont and Alliance, loaded coal trains encounter a grade of 1.55 percent in ascending the Pine Ridge escarpment, a geological boundary between two sub-regions of the Great Plains, the Missouri Plateau to the north and the High Plains to the south. (An escarpment is a steep slope or long cliff that occurs from faulting and resulting erosion and separates two relatively level areas of differing elevations.)

The ascending grade begins three miles east of the town of Crawford, Nebraska, at MP 423, fifty-seven miles west of Alliance. In addition to the heavy grade, the single-track line contained curvature of as much as 10 degrees and a tunnel of 713 feet, the only tunnel in the entire state of Nebraska. To say the least, Crawford Hill was a formidable barrier to efficiency, once a large number of unit coal trains began moving in the direction of Alliance and beyond. A number of derailments occurred due to train dynamics that may have been caused by the combination of a heavy grade and curvature, or buff action that may have been created by a helper locomotive. Imperfections in the track surface in jointed rail might have contributed further to the problem, together with difficulty in maintaining alignment in curves containing several compounds.

Having loaded coal trains moving up the grade at only about 10 to 15 miles per hour on a single-track railroad limited the line's capacity. Because of the heavy curvature and stress that was being put on the track, it had to be taken out of service for some time almost daily for track inspection and general maintenance work. All of these conditions made it necessary for BN to build a second main track over Crawford Hill, reduce the severity of the curvature, install CTC on both main tracks, and signal for operation in both directions. With the improved operating capability that investment provided, train cycle time was reduced, fewer trains had to be re-crewed due to limits of the twelve-hour law, and overall line capacity for the Gillette–Lincoln corridor was increased. With the track rebuilt, the number of derailments was greatly reduced. Overall, BN's operation became much more efficient, safe, and reliable.

The reduction in curvature made by a series of line changes is impressive, including a change from compound and direct reversing curvature to "simple" curves. In the old alignment within the 5.5 miles of the line changes, there were twenty-nine curves, fifteen of which in excess of 4 degrees, including two curves of 10 degrees. One curve contained nine compounds ranging from 5 degrees to two curves of 10 degrees. In the new alignment, there are nineteen curves, all of less than 4 degrees except for one curve of 8 degrees and one of 7

Figure 31.5. The biggest single impediment to the operation of trains moving east from the Powder River Basin was having a single-track railroad on a grade of 1.55 percent (ascending eastbound in the direction of loaded coal trains) and two curves of 10 degrees. A major line-improvement project of 5.2 miles to reduce track maintenance costs and improve the operation at a cost of $13.6 million was completed in 1982. One of the curves in the "horseshoe" was reduced to 8 degrees and the other curve to 7 degrees, 40 minutes. A second main track was built, and the new alignment was built around the tunnel at Belmont. However, Crawford Hill remained helper territory, as the grade was not reduced. BURLINGTON NORTHERN.

Figure 31.6. A view of the high-capacity, high-density line serving the mines in the Powder River Basin that is carrying the highest tonnage of any main track segment of any railway. CHRIS BURGER.

degrees, 40 minutes. In making the line change, the only railroad tunnel ever built in the state of Nebraska was eliminated. It was found more economical to completely remove the tunnel of 694 feet to allow a second main track to be built, rather than widen the tunnel.

The line change of 5.5 miles was completed for $13.6 million in 1982. The new line bypassed the tunnel, and maximum curvature was reduced to 8 degrees. A second main track of thirty-seven continuous miles was built on either side of the grade (between Marsland and Joder) at that time to provide the capacity needed to move the projected number of trains on that route. In subsequent years, this two-main-track territory was extended to fifty-seven consecutive miles. Unfortunately, the grade of 1.55 percent was not reduced, making it necessary to continue using helper locomotives on the new line.

Jerry Pinkepank wrote that when the overall plan was prepared for expanding capacity throughout BN, the possibility of relocating the line far enough to the west to reduce the grade to one percent was considered:

> The justification for this, in addition to elimination of the helper expense, would have been to avoid construction on the southern end of the Orin Line, for which permitting problems were getting more and more serious. Several things killed this idea, foremost of which was that more and more prospective mines were announced at sites farther and farther south on the proposed Orin Line route, which practically eliminated the idea of not completing that line. Secondarily, however, we felt we would have had at least as much trouble to get permits for tearing up the only scenery in Nebraska as we would for the Orin Line.[6]

To facilitate ongoing track maintenance work, the two main tracks were laid on twenty-five-foot track centers, rather than fourteen feet, which had been the general BN standard up to that time. With the installation of CTC, both tracks were signaled for movement in either direction. To provide more flexibility in the operation, a siding of 10,227 feet was built at the top of the grade at Belmont.

The line change at Crawford Hill cost $13.6 million and included 5.5 miles for construction of the grade for the line change and the new second main track. The maximum depth of the cut was 142 feet. Gross tonnage moved over the hill increased from 30 million in 1976 to 185 million in 2009.

LINE CHANGE WEST OF GILLETTE   In the late 1970s a line change to reduce curvature was made on a grade of one percent ascending eastbound a short distance west of Gillette. E. M. (Mike) Martin, superintendent of the Yellowstone Division, recalls a succession of derailments

6   J. A. Pinkepank, email to various recipients, March 30, 1999.

Figure 31.7. Two trains are moving on the new alignment on Crawford Hill built to reduce curvature and allow construction of a second main track. Making this line change overcame the limitations of the old 713-foot single-track tunnel. The tunnel still can be used as an access road for off-track maintenance vehicles and machinery. TIM ZUKAS.

occurred on empty west-bound trains, even when operating at low speeds. There were twelve closely spaced curves of from 2 degrees to 4 degrees. With old jointed rail and poor ballast conditions, the track needed to be upgraded and improvements made in the alignment.

Nine curves were eliminated. The new line contained only two 1-degree curves and one curve of 2 degrees. This line change was yet another example of BN's commitment to improve the coal corridors to the level needed for a safe, reliable, and efficient operation. In this case, the old alignment and track structure had been satisfactory when only the two daily scheduled freight trains of relatively low tonnage were run on this line, with little or no prospect for much growth in business. However, with the new coal business, the track had to be transformed to a much higher standard. Fortunately, BN had the resolve and resources needed to carry out many types of projects to improve its roadway.

Figure 31.8. "Twin spans" over the Missouri River near Plattsmouth, Nebraska. BNSF PHOTO.

LINE CHANGE AT PLATTSMOUTH   At the west end of the bridge over the Missouri River at Plattsmouth, there was a sharp curve of 12 degrees on a grade of one percent ascending eastbound. Due to the curvature and the problem of wheel lift, a 10-miles-per-hour speed restriction was in effect. Wheel lift and derailments were occurring so often that we had to put car inspectors at the site on a twenty-four-hour basis to watch for wheel lift and communicate with train crews by radio to stop the train when necessary to prevent a derailment. We installed floodlights to facilitate inspections and constructed a small building so the inspectors could get out of the weather between trains. This was, of course, a very expensive operation, to say nothing of the congestion and train delay we experienced due to the slow speed on an increasingly busy piece of railroad.

The solution was to undertake a line change of about 4,000 feet and an eighty-foot cut through the bluff at the west end of the bridge to reduce the curvature and grade. Four oil and gas pipelines had to be relocated and additional property acquired. When this project was completed in November 1976, curvature was reduced to 5 degrees, and trains could operate on the new line at 40

miles per hour. A few months before this writing, BNSF announced plans to do additional grading to provide enough space to build a second main track through the cut and construct a second bridge over the river.

LINE CHANGE IN WENDOVER CANYON   On the line between Guernsey and the connection to the new coal connector line (the Gillette–Orin line) at Bridger Junction, a two-mile line change was made in 1984 to reduce high maintenance costs due to heavy curvature. BN invested $5.2 million to improve the alignment, including the purchase of forty-five acres of land. Four curves of 6 degrees were reduced to 3 degrees, and three curves were eliminated. Operating and maintenance savings were reported to be $1.7 million per year.

At the time this line change was underway, the line was carrying 73 million gross tons per year, and an average of twenty-four trains per day. By 1998, the gross tonnage grew to 100 million per year. Six segments of two main tracks totaling 17.7 miles have been built in this 43-mile subdivision to provide the capacity needed to handle the growing business. Construction of one of

Figure 31.9. Four-track main line in vicinity of Logan Hill, Wyoming, on the Donkey Creek–Orin line. Note wide track centers with excellent surfaces and alignment—such a beautiful sight to a railroader. CHRIS BURGER.

those extensions required a 1,928-foot tunnel to be "day-lighted" by removing 800,000 cubic yards of dirt and rock. It was part of the ongoing transformation of this line from handling generally two trains per day for most of its history to a "big time" operation with capability to handle more than thirty trains per day. To further reduce maintenance costs and improve the quality of the track, concrete ties were laid to replace wood ties.

UPGRADING BRANCH LINES

STERLING–WALLACE LINE   To serve a power plant built by the Nebraska Public Power District (NPPD) near Southerland, BN had to completely rebuild 115 miles of its 229-mile branch line between Sterling, Colorado, and Holdrege, Nebraska. The plant is generally known as the "Gerald Gentleman" generating station. NPPD financed the construction of a new eighteen-mile line between Wallace and the plant. While serving a plant that would consume four million tons of coal per year was an obvi-ous enhancement to BN's "top line" (revenue), having to upgrade a branch line of that length took considerable resources of machinery, ballast cars, track material, and management time away from the urgent need to invest as

heavily and rapidly as possible on its main lines. Heavy sections of secondhand welded rail had to be laid, many train loads of ballast had to be placed, and ties had to be replaced at a high rate.

To help drive down the rate NPPD was paying BN to move the coal, NPPD induced the UP to build into its plant. In 1994 the UP completed an expensive ($16.8 mil-lion, or about $2 million per mile) connecting line of 9.2 miles that included a new bridge over Interstate 80, U.S. Highway 30, and the South Platte River, and through a ninety-foot cut. However, the UP had the advantage of a route eighty miles shorter between the source of the coal and the plant. In 1996 BN sold its entire Sterling–Holdrege line to a short-line operator.

LANCASTER (LINCOLN)–NEBRASKA CITY LINE   In 1974 we learned that BN would be running unit trains of coal to a new power plant being built by the Omaha Public Power District (OPPD) along the Missouri River, a few miles south of Nebraska City. This would require complete rebuilding of the fifty-four-mile branch line between Lancaster and Nebraska City, plus six miles to the south on the Schubert branch. These lines were in generally poor condition, with train speed limited to 10 miles per hour over nearly the entire distance. The

Figure 31.10. Line change at Plattsmouth, Nebraska, to reduce grade and curvature. The original line was built on a 12-degree curve on a short ascending grade of one percent for eastbound trains. Wheel lift was a major problem on some eastbound trains, requiring speed to be reduced to 10 miles per hour. Car inspectors had to be on hand to observe trains for wheel lift and advise the engineer by radio to stop if they observed any wheel lift. A line change of 4,000 feet was built in 1976 with curvature reduced to 5 degrees and the grade to 0.4 percent. Train speed could then be increased to 40 miles per hour. It required making a cut of eighty feet through the bluff at the west end of the bridge over the Missouri River. Opening of the new line eliminated the congestion and train delay that was getting progressively worse as business increased. About thirty-five years later, BNSF built a second bridge over the river and widened the deep cut to allow construction of a second track on the line change. BURLINGTON NORTHERN.

Figure 31.11. Members of the extra gang assigned to connect the track at the west end of the bridge over the Missouri River with the newly constructed line change at Plattsmouth, Nebraska, are putting up a banner to celebrate the first "official" train to run on the new line. The date is November 30, 1976, and the snowflakes falling clearly indicate we were at the close of the maintenance season. BURLINGTON NORTHERN.

Figure 31.12. To increase the capacity on its line between Guernsey and Wendover, BNSF carried out the long-awaited plan to "daylight" a 2,480-foot tunnel (as shown on the Alliance Division track charts revised December 31, 1986). Note that the cut was dug wide enough so the Stokes siding could be extended to form 3.1 miles of second main track. It also provided for a road for maintenance vehicles and off-track machinery. MIKE DANNEMAN.

tie condition was poor, with a mixture of 65-, 75-, and 85-pound rail that would have to be replaced with heavier ribbon rail. Curvature was surprisingly severe for a railroad built across the Nebraska prairie. At the top of a 1.4 percent grade, we had a curve with four compounds of six, nine, and two 4-degree curves.

With the demand BN already had for expediting the upgrading of its main lines, it was in no position to allocate funds to acquire property for a line change to reduce the curvature at that point, or to reduce the grade. Nor could a mechanized tie gang and Mannix sled be sent to the Nebraska City line to install new ties and reballast the line on a production basis. Instead, we were told to use two small non-mechanized sleds and pull them with a locomotive or winch cart. We had one small tamper and tie inserter, but all spiking had to be done by hand. Getting the line rebuilt to main line standards with such methods and equipment was a throwback to the days

before track work was mechanized. The tie and ballast work was completed in time for the line to be relaid with secondhand welded rail in 1978, which met the power company's schedule to begin receiving coal.

OPPD and the UP developed plans to build a connection to enable the UP to serve the plant from its Kansas City–Omaha line that ran through Nebraska City. To insure there would continue to be competitive access to its plant, OPPD purchased BN's line between Lincoln and Nebraska City in 1998. UP could then use a connecting track in Nebraska City to get on the former BN line and save the expense of a "build out" at the plant site. OPPD contracted with a short-line operator to serve the grain elevators and other customers on the line. It was disappointing that the line we had worked on so hard to upgrade was used in coal train service for only about ten years.

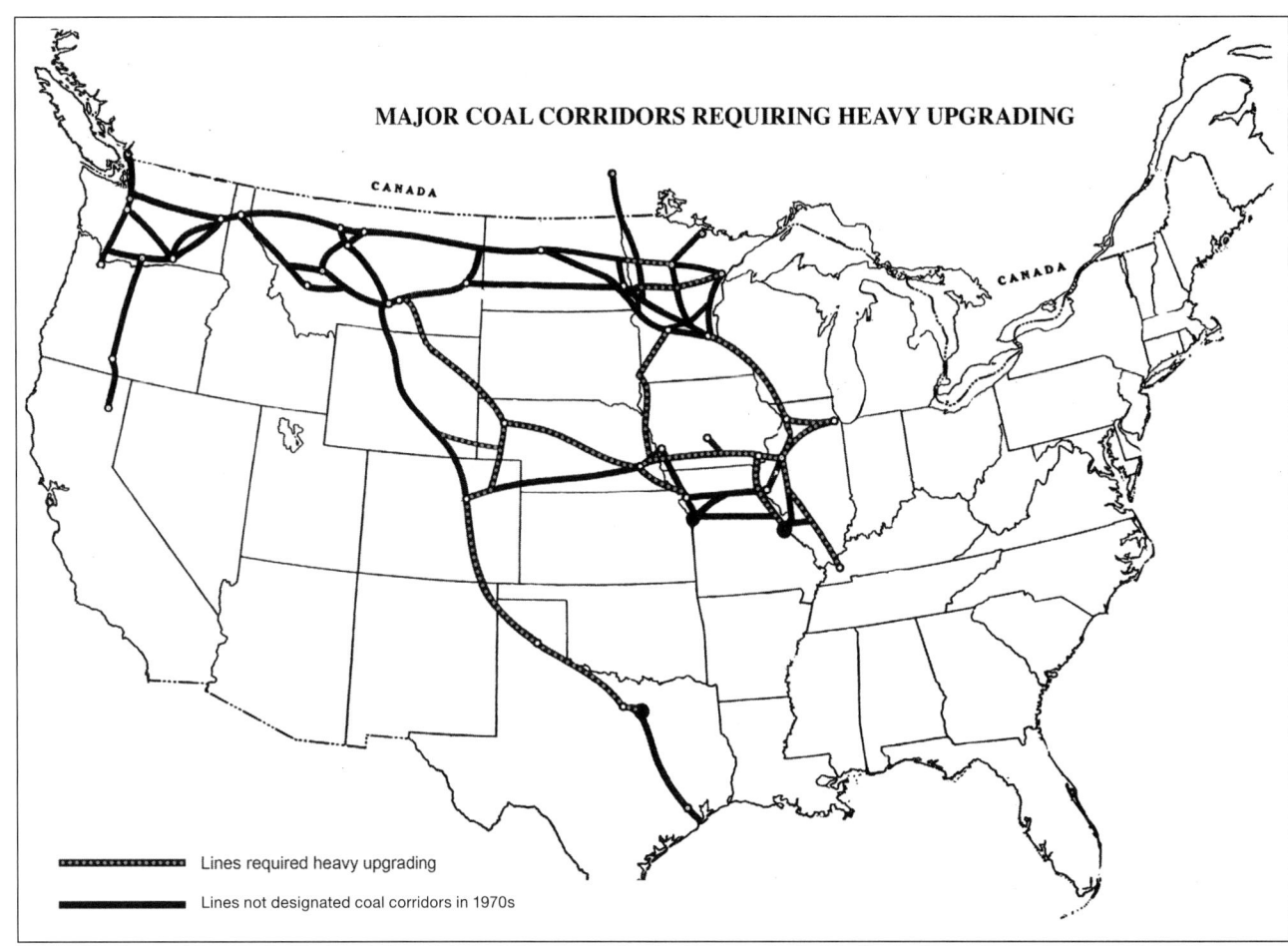

MAJOR COAL CORRIDORS REQUIRING HEAVY UPGRADING

Lines required heavy upgrading

Lines not designated coal corridors in 1970s

Figure 32.1.

CHAPTER 32

IMPROVING TRACK QUALITY
ON THE DIVISIONS

• for officers and employees at the division level, it was thrilling and gratifying to be given
the resources needed to upgrade track to a world-class level for heavy tonnages •
• other than construction of major bridges, yards, line changes, and shop facilities,
the divisions had to gear up to manage rail-, tie-, and ballast-renewal programs, as well as
construction of new or extended sidings and additional miles of second main track •
• divisions were fully accountable for getting the work done
safely, up to standard, and on schedule •

THE DIVISIONS COVERED IN THIS PART ARE those on which heavy trackwork was required to handle the increased tonnage that was projected. The divisions were responsible for completing all roadway projects except for a few large construction projects that were managed by the system Engineering Department's staff: mainly large bridges, construction of new lines, major line changes, and new mechanical facilities. From 1970 and into the mid-1980s, the Alliance, Nebraska, and Colorado divisions reported to the Denver Region; the Yellowstone Division to the Billings Region; and the Chicago, Ottumwa and Hannibal divisions to the Chicago Region.

THE TERRITORY OF DIVISIONS INVOLVED IN HEAVY UPGRADING PROGRAMS IN THE 1970S

ALLIANCE DIVISION   Alliance–Ravenna, Alliance–Sheridan, Donkey Creek–Orin, Alliance–Northport–Guernsey–Casper

NEBRASKA DIVISION   Lincoln–Ravenna, Lincoln–Hastings, Lincoln–Omaha–Oreapolis, Lincoln–Pacific Junction via Louisville, Lincoln–St. Joseph

COLORADO DIVISION   Hastings–Denver, Sterling–Brush, C&S Railway (Wendover–Denver–Texline)

YELLOWSTONE DIVISION   Laurel–Mandan, Huntley–Sheridan

OTTUMWA DIVISION   Galesburg–Pacific Junction, St. Joseph–Kansas City, Terminals at Galesburg and Kansas City

HANNIBAL DIVISION   Galesburg–Kansas City, St. Louis–Burlington, Bushnell–Metropolis

CHICAGO DIVISION   Chicago–Galesburg, Aurora–Savanna–St. Croix Tower, Savanna–Galesburg

The above listing includes all main-line segments managed by each division. Most (but not all) of the lines listed above were in the heavy-track-upgrading programs of the 1970s.

To be able to handle the very large tonnage of coal to be mined in the Powder River Basin and moved to electric-power-company plants at widespread destinations required BN to make a major financial commitment. In 1974 the BN board of directors approved an expenditure of $2 billion to expand the capacity of the railroad, with new track construction, upgrading the train control (signal) system, acquiring additional locomotives and cars, and building new car- and locomotive-maintenance facilities and higher-capacity locomotive-fueling facilities. Also required would be the upgrading of hundreds of miles of track on main lines and some secondary lines to the standard needed to handle unit coal trains with ten thousand tons of coal, loaded in 100-ton capacity cars.

After the board and the Finance Department had done their part—meeting the challenge of raising the money for this work—came the unprecedented challenge of getting the work done in the field. The extensive program for upgrading the track would have to be done under traffic. The volume of coal being mined was increasing rapidly, before the work of expanding capacity and improving the track had even begun. Most of the new coal business was to be moved on lines of the former Burlington. The lines radiating out from the area to be mined had been built and maintained only to the standard needed for handling light tonnages. Heavy upgrading amounting to a complete rebuilding of the track structure had to be done on all of the lines in Wyoming and western Nebraska.

With the very high and immediate demand for coal that developed in the early 1970s, a gradual, incremental program for track maintenance would not be acceptable. Instead, an all-out effort to improve the railroad had to be undertaken on nearly all of the lines of the former Burlington, as well as completed on all of them within a very few years.

UPGRADING THE SKILLS OF
TRACK FOREMEN AND TRACK WORKERS

Getting the heavy track upgrading and capacity expansion programs started required much more than approval by the board. Because there had been only light maintenance done on the lines of Burlington (CB&Q) for several years, the kind of management organization needed to plan and carry out the work had to be formed. There were too few maintenance engineers and supervisors in place. For the most part, they did not have the level of skill or experience needed at the outset. The same was true for the foremen needed to run the gangs set up to do the work. Because little modern track-maintenance machinery had been acquired for some time, there were few people qualified to operate the kinds of technically

advanced machines that would have to be acquired. The local track-maintenance forces had not been trained to perform their work to the standards or quality needed for safe movement of large numbers of heavy coal trains at 40 miles per hour, the authorized speed at that time.

Large numbers of people had to be hired and trained for all types of work. The high level of wages paid by BN made it an attractive employer to residents of the sparsely populated territory. A large number of employees (mainly conductors, engineers, and brakemen, and a number of machinists and electricians) with some amount of experience in railroad work moved to Nebraska and Wyoming from the Midwest, where they had been laid off by railroad companies that had been shut down or were struggling to survive in the changing times of the rail industry. Developing competency in first-line managers to supervise the rapidly growing labor force also was a big challenge. During the time the railroad's infrastructure and its personnel were being upgraded and qualified to handle the work, an ever-increasing number of trains had to be run to deliver the coal at a standard of service acceptable to the electric-power companies.

To overcome these deficiencies required a massive effort to train and upgrade the skills of roadway-maintenance people at all levels of the organization, from the engineering and maintenance officers on the region and division staffs on down to local supervisors, gang foremen, track inspectors, machine operators, and track laborers. A large number of workers had to be hired. Many of the foremen had only limited experience, even as machine operators or laborers. Fortunately, GN and NP had a strong cadre of maintenance officers and supervisors who could be transferred to management positions at the numerous locations on Burlington territory as the heavy work was undertaken.

As might be expected at the start of this effort, the heavy work needed to upgrade the track required considerable interruption to the train operation. The track had to be taken out of service for several hours each day so the work could be done efficiently and kept on schedule. Increasingly frequent derailments caused by the growing weakness in the track structure also disrupted the flow of traffic. It was not always possible to get the work done fast enough to overcome all of the weak spots as fast as they developed.

MEETING THE STANDARDS FOR THE QUALITY OF TRACK WORK

Some of our maintenance supervisors were reluctant to apply the new standards for maintenance and new construction that they had not grown up under. Others made some mistakes along the way. Patience was shown with those who were trying to adapt to changing times. Under our policy for evaluations at that time, we did not make it a case of "one strike and you're out" for the superintendents and their staffs who showed they wanted to "get on the team" and apply new methods and standards to the work for which they were responsible. For those showing promise and delivering results, we helped make sure they were getting the experience and education they needed in all aspects of their work. Those who were the best performers were placed on the company's list of promotable candidates for higher-level positions. As a result, we developed some highly proficient teams on each division. Many of those who were developed under this tutelage moved to higher level positions, including that of vice president, over the next ten years.

We did not want our people to feel their job was something they had to simply endure day after day, rather than an experience in which they could develop and use their skills, and also feel they were making a contribution. As a result, most of our people were glad to be working for BN. Even though the work was often strenuous and taxing, most people gladly put their talents to work, whatever their job responsibilities were at the time. The style of leadership we applied in those years fit in well with that of nearly all of the senior executives BN had at that time,

especially Bob Downing, Norman Lorentzsen, Ivan Ethington, and John Hertog. They were good coaches and good communicators, affirmative in their dealings with people at all levels. All of them gave us the authority, responsibility, and resources we needed to get the work done. Accountability for results also was a part of the process.

At that time in BN's history, there were no politics in the operating organization, and no "in groups" or "out groups." There was just one team, the first team, and everyone was on it, as long as you delivered on your responsibility. All of this brought out the best in people. At the same time, those who wanted only to ride out their time until retirement or otherwise not carry their load had to be moved out of the demanding jobs on the coal corridors. This was a "game" only for those with the energy, commitment, desire, and ability to get the job done under very demanding conditions. Many of us who worked on the division level in those years moved on to great careers and developed self-confidence and professional competency. We were not held back in the use of our talents or ideas for running a good railway operation. We had a life that was hard, but good. As I look back on those years, I find it amazing that so many people (among them neighbors, friends, and former classmates) had written off the rail industry as a place for people to invest their talents and education. Instead of being part of a losing effort, we established ourselves with pride, reward, and satisfaction in working for a successful company, with an even greater future. We felt we were part of something great in those years.

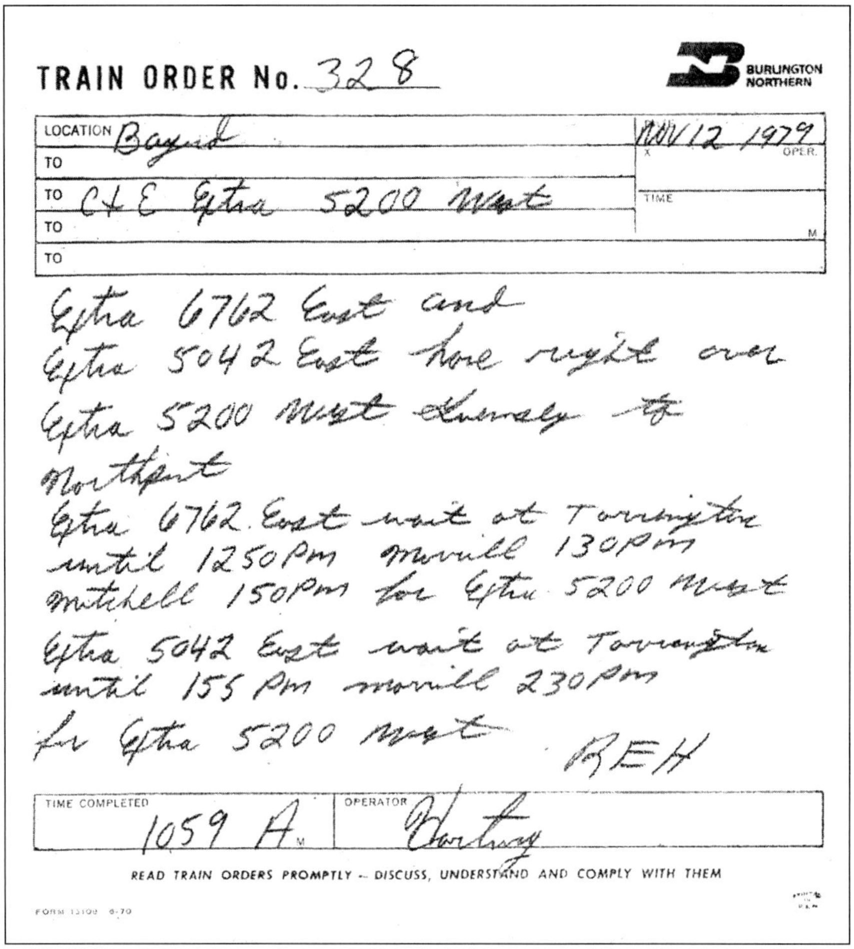

TRAIN ORDER No. 328

BURLINGTON NORTHERN

LOCATION Bayard   NOV 12 1979

TO

TO C&E Extra 5200 West   TIME

TO

TO

Extra 6762 East and
Extra 5042 East have right over
Extra 5200 West Laramie to
Northport
Extra 6762 East wait at Torrington
until 1250 Pm morrill 130 Pm
mitchell 150 Pm for Extra 5200 West
Extra 5042 East wait at Torrington
until 155 Pm morrill 230 Pm
for Extra 5200 West
                              REH

TIME COMPLETED 1059 A M   OPERATOR

READ TRAIN ORDERS PROMPTLY — DISCUSS, UNDERSTAND AND COMPLY WITH THEM

FORM 13108 6-70

Figure 33.1. Author's collection.

CHAPTER 33

THE ALLIANCE DIVISION

• transforming a railroad from a low-tonnage, low-density
operation to lines handling over one hundred million gross tons per year •
• made investment of $716 million[1] from 1972 through 1988
to upgrade and expand capacity •

MAJOR TRACK WORK ON THE ALLIANCE Division started with replacement of 90- and 112-pound jointed rail on 997 miles with new 132-lb. welded rail, replacing 1.6 million ties and upgrading from fouled ballast of inferior quality with heavy crushed granite or other hard rock ballast. Simultaneously, sidings were added or extended, second main track was built, CTC was installed, a new line of 127-miles was built, and a new car and locomotive shop constructed. Hundreds of new people were hired and trained to work in all crafts. While all of this work was underway, increasing numbers of trains had to be kept moving.

A total of 4,311 miles of road on the BN system (including the C&S and FW&D railways) had to be substantially upgraded or rebuilt to handle the high tonnages of coal, ranging from about 25 to 250 million gross tons (MGT) or more. The only major coal corridors not requiring heavy work were the line between Huntley, Montana (13 miles east of Billings), and Minneapolis, the 137 miles between Superior, Wisconsin and Coon Creek Junction (7 miels west of Minneapolis) and 93 miles between Willmar and Minneapolis. In addition, 170 miles of new track construction was required to serve new mines.

The largest part of the investment BN made in upgrading and expanding the capacity of its rail network had to

be made on its Alliance Division. Most of the new mines were located within the division, and the rail lines that would be handling the coal had been built and maintained to handle only a light density of traffic. The track had been maintained at a level satisfactory for a low-tonnage operation, but it was not at the standard needed to handle large numbers of heavy-unit trains. For most of the division's history, only two daily scheduled freight trains (one in each direction) had been operated over the entire 829 miles between Alliance and Laurel, and east of Alliance to Ravenna, 238 miles. Local train service was provided on a seasonal or as-needed basis to handle mainly bentonite clay and grain at scattered locations.

To handle the tonnage forecasted in the early to mid-1970s required that all lines of the Alliance Division be transformed to a very high standard. The only exception was the line of 382 miles between Bridger Junction and Laurel (the line via Casper) was not designated a coal route. Not long before the "coal boom" began, the McCook Division, responsible for the 385 miles of main line between Hastings, Nebraska, and Denver, was consolidated with the Alliance Division.

Some heavy-track-maintenance work was started on the Alliance Division as far back as 1972. With the new projections for the demand for coal made in 1974–75, the maintenance program had been rapidly expanded. Later revisions to the Clean Air Act in 1990 caused even more electric power companies to decide to use Powder River coal, requiring BN to invest even more to expand the

---

1 Includes investment made for track improvements and the installation of CTC on the Gillette–Huntley line of the Yellowstone Division.

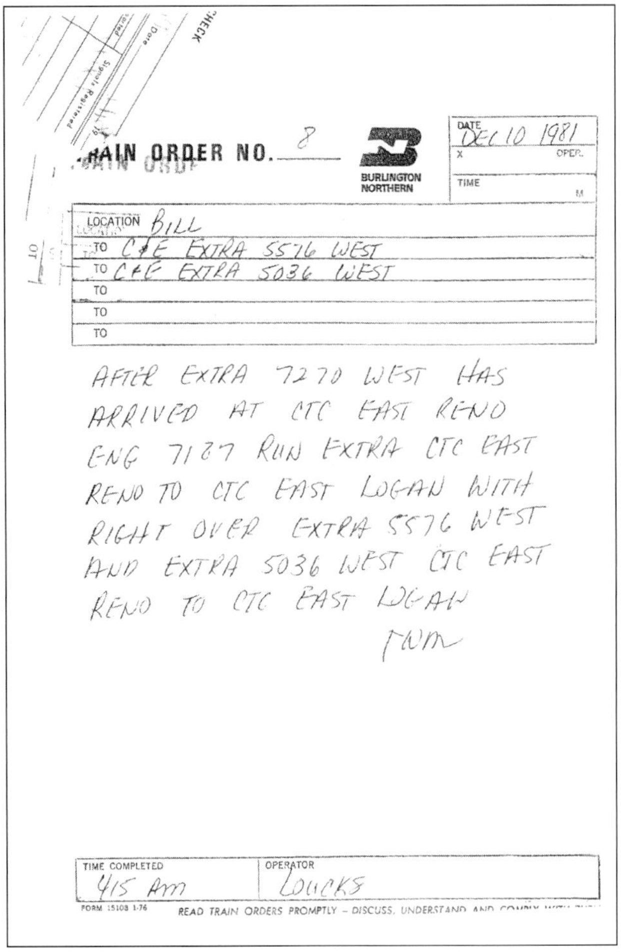
TRAIN ORDER NO. 8

BURLINGTON NORTHERN

DATE DEC 10 1981

LOCATION BILL

TO C&E EXTRA 5576 WEST

TO C&E EXTRA 5036 WEST

AFTER EXTRA 7270 WEST HAS ARRIVED AT CTC EAST RENO ENG 7127 RUN EXTRA CTC EAST RENO TO CTC EAST LOGAN WITH RIGHT OVER EXTRA 5576 WEST AND EXTRA 5036 WEST CTC EAST RENO TO CTC EAST LOGAN

TWM

TIME COMPLETED 415 AM | OPERATOR LOUCKS

FORM 15108 1-76 READ TRAIN ORDERS PROMPTLY – DISCUSS, UNDERSTAND AND COMPLY WITH ...

Figure 33.2. AUTHOR'S COLLECTION.

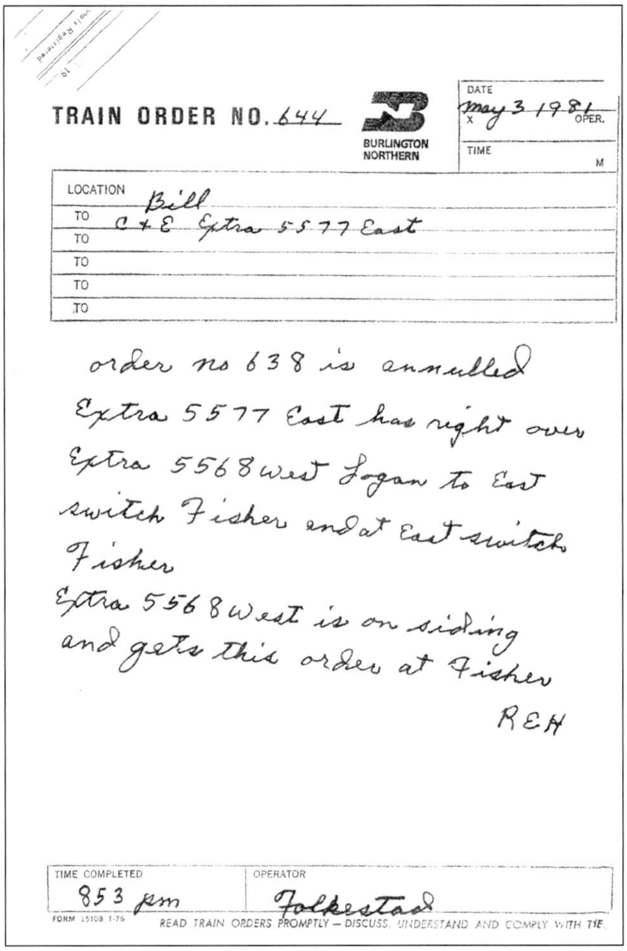
TRAIN ORDER NO. 644

BURLINGTON NORTHERN

DATE May 3 1981

LOCATION Bill

TO C&E Extra 5577 East

order no 638 is annulled

Extra 5577 East has right over Extra 5568 West Logan to East switch Fisher and at East switch Fisher

Extra 5568 West is on siding and gets this order at Fisher

REH

TIME COMPLETED 853 pm | OPERATOR Folkestad

FORM 15108 1-76 READ TRAIN ORDERS PROMPTLY – DISCUSS, UNDERSTAND AND COMPLY WITH THE

Figure 33.3. AUTHOR'S COLLECTION.

capacity of some of its major coal corridors. Construction of additional miles of second main track and extensions of sidings continues at the time of this writing, in 2013. By 1976 BN found that the size of the territory and the rapidly increasing complexity of the Alliance Division was far too extensive for one superintendent to oversee, and so established a new division, the Colorado Division, to include the lines of the former McCook Division (the main line between Hastings, Nebraska, and Denver and several branch lines in western Nebraska), plus the 465 miles owned by the Colorado and Southern Railway. The new division also had responsibility for the Denver terminal.

Even with this reduction in territory, the job of superintendent of the Alliance Division was considered by most BN people who understood operations and maintenance to be the heaviest job on the BN in those years. Norman Lorentzsen put his confidence in Glenn Saylor as the best-qualified officer to serve as superintendent at Alliance in the toughest years, from 1974 into 1979. At that time, the Alliance Division outranked even the Chicago Division for the complexity and demands of the operation. However, it should also be recognized that in those years, the Spokane, Nebraska, and Chicago divisions could have been ranked as close seconds. In 1974 Wayne Arntzen was moved in as assistant vice president–Operations for the Denver Region, which covered the Alliance, Nebraska, and Colorado divisions. Wayne's appointment as the head of operations and maintenance on the region turned out to be one of the best decisions BN made on personnel in those times.

In addition to upgrading the track, line capacity had to be expanded to handle the increasing number of trains

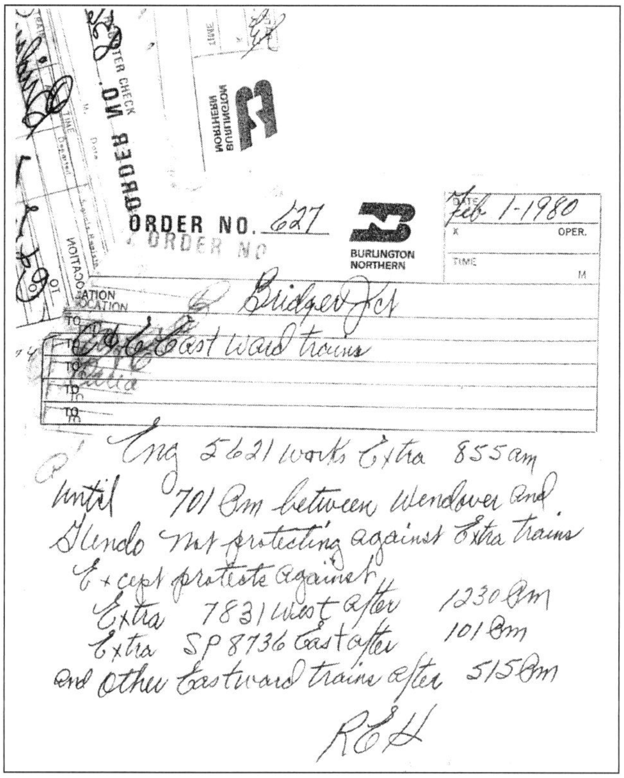

ORDER NO. 627
BURLINGTON NORTHERN
DATE Feb 1-1980
X OPER.
TIME M
STATION Bridger Jct
TO C&E East ward trains

Eng 5621 works Extra 855 am until 701 pm between Wendover and Glendo not protecting against Extra trains Except protects against Extra 7831 West after 1230 pm Extra SP8736 East after 101 pm and other Eastward trains after 515 pm

REH

Figure 33.4. AUTHOR'S COLLECTION.

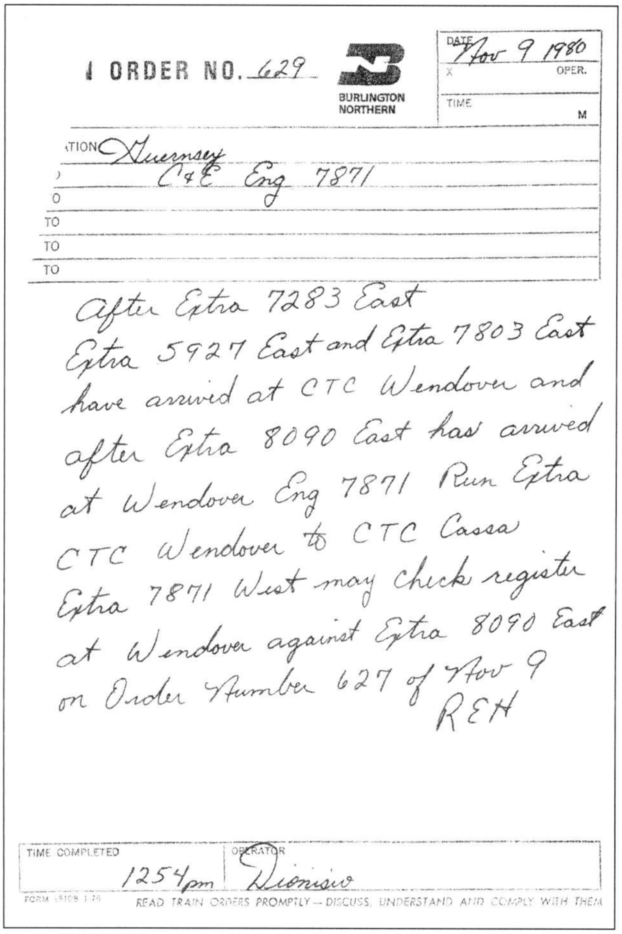

ORDER NO. 629
BURLINGTON NORTHERN
DATE Nov 9 1980
X OPER.
TIME M
STATION Guernsey
C&E Eng 7871
TO
TO
TO

After Extra 7283 East Extra 5927 East and Extra 7803 East have arrived at CTC Wendover and after Extra 8090 East has arrived at Wendover Eng 7871 Run Extra CTC Wendover to CTC Cassa Extra 7871 West may check register at Wendover against Extra 8090 East on Order Number 627 of Nov 9

REH

TIME COMPLETED 1254pm OPERATOR Dionisio

READ TRAIN ORDERS PROMPTLY — DISCUSS, UNDERSTAND AND COMPLY WITH THEM

Figure 33.5. AUTHOR'S COLLECTION.

being run by extending sidings and building additional ones, constructing second main trackage, and installing CTC on all main-line subdivisions. A major line change had to be made to reduce curvature on the 1.55 percent grade over Crawford Hill. The single largest project was the construction of the 127-mile line of railroad between Donkey Creek and Bridger Junction to provide access to the new mines being open, and to reduce the distance for coal trains moving to Colorado and Texas.

A large locomotive- and car-maintenance facility had to be built at Alliance, along with a new office building to house the growing force of dispatchers, crew callers, and the division management staff. These impressive new facilities stood out at a real credit to the community. They were symbols of the kind of company BN had evolved to within only a few years. Employees could take great pride in working for a company that had a strong, positive impact on the community in many respects, including certainly, by creating hundreds of well-paying jobs. A fine spirit of respect and appreciation was built between BN and the community in those years. It has carried forward to this day.

EXAMPLES OF TRAIN ORDERS ISSUED BEFORE INSTALLATION OF CTC ON ALLIANCE DIVISION

It is quite commendable that dispatchers, conductors, and engineers with only limited experience (often with less than a year of service) had been trained well enough to comply and operate safely and efficiently with train orders as complex as the examples given in this chapter.

Until the completion of new installations of CTC, trains on the high-density lines on the Alliance Division and parts of the Nebraska Division had to be dispatched under the historic timetable and train-order system for movement authority. The train orders shown above are examples of the orders issued every day to train crews operating heavy trains of about 14,000 gross tons. Each train consisted of newly acquired cars and locomotives valued at $50 million or more, operated by four-person crews on which even the conductor and engineer may have had less than one full year of experience. Even for "old head" crew members, the orders were complex enough to be challenging to understand and comply with.

Keep in mind also that many of the train dispatchers also were new on the job. The train crews had to bet their lives on each trip that the dispatcher had not overlooked the presence of a train or issued overlapping or conflicting movement authority with no automatic-block-signal system in service as a "fall back" under which a signal would display red or stop as an indication of another train or engine fouling the main track.

On top of issuing such a set of orders to crews when they were going on duty, the dispatcher would have to issue additional orders as the train progressed. Those orders might change the location at which opposing trains were to meet on the single-track railroad, or require the train to run a designated number of minutes later than the times shown on the original set of orders,

thereby requiring the dispatcher to issue additional orders. Quite likely, there would be additional trains they would have to meet along the way. The dispatcher might also have to change the designation as to which train was to "take siding" at the meeting point.

It is amazing the operation ran as well as it did in the 1970s and into the early 1980s, when train movements on high-density lines could finally be made under the rules for CTC.

BN and its investors bet their fortunes on the ability of the managers of the Alliance Division and their counterparts on neighboring divisions, and those who worked on the Chicago Region, to complete all of the capital projects authorized on schedule. It was vital that enough coal be moved to meet the goals for increased revenue and higher net income, plus a higher rate of return on invested capital needed to help drive up the price of BN stock. Major projects carried out on the Alliance Division are covered in greater detail in chapters 31–33, 38, and 42, and in the series of reflections in chapter 41 on the lives and work of several officers and employees who served on the Alliance and Nebraska divisions in the years of rapid growth and transformation into a "big time" operation.

CHAPTER 34

THE NEBRASKA DIVISION

• the hub for five major coal corridors •
• challenge to upgrade 718 miles of main-line track for high-tonnage service •

THE TERRITORY OF BN'S Nebraska Division was named the Lincoln Division until about the middle of 1976. It connected with the Alliance Division at Ravenna, with the Ottumwa Division at Pacific Junction, and with the Colorado Division at Hastings. Lincoln served as a hub for the main lines that ran to Denver, Chicago, Kansas City, and Minneapolis (via Sioux City). Coal loaded on the Alliance Division en route to Chicago and Kansas City moved east on the main line to Lincoln, where the two main coal corridors split.

The operating and maintenance challenges and the extent of the programs to upgrade the roadway on the Nebraska Division in the 1970s were not unlike those that BN faced on the Alliance Division. At the start of the "coal boom," only the 100-mile subdivision between Lincoln and Hastings was in good condition. It had been relaid with new 115-pound welded rail in the early 1970s. But even with that work completed, that line still was not in as good shape as it should have been, since a great deal of the necessary work connected with a rail relay or for a general upgrading of the track was not done at the same time. Senior executives who were knowledgeable on track work were dissatisfied when they noted on an inspection trip that the switches had not been raised, road crossings had not been rebuilt with new ties and ballast, drainage problems had not been corrected, insufficient ballast had been placed on curves, and not all of the track material recovered in the relay had been picked up, even two years after the new rail was laid. This situation gave the Nebraska Division a poor reputation for its

ability to do quality work, at the time the all-out effort to upgrade its main lines was underway.

When I was assigned the job of division superintendent at Lincoln in October 1975, I was told in no uncertain terms that the Nebraska Division forces were not showing they were capable of getting the work done on schedule and to the standard BN expected, even on main lines. As was the case on the Alliance Division and on all of the Chicago Region, completing the vast number of projects needed to upgrade the track and expand its capacity was vital to BN's future. We were constantly being compared to the Union Pacific and the "gold plate" standard one could see on UP's track at the numerous locations in Nebraska where lines of BN and UP either crossed or ran in close proximity to each other.

The challenge to the people of the Nebraska Division was formidable, but it was also seen by them as an opportunity few railroaders could have expected to have in their careers. To have the resources needed to overcome years of substandard maintenance and to move tonnages of a magnitude we had never envisioned was indeed a unique opportunity. In looking back over my career, I believe that my two-year term as superintendent of the Nebraska Division was the most rewarding position I held in my career.

Every main-line subdivision on the Nebraska Division had to be upgraded as required, per the plan for current and future coal movements. In addition, we would have to upgrade two branch lines, one of fifty-four miles between Lancaster (three miles south of Lincoln yard) to a new power plant five miles south of Nebraska City, and

Figure 34.1. Unloading CWR (continuous welded rail) from a train of rail "ribbons" of about one-quarter-mile in length. When the new rail is seated in the tie plates, it will be spiked and rail anchors will be applied, followed by the unloading of additional ballast. A track-surfacing and tamping machine will surface and line the track and raise it on the new ballast. Finally, a ballast regulator will shape the ballast on the shoulder up to the standard required for CWR and move excess ballast out from the center of the track and to the shoulder at the ends of the ties. BURLINGTON NORTHERN.

the line between Aurora and Hastings. By using the latter line, we gained the equivalent of a double-track railroad for the seventy-three miles between Lincoln and Aurora.

The following is a summary of work that had to be done between 1974 and 1980 to bring the track condition and capacity of the division to the standard needed to handle the tonnage projected.

LINCOLN–RAVENNA (126 MILES)   This line had been maintained at the standard needed to handle the four daily freight trains operated between Lincoln and Laurel, plus the four trains BN operated jointly with the UP between Lincoln and North Platte. The line was operated under train-order authority, with an automatic block-signal system. The rail, ties, and ballast conditions were not in the condition of standard needed to handle the heavy tonnages of coal that began to move in 1972. Following the hi-rail inspection trip Lorentzsen made over several lines of the Nebraska Division in August 1975, a letter he sent to Ethington left no uncertainty as to the condition of the line: "Tie condition from Seward west is very poor. Ties are broken and badly plate cut. Ballast badly fouled. This area should be plowed when relay is done instead of smoothing lift as proposed. Track through Tamora is laying in nothing but mud and this continues to the west for several miles. Tie condition continues poor."[1] At the time this report was written, four or five loaded coal trains were already being run on this line every day. The opportunity to at least begin to upgrade the track before the start up of the coal-train movements had passed. A major upgrading program had to be set up to bring the tie condition up to standard, to correct numerous drainage problems, to improve the condition of the sub-ballast section, replace the 112-pound jointed rail with new 132-pound welded rail, and to remove many miles of contaminated ballast with a Mannix sled and plow.

To have adequate capacity for the number of trains to be operated, we installed CTC and extended several sidings. We began to add segments of two main tracks in a "10-10" configuration. Under this concept, we connected sidings with the new second main trackage we built between them, thereby getting a ten-mile section with two main tracks, followed by ten miles of single track, over the entire subdivision. The same plan was used on the 238-mile Alliance–Ravenna line of the Alliance Division, resulting in a 10-10 configuration for the 364 miles between Lincoln and Alliance. BN found that the tonnage it had projected by 1980 was understated, requiring a second main track to be built in many of the ten-mile single-track sections.

LINCOLN–PACIFIC JUNCTION (61 MILES)   At Lincoln, the main coal corridor from Ravenna splits into the line that runs east toward Chicago and the one that runs southeast toward Kansas City. The line running toward Pacific Junction and Chicago was of heavier construction and had been better maintained than the Ravenna line, since it had been the main line for the Chicago–Lincoln–Denver passenger trains and the priority merchandise trains operated to and from both Denver and Grand Island.

There were subgrade problems to deal with on the alternate routes via Omaha and via Louisville. The tie-and-ballast condition had to be upgraded. There were twelve miles of 136-pound welded rail on both main tracks east of Lincoln, but on the "preferred route" via Louisville, the rail was the 129-pound section laid in 78-foot lengths. That rail was curve worn with heavily battered welds where the two 39-foot rails had been welded to form rail laid in 78-foot lengths and needed to be replaced. We started a rail program by relaying the curves on all main-track segments with new welded rail, and on the tangents between curves only where the tangents were less than a quarter mile in length. Although short-curve relays cannot be done as efficiently as continuous (out-of-face) rail relays, getting just the curves relaid got the worst rail out of the track. The tangents would be relaid when funds became available, and after higher priority relays had been completed elsewhere. It was the best way to deal with the fact that not enough money was available to program out-of-face rail relays everywhere the rail needed to be replaced.

LINCOLN–ST. JOSEPH (VIA TABLE ROCK) (147 MILES)   No unit trains of coal destined for Kansas City or beyond could be run on the most direct route (via Table Rock) until the bridge over the Missouri River at Rulo was replaced. Although the new bridge would not be in service until 1978, we undertook heavy-maintenance and capital-expansion programs starting in 1974. At that time, this line was handling only four to

---

1   N. M. Lorentzsen to I. C. Ethington, letter, August 1975, personal papers of N. M. Lorentzsen.

six general freight trains per day plus local service. Due to weight restrictions on the bridge, all loaded 100-ton capacity cars had to be moved on a circuitous route via Pacific Junction, twenty-eight miles out of route.

Between Lincoln and Napier, the Table Rock line was of light construction, with only fair tie condition. Very little heavy crushed-rock ballast had been used. There was an automatic block-signal system in service on the forty-eight miles between Napier and Table Rock, with non-ABS territory for the remaining sixty-four miles into Lincoln. This line was under CTC for the thirty-five miles between Napier and St. Joseph. That part of the line was in far better maintenance condition than the segment between Lincoln and Napier. We installed CTC on the entire line and re-ballasted it by use of the Mannix sled to plow out the old, fouled ballast and replace it with new heavy crushed-rock ballast. Several sidings were extended. In time, nine miles of second main track were built south from the yard at Lincoln. By the 1990s, the tonnage had increased so much that concrete ties and new 136-pound welded rail was installed between Lincoln and Kansas City.

By the time the new bridge at Rulo was completed in 1978, the entire Table Rock line was in excellent condition. Other than the new line built between Donkey Creek and Orin, Wyoming, and the Nebraska City line, the Table Rock line was the only one where we had the opportunity to get it in shape before the coal began to move. After several maintenance seasons of heavy work to build this line up to the condition to handle a heavy tonnage, it was gratifying to receive this report from George Lamphier:

> All the (rail) relay was completed in a good manner with gauge good, plates square on ties, spikes driven down, anchors placed tight against ties and placed to BN standard and a good cleanup of both old and new material. . . . Work all completed in very good manner with excellent line and surface. Cross level on line is good. Ballast section over the entire line is excellent with good shoulders and full cribs. Dressing of track is very good. . . . No slow orders on the entire line. . . . This line going into winter in good condition.[2]

Clearly, program work was now being completed to standard. The lessons taught repeatedly by George, Don

Rogers (the region's track engineer) and senior executives who had made trips over the railroad had taken hold. The big investment made in material, machinery, and training for maintenance personnel was paying off.

With the new bridge at Rulo in service, and the improvement in the quality of the track, it was possible to establish an interdivisional run for train crews between Lincoln and Kansas City. This change enabled us to eliminate the use of crews on a short turnaround basis between St. Joseph and Kansas City, a round trip of only 120 miles. A major benefit of the merger that created BN and the heavy tonnage of coal projected was having the resources to replace a bridge as large as the one at Rulo.

PACIFIC JUNCTION–NAPIER (77 MILES)   Since this line was not included in the system plan for upgrading to the standard of a coal route, no major capital-improvement programs were established for it. It was anticipated that in getting the Table Rock line upgraded to the standard required for a major coal corridor, we would not need to upgrade the alternate route on the line between Pacific Junction and Napier. This line generally handled only two to four daily general-freight trains and was of light construction. It was laid with 112-pound jointed rail and was in need of a tie-replacement program. There were numerous mud spots due to poor drainage and weak ballast and subgrade conditions.

When we found that coal trains for Kansas City would start moving early in 1977, several months before the completion of the new bridge at Rulo, there was no choice for the Nebraska Division but to handle them on the alternate route moving east from Lincoln to Pacific Junction and then south to Kansas City. The Pacific Junction–Napier segment was woefully lacking in rail anchors, and Merle Fivecoat, our roadmaster for that line, and Larry Ficke, the assistant superintendent–Roadway Maintenance for the division, argued strongly for a non-budgeted program for additional anchors to be placed in 1976. Unfortunately, our request was not progressed at the system level. It was felt by some of the administrative people (but not by the region or system engineering departments) that we could get by without the additional anchors since the new bridge would be open for service in the not-too-distant future, and because upgrading work on this part of the alternate route had not been included in the "master plan." However, we knew we could not wait that long.

---

2   Report dated November 12, 1980.

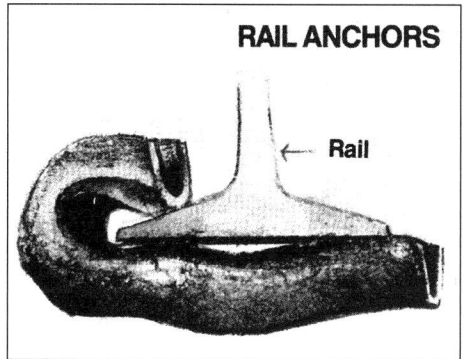

RAIL ANCHORS

← **Rail**

Figure 34.2. Rail is kept in proper alignment and good surface by the combination of ballast, ties, tie plates, and good drainage. Rail can "run" lengthwise with the flow of train traffic if it is not properly anchored. Without adequate anchorage, it may expand or contract with fluctuations in temperature. Rail anchors are applied under compression to the base of the rail and set against the ties to restrain such lateral movement. In extreme cases, rail that is moving or expanding in hot weather may become distorted into a "kink" if not properly anchored, a definite hazard to safe operation. BN NEWS.

After only two loaded coal trains had been run on that line, the rail "ran" so much due to inadequate anchorage that it kicked out where it ran into a switch and derailed twenty-five or thirty cars of the third train. The next coal train derailed from the same cause. At several locations, rail was found to be moving between thirty and sixty-five inches. In jointed rail, even three inches is considered excessive. This brought about immediate approval of the AFE we had submitted for 120,800 additional anchors, and an all-out effort was started to apply the anchors needed to hold the rail. We called back about thirty-five track workers who had been laid off for the winter months, to put anchors on the entire line in winter weather conditions and get them applied within only a few days. Until the anchors were applied, all trains had to be restricted to 10 miles per hour. When the work was completed, coal trains could be run safely, but speed was limited to 30 miles per hour on that line.

A day or two after the two derailments occurred, George Lamphier made an inspection trip over the line to determine if the amount of rail movement was as severe as reported. In his report to senior management, George wrote:

> As first reported, rail was showing a movement of three to four feet and the report dated January 28 showed this to be not only true, but we found additional movement at some locations up to six or seven feet and was starting to build up some high pressure in the rail at various locations where rail was being stopped by the grip type anchors. . . . Due to the roadbed being froze up from the extremely cold weather this winter, and with the heavy tonnage on the line, we know the effect of trains on rail

is primarily a wave motion set up in the rail which moves the rail forward in the direction of the train movement. With the roadbed and ties frozen in the ballast, all this rail movement was between the tie place and the base of rail, and the clip anchors just did not hold the rail in place.[3]

As the volume of traffic increased on the Lincoln–Kansas City corridor, the Pacific Junction–Napier line was used more and more as an alternate route for empty coal trains, thereby giving double-track capability. It is also used for loaded trains when heavy maintenance work is underway on the shorter, more direct route via Table Rock. South of Napier, the line had been maintained to a higher standard, since it had been carrying tonnage for both the Table Rock line and the line between Napier and Pacific Junction. It was a big relief for all concerned when the new bridge at Rulo was completed. Having the Table Rock line fully upgraded gave us the capacity and quality of railroad needed to handle with the high tonnages of coal safely and within the cycle time specified in the operating plan. The alternate route between Pacific Junction and Napier has been used as the route for empty coal trains at times when congestion is building on the preferred route via Table Rock.

NEBRASKA CITY LINE (60 MILES)   The rate of defective ties was very high, and almost all of the line was restricted to 10 miles per hour. As a result, it had become a very expensive line to operate. Very little

---

3   George Lamphier, lette to B. G. Anderson, assistant vice president–Engineering, and R. G. Brohaug, chief engineer–Maintenance.

grading had been done before the original track was laid. Even though the line was built on the Nebraska prairie, there were curves of 9 degrees and a grade of 1.4 percent ascending eastbound, in the loaded direction. There were a few short "hogbacks" or humps that had never been leveled or at least reduced. Derailments were common, although not severe, since only small local trains were being run on the line and at a low speed. To bring the line up to the standard needed to handle coal trains required complete rebuilding of the line. Even one coal train could not have been run more than a mile or two from Lancaster without derailing.

We started to upgrade the line by using two "small sleds" to raise the track out of the mud, to permit replacement of defective ties, to place a new ballast section of crushed rock under the ties, and to form a ballast shoulder. Bank widening was required to hold the ballast shoulder. We pulled one sled with a locomotive and the other with a winch cart. Due to the large programs we had for re-ballasting on several main-line subdivisions at that time, a "big" Mannix plow and sled was not available for work on a branch line. We were told we would have to use the small sleds and still get the job done in time for the first scheduled movement of coal to the new power plant. With the heavy tie renewal programs we had on the main lines, it was not possible to schedule a mechanized tie gang for this branch. New ties had to be placed with hand labor, including the driving of track spikes. When the sled program was completed, we relaid the rail with second hand-welded 112-pound rail. No money was allocated for the top of the "hill" where we had a curve with four compounds (curvature of 9, 9, and two 4-degree curves) and a grade of 1.4 percent. Additional property would have been required, together with a contract to move enough dirt to build the grade for a new alignment. It was too big a project for the division to undertake "informally" under its own authority, without the benefit of an approved AFE.

Having the opportunity to lift an entire line of railroad out of the mud and upgrade it to the standard of a main line was fun and rewarding. We got the job done, and the track looked as good as one could find anywhere. A few years later, the compound curve at the crest of the 1.4 percent grade was flattened out to some extent.

Omaha Public Power Development (OPPD) and UP developed plans to build a connection so UP would be able to serve the plant from its Kansas City–Omaha line that ran through Nebraska City. To ensure there would be competitive access to its plant, OPPD purchased BN's line between Lincoln and Nebraska City in 1998. OPPD was then able to allow UP to use a connecting track in Nebraska City so its trains could get on the former BN line and save the expense of a "build out" at the plant site. OPPD contracted with a short-line operator to serve the grain elevators and other customers on the line.

RULO BRIDGE   With the weight restrictions on the bridge at Rulo, no cars in excess of 210,000-pound gross weight could be moved on the most direct route (via Table Rock) between Lincoln, St. Joseph, and Kansas City. Before the merger and the onset of the coal movements, not enough capital was available for a project of the magnitude of replacing a large bridge such as the one over the Missouri River at Rulo, even though a replacement bridge had been needed as far back as the 1950s, when it became clear that more and more bulk materials would be moving in 100-ton capacity covered hopper car.

SIOUX CITY BRIDGE   At the time of the 1970 merger, BN had operating rights on the C&NW's bridge over the Missouri River a few miles south of Sioux City. Due to its light construction, the weight limit for cars was 210,000 pounds. With the rapid switch to 100-ton cars for loading grain and fertilizers, replacement of this bridge became a high priority. On most days, there were enough loaded 100-ton cars for movement between Willmar and Kansas City for a full train that had to be run via Minneapolis, Savanna and Galesburg, which was 319 miles out of route, much slower, and required three additional crews.

A large program had to be undertaken to upgrade the track on the entire line between Ashland and Sioux City. Between Sioux City and Willmar (a line on the Minnesota Division) the 90-pound rail had been satisfactory, since only four-axle locomotives were being used, and no 100-ton cars could be moved for destinations south of Sioux City. The entire corridor was relaid with heavy welded rail, several timber bridges were replaced with steel and concrete structures, and the ballast section was upgraded with crushed granite rock. The entire line of 332 miles between Ashland and Willmar was transformed from a secondary status to a high-grade main line.

Figure 34.3. A device appearing to be mounted on sled runners, inserted under the track. A hydraulic jack is used to raise the track enough to insert the sled. It is then pulled by a locomotive or winch cart so the fouled or otherwise inadequate quality of old ballast can be plowed out from under the track and shoved or conveyed to the side. Any defective ties will be removed while the track is lifted out of the ballast. The track will then be lowered back on the grade and new ballast is unloaded by a work train to about the top of the rail. A tamper will raise the track enough to get enough new ballast under the ties to meet the usual standard of about six inches. Finally, a combination ballast regulator-broom will fill the cribs with ballast that is still on the tops of the ties, and plow any excess ballast out from between the rails and onto shoulder of the ballast section, which generally is to extend twelve inches beyond the ends of the ties. JOHN PHILLIPS III COLLECTION OF PHOTOS OF OPERATIONS ON THE NORTHERN PACIFIC RAILWAY (PRE-1970).

BYPASS TRACKS AT LINCOLN YARD   New bypass tracks were built around both sides of the yard at Lincoln in order to keep the loaded and empty coal trains out of the congested yard. Getting those tracks in service greatly reduced the amount of time that trains had to be detained at Lincoln to change power and crews and for the required 500-mile intermediate inspections. In time, fueling stations were built on the bypass tracks. The result was greatly improved flexibility in the operation of the Lincoln terminal and reductions in the cycle time for the coal trains.

CONSTRUCTION OF SECOND MAIN TRACK, WAVERLY–GREENWOOD   At Ashland, twenty-four miles east of Lincoln, the main line divides in three directions. East from Ashland, main-line trains used the line via Louisville due to its shorter mileage and lower grades. The second "split" at Ashland is the line to Omaha; it is used mainly for service to local business in that area, access to the yard in Omaha, and Amtrak trains. The line via Omaha also serves as a "second main" to take some trains off the Louisville line in times of heavy traffic and when maintenance projects are underway. The third line runs north from Ashland to Sioux City, Willmar, and Minneapolis.

Between Ashland and Lincoln, a high-density territory, there was a 5.4-mile segment of single track that became a major bottleneck in handling trains running to and from the three lines just described. Fortunately, no large bridges had to be built for a second main track. Because the line between Lincoln and Ashland was handling about fifty million gross tons per mile per year, making it one of the heaviest on BN at that time, we put the addition of this segment of second main track close to the top of the priority list. It was opened for service in 1977.

CURVE RELAYS   The replacement of old jointed rail was needed with the rapid increase in tonnage we were handling. Replacement of much of that rail on an out-of-face basis had to be spread over several years. There was not enough new 132-pound rail available on the market to do everything at once. Therefore it was best to set up programs to first relay the curve-worn rail on curves of 2 degrees and over with small gangs of thirty-five people. If the length of the tangent track between any two curves

to be relaid was less than one-quarter mile, we relaid the tangent between the curves at the same time.

This program enabled us to get the worst rail out much sooner than if we had waited until a big out-of-face rail gang could be scheduled to relay several consecutive miles, both curve and tangent. Because the productivity of the smaller curve relay gangs was lower than the large rail gangs, criticism came from some managers who came into BN with the Frisco merger. However, George Lamphier and Don Rogers, two "ace" track people, stuck to their guns and made it possible to continue the practice of relaying the curves with heavily worn rail, to buy us some time. Don Rogers went so far as to say that George's curve relay program was the "savior" to keep many lines in safe operating condition.

CHALLENGE AND REWARD

Many of our roadway-maintenance supervisors and employees had had little experience with modern, high-productivity machinery before the 1970 merger in applying the best practices for organizing large-scale work. With the large number of gangs we were working, we were short of experience, qualified foremen and machine operators. BN's System Engineering Department helped us a great deal in conducting training programs and providing on the job instruction as to best practices and in doing the work at the new standards. The people we had were the salt of the earth, as track people usually are, no matter what railroad you are working. But as we geared up for the work, it was a challenge to get the work completed at the standards of quality expected in the massive projects we had underway.

We experienced many ups and downs through the process of getting the work done. The task of coordinating the work among the track, bridge, signal and operating people was complex and difficult at times. Some mistakes were made through this massive undertaking, but for nearly everyone, any feeling of discouragement or frustration along the way never lasted very long. We kept our sights on the big picture. We felt pride, reward, and satisfaction as one big project after another was completed. We moved on to new projects with enthusiasm.

The board and senior management maintained faith that the tonnages of coal forecasted by the electric power industry and BN's experts in coal marketing were "real." As it turned out, those forecasts were understated. By 1977, still in the early stages of growth in business originating in the Power River Basin, we built up to about 100 million tons. Little did we know at that time that some segments of the Gillette–Orin line would one day carry over 440 million gross tons, or that the density would grow to over 200 million gross tons per mile per year on the Alliance–Ravenna–Lincoln segment by then. No other railroad in North America had ever moved tonnages of that magnitude.

For hundreds of people, these challenges built professionalism, skill, and quality into all of the activities managed at the division level. Few people in the rail industry have had the opportunity for a part in a transformation of this magnitude. Several workers who began their careers on these major track projects—some coming to BN right out of engineering school and others starting as track laborers—advanced to leadership positions; some are still working in today's BNSF Railway. The investments the company made in people, materials, equipment, and in teaching the best practices in the 1970s have been validated many times over in the financial and operating performance that BN and BNSF have achieved in subsequent years, continuing to the present.

This success provided some of America's best jobs for many people. Because the coal we deliver is an inexpensive form of energy, the rates for electric power are lower for millions of people than would otherwise be the case. The low-sulfur coal has helped in reducing the amount of nitrous oxide and sulfur dioxide where high-sulfur eastern coal had to be relied on in the past for electric power generation. It has lessened our country's dependence on foreign sources of energy. Having to move these unprecedented tonnages has fostered the development of higher quality steel for rail and turnouts, the use of concrete ties, and far more efficient methods of inspecting and maintaining all of the components of the track structure.

CHAPTER 35

THE COLORADO DIVISION

• created in 1976 to give some relief to the Alliance Division •

THE COLORADO DIVISION WAS CREATED AS a new operations and maintenance unit in 1976 to give some relief to the Alliance Division, which had become overloaded with the many large projects underway on all of its lines, over a very wide expanse of territory. The new division was made up mainly of the lines of the former McCook Division, which had been consolidated with the Alliance Division not long before the coal boom started. In addition, the Colorado Division included the 585 miles of main lines of C&S (Colorado and Southern Railway, a subsidiary of BN). C&S remained a corporate entity until it was merged into BN in 1982. It was headed by a president who also served as president of a C&S subsidiary, FW&D (Fort Worth and Denver) Railway. Under this arrangement, the Colorado Division superintendent reported to both the assistant vice president–Operations of BN's Denver Region and to the president of the C&S.

The Colorado Division was also responsible for managing the Denver terminal, which included a large yard and locomotive-maintenance facility. It included part of the new major coal corridor between Sterling, Brush, and Denver. To the south from Denver, the new coal corridor used the trackage rights that C&S had to Pueblo on a 118-mile double-track line jointly owned by the Santa Fe and Rio Grande. When the heavy flow of coal trains to Texas and Oklahoma started, C&S became by far the heaviest user of that line (refer to chapter 39 for details of track work and major projects completed on C&S).

The next segment of the coal corridor to the Southwest was the C&S line of 230-miles between Pueblo and Texline, where it connected with the FW&D's line to Fort

Worth. Chapter 39 on the C&S covers in detail the heavy workload undertaken by the Engineering Department of the C&S and the Colorado Division to transform that line from a lightly used secondary line to a highotonnage main line, capable of handling annual tonnage of 60 million gross tons. It was especially challenging because of the extremely heavy curvature in several locations that required major line changes to be made within a short span of only about three years.

To serve a power plant built by the Nebraska Public Power District (NPPD) near Sutherland, BN had to completely rebuild 115 miles of its 229-mile branch line between Sterling, Colorado, and Holdrege, Nebraska. The plant is generally known as the "Gerald Gentleman" generating station. NPPD financed the construction of a new eighteen-mile line between Wallace and the plant. While serving a plant that would consume four million tons of coal per year was an obvious enhancement to BN's "top line" (revenue), having to upgrade a branch line of that length took considerable resources (machinery, ballast cars, labor, material, and management time) away from the urgent need to invest as heavily and rapidly as possible on its main lines. Heavy sections of secondhand welded rail had to be laid, many train loads of main line standard ballast had to be placed, and the ties had to be replaced at a high rate.

Also during this time, the yards that were operated separately by BN and C&S in Denver were consolidated. To add to the complexity, C&S trains were manned by Santa Fe crews between Denver and Pueblo. C&S crews handled Santa Fe switching and train makeup in the Denver terminal.

As one would expect, the division superintendent and staff of the Colorado Division had to manage what could have been a touchy, sensitive situation by having to report to two bosses. However, with the common goal of getting the track upgraded to handle the heavy tonnage that was already moving, officers of both companies set aside the rivalries of the past on administering joint-facilities contracts and the apportionment of expenses between the two companies. The C&S was given all of the authority it needed to carry out its programs for major track upgrading and to expand capacity. It was no longer treated as an errant step-child.

As the work was completed, safety improved, the coal-train cycle time went down, and operating costs were reduced to an acceptable level. Having seasoned division superintendents with prior experience in managing large-scale maintenance and operating activities provided the strength of leadership needed to show BN's senior management that they had made a good decision in allowing a new division to be formed at this time of a very heavy work load.

The major line changes and heavy upgrading on C&S and FW&D are covered in chapters 39 and 40.

CHAPTER 36

THE YELLOWSTONE DIVISION

• track on the main lines of the former Northern Pacific was in excellent condition,
but it still required some heavy maintenance work to handle the large increases in tonnage[1] •

THE 232-MILE LINE BETWEEN GILLETTE and Huntley was in a state of maintenance and utility very similar to other main lines of the Alliance Division at the time of the 1970 merger. Upgrading soon became necessary when new mines opened on the Decker spur in Montana in 1973, where annual gross tonnage of 17 million tons was projected annually. The tonnage actually reached 32 million by 1984. A program to relay the 115 miles of 90-pound rail with heavier new welded rail was urgently needed. Burlington (CB&Q) had upgraded the rail on the entire line from Lincoln to Sheridan with mostly 112-pound TR (torsion resisted) jointed rail in the 1950s, but at the time BN was formed, there were still 115 miles short of replacing all of the 90-pound rail to where it connected with the NP main line at Huntley. No continuous welded rail had been laid on any part of this line segment by the startup of heavy unit train movements of coal and grain.

The line through Sheridan and Alliance was designated the preferred route for traffic moving between the Pacific Northwest and Kansas City in the operating plan for the merged company. Business for Kansas City and beyond that had been moving by way of Minneapolis or Sioux City before the merger was then shifted to the Sheridan–Lincoln line. Even with that increase in business, there still were only four regular trains operated on the Sheridan line before the new mines were opened at Decker.

A new connection had to be built at Huntley to allow the coal trains to make a direct move from the line from Sheridan to the former NP line to Glendive and beyond. Having that connection eliminated the need for a coal train to stop at a siding or yard track west of Huntley to run around the train before heading east toward Glendive. Also, a new interdivisional run for train crews was established between Sheridan and Forsyth.

In addition to relaying the light rail, the line through Sheridan had to be reballasted with crushed rock and upgraded with bankwidening and a heavy tie renewal program. It was vital to remove the scoria, a lightweight, low-density volcanic rock that Burlington had used for ballast on a large part of the line. For the amount of traffic that moved on this line before the new coal mines were opened, scoria had proved adequate, but it was not of the quality needed to support the track, once the heavy tonnages began to move.

Fortunately, the design of the 112-pound TR rail section used on many lines of the former Burlington was stronger and had a longer service life than the 112-pound RE rail section that had been widely used throughout the rail industry, including the NP's Yellowstone Division. The design characteristics of TR rail made it possible to keep it in service until the 1980s, when enough money became available for a program to lay new 132-pound welded rail on the entire line between Sheridan and Huntley. The advantages of the jointed TR rail in service made it possible for BN to delay replacing it for several years and, instead, concentrate first on replacing rail in locations with poorer-quality rail in those early years of the coal boom.

---

1   Refer to page 195 for information on a line change made near Gillette and page 188 for construction of new lines to the Decker, Big Sky, and Absaloka mines.

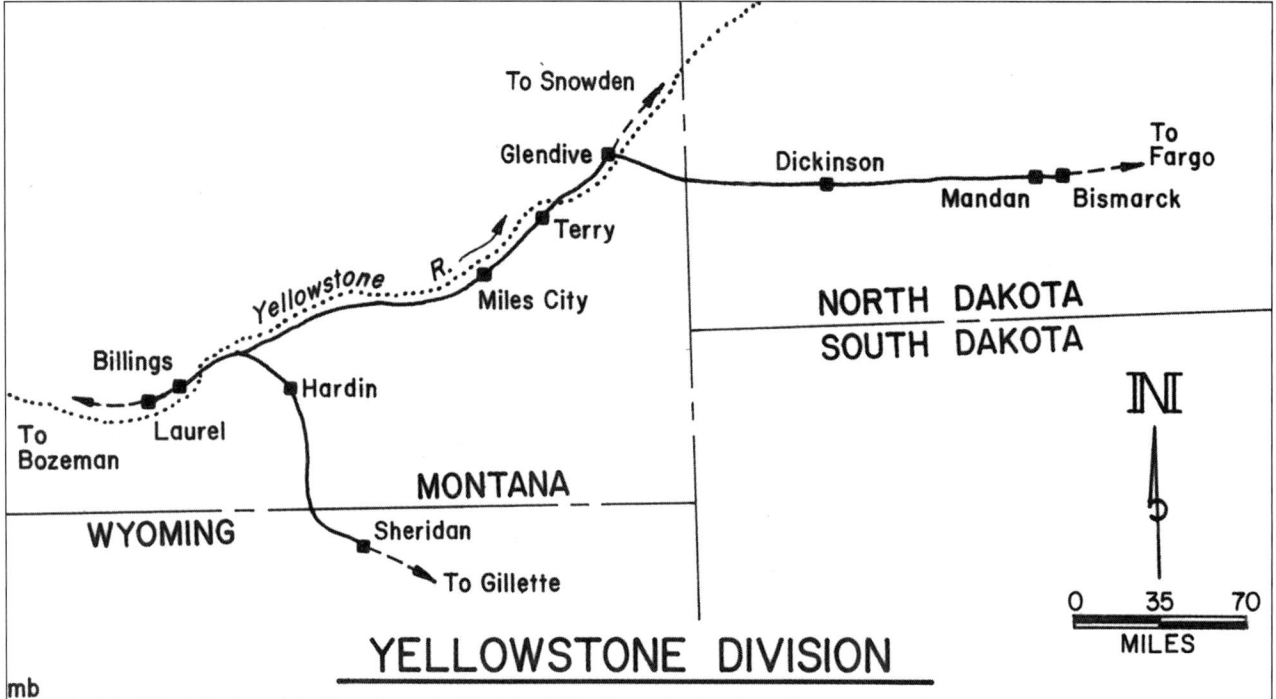

Figure 36.1.

Capacity on the Sheridan line was increased by extending several sidings to at least 7,000 feet, long enough to handle trains of 100 to 110 cars. Getting the number and length of slow orders reduced as the old ballast section was plowed out by the sled method and replaced with good crushed rock ballast helped reduce the running time for trains. In 1984 the installation of CTC was completed from Gillette to Huntley, allowing more trains to be run than was possible under the old train-order system of dispatching.

The grades on both sides of Sheridan were quite heavy, amounting to 1.25 percent in each direction. For a while in the early 1970s, radio-controlled helper units were placed in trains moving on these grades. However, with a number of difficulties experienced in maintaining continuity between the lead locomotive and the remote units, the use of Locotrol equipment was suspended and replaced with manned helpers. Today's use of "distributed power" is an outgrowth of the early Locotrol system concept for remote control of the locomotives placed at the rear of a train or cut into the train consist.

One might have thought the former NP's main line east of the junction at Huntley was so well maintained that it would not have needed much heavy maintenance for a long time, even with the addition of heavy coal tonnage from the new mines at Decker, Colstrip, Big Sky, and Kuehn. NP had maintained its track for 75 miles per hour for passenger trains and 50 miles per hour for freight, and kept the surface and alignment of its main tracks and sidings in good condition right up to the time of merger in 1970.

Mike Martin, who was superintendent of the Yellowstone Division in much of the 1970s, and Don Rogers, assistant superintendent–Roadway Maintenance in those years, recall how the heavy-unit trains soon began to take their toll, especially on the jointed 112- and 115-pound rail that had been laid between the early 1940s and late 1950s, before NP began to install ribbon rail in its main-line rail-replacement programs. More welding had to be done to repair the increased batter on rail ends on the jointed rail. Thousands of additional anchors had to be applied over hundreds of miles to keep the rail from "running" (i.e., moving longitudinally) and thereby increasing the stress at the point where the running rail would hit a "stone wall" at switches and the ends of bridges.

Don Rogers pointed out that the problem of elevations in curves was especially severe on the thirteen miles of one percent grade in the vicinity of Fryburg, east of Glendive. There was a series of twenty-three curves of between 1 and 3 degrees on the grade, on which loaded coal trains could run at only about 10 to 15 miles per hour. Having to maintain superelevation for a passenger-train speed of 60 miles per hour on those curves made it difficult to maintain surface and line, to say nothing of the accelerated wear on the low rail on the curves. Mud spots began to develop, which required fouled ballast to be dug out to improve drainage from the track section, often in areas where there had been no such deterioration in the ballast condition in the past. All of this placed challenges on a part of the BN network (between Laurel and Casselton) that had not been expected when the merger plan was developed ten to fifteen years earlier.

Mike Martin provided a good description of the tension that built up in having a large number of new employees at work on safety sensitive jobs in those times: "New recruits were young and received their baptism rather harshly. Remember, we not only had young inexperienced Operators out in the field—but perhaps more worrisome, we had young, inexperienced Dispatchers putting out train orders to young, inexperienced Operators which were then to be executed by young inexperienced train service employees."[2]

The annual gross tonnage on the former NP main line increased from about 14 million to 30 million tons per mile by 1975, and by 2009, to 74 million gross tons per year. The number of employees required for train and engine service nearly quadrupled in the 1970s. In the mechanical facilities at Mandan and Glendive, employment tripled. Having such a large increase of new business was of great benefit to the operating and maintenance employees of the former NP, who had stood to lose considerable work under the plan to reroute long-haul traffic over the former GN line through North Dakota and Montana upon merger.

As happened elsewhere on BN, once the major track-improvement projects were completed on the Yellowstone Division, the operation improved immensely. Mike Martin wrote, "By the early 1980s, the division was finally at a place where the track infrastructure and capacity could efficiently accommodate the increased coal traffic."[3] It became possible to run trains within the cycle times established in the plan for coal service. With that improvement, fewer sets of equipment were required, thereby reducing congestion on the line and through terminals. The number of derailments was reduced, service became more reliable, and operating costs decreased. The large investment made to upgrade the track began to pay off. These improvements were highly gratifying to the hard-working employees, and supervisors at the field level could begin to see the results of the effort they had put in over the past several years to upgrade the railroad and improve its performance.

2   E. M. Martin, message to author, March 2, 2014.

3   E. M. Martin, message to author, March 2, 2014.

CHAPTER 37

THE CHICAGO REGION

THE OTTUMWA, HANNIBAL, AND
CHICAGO DIVISIONS

• heavy upgrading required on over 250 miles of double-track and second main trackage
between Savanna and the Wisconsin-Minnesota state line, and 125 miles of two-main-track
territory between Aurora and Galesburg • challenge of heavy upgrading in the
thirty-eight miles of three and four main tracks in the high-density
Chicago suburban territory • also required heavy work on
245 miles of single track: Aurora–Savanna and Galesburg–Savanna
recovery from deferred maintenance; restored to excellent condition •

THE CHALLENGE FOR THE THREE DIVISIONS of the Chicago Region was to upgrade the main tracks and sidings on 1,562 miles of road. All but 200 miles of the region's main track became a high-tonnage coal route. There were 776 miles of road containing two main tracks, including a third and fourth main track for five miles east of Cicero, within 38 miles between Chicago and Aurora. The work required to bring those lines up to standard amounted to far more than putting in the standard number of ties that would be anticipated in a seven-year cycle for tie renewal.

The tie condition had become so poor on much of each division that from about 1,100 to 1,800 ties per mile had to be put in so that train speeds of 79 miles per hour for passenger trains and 60 miles per hour for "regular" freight trains could be restored. There were only a few miles of continuous welded rail (CWR) on any lines on the Chicago Region at the time of the 1970 merger. All of the slag-and-limestone quality ballast had to be replaced with granite rock moved in long hauls from ballast sources near St. Cloud and Granite Falls, Minnesota. This required hauls of 525 miles for a unit

train of ballast to be run from St. Cloud to Creston (via Galesburg), for example.

It took far more than one maintenance season to bring these lines up to standard. We started out by replacing between 800 and 1,200 of the defective ties in each mile, and laid new welded rail on curves of 2 degrees and over where the rail had a high defect rate and/or was head worn. Later, after the worst of the curves had been re-laid and funds were made available, we replaced the rail on the tangents between those curves on an out-of-face basis. Mud had to be dug out at road crossings and in most switches. In some areas, the ballast was so badly fouled with mud that it had to be plowed out with a sled and replaced with a new ballast section twelve inches in depth. If the old ballast was not that badly fouled, we would give the track a two-inch or four-inch lift, using new granite ballast when it was surfaced behind a tie or rail gang. That gave us a chance to improve the surface and line for a while, but all too often, using only that much new ballast amounted to a "camouflage" job: it was enough to help correct the problem for a while, and made the track at least look better. But within a year or so, the mud would begin to work its way up through the thin layer of new ballast and, again, obstruct drainage and cause the restored surface and line to deteriorate.

We would have to come back into these areas with spot undercutting or use the ballast sled to get enough good-quality ballast in place to restore drainage so the track could withstand the impact of heavy tonnage carloads. Often, a fabric-type material had to be placed to keep the mud from working its way up through even the twelve inches of new ballast. The shoulders of the ballast section had to be widened further off the ends of the ties to provide enough strength to keep the welded rail in place, especially on curves. Often, that required us to widen the embankment enough to keep the additional shoulder ballast we placed from simply running down the embankment and thereby being wasted. On our first pass with a tie gang, the replacement of from 800 to 1,200 ties per mile often left so many defective ties in the track that we could not put the track on the standard of seven years between tie renewal cycles. Instead, we would have to bring a tie gang back after only three or four years after the first pass.

The point of listing all of these very basic types of work is to show that it took much more than simply replacing some ties and laying new rail to restore these lines to a standard strong enough to handle the ever-increasing tonnages that would be run over them for a long time to come. This tested the patience of some senior executives and department heads who had little or no experience in track maintenance, or did not have the opportunity to get out on the railroad to see for themselves—and listen to the maintenance personnel on the magnitude of the challenge to finally get enough work done so the track could be put on a normalized maintenance basis. It actually took five or six maintenance seasons to get the job done on many of the line segments involved in these massive upgrading programs. No one could realistically complain that we did not have enough resources for labor, material, and machinery to do the work, but having so few experienced gang foremen, mechanics, and machine operators made it challenging to get the work done at the rate of productivity for which it was programmed and to the level of quality that was necessary.

By 1980 the heaviest upgrading work had been completed. Of course, that was not the end of the job of maintaining track. It still required diligence and attention to correct small problems before they expanded into big problems that required major remedial work, or that would soon require temporary speed restrictions. It became necessary to program surface correction on at least an every-other-year basis. Programs for spot undercutting and shoulder-ballast cleaning had to be carried out to keep the track in shape, and in some cases, to remedy problem areas that were not given enough attention in the first pass of upgrading the railroad. There were enormous quantities of scrap and recovered track material to be picked up after of years of heavy-work programs and not always requiring the gang to remove it as part of the program to finish the job before moving on to its next work location.

In addition to the work of upgrading the track, there were some large projects carried out to expand capacity on the Chicago Region in those years. Examples were the installation of CTC on the 96-mile Galesburg–Savanna line, and the extension or construction of sidings to be used for meeting trains on some single-track lines on which the number of trains operated increased significantly. It was necessary to convert a large number of bridges from open deck to ballast deck bridges. Some yard tracks at Creston, Ottumwa, and Galesburg had to be upgraded and extended enough to handle loaded coal trains when they had to clear the main track to be overtaken by faster trains, or to hold them back when a line was to be shut down for several hours for trackwork, or when it was necessary to stage the trains for delivery to unloading points. Considerable trackage had to be constructed to serve rail-to-water transfer facilities on the Mississippi River at North St. Louis and Montrose, Iowa.

Bob Downing recalled many times how he was surprised that a main line of the Chicago Region would have deteriorated to that extent. The cuts made in normalized maintenance had not caught up with the Burlington before the effective date of the merger. Indeed, the cuts in heavy maintenance work such as tie renewal programs were "well-timed." The territory of the former Burlington soon was generously paid back for having made dividend payments it really could not afford to be making in the years just preceding the merger. The amount of money and commitment of senior management to restore these lines to a high quality of maintenance was impressive and long lasting. No reasonable person could complain after the merger that the former Burlington property was neglected in any way. As major track-maintenance projects were completed, money was made available to do more work. By the late 1970s, the track had been upgraded to the condition needed to handle heavy tonnages of coal. The property had been cleaned up, all gangs (including section gangs) were properly staffed with trained people and the best available machinery for light and heavy maintenance. Roadway-maintenance supervisors were given the resources needed to keep the track in good condition for the long term.

In the early years of BN, we could not wait until skills in the entire workforce and all supervisors had been upgraded to the level needed to move ahead with heavy work programs that were so badly needed. Under the decentralized organization structure in effect on BN in those years, the regions and divisions were expected to manage the day-to-day operating and maintenance matters, plus large projects such as siding extensions, line changes, construction of second main tracks, hundreds of miles of re-ballasting, and large programs to replace rail and ties. We had all the responsibility and accountability for organizing and carrying out the work. With several projects simultaneously underway on each main line subdivision, the challenge seemed formidable at times. However, the rewards from meeting those challenges were great. It was exciting to visualize how much better our operation would be when all the heavy track work was completed.

The next three chapters contain examples of the work done on line segments of the Chicago Region to bring them up to a satisfactory level of maintenance. Except for the line of 200-mile between West Quincy and Kansas City, all main line subdivisions had to be upgraded, a total of 1,550 miles of road.

THE CHICAGO DIVISION

CHICAGO–AURORA   A tremendous amount of expensive detail work had to be done under traffic, since operations could not be shut down during daylight hours due to having to run suburban passenger service, even outside of the rush hours. Also, there were several priority freight trains that had to be run in and out of Cicero that could not be held until the end of the hours worked by the maintenance crews. All three main tracks had to be rebuilt, and extensive undercutting had to be done at station platforms, crossover switches, and through numerous road crossings. All rail had to be replaced, together with tie and ballast replacement.

AURORA–GALESBURG   A two main track with undercutting required through stations, crossings, and switches. All rail had to be re-laid except for the sixteen miles of 136-pound welded rail laid in 1961, plus heavy-tie-and-ballast renewal over several years.

AURORA–SAVANNA   A line of 105 miles of single track plus eight miles of two main tracks on which passenger trains had been authorized to run 90 miles per hour until 1968. It required complete renewal of ballast, extensive tie renewals and relay of 131-pound rail that deteriorated under heavy tonnage.

GALESBURG–SAVANNA   A single-track line of 98 miles that by 1970 required total rebuilding, with replacement of several miles of 100-pound and 110-pound rail laid secondhand in 33-foot lengths, except for 21 miles of secondhand 112-pound welded rail laid in 1966. Extended sidings and installed CTC.

ST. CROIX–SAVANNA   A 267-mile line of double-track and two-main-track CTC on a river grade (including 13 miles of operation on track owned by the Illinois Central Railroad), with three short segments of single track on bridges over large rivers, as well as through the yard at North La Crosse and three miles of single track at East Winona converted from double track due to flood damage in 1964. Much of the rail on this line was 112-pound TR rail laid in 78-foot lengths. We had to reballast the entire line, with a high rate of tie renewals, in the range of 1,800 to 2,000 per mile over several years.

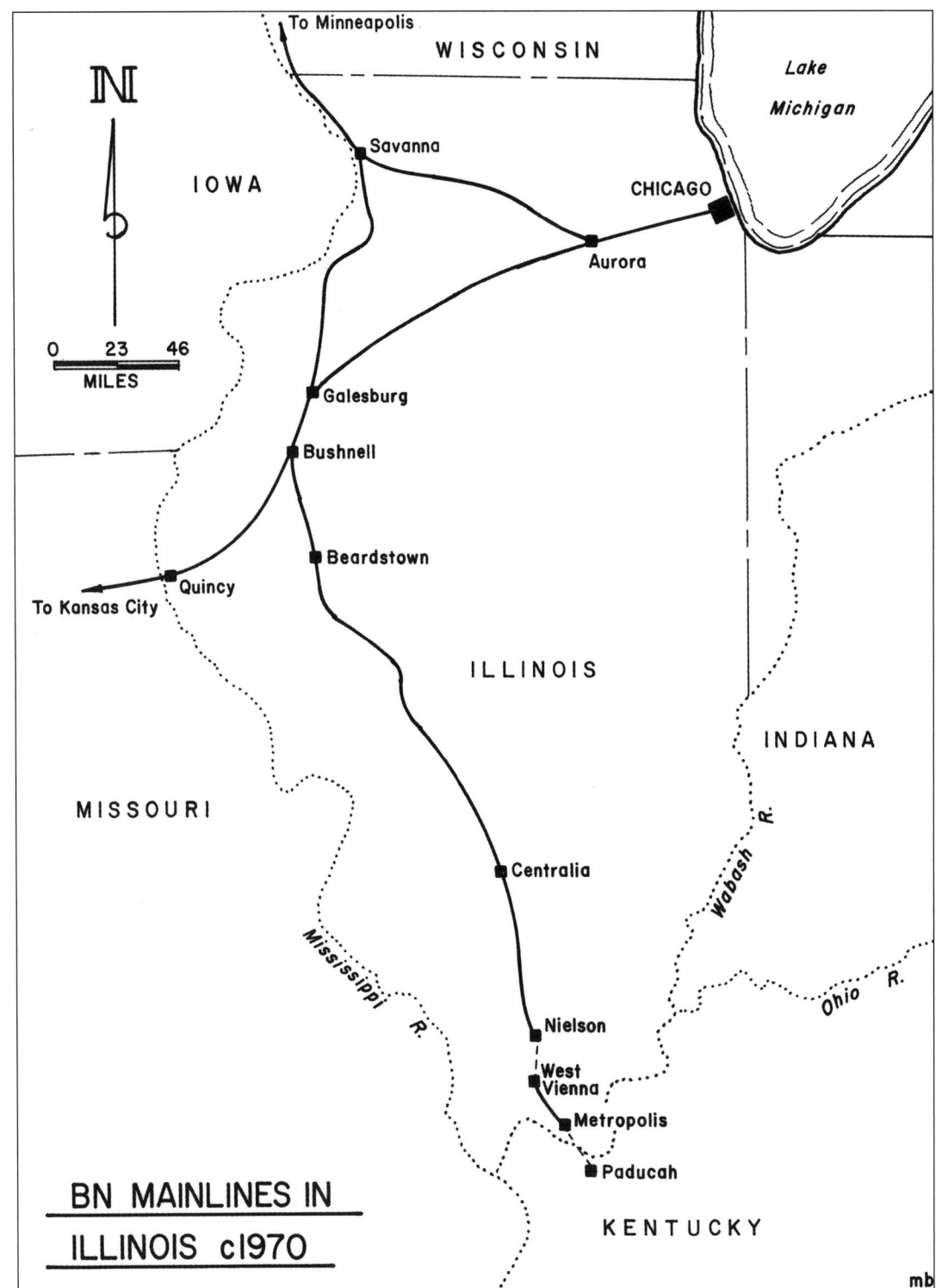

Figure 37.1.

The problem of deficiency in track conditions on the Chicago Division came to a head in the fall of 1972 on an inspection trip headed by George Lamphier when it was determined that a speed restriction of 40 miles per hour should be placed on both main tracks over the entire 127-mile Aurora–Galesburg subdivision. This caused great consternation among the executive officers in St. Paul, who had not yet focused on how much deferred maintenance there was on large parts of the former Burlington, in addition to what they had learned about the generally poor condition of the Galesburg–Savanna line.

Senior officers from headquarters, including Bob Downing, president, and Bruce Anderson, AVP–Engineering for the BN system, made a follow-up inspection trip over the line. None of them concluded that George's assessment should be overruled. Instead, two tie gangs, one each from the Montana Division and the Yellowstone Division, were moved in to supplement the division gangs, in order to get as many additional ties placed as possible in the few weeks remaining before freeze-up. These gangs were reputed to be the best-performing tie gangs on BN at that time. To get this very large additional tie program started rather late in the maintenance season, it was necessary to have new ties picked up that had already been unloaded for placement on lines in Montana and shipped to Illinois. Going to work in Illinois was far from the limits of those gangs' seniority district, but the situation was so desperate that it had to be dealt with by any means before the onslaught of winter. Further, since there had been no track-maintenance employees on the Chicago Region had been laid off yet that fall, it was felt there would be no basis for a penalty claim that work had been taken away from any of them.

One of the gangs from Montana was run by Joe Stiffarm, general foreman. Joe soon became a legend across the Chicago Region for the amount of work done by his gang within a few weeks. Joe was promoted to Roadmaster on the "high line" (the former GN main line across Montana) not too long after that work was done and the gang returned to Montana.

OTTUMWA DIVISION

GALESBURG–PACIFIC JUNCTION   Consisted of 228 miles of double-track and two-main-track territory between Galesburg and Creston, and 44 miles of two main track and 38 miles of single-track CTC between Creston and Pacific Junction. Required a major effort for embankment stabilization, bank widening, tie renewal, undercutting, and rail replacement over several years. As an indication of the amount of work required in embankment stabilization, Dave Nelson, division superintendent, reported in 1980, "Since 1972 we've driven 137,000 linear feet of stabilization pilings, and that translates, if all the piling were laid end-to-end to some 26 miles. And we're still driving them in soft spots to stabilize hills; the soil in southern Iowa isn't the most stable in the world."

ST. JOSEPH–KANSAS CITY   A subdivision of 58 miles requiring complete reballasting and heavy-tie replacement.

HANNIBAL DIVISION

BUSHNELL–METROPOLIS   A 280-mile secondary line (including 16 miles of trackage rights on the Chicago and Eastern Illinois Railroad [C&EI]) operated under timetable and train-order authority. Except for 19 miles of secondhand welded 112-pound rail laid in 1966, most rail was 100-, 110-, and 112-pound jointed rail laid second hand. Required complete rebuilding, including major tie renewal, replacing 302 miles of rail on main tracks, and extending sidings to handle heavy coal tonnage to a rail-to-water transfer facility on the Ohio River near the end of the line at Metropolis, plus increasing merchandise tonnage that BN interchanged with the Southern and L&N (Louisville and Nashville) railroads. This line crossed the tracks of 17 railroad lines that were in service to East St. Louis or for access to the mining areas in southern Illinois.

At one time, this line had been the artery for moving coal from mines in southern Illinois to Chicago. However, it was allowed to deteriorate with the reduction in the amount of coal being moved for home heating and industrial purposes. Also, with its high sulfur content, this coal could no longer be used by electric power plants. In the early 1970s, we learned we would soon be handling large tonnages of low sulfur coal mined in the

Powder River Basin to a rail-to-water transfer facility at Metropolis on the Ohio River. After recently moving our interchange points with the L&N (Louisville and Nashville) Railroad and the Southern Railway from East St. Louis to junctions on this line (Woodlawn for the L&N and Centralia for the Southern). Even with so many miles of sub-standard track, our service had already improved so much from this change that at least one full tonnage train was being run daily to and from each of those roads. In addition, we started to run unit trains of taconite to a steel mill at Granite City over part of this line.

The projected tonnage for low sulfur coal was so high that we had to relay all of the rail with new 132-lb. CWR except for the 18 miles of secondhand 112-lb. welded rail that had been laid at the north end of the line in 1966. We replaced an average of 1,500 ties per mile. To reballast the line, we used a ballast sled and plow. We had to build some new sidings and extend others. Nowhere else on the Chicago Region was the work as extensive as on the long north-south line through central Illinois.

BURLINGTON–NORTH ST. LOUIS   Also required major tie-renewal programs and replacement of nearly all rail, plus undercutting and sledding to remove ballast fouled by flooding over the years, to bring track back to standard to handle coal to barge transfer facilities at North St. Louis and Montrose, Iowa.

GALESBURG–PEORIA   A branch line of 52 miles with 90-, 100-, and 110-pound rail that handled only light tonnage until 1971, when several unit trains began moving to barge transfer facilities on the Illinois River near Peoria. Required complete rebuilding of the line over the next ten years.

In addition to the heavy restoration work required on main lines, it was necessary in those years to also carry out large programs to overcome track deficiencies in the yards at Galesburg, Cicero, and Eola (Aurora).

GREAT IMPROVEMENT MADE BY 1980, BUT TRACK WORK IS NEVER ALL DONE

Ten years of heavy work were required to overcome the deficiencies that we had developed over the entire Chicago Region from the lack of adequate maintenance in the 1960s. In addition, having conventional bolted (jointed) rail on nearly all of the mileage involved resulted in accelerated wear on rail joints and ties. The slag-and-limestone-type ballast could not withstand the impacts from the large number of heavy-tonnage trains moving on all lines by the mid-1970s. The work required was so widespread that we could not have concentrated on complete rebuilding of one subdivision at a time. Instead, heavy work had to be undertaken on every line at the same time. Because the coal was moving everywhere, we could not delay the start of rebuilding on any line until all necessary work was completed on one or two lines or subdivisions. By 1980 most of the heavy work was done, but even after that, continued surfacing, drainage-improvement programs, rail grinding, and shoulder-ballast cleaning were necessary to keep the track in good shape. Fortunately, improved technology and work methods for maintenance and inspection have made that possible, together with the strong year-to-year earnings generated by BN and BNSF.

CHAPTER 38

EXPANDING ROADWAY CAPACITY

• building the capability to handle a high-density operation •

THE INSTALLATION OF CTC ON THE COAL routes on the Alliance, Nebraska, and Yellowstone divisions was a vital part of the program to add capacity to those lines. However, by 1974, the line-capacity model run by the Industrial Engineering Department showed that we would also have to build considerable second main trackage to provide enough capacity to handle the tonnages forecasted for 1980 and beyond. It was determined that we could stay within the limits of "allowable delay" (i.e., the time trains will be delayed in meeting trains moving in the opposite direction on a single-track line) if we built a 10-10 configuration between Gillette and Lincoln, which was to become the corridor with the heaviest tonnage.

We constructed ten miles of second main track followed by segments of ten miles of single track. An exception to this standard was the heavy-grade territory over Crawford Hill, where thirty-seven miles of two-main-track territory were built on the grade and on both sides of it, between Marsland and Joder. Where the computer simulations showed that the threshold level of delays with a 10-10 configuration would be in excess of an acceptable level, additional second main track would have to be built to gain enough additional capacity.

There were many locations at which two existing or recently extended sidings were connected to form the ten-mile segments of two main tracks. This concept worked well until the number of trains increased well beyond the projections we worked with at the time the line capacity model was run. BN then began to build even

more miles of second main track by connecting some of the existing ten-mile segments. By 2010, 81 percent of the mileage of the Donkey Creek (the junction with the new line built to Bridger Junction)–Alliance segment had two main tracks, 91 percent between Alliance and Ravenna, and 84 percent between Ravenna and Lincoln. On other subdivisions on the Alliance and Nebraska divisions, CTC was installed, additional sidings were built, and existing sidings were extended to about 7,000 feet to provide enough capacity to run trains within the agreed upon cycle times.

NEW INSTALLATIONS OF CTC

CTC is defined in the Consolidated Code of Operating Rules, edition of 1967, as "A block-signal system under which train or engine movements are authorized by block signals whose indications supersede the superiority of trains for both opposing and following movements on the same track." The Consolidated Code is used for reference since it was the book of rules in effect in the years covered by this writing.

Installing CTC on a segment of railroad line increases its capacity and reduces train delays. Operators are no longer needed at intermediate stations to copy train orders issued by a dispatcher, which would be handed up or delivered to the crews of trains. Instead of the movement of trains being governed by train orders and a timetable, trains either proceed or are restricted according to the indications of wayside signals controlled by a dispatcher.

The dispatcher, or a control operator working under his or her direction, may also have remote control of switches at junction points and crossovers and at sidings used to meet or pass trains. Having electrically powered switches controlled remotely eliminates the delay to trains from having to stop and have an employee get on the ground to line switches by hand. The dispatcher also is able to block segments of track between control points to authorize and protect the maintenance or inspection work of signal-, track-, and bridge-maintenance forces.

In a typical installation, CTC will expand the capacity of a single-track railroad by about 80 percent. It also gives the dispatcher the flexibility needed to allow priority trains to pass "inferior" trains, or trains that are operating with a lower ratio of locomotive horsepower to a train's tonnage.

CTC may also be installed on a double-track railroad to expand its capacity by signaling both main tracks for movement in either direction with powered crossover switches at intermediate points. If a railroad no longer needs all of the capacity of a double-track line, some or all of it may be reduced to single track by installing CTC and controlled sidings. Depending on the spacing between sidings and the amount of two-main-track territory—if any—that is retained, a single-track railroad with CTC will have 70 or 80 percent of the capacity of the double-track railroad it replaced. By being able to reduce a substantial amount of double-track, track-maintenance costs and the cost of replacing rail, tie, and ballast over the long run are greatly reduced.

CTC (CENTRALIZED TRAFFIC CONTROL)

With the light density of traffic on the Alliance Division before the opening of the new mines starting in the early 1970s, operating trains under the timetable and train order system was perfectly adequate. The only line with CTC or even an automatic block system (ABS) in effect anywhere on the division was a "poor man's CTC" system of 238 miles between Alliance and Ravenna. There was a short segment of 13 miles of ABS over Crawford Hill and in the 4 miles adjacent to Casper, Wyoming.

The type of CTC in service between Ravenna and Alliance provided only limited capability to dispatchers to move trains once traffic began to increase to a high

level. The sidings used for meeting or passing trains had a power switch only at the east end of each siding and a spring switch at the west end. If the dispatcher wanted an eastbound train, rather than the westbound train, to take the siding to meet an opposing (westbound) train, the eastbound train would have to stop and line the spring switch by hand. This somewhat primitive version of CTC was adequate for the light density of train traffic in the 1950s when it was installed, and it allowed the company to eliminate a large number of operator positions at intermediate points to copy and deliver train orders. However, as train traffic built up to twenty or more trains per day by the mid-1970s, the old system had to be upgraded by installing power switches at both ends of the sidings, extending existing sidings, building additional sidings, and replacing the old code line that had become far too slow in sending indications of the position of switches and signals back to the dispatcher. At times, a dispatcher would have to wait twenty minutes for the field indication to come back.

Between 1974 and 1980, 623 miles of new CTC were installed on the Alliance Division, including 127 miles on the new line between Donkey Creek and Bridger Junction. See table 38.1.

On the Nebraska Division, CTC was put in service on the entire line between Lincoln and Ravenna in 1979 and the Lincoln–Napier line in 1983. On the Yellowstone Division, the entire line between Gillette and Huntley was converted to CTC in 1984.

On the Chicago Region, one major installation of CTC was made on the 98-mile line between Galesburg and Savanna, completed in the mid-1970s. All of the other coal routes on the Chicago Region (except the branch line between Galesburg and Peoria and 267 miles of the 283-mile between Bushnell and Metropolis, Illinois) were under CTC or had double track before the "coal boom" began.

INCREASING CAPACITY ON THE LINCOLN–KANSAS CITY LINE

To expand line capacity and reduce operating costs, BN built nine miles of second main track between its yard in Lincoln and Saltillo, on the corridor used by trains enroute to or from Kansas City and beyond. This

Table 38.1. New CTC Installed on Alliance, Yellowstone and Nebraska Divisions

Alliance–Gillette–Huntley	464 miles
Alliance–Northport	34
Northport–Guernsey	95
Guernsey–Orin	40
Donkey Creek–Bridger Junction	127
Lincoln–Napier	112
Lincoln–Ravenna	126
Dutch Creek–Decker	15
Campbell–Eagle Butte Junction	10
Reno–Black Thunder Junction	3
TOTAL	1,026 miles

Table 38.2. Gross tonnages handled between Alliance and Ravenna[1]

Years[2]	Gross tons handled[3] (millions)
1960s[4]	8
1975	27
1981	92
1983	118
1986	122
1988	200
2009[5]	226

1 Heaviest tonnage line except for the Gilette–Orin line on which 402 million tons were handled in 2009.

2 The years shown are those that were available throught the author's collection.

3 Combined eastbound and westbound tonnage.

4 Tonnage reported in the application for the 1970 merger.

addition was necessary to eliminate the congestion that occurs when only a single main track is in place adjacent to a terminal such as Lincoln. Trains will tend to "bunch up" where they must reduce speed and stop to either change crews and motive power or be switched. Also on that line, a new second main track was built on a grade of 0.4 percent over Firth hill, twenty-two miles southeast of Lincoln. The grade of 0.6 percent on the original main track restricted line capacity and required a third unit (SD40-2 or equivalent type) on only the 2.6 miles of 0.6 percent grade out of the 570 miles between Alliance and Kansas City. By retaining the old line, primarily for empty trains, BN gained an additional 7.7 miles of two-main-track territory.

Between Napier and Lincoln, empty trains are often run on an alternate route via Pacific Junction instead of on the more direct route through Table Rock. Having this flexibility adds greatly to line capacity between Lincoln, Napier, and Kansas City. However, the alternate route is twenty-five miles longer, if trains are run between Pacific Junction and Lincoln via Louisville, the most direct route, or forty-five miles, if run via Omaha (refer to figure 38.1). The track between Napier and Pacific Junction had to be upgraded with second hand-welded rail, crushed-rock ballast, and a heavy program of tie renewals to allow a speed of 40 miles per hour for the empty trains. It is operated under TWC (Track Warrant Control).

ADDITIONAL PROJECTS TO
INCREASE LINE CAPACITY

In 1977 a 5.4-mile segment of single track was converted to two main tracks between Waverly and Greenwood, Nebraska, a short distance east of Lincoln. Elimination of this restrictive piece of track greatly facilitated the movement of trains between Lincoln and the "three-way" junction at Ashland, which had become the heaviest tonnage line on the Nebraska Division. BNSF has added 9.4-miles of second main track east of Ashland on the Louisville line to increase line capacity on the preferred route between Lincoln and Pacific Junction.

In 2013 BNSF built a second bridge over the Missouri River at Plattsmouth and widened the 100-foot deep cut at the west end of the bridges enough to permit construction of a second main track. The additional 2.2 miles of second main track use the old bridge to form what has been designated the "River siding."

Hopefully, BNSF will soon be able to extend second main trackage on the 2.9 miles of single track that remain to form a combined route of seventy-eight miles of two main tracks between Lincoln, Pacific Junction and Balfour, Iowa.

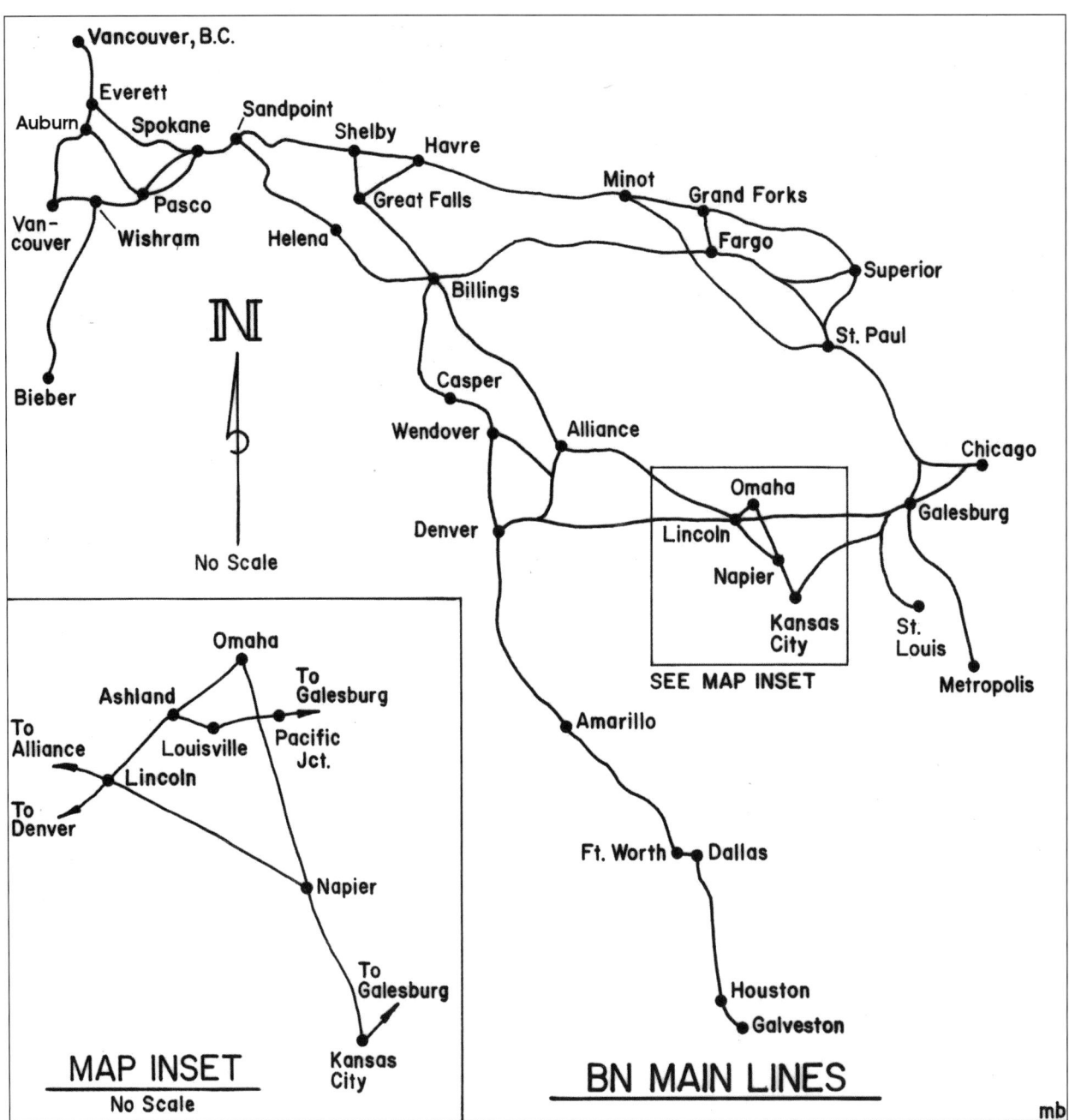

Figure 38.1.

By upgrading the twenty-seven-mile branch line between Hastings and Aurora to main-line standards, BN gained the capability of an alternate route for seventy-six miles between Lincoln and Aurora. Although the route using the Aurora–Hastings line is forty-seven miles longer than the direct route via Seward, it can be used to great advantage at times of traffic interruption for maintenance, accidents or adverse weather conditions. The portion of the alternate route between Lincoln and Hastings is operated under CTC, and the branch line between Aurora and Hastings, under TWC rules.

At Lincoln, two by-pass tracks were built around each side of the yard. Complete fueling and servicing capability was built at both ends of each track. Having the new by-pass tracks made it possible to keep the coal trains out of the yard and interfering with the switching and makeup of trains handling business other than coal.

One critical location that remains a single-track operation is through Grand Island, where BN built an overhead bridge to eliminate crossing the UP at grade, where the two railroads were running a combined total of about a hundred trains per day. In 2013 the 10.3 miles of single track through Grand Island were reduced to 2.2 miles by building an additional main track up to the start of the embankment approaching the bridge that crosses the UP. Since it appears the tonnage of coal to be mined and moved out of the Powder River Basin will gradually decline rather than grow any further, it seems unlikely it will become necessary to close up this short piece of remaining single track or any of the few remaining segments of single track on other major coal corridors.

On the Ottumwa Division, 10.5 miles of single-track CTC were upgraded to two-main-track capability by connecting the sidings at McPherson, Emerson, and Hastings, Iowa (between Pacific Junction and Creston). Also, powered crossovers were installed in two long stretches of two-main-track CTC. East of Creston, on the main line to Galesburg, there are 235 miles of double track operated under ABS and TWC, except for 18.6 miles in three short segments with two-main-track CTC. Although having so many miles of double-track and two-main-track capability with CTC gives this line a high capacity, it was difficult for dispatchers to deal with the problem of having to run trains with a wide range of allowable speeds. There were two Amtrak trains authorized to run 79 miles per hour, a few merchandise trains that may operated up to 60 miles

per hour, and a large number of coal trains running at much lower speeds of 20 or less, on the grades of 0.66 percent ascending eastbound at several locations on this line. With no powered crossovers outside of the short segments of CTC, it was difficult to arrange for faster trains to overtake the loaded coal trains, making it necessary at times to hold the slower trains for some time at Creston or Ottumwa until the faster train has passed.

The modern communications and signal equipment developed in recent years has provided the opportunity to build short "islands" of CTC at crossover switches to facilitate such overtakes out on the line of road without having to invest in a continuous system of CTC on hundreds of miles of double track. Plans to install several CTC islands in the double-track territory dispatched under TWC rules were announced in 2013.

That, in turn, enabled the power companies to sell electricity to their industrial and residential customers at reasonable rates. This has been possible even with the low BTU (British Thermal Unit) content and relatively high moisture content of the western coal. BN's efficiency kept the delivered price of the coal very low and kept the power companies from having to procure more expensive fuels for power generation.

Completion of most of the large track-maintenance and upgrading projects also had a highly favorable impact on the running times of loaded and empty coal trains. Eliminating many miles of slow orders that had been placed as temporary speed restrictions either for track-improvement work in progress or because of deteriorated track conditions awaiting corrective action reduced train delay, which in turn produced improvement in the utilization of train-operating crews and the number of cars and locomotives needed in coal service. Another benefit was that more trains could be run in a twenty-four-hour period on a given line segment when train speeds could be increased. Getting the track upgraded reduced the need to take track out of service for upwards of eight to ten hours when large track gangs were working. The "spot" maintenance work, needed to keep good track up to standard from then on, did not require taking the track out of service for so many hours. The overall operation improved enough to substantially reduce the cycle time on equipment, which then made it possible to move more coal with the same equipment.

CHAPTER 39

BRINGING THE C&S UP TO MAIN-LINE STANDARDS

THE COLORADO AND SOUTHERN RAILWAY

• *a small subsidiary railroad taking the initiative to reduce*
heavy curvature with major line changes •

THE COLORADO AND SOUTHERN RAILWAY (C&S) and Fort Worth and Denver (FW&D) were two smaller subsidiary railways of Burlington Northern that rose to prominence in the 1970s with the start of large coal movements to power plants in Colorado and Texas. Up to that point, they had been low-density operations, handling in the range of 5 million to 10 million tons per mile. Their lines were maintained only to the standard needed to support that level of business, and not at the standard required to handle the 40 million–plus tons that were moving over the C&S and FW&D by the late 1970s.

Both lines operated as separate companies, each with a president and board of directors. They had their own budgets for operations and maintenance, and for capital expenditures. At the time of the 1970 merger, BN owned 80 percent of the stock of the C&S, which in turn owned 99 percent of the FW&D. To keep from adding to the complexity of the application to merge the northern lines and the Burlington, the decision was made to not include the C&S and the FW&D in BN at that time. This avoided the need to negotiate with minority shareholders, which might have delayed the merger transaction.

As already mentioned, the density of traffic on both subsidiaries was light. A new schedule was set up for freight service (Nos. 77 and 78) between the Pacific Northwest and Houston soon after the merger. These trains operated via Missoula, Laurel, Casper, Denver and Fort Worth. The volume of cars moving end to end was small, making it necessary to add intermediate business to these trains to make a "full" train.

Control of the C&S was acquired in 1908 through the Burlington (CB&Q). James J. Hill wanted to move cotton westward from the South to the Pacific Northwest, where it would be loaded on ships of the Great Northern Pacific Steamship Company headed for Japan. To balance traffic and reduce empty car mileage, Hill worked with lumber companies in Washington and Oregon to develop new markets in Texas. There never was much success in developing westbound business, so for the most part, No. 77 consisted mainly of empty box cars and flat cars returning to the western part of the system. Besides Nos. 77 and 78, trains were operated as needed to handle local business and large volumes of grain at harvest time.

The C&S and FW&D never were big money makers. The operating ratio of the C&S ran in the range of 65 to 79 in the years before the heavy coal movement began in the early to mid-1970s. On the FW&D, the ratio fluctuated widely, but for the most part, it ranged between the mid-70s and mid-80s. Both companies had a high-quality fleet of motive power, consisting mainly of SD7 and SD9 units acquired in the late 1950s. The SD-type units of that generation of motive power produced high tractive effort at low speeds and could operate at a low minimum continuous speed of less than 10 miles per hour without getting below the short-time rating, which could damage the traction motors, as could occur with the four-axle GP-class units. The SD units were well suited for moving heavy tonnage trains of grain and ore on the fairly steep grades that were on parts of each railway.

The track of the C&S south of Denver and on the FW&D as far as Fort Worth was satisfactory for the amount of business handled up to the mid-1970s, when the heavy movement of coal began. On most of each railway's main lines, the rail consisted mainly of 112-pound TR jointed rail. Ballast was local material, generally scoria or slag. The tie condition was fair, adequate for the amount of tonnage being moved at that time and for the speeds (generally 40 miles per hour) at which freight trains were authorized to operate.

Grades were fairly heavy on several areas of the C&S. On the joint line between Denver and Pueblo (with C&S operating on tracks owned by the Santa Fe and Rio Grande railways), the maximum grade in the direction of the loaded coal trains was 1.42 percent, approaching Palmer Lake. In the opposite direction, the grade was 1.4 percent. Both the Santa Fe and the Rio Grande maintained their track to a high standard, fully satisfactory for the operation of the C&S–BN heavy-tonnage coal trains. The cost of using their tracks was based on the percent of the gross tonnage moving on the line. The only deficiency besides the heavy grade in this segment of the route between Denver and Pueblo was having a 32.4-mile section of single track between Palmer Lake and Crews. As the number of coal trains operated by the C&S increased, this segment became a bottleneck at times. The Denver–Pueblo joint line was dispatched by the Santa Fe. All C&S trains were operated with Santa Fe crews over the 115-mile joint line between South Denver and Pueblo Junction.

HEAVY CURVATURE AND GRADES

Between Pueblo and Trinidad, the maximum ascending grade in the loaded direction was 1.3 percent, and 1.17 percent in the direction of the returning empty coal trains. Between Denver and Pueblo, the maximum grade in the loaded direction (southbound) was 1.5 percent. Going south from Pueblo, the grade was 1.2 percent, and between Trinidad and Texline, 1.3 percent. Curvature was quite severe, with 102 curves of between 6 and 7 degrees and 26 curves between 5 degrees, 6 minutes, and 4 in excess of 7 degrees. This kind of alignment would

indicate that the C&S was built to minimize the cost of grading and not to obtain the efficiency of a low-grade, minimum-curvature railway line. Until the 1970s, no programs had been undertaken to make line changes to reduce the grades and curvature of the original line. Also, most curves were laid out as "simple" curves with no spirals. The alignment of the FW&D between Texline and Fort Worth is discussed in the next chapter.

Jim Daume, who served as chief engineer of the C&S from 1976 to 1983 and of the FW&D starting in 1982, provided a good summary of the standards to which the C&S had been maintained up to the time when programs were started in the late 1970s to upgrade the track:

For decades the C&S and the FW&D were ballasted with material from Twin Mountain, New Mexico. This ballast was scoria, a red volcanic cinder, the same material used for landscape purposes all over the country. It served well until the advent of heavy-tonnage unit coal trains. We tried using slag from the blast furnaces at CF&I Steel at Pueblo. That material was satisfactory, but supply and dependability were not good for the long term. We found several sources of basalt, a heavy, dense, volcanic material that tested very well. However, these sites were not pursued by the System Engineering Department of BN. Ballast from existing BN sources (Guernsey, Wyoming) was implemented but required considerable haul distance. A smiling side note concerning scoria cinders: when tested for ballast specifications, the material floated in one of the test chambers. Floating ballast from the C&S was a first for the lab![1]

Dave Burns was assigned as division superintendent of BN's Colorado Division (which included the C&S lines) in 1979. Dave has written about the situation he encountered at that time: "Being in Chicago, Missoula and Seattle during the '70s, I wasn't in on the coal action. Got to Denver in '79 just after the BN routes had been beefed up, but did live through two hellish years of running on the C&S with worn out 112-pound bolted rail on soft rock ballast—lots of 10 miles per hour track, fighting to avoid crews having to tie up on line. One aspect I remember well was the cash crunch and the concerns over paying our bills on a delayed basis."[2]

---

1   Conversations and messages exchanged between the author and James Daume, chief engineer of the C&S Railway.
2   Conversations and messages exchanged between the author and Dave Burns.

A RASH OF DERAILMENTS

Soon after BN, C&S, and FW&D began to run coal trains to electric power plants in Texas, the weakest spots in the track structure soon showed up. Jim Daume wrote:

> Unit train derailments were frequent and always involved multiple cars. It was not always the case, but more than likely the "bathtub" design coal cars would be involved. The 100-ton plus cars in units of a 100/105 car consist, on multiple curves and descending/ascending grades is a complexity of forces, to say the least. . . . The weakest fixture at a particular point could be triggered for failure by almost any sudden slack, gauge, variance, elevation, rail, and any number of other factors. The rail did its job, to a point, in that you could see it flex, yet maintain its girder strength characteristics. We will not belabor the cause issue. Everyone whoever investigated or picked up a derailment, without there being an obvious cause, have their own methods and theories for cause. . . .
>
> The only coal train derailment I ever witnessed was at Branson, Colorado. We were following a unit train by hi-rail auto, in the company of the system derailment team. This train was at near stall speed, when the cars began to scatter and pile up. Almost half of the train was on the ground before the dust settled. The derailment team could only say, "I don't believe it."

They left the next morning, Jim Daume recalls. Clearly, there was much work to be done to bring the railroad up to standard.[3]

Dave Burns commented on the operating and maintenance practices of the Santa Fe and Rio Grande between Denver and Minnequa on which the C&S had trackage rights, and for which the Santa Fe provided the train crews:

> The Santa Fe had done a good job of upgrading its track. The only real trouble we had was with the Santa Fe crews on weekends when the Denver Broncos were playing football, and we would run out of crews which caused us no end of trouble, having to hold them with no good places to do so. I never understood why the Santa Fe couldn't do a better job of controlling layoffs, but we had few derailments on their track.
>
> Not so on the C&S, especially south of Trinidad. On my first day on the Division, I was told to expect around two derailments per month. That didn't sit well with me, but on my first hi-rail inspection trip three days later, I saw first hand what coal trains could do to an otherwise nice 112-pound bolted railroad, especially on curves. . . . There was way too much superelevation left from the days of passenger-train operation. There was even a long 13-degree compound curve entering Trinidad where the superelevation had been taken out, but the high rail was braced with tie plates because wooden bracing simply shattered. I'd never seen stresses like this.

Dave requested that the program for rebuilding the railroad northward from Texline, scheduled to begin in the spring of 1980, be moved up and started sooner. During a meeting on that subject, those present were notified of a thirty-car derailment at Folsom, New Mexico. In going over the data for that derailment and others before it, Dave found that derailments were occurring on curves of 6 degrees and greater, on the high side of the curve and mainly in a sixty-mile stretch south of Trinidad where the track speed was 25 and 30 miles per hour. A slow order of 10 miles per hour was placed on the entire sixty miles. This created great problems in getting trains over the road, with "dark" (non-ABS) territory, and with 24-hour operators only at Trinidad and Texline, at the end points of the 137-mile subdivision, and at a station one-third of the way down that line. Having such a restricted operation with eight to twelve trains per day running on that line was very costly in all respects.[4]

A large part of the operating ratio was the result of the changes noted in increased spending for activities in the

---

3 The System Derailment Team was a small group of technical experts representing each discipline of the Operating Department—Engineering, Mechanical, and train operation—appointed to determine the root causes of derailments, examine maintenance practices and train handling techniques in place at that time, and determine what changes needed to be made in those practices and standards to prevent derailments.

4 "Dark territory" refers to a part of the railroad on which there are no block signals in service to govern the spacing between trains. At this time, train movement was authorized by written train orders issued by a train dispatcher and delivered manually to train crews at the start of their trip, and supplemented by additional orders delivered en route. Also known as "non-ABS," meaning there is no automatic-block-signal system in service between locations specified.

Table 39.1.

C&S "FACTS"

Miles of road—owned: 465 (main line)
Miles of road—under trackage rights: 135
Miles of road—branch line: 78
Merged with Burlington Northern December 31, 1981

Table 39.2.

COMPARISONS 1970 AND 1980

	1970	1980
Revenue	$87.7M	$186.1M
Operating income	$1.931M	$4.719M
Operating ratio	68.9	96.3
Transportation	33.7	48.5
W&S (Way and Structures)	12.6	22.0
M of E (Maintenance of Equipment)	15.3	22.1
G&A (General and Administrative)	7.3	5.7
Locomotive units owned	60	235
Revenue ton-miles	1.365B	7.230B
Revenue freight density*	1.973M	10.663M

* Gross ton-miles per mile of road
Source: *Moody's Transportation Manual*

Table 39.3.

C&S—REVENUE TON-MILES

Year	RTMS
1970	1.365 B
1971	1.581
1972	1.784
1973	2.289
1974	2.504
1975	1.911
1976	2.072
1977	3.322
1978	4.165
1979	5.606
1980	7.230

Revenue ton-miles increased by a factor of seven from 1970 to 1981. This level of increase put great stress on a track structure that was neither built nor maintained for tonnage reached by 1974, especially since nearly all of it was being handled in unit trains of 100-ton capacity cars, and in trains of about 14,000 gross tons.

From 1970 up to the merger of C&S and Burlington Northern, the volume of business handled on the C&S increased dramatically. Its lines became a key component of BN's primary corridor for unit coal trains moving to electric power plants in Colorado, Texas, and Oklahoma.

maintenance-of-way function. Note the increase from the very low level of 12.6 in 1970 to ratios to well over 20 starting in 1976.

What was memorable for me and got me sold on QIP [Quality Improvement Process] through involving employees in problem solving was the notice to crews I used to explain why we had put on this speed restriction and put them up against the clock. I also asked for their alertness in observing and reporting anything they would hear or feel as they transited the curves. Many

Head Brakemen began to ride out on the front platform of the locomotive, to give them a better chance to hear or see anything unusual occurring under the lead locomotive at they moved along. Interestingly, back came reports of hearing a cracking or groaning sound out of the low rail side. Section crews started laying along the low rail side to watch the track under load—even the local officers and I did, too, although in hindsight that might not have been the safest thing. But now came reports of noticing a small flex in the web on the low side, with the rail base angling slightly upwards simultaneously. This resulted in the high side wheel would be close to dropping in and putting overwhelming stress on the high rail.

I believed the reports and so did "super gandy" George Lamphier (no one else would) who, though mystified, sent carloads of compression clip anchors and emergency funding to get them installed quickly on every tie on the low side of the curves in the entire territory. And, the

Table 39.6.
OPERATING RATIO

Year	C&S
1970	68.9
1971	66.1
1972	68.6
1973	75.3
1974	81.6
1975	82.5
1976	80.2
1977	78.9
1978	104.3
1979	108.8
1980	96.3
1981	95.9

In the early 1970s, the c&s had a respectable operating ratio in the '60s, far better than the BN or its predecessors had achieved as far back as the 1920s. The operating ratio increased dramatically starting in 1973, when the heavy upgrading work was undertaken. It actually exceeded 100 in 1978 and 1979. Before 1981, under the accounting rules established by the ICC (the so-called "betterment" method) for trackwork, major track renewals such as tie, rail, and ballast replacement were chargeable to operating expense, rather than being capitalized. Once such major work was completed, together with the major growth in revenue with the increased coal business, the operating ratio would have declined dramatically. However, with the c&s being merged into BN in 1981, such separate numbers were no longer reported for the c&s.

Table 39.5.
C&S—MW&S (MAINTENANCE OF WAY AND STRUCTURES) RATIO

Year	C&S
1970	12.6
1971	10.1
1972	11.2
1973	13.2
1974	20.8
1975	16.8
1976	22.3
1977	20.0
1978	25.1*
1979	29.3*
1980	22.2*
1981	N/A

* Based on all operating expense related to MW&S in accordance with new ICC expense groupings effective January 1, 1978; not comparable to 1977 and prior years.

derailments stopped! The clip anchors seemed able to keep low rail bases from lifting up and letting the ball of the rail tip slightly outward, with or without web flex.

EXERCISING SELF-HELP

"With the rebuild came the standard 132-pound CWR, new ties, flatter curves, good granite ballast, and ultimate success regarding derailments. But what impressed me here was the initiative of Frank Thurston, Roadmaster, who used the rebuild process to 'boot leg' in shoulders wide enough to drive on for much of the entire territory

(most of which had no real road access). I took Grayson and Thompson over it in a business car, and they were impressed too. . . .[5] I never forgot the impact and the reception we got from actively involving those doing the work in ways to improve."

Carl Peglow, who worked as assistant superintendent–Roadway Maintenance on the Colorado Division from 1976 through 1981, remembers Frank Thurston as "the best Roadmaster ever." Carl reported that Frank was innovative and had the courage, resourcefulness, and desire to do whatever was needed to improve the railroad. Since the c&s was a small company, unorthodox managers such as Frank were able to get more work done and more quickly than often is the case in a large company in which many protocols and regimented

5   This refers to Richard Grayson, president of BN, and W. F. Thompson, senior vice president–Operations.

procedures must be followed before new projects or new methods can be undertaken.

As an example, Frank hired the Van Mytre brothers and their cousin who had run road-construction gangs in mountainous areas. They knew how to blast out the rock where sharp curvature needed to be reduced, and then get the loose rock material moved to locations where it could be used as fill for siding extensions and bankwidening. Carl said the brothers and their cousin became known as "the powder monkeys." Since Jim Daume had the authority from George Defiel, the president of the C&S, to buy whatever equipment and material were needed for such heavy work, it got done quickly, and without delay for scrutiny and evaluation from the expert staffs at BN headquarters. Frank Thurston declined promotion, as he did not want to leave the Trinidad area, where he had many ties in non-railway activity, such as membership on the board of a local hospital.

Peglow and Thurston convinced George Lamphier that he should authorize the rail-welding plant at Pueblo to weld recovered 90-pound rail to replace the 65-pound rail on sidings and some yard tracks that were needed to hold loaded coal trains. This bought some time until heavier second-hand 112-pound rail recovered in main-track-relay programs would become available as replacement rail in tracks other than main tracks.

In addition to the C&S main line south of Pueblo, Carl had responsibility for maintenance of BN lines that were under the Colorado Division. To give some relief to the Alliance Division, the Colorado Division was set up in 1976. The new division consisted of the main line of 386 miles between Denver and Hastings, and several branch lines in western Nebraska and eastern Colorado, plus the lines of the C&S. R. L. (Randy) Beem was named superintendent of the Colorado Division when it was established. In the Denver terminal, new yard tracks had to be built and some yard tracks extended and upgraded to handle the coal trains. Dave Hestermann, a general foreman on the Nebraska Division, was promoted to roadmaster in the Denver terminal of the Colorado Division and put in charge of these major projects. Dave rose to a high position in the Engineering Department of BNSF, that of assistant vice president–Chief Engineer of the north lines.

To be able to move coal to the Gerald Gentleman power plant near Wallace, Nebraska, Carl had to set up two steel (rail-laying) gangs to install second-hand

continuous welded rail on 116 miles of the 230-mile branch line between Sterling and Holdrege. The work of rebuilding this line with high-quality ballast, bankwidening, replacement of a high percentage of the ties, and replacing old, lightweight rail had to be done on those 116 miles in only two work seasons.

These demands put some tension on the division officers, as to the amount of time they would spend on improvements to the line of the C&S, compared to the BN lines that were under the Colorado Division. Having a division report to higher-level management in two companies at the same time was a most unusual arrangement, but Carl remembers that Dave Burns and his predecessor, Randy Beem, handled it well. They managed to prevent this tension from becoming a distraction and frustration to those who were "out there doing the work." Also, as many other maintenance officers felt, Carl appreciated the support and expert advice he received from George Lamphier and Don Rogers in the late 1970s and into the '80s—"the best railroaders on BN," in Carl's opinion.[6]

A HEAVY UPGRADING PROGRAM

As was the case on much of BN in those years, the lack of experienced foremen to run track gangs among the most difficult situations Carl and his counterparts had to deal with. There were gangs on which the foreman had seniority of barely one year. The training programs BN had at that time were good, but they could not keep up with the rate of turnover and the increasing number of gangs that had to be set up. Due to inexperience and lack of training for machine operators and track laborers, the number of injuries was quite high on some gangs in those years. These new employees were under exposure to injury since their work involved the use of heavy hand tools and handling material such as ties and rail. Hand tools still had to be used since not all trackwork was fully mechanized at that time. There was no program in place to train newly hired employees on safe work methods. They learned the work on the job through coaching and oversight from the gang foreman and their fellow workers. The turnover among these employees was high.

Heavy upgrading work was started north from Texline

---

6   George Lamphier served as director–Maintenance Programs for the BN system and Don Rogers as engineer–Track for the Denver Region.

Figure 39.1. The rail profile at the near right shows the extent of head wear that had become all too common on many curves in the track of the former C&S Railway. DAVE BURNS AND AMY JO STOCKINGER.

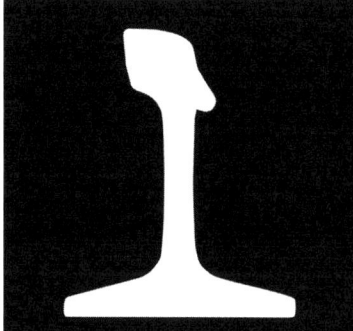

Figure 39.2. Comparison with the profile of the new rail to the right shows how severe the loss of metal in the head resulted from the ever-increasing tonnage of coal moving over the line. Improved rail lubrication and rail-grinding practices helped extend the life of new rail laid to replace the heavily worn rail on this line and virtually all other lines on BN's coal corridors. BN NEWS.

in the spring of 1980 with the installation of 132-pound welded rail, the standard section for rail laid on main tracks carrying coal. By then, enough tie-replacement work had been done to bring the tie condition up to a level of fair to good. Also, some of the most heavily worn rail on the curves had already been replaced with new continuous welded rail. Once the decision was made to undertake upgrading the C&S track, Dave recalls how quickly approvals for individual projects moved through the review and evaluation process BN used for major capital projects:

> One other facet of C&S operation worthy of note was the quickness with which line-change authorization and funding could be obtained, compared to the way it had been done on BN for similar projects.
>
> An example was the quick approval given for a major line change between Pueblo and Minnequa, where there was a grade of 2.2 percent, curves of 6, 8, and 9 degrees and on which there was a problem of derailments involving the use of helper locomotives. Dave wrote, "Jim Daume and I begged Jim Brown, the C&S president, to stick his neck out and authorize fast engineering and land acquisition to build a new line (with a lower grade) that wouldn't require helpers on coal trains. Within two

months the dirt was flying. I used this display of expeditious elimination of customary bureaucracy to highlight one of the nimbleness advantages of a small organization.

Jim Daume served as chief engineer in those very challenging times. In recalling the projects undertaken to improve the track in those years, Jim wrote,

> At every opportunity a sharp curve was reduced, an embankment was widened, superelevation adjusted, along with ballast stabilization. A machine was seldom idle, there was always another project. The "locals" embraced this approach and accomplished many roadbed improvements. Much of the work, or at least some, circumvented the capital accounting procedures. If we were in the area on routine work and the opportunity arose to combine the task with embankment work, it was done! You will note in many curve areas that the fence line may be 20 feet from the center line of the track on one

side and 80 feet on the other. This was a curve reduction we made while staying within our property boundaries.

A line change at Pueblo was 3.1 miles long and eliminated a helper district for loaded coal trains . . . This was the project that was approved, on the spot, by the Board of Directors in their meeting at the Brown Palace Hotel in Denver. The verbal authority to proceed, for a proposal of this scope, was indeed management at its best.

This project was especially significant because of the reduction in grade from 2.2 percent to one percent. This allowed the elimination of helper locomotives on the coal trains. The line was shortened by 0.53 miles. Curvature was reduced from three curves of 6, 8, and 9 degrees on the original line to four curves of 3 degrees, 41 minutes, or less.[7]

All of the property required for the line change at Pueblo was secured from the Colorado Fuel and Iron Company (CF&I), which had produced large quantities of new rail for the C&S, BN and the former CB&Q railroad.

Many more ambitious projects were undertaken by the C&S management in the early 1980s to bring the railway closer to the standards needed for a high-tonnage route for unit coal trains. These projects included both upgrading the track and increasing the capacity of the line. In addition to replacing rail on the entire line, fifteen sidings were extended for the meeting of trains moving in opposite directions. Curvature was reduced on fifty-three curves. Five major line changes were made to reduce curvature. The use of welded rail required placement of a much stronger ballast section, with higher-quality rock, and at the depth and shoulder width needed to support the new rail and tie structure. That in turn

required widening the embankment on which the track was laid. All of this work required training and close follow-up with maintenance supervisors and the foremen of gangs who in many cases had little or no experience in upgrading or building a railway to these standards.

It was fortunate that nearly all of the rail on the C&S south of Denver had been the 112-pound TR design developed by the CB&Q's Engineering Department in the 1940s. The extra girder strength and additional metal in the head of the TR design made it more durable, giving it a longer service life than the 112 RE section adopted by many other railways. The 112 TR in service on the C&S was laid in the early to mid-1950s. Assuming gross tonnage of about 8 million per year, this rail probably had only about 150 to 200 million accumulated gross tons by the late 1970s when the coal tonnage began to hit. At that time, the anticipated life of jointed rail on tangent or light curvature track was between about 400 and 500 million gross tons. On that basis, the 112 TR rail had enough life remaining to carry the tonnage until funds could be made available to relay the rail, hopefully, by the early 1980s. As a result, the TR rail lasted longer under the greatly increased tonnage than the RE rail would have.

To further buy some time until enough money became available (starting in the early 1980s), George Lamphier set up a program to start replacing the 112-pound jointed rail with 132-pound welded rail on the sharpest curves and, if enough rail was available, on tangents between curves if less than one-fourth mile in length. That approach made it possible to get the most heavily worn and highest-defect rail out, making the track safer, reducing short-term maintenance costs, and eliminating or preventing many miles of slow orders.

MAJOR LINE CHANGES UNDERTAKEN

In addition to the line change at Pueblo, discussed earlier in this chapter, major line changes were made at these locations:

TRINIDAD (BETWEEN MP 210.32 AND 214.65)
- Length: new construction of 1.23 miles
- Line shortened by 4.33 miles
- Curvature reduced from twenty curves with a maximum of 11 degrees, 54 minutes (and compounded to a 3-degree curve), on the old line to three curves

---

7   Author's note: I recall several occasions when Bob Downing related the decisions made in this meeting that would expedite the work needed to upgrade the C&S. At that time, Downing served as chairman and CEO of the C&S. Jim Daume was asked to remain just outside the meeting room in case his input was needed when this project was to be discussed by the board. However, once Downing and other board members understood the benefits of the project and the engineering work it would involve, they wasted no time in decreeing it approved. Downing then stepped out into the hall and told Daume it was approved and to get the work underway. No detailed paperwork had to be run through the approval process, based on Downing's interest and support of the project.

of 2 degrees, 10 minutes; one curve of 1 degree, 57 minutes; and one curve of 3 degrees, 23 minutes.

- Property for this line change had to be purchased from adjacent land owners.

TRINCHERE (MP 255.3–257.48)
- Length: 1.7 miles
- Line shortened by 0.31 miles
- Curvature reduced from seven curves (with four curves between 6 and 7 degrees) to one curve of 1 degree and one curve of 2 degrees.

BRANSON (MP 261.7–263.6)
- Length: 1.8 miles
- Six curves of 6 degrees, one curve of 4 degrees, and one of 3 degrees were reduced to one 3-degree curve; one curve of 3 degrees, 50 minutes; and one curve of 1 degree (on the old alignment, there was less than 100 feet of tangent between two reversing 6-degree curves).

ALPS (MP 270.1–274.0)
- Length: 3.9 miles
- Reduced from 13 curves of 4, 5, and 6 degrees to seven curves (five of 3 degrees and two curves of 2 degrees).

These major line changes involved much more than simply lining curves over a few inches to take out small deviations from "true" curvature.

In addition to the reduction in curvature gained with the major line changes, curvature was reduced at many other locations, by the means described by Jim Daume and Dave Burns earlier in this chapter. Between Pueblo and Texline, there were 102 curves of 6 degrees and over, plus two curves of 8 degrees, one of 9 degrees, and one of 10 degrees, 16 minutes. A total of fifty-three curves of 6 degrees and over were reduced to 4 degrees or less, with many at only 2 or 3 degrees. There still are some curves of between 5 and 7 degrees remaining in this line, but the large reduction in curvature that was made in upgrading the railway in the 1970s and early 1980s has paid off handsomely through reduced maintenance, longer rail life and less need to delay trains for heavy maintenance work.

The only major line change proposed that was not made in these years was at Folsom, New Mexico. Curve 281 at Folsom is about 0.6 miles in length, a curve with five compounds between 5 degrees and 6 degrees, 9 minutes.

The Engineering Department of the C&S was ready to undertake this project in 1981 when BN became ready to officially merge with the C&S. That plan, together with a major reorganization that would put the C&S line under the Fort Worth Division of BN, caused this project to be set aside. BN's priorities and standards for evaluating capital expenditures were changing at that time. Only those projects with a short-term payback and return on investment (ROI) of 25 percent or higher were looked upon favorably. The ROI of an individual line change generally was quite small. Unless one was willing and able to look at a line change as one of several projects needed to improve the quality of track on a given corridor over several years, it would likely not be approved. Hence, the line change proposed at Folsom is yet to be done. With directional running now in place on this corridor, the need for it has diminished.

Sidings were extended at fifteen locations to between 7,500 feet and 9,000 feet, for an average spacing of 15 miles between sidings over the 228 miles between Pueblo and Texline.

A VERY SUCCESSFUL EFFORT

What was accomplished in these years through the initiative of the Engineering Department and division-level officers, together with support of higher-level managers, is a clear demonstration of what can be gained when a team of resourceful, hard-working, and committed leaders is given the resources and the authority to take on ambitious projects that may have seemed by some as too formidable to take on at a particular time. The successful efforts of this group of officers brought the coal-carrying lines of the C&S up to the standard to which BN lines also had been upgraded.

Efforts to get electric-power companies in Texas and Oklahoma to agree to rate increases high enough to give BN a reasonable return on the investment were rejected. Even though the ICC approved at least part of the increases BN requested, the power companies sought relief through the courts. A protracted process of appeals ensued for several years. In time, BN obtained most of the rate increases it sought, and long-term contracts were finally negotiated and agreed upon between the parties. It was tough for BN to have to be "at war" with its customers, but it was mandatory for BN to be able to earn enough from hauling coal to pay for the massive investment that had to be made in infrastructure on lines with the challenges of grades, curvature, and sub-standard track it had on the C&S.

The accomplishments of the C&S team in those years went generally unheralded except for one issue of the Denver Region newsletter for December 1980. In other company publications, the projects done on the C&S were not given as much recognition or publicity as those carried out on BN's Alliance and Nebraska divisions and on the Chicago Region. In the newsletter mentioned, Jim Daume provided a summary of the work completed on the C&S in 1980:

> This has been a good year for the C&S maintenance-of-way programs, particularly on our first subdivision which runs from Pueblo to Texline. We have an accelerated line change and curve reduction program on this subdivision, calling for several million cubic yards of earthwork. So far this year, we have completed a 1.8-mile line change, reduced 24 curves, and completed three unit coal train sidings.
>
> During 1980, there were 34 miles of CWR relay on the first subdivision, and system-wide, we utilized 200,000 tons of ballast in an aggressive surfacing program. . . . We'll see nearly 100 percent completion of our projects by the end of this year and much of the credit must go to our employees, along with the employees of the other departments we work with so closely. They have given us an all-out effort in getting all this accomplished. I'm very proud of all our people. . . . Our system is in the best shape we can remember and we're all quite optimistic about the C&S and its progress.

The article went on to say that 1981 promised to be another busy year, with forty miles of new relay and the spending of about $2 million on more line changes and curve reductions.[8]

***

8   "C&S Engineering Closes a Busy Year," *Denver Region News*, December 1980, page 4.

The C&S was run as a separate company even though nearly all of its stock was held by BN. In the years that so much work was being done, a railway as small as the C&S could only finance a small part of the cost of the many large projects that had to be undertaken. The same was true for the FW&D, which in turn was owned by the C&S. Nearly all of the members of the C&S board were senior executives from BN. The board did not get involved in the day-to-day matters on the C&S. Responsibility for running the railroad was left to officers assigned to the C&S.

It was fortunate that, in the years the C&S had to undertake the rebuilding of the railway, Bob Downing and Tom Lamphier were on its board. With their background in engineering and maintenance Downing and Lamphier understood and supported the need to authorize major expenditures for line improvements. They also were able to give coaching and direction to the president and department heads on the C&S and the FW&D. Without that support, the presidents in those years, some of whom had no background in railway maintenance and operations, might have been hesitant to move ahead with the expensive projects needed to overcome the danger of derailments and to upgrade the track to the same standards set for the coal corridors on BN.

BN officers at all levels (including those on the board of the C&S) really had their hands full in managing the expansion and improvement programs on BN in those years. It is good that managers on the C&S used this as the opportunity they needed to take hold of things and get them done, and not wait for direction or micro-managing on the part of BN personnel. The process of getting the work done would have been slowed down if they had been expected to wait for orders from their counterparts at BN headquarters. As a result, innovation and a spirit of entrepreneurship flourished on the C&S. We were fortunate to have had a group of "can do" people holding responsible jobs on the C&S in those years.

In 1981 the C&S was merged into BN and put under the Fort Worth Division.

CHAPTER 40

UPGRADING THE FORT WORTH AND DENVER RAILWAY

• a small railroad upgraded to heavy-haul capability
to meet the energy needs of Texas •

THE COLORADO AND SOUTHERN RAILWAY acquired ownership of the Fort Worth and Denver Railway in 1898. On December 31, 1982, the FW&D was merged into Burlington Northern. BN and C&S merged on the same date in 1981. In 1908 James J. Hill had directed the Burlington to acquire the lines of the C&S and FW&D to gain access to the cotton-growing areas of Texas. Hill believed there was a huge market potential in Asian countries for clothing made from cotton. To provide loads for cars made empty at Seattle when the cotton was transferred to ships, Hill set favorable rates on lumber needed for construction in Texas.

Until coal trains started to move to power plants in Texas and Oklahoma in the mid-1970s, the density of traffic on the FW&D was light. As on the C&S, the FW&D had to undertake major track-upgrading programs. Most of the work was completed in the late 1970s.

The amount of trackwork done in those years was far greater than what would have been needed to maintain it at the standard that was adequate to handle only the non-coal business moving on the C&S and FW&D in those years. The rates charged for hauling coal to the power plants had to be set at levels high enough to provide BN and its investors a reasonable return on the money spent to upgrade the railroad and expand its capacity. BN's actions in raising rates resulted in protracted disputes before the ICC and, later, in the courts.

The trackwork required on the FW&D was major, although not as formidable as what had to be done on the C&S. Curvature was much less on the FW&D, which allowed the accumulated tonnage to be much higher before rail became so heavily worn that it had to be replaced. When the heavy coal movements started, nearly all of the north end of the FW&D, on the 120 miles between Texline and Amarillo, was laid with 90-pound rail—too light and too heavily worn to be kept in service once the unit-coal-train movement started. Further south, between Amarillo and Fort Worth, the rail consisted almost entirely of 112-pound TR jointed rail, with a few miles of 110-pound and 112-pound RE rail. The heavier 112-pound rail was much more serviceable than the 90-pound rail north of Amarillo, but still had to be replaced.

With some sidings added and a number of them lengthened, sixteen long sidings were available for meeting and passing trains on the 218-mile subdivision between Valley Junction (118 miles north of Fort Worth) and Amarillo, for average spacing of about 15 miles. On the 117-mile subdivision between Amarillo and Texline, seven sidings of sufficient length were put in service, at an average interval of 17 miles. Between Valley Junction and North Yard in Fort Worth (118 miles) eight "good" sidings were in place, with spacing of 15 miles. This siding configuration was fully satisfactory for the number of trains needed to deliver the coal, and for the amount of non-coal business projected.

A major line change at Decatur, Texas, made it possible to eliminate the use of helper engines on loaded coal trains.

The rail on the entire line of 450 miles between Texline and North Yard in Fort Worth was relaid between 1978 and 1985. Together with heavy tie renewals and the use of high-quality ballast, the relay of rail on the entire line produced track of the quality needed to handle very heavy tonnages of coal. That permitted the establishment of interdivisional runs and the reduction from five crew districts to three between Trinidad and Fort Worth. With the completion of these major track-improvement projects, the cycle times on unit coal trains improved greatly, making it possible to reduce the number of train sets and locomotives in service to move a given level of tonnage.

Tom Lynch, who became division superintendent at Fort Worth in 1982, recalled,

> They were still using motor cars to patrol track. The shops at Childress were still open. In the summer of '83, Wayne Arntzen and the Denver Region took over the Fort Worth Region and Brown [Jim Brown, regional vice president at Fort Worth] was leaving. Updates to dispatching, communications and maintenance quickly followed. . . . with the new technology BN brought to Texas we cut maintenance expense, clerks and operators everywhere from Galveston to Pueblo.[1]

Track Warrant Control (TWC) was implemented in all non-CTC territory, and all maintenance and operating practices were integrated into the BN system.

The wasteful, redundant one-on-one (with the regions having only one division reporting to it) Fort Worth Region/Fort Worth Division Organization that had been set up shortly after the Frisco merger was broken up when its territory was taken over by the Denver Region. This change saved many managerial and staff positions and eliminated the micromanaging and unnecessary complexities in organizational processes that stifled the initiative of those who wanted to make improvements in operating and maintenance practices.

---

1   Tom Lynch, correspondence with author.

Table 40.1.

Fort Worth–Dallas (on Rock Island)	36 miles
Dallas–Waxahachie (on Missouri–Kansas–Texas Railroad)	28 miles
Waxahachie–Teague (on Joint Texas Division*)	67 miles
Teague–Belt Junction (on Joint Texas Division*)	147 miles
Belt Junction–New South Yard (on Houston Belt and Terminal Railway)	12 miles
New South Yard–Galveston (on Sante Fe)	48 miles
TOTAL	338 miles

* Track of the former Burlington–Rock Island Railroad Company, owned in equal shares by BN and RI, and later, operated as the Joint Texas Division. After the RI ceased operation in 1980, BN became the sole owner.

In addition to the 450-mile "coal corridor" between Texline and North Yard at Fort Worth, the FW&D operated on the 338 miles between Fort Worth and Galveston, much of it on trackage owned by other railroads or owned jointly with the Rock Island. See Table 40.1.

By having trackage rights on several railroads, the FW&D was able to reach Dallas, Houston, and Galveston. Having trackage rights of this extent was expensive, especially with volume as small as the FW&D had. After the Rock Island ceased operation, and with new contracts that were negotiated with the Union Pacific after its consolidation with the Missouri Pacific and the Missouri-Kansas-Texas Railroad, BNSF was able to simplify these arrangements.

An example of the kind of strife and maneuvering for advantage that sometimes occurs in operations on joint facilities occurred on the Joint Texas Division (JTD) in the early 1970s. This was during one of the five-year intervals in which the Rock Island was in charge of managing the JTD (management of the JTD rotated every five years between BN and the Rock Island). Bob Downing, who was serving as president of the C&S and FW&D in addition to being president of BN, recalled that BN had unloaded several miles of new continuous welded rail on the JTD where it was to replace the old heavily worn

Table 40.2.

FW&D FACTS

COMPARISON: 1970 AND 1981

	1970	1981*
Miles of road owned	995	926
Revenue freight density**	1.217M	8.329M
Revenue	$19.5M	$177.1M
Revenue ton miles	1.493B	9.837B
Train miles	1,158,135	2,743,112
No. Employees	918	1,671

* Last full year before merger with BN

**Defined as the number of revenue tons moved over a defined
   area in one year

Source: *Moody's Transportation Manual*

Table 40.3.

FW&D OPERATING INCOME, 1970–1980

Year	Operating Income
1970	d $.094M
1971	d .236
1972	d 1.585
1973	d .197
1974	d 3.418
1975	2.072
1976	.650
1977	1.811
1978	6.008
1979	17.073
1980	19.410

(d) indicates a deficit.

Source: *Moody's Transportation Manual*

rail that had a high defect rate. The new rail had been unloaded in the area where it was to be placed in the track, but the work had not yet been started when the five-year term for BN to manage the line expired. Shortly after the Rock Island began its next term of managing the JTD, its management ordered that the new rail be picked up and moved to a location somewhere on the Rock Island where it would be laid in their track. Apparently the Rock Island interpreted that the joint-facilities agreement for the JTD allowed it to appropriate material from the JTD for its own use without approval of the BN, even though BN had furnished the material.

In several conversations I had with Bob Downing in later years, he recounted the pressure he put on the Rock Island to either return the rail to the JTD where it was supposed to be laid or pay BN for it. In desperate financial condition by this time, the Rock Island made no response either to return the equivalent amount of rail to the JTD or to compensate BN for its share of the expense. BN simply was out the money it paid for the rail.

Tom Lynch recalls that after Robert VanMatre, the contractor Frank Thurston used so successfully in making line changes on the C&S, completed those projects, he came to Teague to oversee the clean-up of scrap, old buildings, and unused track material that had accumulated for many years on the JTD. It took two years to get

The FW&D reported a deficit in operating income from 1970 through 1974. As the coal tonnage increased, operating income rose dramatically by 1980, the last year before the FW&D was merged into BN.

the job done. This was another example of how poorly managed the JTD had been under the joint ownership and rotating management of the two owning railways.

The FW&D was not able to compete well for non-coal business in Texas with the two dominant railroads, the Southern Pacific and Missouri Pacific. The FW&D did not have direct access to the chemical plants and oil refineries in the Houston area along the Gulf Coast. Instead, the FW&D had to depend on locally originated business from the agri-business sector for most of its revenue. Of course, when some of the large utilities in Texas and Oklahoma had to turn to low-sulphur coal mined in the Powder River Basin, the standing of both the FW&D and the C&S rose considerably. With the amount invested in the track of both companies, they became more respected and more widely known as prominent providers of rail transportation service.

Table 40.4. FW&D OPERATING RATIO, 1970–1981

Year	Operating Ratio
1970	83.8
1971	78.2
1972	83.1
1973	74.5
1974	85.7
1975	72.9
1976	76.8
1977	72.0
1978	90.5
1979	81.7
1980	85.8
1981	85.5

Source: *Moody's Transportation Manual*

The FW&D's operating ratio fluctuated widely from 1970 through 1977, depending on the amount of grain originated and the amount of track-maintenance work that was done. Beginning in 1978, the operating ratio increased to a high level, due mainly to the large amount of track maintenance that was chargeable to operating expense under the accounting rules in effect at that time.

Table 40.5. MW&S* RATIO

Year	MW&S Ratio
1970	15.0
1971	14.1
1972	15.8
1973	14.9
1974	17.8
1975	17.9
1976	18.5
1977	20.3
1978	23.0**
1979	14.7**
1980	17.9**
1981	16.0

* Maintenance-of-Way and Structures

**Based on all operating expense related to MW&S in accordance with new ICC expense groupings effective January 1, 1978; not comparable to 1977 and prior years.

Source: *Moody's Transportation Manual*

The MW&S ratio increased considerably beginning in 1974 when major track upgrading projects were undertaken. The ratio increased above 20 for two years, and almost unheard of level, reflective of BN's desire to quickly bring the track on the FW&D's coal-hauling corridor up to the standard needed for a safe and efficient operation.

CHAPTER 41

REFLECTIONS ON THE WORK

"HOW FAR WE'VE COME"

I've seen it go from a local line in serious disrepair to an
ultra-modern state-of-the-art segment—the busiest of stretch of railroad in the world.

BILL CLARK, TRAINMASTER/ROAD FOREMAN,

QUOTED IN *BN NEWS*, SEPTEMBER–OCTOBER 1986

Track upgraded to handle coal tonnage far greater than anticipated,
reaching annual tonnage of from 200 million to 400 million gross tons, on some line segments.

. . .

Since 1970, we have invested $2.7 billion to improve Burlington Northern's performance.
The largest portion of this investment has gone to increase our railroad's capacity to do business . . .
and by rebuilding and refurbishing our rail plant.

. . .

In 10 years, Burlington Northern's consolidated net income has grown at
an average annual rate of 22 percent. Consolidated revenues have increased
at an average rate of 13 percent annually.

BURLINGTON NORTHERN 1979 ANNUAL REPORT,

LETTER TO SHAREHOLDERS, PAGE 3

A NUMBER OF EMPLOYEES WHO BEGAN their careers on the coal corridors in early years of the coal boom have shared some thought-provoking reflections on the part they have played in raising the railroad's operational effectiveness to a high level over the past thirty to forty years.

The remembrances of several employees who were working on the day the first train operated out to the end of the new Gillette–Orin line were included in an article, "A Memorable Moment in Time," in the December 2009 issue of the *Powder River Division Reflections* newsletter. After the passage of thirty years since the line was opened, those employees revealed the significance the opening day had in the early years of their careers, and in how much more capacity has been built into the Orin line since it was opened, far more than was envisioned by BN when the decision was made to build the line.

A number of those remembrances follow:

Recalled T. J. Smith, locomotive engineer: "Railroaders have masters, doctorates, GED diplomas or certificates and everything in between and we've all helped build the coal corridor. . . . It is simply the class of people BNSF has in the Powder River Division."

On the day the Orin line was put in service, Joleen

McIlravy was working as an operator at Reno and copied the train orders that sent the first train down the new line. She watched the first train travel down the track through the traditional banner. "It was a wonderful, exciting experience. I had been part of the congested railroad prior to this. I only saw opportunity. The present-day technology is impressive. To see how we evolved is almost unbelievable."

John Ford, conductor, remembered: "It [BN's success in moving the coal] was all because of the Orin line. It has been amazing to see the progress. It is hard to believe I've been a part of this so long. Thirty years went by fast."

Roadmaster Charlie McCoy looked back: "I am amazed at how much work we've got done and how it went. I have experienced it.... I am very proud to be part of the Orin line. It is in my blood. I don't want to leave it."

Chuck Garrett, locomotive engineer, was at the controls of the second empty train leaving Guernsey for loading at a mine on the new Orin line. "At the time I knew it was an important event, but I probably didn't take it as seriously as I should've looking back. It was an important step for myself and the railroad. I feel I have contributed greatly to its development. I am very glad to have been and still be around to be a part of it."

Jake Greeling was serving as assistant superintendent at Gillette on that day and set up the special event to commemorate the operation of the first trains to use the new line. "We had to build the yard (referring to the new yard and support facilities at Guernsey) for coal service. It was a monstrous task.... The whole division had to be rebuilt to take the stress and burden of the coal freight. ... It was a great sight to see that long unit train (the first loaded train) coming around the connection at Bridger Junction. It was a special event, a historical moment."[1]

By the end of 1979, the main lines of the Alliance Division had been upgraded from a light-density secondary railroad to the standards needed to handle annual gross tonnage nearing 100 million tons per mile. These tonnages were by far the heaviest of any lines on the BN system. The lines of the coal corridors in Wyoming,

Nebraska, Missouri, Iowa, Illinois, Wisconsin, and North Dakota handling heavy tonnages had also been upgraded substantially by the end of BN's first decade of operation. This turned out to be only the beginning of spectacular growth in coal tonnage that reached 400 million gross tons (combined tonnage of BN and UP) handled by on parts of the new Gillette–Orin line by 2010.

While not all of the work required to expand capacity and raise the quality of the track to higher standards had been completed by 1979, the lines on the coal network had been improved so much that we could pause for a moment to take stock of what we had accomplished in only five years. BN's annual report for 1979 contained many impressive numbers showing the magnitude of work completed, along with year-to-year comparisons of financial and operating performance. These numbers show where we stood in 1979 as we looked ahead to the new challenges we would face in the next decade.

In 1978 an average of 1,151 miles of track were under temporary speed restrictions (slow orders, in railroad terminology) due to work in progress, or where necessary maintenance or upgrading work had not yet been completed. At the end of 1979, that number was reduced to 621 miles. In 1975 BN had reported 2,369 miles were under slow order (these mileage figures cover the entire BN system, not only the miles designated as "coal corridors").

Other interesting comparisons between 1979 and 1978 are the following:

- Coal tonnage increased 44 percent in just one year.
- Railroad pre-taxed income increased from $9.1 million to $41.0.
- Consolidated earnings rose 53 percent, from $114.5 million to $175.6 million.
- Return on average net investment in the railroad increased from 2.7 percent to 3.8 percent, still substandard, but an improvement.
- Corporate return on equity increased from 6.0 percent to 8.7 percent .

Comparisons of BN's performance from 1971, the first full year in BN's formation, to 1979 are even more impressive. See Table 41.1.

---

1   The Powder River Division was established after the merger that formed BNSF Railway in 1995, when the territories of the Alliance and Colorado divisions were realigned.

Table 41.1.

	1971	1979
Operating revenue	$1.1B	$3.3B
Operating income	$80M	$256M
Dividends per share	$1.50	$1.95
Capital expenditures	$146M	$542M
Revenue ton-miles	65M	135M
Ties inserted	1.4M	3.6M
New rail laid (mi.)	159	768
Track-mi. of CWR* in service	3,202	6,682

* Continuous welded rail

Source: "Ten-Year Summary of Operations; Ten-Year Financial
and Statistical Summary," Burlington Northern *Annual Report
for 1979*, page 31.

SUMMARY OF BENEFITS GAINED FROM
BN'S LARGE INVESTMENTS
IN CAPACITY AND
TRACK-IMPROVEMENT PROJECTS

The immediate pay-off for the large investments BN made
in additions and improvements was to gain the capacity
to handle the rapidly increasing demand for low-sulfur
coal. At the same time, it provided ability to increase the
efficiency of the operation, as represented by a reduction
in cycle time for the unit train equipment provided by a
combination of BN, the electric-power companies, and
some third-party investors. Bill Greenwood, division
superintendent, reported that cycle time was cut 24 percent from November 1979 to June 1980 by the opening of
the new Donkey Creek–Orin line alone. The installation
of four hundred additional miles of CTC was in progress,
which would soon provide additional line capacity, by
overcoming the limitations of the old system of dispatching by train-order authority for movement.

Jerry Pinkepank added, "Just 10 years ago, we saw
barely 15 trains a day over this division, and very few of
them were coal trains." A writer of BN *News* noted that
ninety trains were being operated daily on the Alliance

Division, and that sixty-eight of them were hauling coal
or returning empty for their next load.[2]

During the early 1980s, additional improvement
in cycle time became quite dramatic, as revealed by
John Hertog in an interview with *Railway Age*. Hertog
reported that in August 1982, BN loaded 1,061 coal trains,
a record high. He added, "The other side of the coin is
that BN could still haul even more coal. During most of
last year and this year, we have had thousands of coal
cars and hundreds of locomotives sitting idle.... At last
count, we had 5,800 shipper and railroad-owned coal
cars in storage."[3] That apparent "surplus" of equipment
made it possible for more power companies to purchase
coal mined in Wyoming and Montana, without having
to invest in still more cars. The number of idled cars
and locomotives was soon worked down, as business
increased. It should be pointed out that many locomotives were idled at that time due to the general economic
recession affecting other lines of business and because of
improvements made in utilization of locomotives used
in other than coal service.

The work done by BN to build a high-capacity and
highly efficient network of coal corridors has made it
possible to deliver huge tonnages of coal that is in compliance with Environmental Protection Agency standards for emissions, and at very low rates. The power
companies had to shift a large part of their sourcing for
coal from mines in the Midwest and Appalachians to the
low-sulfur coal produced in Wyoming and Montana. The
length of haul for that coal would be much greater, and
it was vital for the power companies that BN develop a
very low-cost, high-capacity delivery system to plants
scattered throughout the states of the Midwest and
southwestern parts of the country. Not too many years
later, coal mined in the Powder River Basin was being
shipped to the southeastern and northeastern states and
Ontario as well.

That, in turn, enabled the power companies to sell
electricity to their industrial and residential customers

2   "Alliance Thrives Amid Tough Challenges," BN *News*, July
1980.

3   "New Records for BN Coal," *Railway Age*, October 11, 1982,
p. 12.

at reasonable rates. This has been possible even with the low BTU content and relatively high moisture content of the western coal. BN's efficiency kept the delivered price of the coal very low and kept the power companies from having to procure more expensive fuels for power generation.

In ten years, consolidated net income for all BN units—transportation, resources and other non-rail units—grew at an average annual rate of 22 percent. Capital expenditures totaled $2.7 billion, with $1.1 billion of that amount spent on BN's railroad company to increase roadway capacity and $590 million for coal-hauling cars and locomotives.

Overall, it was a good showing, one in which every member of the BN team could be proud. But this was not the end of the journey to improve BN's operating and financial performance. Another $620 million of capital expenditures in rail projects was slated for 1980. BN's people would be challenged to maintain the thousands of miles of track that had been upgraded to safely carry record-high tonnages at high standards. Increasing the rate of return on the company's massive investment would require a greater reduction of operating expenses by improving productivity and by raising rates on coal shipments that had been set too low in the early years of what had developed into the "coal boom."

CHAPTER 42

MECHANICAL DEPARTMENT UPGRADES

This project was completed in 1979 at a cost of $47 million. The new locomotive shop was designed to handle routine maintenance on six hundred assigned units and to repair an average of seventy cars per day in the new car shop. The locomotive shop was equipped to handle both running repairs and heavy repairs. The portion of the shop built for running maintenance and the car shop were open first, in September 1978. The heavy-repair section was opened in 1979. The new facility included a servicing facility located outside the locomotive shop for fueling, sanding, cleaning, and inspecting locomotives. The new shop had the capacity to rebuild older units built by GE, usually completing two units per month. Using new employees to overhaul locomotives was found to be a good method for training of apprentices, and it was preferable to sending all units needing heavy repairs to the system shop at West Burlington.

As the number of units assigned to Alliance for running maintenance was increased from the planned capacity of six hundred units to seven hundred units (23 percent of BN's total fleet), it became necessary to reduce the amount of space allocated in the shop for heavy repairs. Also at this time, the Federal Railroad Administration allowed the railroads to extend the interval between mandatory inspections from thirty days to ninety-two days. This reduction in the frequency of inspections had the effect of increasing the capacity of the shop. In time, the equivalent of nearly a thousand units could be maintained at Alliance. In addition to a large shop for running maintenance and overhauls, a separate new facility was built outside the shop with two tracks for fueling, sanding, and inspections.

In 1982 the capacity of the servicing facility was doubled, to allow sixteen locomotive units to be serviced at one time. The two-phase project cost $3.1 million. Locomotives could be refueled at a rate of two hundred gallons per minute on each unit. Figure 42.1 contains a drawing of the exterior of the new shop and the tracks leading to or from it. Following are its specifications:

1. Car shop: two tracks for light repairs, mainly wheel changes on coal cars, and two tracks for heavier repair of freight cars.
2. Locomotive shop: consisted of five tracks for light repairs requiring four to twelve hours of shop time, plus units due for the required three-month and six-month inspections; three tracks holding four units each for heavy repairs and rebuilding; and two tracks with a 35-ton crane; and a drop table for changing wheels and trucks and truck repair. A wheel-truing machine was located in an attached building.
3. Locomotive servicing facility: for fueling, sanding, and inspection.
4. Buildings for material storage, wastewater treatment, and a heating plant.
5. An area outside the shop building designated for re-starting, load testing and making adjustments to locomotives on which heavy repairs have been completed.

Figure 42.1. Engineer's drawing of the combined locomotive- and car-maintenance facility completed in Alliance, Nebraska, in 1979. The locomotive shop was built to handle both light and heavy repairs, with a capacity for running maintenance of eight hundred locomotive units assigned to BN's coal service. A new division office building, dispatching center, and yardmaster's tower (not shown) also were constructed as part of the complex BN needed for Alliance to serve as a hub for its coal network. BURLINGTON NORTHERN.

A NEW CAR AND LOCOMOTIVE
SHOP PLANNED AT MANDAN

In an AFE approved by President Tom Lamphier in 1979, authorization was given of $46.6 million for construction of a combined locomotive and car shop at Mandan, North Dakota, on the major coal corridor leading to the Twin Cities, Cohasset, and the rail-to-water transfer facility in Superior, Wisconsin. The shop building was constructed, but only the portion designed for making light repairs to coal cars was ever put in service.

Instead of finishing all the work needed to outfit the new shop building for maintaining locomotives used on

that corridor, BN decided to make modest improvements to the existing shop in Glendive, located 205 miles west of Mandan. At a cost of about $10 million, the remodeled shop at Glendive could handle the maintenance of three hundred units. Two new servicing tracks were built, together with a fueling station on the main track. The wheel-truing machine at the shop in Livingston was moved to Glendive, a major improvement to the capability for maintaining locomotives. Together with the increased capacity that was built into the shop at Northtown at that time, there was enough capacity on the north corridor to maintain locomotives needed for this service. Having that much capacity made it unnecessary to complete the locomotive shop at Mandan.

At the time plans were made to build a car shop at Mandan, it was a 500-mile inspection point for through trains. When the requirement for inspections was extended to a thousand miles, Mandan was eliminated as an inspection point, which reduced the need to have car-repair capability at Mandan. Also, more and more of the cars being acquired for coal service were owned or leased by electric-power companies, and maintained at private car-company shops rather than in railroad-owned shops.

Having fewer BN-owned cars in service than planned further reduced the need for a new car shop at Mandan.

NEW MECHANICAL FACILITIES
AT GUERNSEY

A new facility with two servicing tracks was built for locomotive fueling at both ends of the yard at Guernsey. It was necessary to have the capability for fueling locomotives on trains that entered and departed from the lower end of the new Donkey Creek–Orin line, in moving coal to and from power plants in Colorado and Texas. Since those trains would not pass through Alliance, it was necessary to build a fueling facility at a point on the route on which they moved. If no problems were indicated in BN's system for tracking the performance history of a locomotive's components, the 92-day inspection could be made at Guernsey instead of having to move it 129 miles back to Alliance for the prescribed inspection. This capability enhanced locomotive utilization and eliminated train delay from having to swap units with another train at an intermediate location.

CONSTRUCTION OF A NEW SHOP
AT LAUREL FOR HEAVY REPAIR
AND PREVENTIVE MAINTENANCE
ON BN-OWNED COAL CARS

With the opening of the new shop at Laurel, cycles were established for reconditioning car components based on actual mileage. This was a shift to preventive maintenance as opposed to the historic practice of waiting until the condition of a car required it to be placed in bad-order status, and then held out of service for several months until there were enough cars in the same series for a program for heavy repair. Instead, cars used in unit trains that still were in serviceable condition were run through the shop on a pre-scheduled preventive basis, thereby reducing the number of in-service failures and the length of time cars had to be held out of service for repairs. Also, the change-out of components on a life-cycle basis replaced the old practice of a 100 percent change-out at had been done in overhauls. BN projected the useful life of cars maintained under the new program could be extended up to fifty years from the heretofore anticipated life of twenty-five to thirty years.

COST AND BUDGET ISSUES

Just as the expenses for maintenance-of-way work increased in the 1970s with the rapid growth in the tonnage being move, the cost of maintaining cars and locomotives increased dramatically. For example, in 1975 and 1976, the expense for equipment labor and material increased 26 percent. From 1978 to 1979, equipment expenses rose 24 percent. In addition to the increase in work load due to having a larger fleet of cars and locomotives to maintain, the number and severity of derailments caused expenses to rise significantly in several years. Also, with the need to apply capital-expenditure dollars to move the coal, there was not enough money available to invest in the types and quantities of new tools and equipment needed to improve productivity and reduce costs elsewhere on the system. It was not possible to retire enough of the old cars and locomotives with high maintenance costs as soon as they should have been taken out of service and replaced with more efficient equipment. Finally, with the responsibility for operating expenses in the mechanical function being decentralized to the regions, costs were not always managed as tightly as might have been the case with a top-down system of command and control.

CHAPTER 43

PROJECT YELLOW

INCURSION BY C&NW AND UP

R. W. Downing [president of BN] to Larry Provo [president of C&NW] in 1973:
"How do you expect to pay for the construction?" Provo: "I guess we'll use cash."
EUGENE M. LEWIS, *12,000 DAYS ON THE NORTH WESTERN LINE*, PAGE 700

J. E. Wolfe [president of C&NW]: "BN management has had the feeling that
they own Wyoming and that God gave it to them."
BUSINESS WEEK, NOVEMBER 3, 1980, PAGES 116 AND 121

To use the government's credit to give the Union Pacific all that tax shelter is an outrage,
especially when the UP could just write out a check for the whole project.
AN UNNAMED VICE PRESIDENT OF BN, *BUSINESS WEEK*, NOVEMBER 3, 1980, PAGE 121–122

We think it's wrong for the UP and C&NW to use taxpayers' money to build
a new and unnecessary rail line into the [Powder River] Basin. Congress authorized
the use of public money to rehabilitate worn-out lines—not to build new ones.
RICHARD M. BRESSLER, PRESIDENT AND CEO OF BN, 1980,
IN A LETTER TO EMPLOYEES, DECEMBER 1980

Those who want to compete are welcome to the party—but bring money. . . .
We do not object to competition as long as the funding is obtained from the private sector,
just as Burlington Northern did. . . . We have been waiting to see the C&NW money since 1975.
RICHARD C. GRAYSON, PRESIDENT AND CEO OF BN RAILROAD COMPANY, 1981–1983,
IN " 'BRING MONEY,' BN TELLS WOULD-BE COMPETITORS," *TRAFFIC WORLD*, OCTOBER 19, 1981

AS MENTIONED NUMEROUS TIMES IN THIS history of BN, the company took tremendous risks in deciding to upgrade and expand the capacity of hundreds of miles of track to safely and efficiently handle the large tonnages of coal to be mined and shipped from the Powder River Basin. This commitment required the hard work of planning, raising money, constructing new lines and facilities, and getting set up to operate to the standards expected by the mining companies and electric-power companies. Since the coal to be mined was a low-value commodity, it could be moved only if low rates were in place. In line with the principles of economics for running a railroad, a high density of traffic had to be reached to make coal a profitable commodity.

A large part of the $2 billion invested in the coal operation was the construction of 116 miles of new railroad to serve the new mines to be opened between Gillette and Orin, Wyoming. The cost of building a single-track railway over that distance was $110 million. Other than BN, the only railroad company that had a line even close to the proposed new line was the C&NW, which had a branch line of 544 miles running across the entire state of Nebraska, and into Wyoming to Lander. Its line was located within a short distance of the lower end of the line BN proposed to build to serve the new mines.

By 1973 the C&NW let it be known that it was determined to get access to the new mines. It was planning to build its own line to reach the mines in competition with BN. At the start of the C&NW's quest to share in the "wealth," it seemed totally unrealistic, almost laughable, that the C&NW could believe it had any chance to become a big-time operator in any respect. The C&NW had been a marginal, struggling railway company for as long as any of us could remember. Certainly it could not expect to run unit trains of 100-ton capacity cars across a line with light rail, on which the short, low-tonnage, infrequently operated local trains were not authorized to exceed ten miles per hour over nearly every mile of track.

However, the C&NW was a tough-minded company that had survived in the face of the failure of other marginal railways in the Midwest. It seemed to have been energized by the success it had in upgrading one track of its double main-line route across Illinois and Iowa with government loans. Also, the UP had designated the C&NW as its preferred connection for interchange at the Missouri River after failing to acquire the Rock Island for access to Chicago. The C&NW had acquired other marginal midwestern railroads and successfully merged them into its system. It was a survivor and with no apparent merger partner out there, it was ready to move ahead on its own. Gaining access to the new mines to be opened along BN's new Gillette–Orin line would be its next frontier or challenge.

On May 5, 1973, the C&NW filed an application to the ICC to construct a line into the new coal fields, seven months and seventeen days after BN had filed its own application. BN promptly announced it would fight the C&NW's application. In July of that year, R. W. Downing met with Larry Provo, his counterpart on the C&NW, and heard the C&NW's proposal to build a new line jointly with BN. When Downing asked Provo how he intended to pay his share, Provo answered, "Cash," without hesitation.

ICC DIRECTIVE: ONLY ONE RAILROAD TO BE BUILT

The ICC advised it would not approve the construction of two parallel, competing lines. That ruling of course required BN and the C&NW to prepare a joint application. The "joint-line accord" was filed with the ICC in February 1974. Having to come to such terms with a competitor was hard for BN to take. BN had taken all the risk, done all the planning, and negotiated with land owners for the purchase of the right-of-way needed to build the railway. To then be forced by a government entity to allow a competitor to share in the wealth it had done nothing to create was disheartening, even more so when it was a company we thought of as weak, on the verge of going out of business, and struggling to stay out of the bankruptcy court. But knowing there were many other instances in which the owner and builder of a line had been forced by the ICC to allow a competitor to use its track (for example, Southern Pacific being forced to grant trackage rights to the Great Northern on seventy-five miles of the new line it built in central Oregon in the late 1920s), there was no choice but to proceed to build the line shortly as directed and prepare to have competition in serving the new mines.

It was nearly two years (January 1976) until the ICC approved the joint application. During this time, the C&NW had failed to obtain private financing for its 50 percent of the cost of building the line. BN did not object to granting an extension to the C&NW to obtain financing because as long as the C&NW was in the picture, BN felt the UP would not try to move in. During that time, the C&NW worked with the FRA to obtain funding through the 4-R Act, using the federal government at the lender of last resort. With pressure building up to have the line completed and in service by the time the new mines would be ready to ship coal, the ICC approved BN's request to begin construction. Also during that time, the C&NW decided it had to abandon its plan to upgrade the 544-mile line (the "cowboy line") across Nebraska and

Wyoming. The cost to upgrade that line was estimated at $530 million, far more than any group of investors would agree to finance. Further, the FRA advised this amount was more than it would commit to a single project. Provo then approached John Kenefick of the Union Pacific with the possibility of building a connection with the UP's North Platte branch at Joyce, Nebraska. The UP's initial reaction was that it would build into the Powder River Basin on its own, independent of the C&NW.

Late in 1978, the C&NW and UP worked out the basics of a joint plan for the financing of 50 percent ownership of the joint line and for building a line of fifty-six miles to connect the UP's North Platte branch with the C&NW's line. Under this plan, the C&NW would have to invest less than half the cost of upgrading its own line and buying into the joint line on a 50-50 basis with BN. During these joint discussions, the C&NW and UP came up with the name "Project Yellow" to designate the solid, joint commitment they made to get access to the new mines. Following were the components of the C&NW–UP project:

- C&NW to build a six-mile connection between the joint BN–C&NW line at Shawnee and Shawnee Junction, on the exisiting C&NW line.
- C&NW to upgrade forty-five miles of its existing line between Shawnee and Crandall, Wyoming.
- A new line of fifty-six miles to be built between Crandall and Joyce, where it would connect with the UP's line that ran between Joyce and O'Fallon, Nebraska (the junction on UP's main line, seventeen miles west of North Platte).

No upgrading was needed on the UP's North Platte branch to handle the coal tonnage forecasted for the short term. About 115 miles of that 162-mile line had already been upgraded enough to handle the coal interchanged by BN to the UP at Northport for delivery to a power plant located on the UP.

The C&NW formed a subsidiary called Western Railroad Properties, Inc. (WRPI) to purchase one-half interest in the joint line BN had just completed. The UP worked out all of the arrangements for financing of $387.2 million with a group of banks led by Manufacturers Hanover Trust Company. With that done, the C&NW withdrew its application for federal financing. Construction of the two connector segments and upgrading the

existing segments of C&NW track got underway in June 1983. It took only fourteen months for all the UP and its contractors to complete the work. Due to a general slowdown in construction activity at that time, the project was completed for about $300 million, much less that the $387.2 million that had been raised. On August 13, 1984, the first C&NW train was loaded.

Soon after the C&NW began to serve mines on the joint line, between Shawnee Junction and Coal Creek Junction (the limits of C&NW access agreed upon in the contract for operating the joint line), pressure mounted on BN to allow access to the three mines located beyond Coal Creek Junction. According to Robert Downing, Larry Provo had asserted that the C&NW would not demand an extension of ownership beyond what was defined at the joint line as the time the agreement was negotiated. Provo died while in office, and his successors soon disregarded that commitment. They sought an eleven-mile extension of joint ownership to East Caballo Junction, to reach the additional mines. When the C&NW received authority from the ICC to construct its own line to serve those mines, BN agreed to sell one-half ownership in that part of the line for $27 million, effective December 15, 1986.

On the day BN was to receive the check for $76.2 million from the C&NW, I recall some BN executives stating that it might be better for us to have that cash in hand rather than hauling the coal that would soon be moving on the C&NW. They were concerned the profit from transporting coal would soon become so low that selling one-half interest in part of the Gillette–Orin line was a better deal for us. At the time, this seemed to be an extension of the attitude on the part of some executives who had come to BN from outside the industry, and questioned the wisdom of whether BN's decisions to invest in the large projects we had completed. Although some of us found this view disappointing, we had to recognize there was good reason on their part to fear that the C&NW would be undercutting our rates just to establish a "beachhead" in the Powder River Basin. That would require us to offer reduced rates to hold the business on which we were trying to negotiate long-term contracts at higher rates. Further, it might become difficult for us to maintain the higher, more compensatory rates we had

J. C. KENEFICK: A SURPRISE GUEST

IN MAY 1976, WHILE I WAS THE DIVISION SUPER-intendent at Lincoln, Nebraska, the annual inspection trip of the board of directors was to originate in Lincoln. The special train of ten or twelve business cars was to run west to Denver and then over the lines of the Alliance Division being upgraded to heavy-duty, main-line standards for unit coal trains. On the morning of the departure from Lincoln, it was most surprising to see John Kenefick, president of the Union Pacific, on the station platform visiting with some of the board members and BN officers about to board. I asked one of the senior BN officers why our number one competitor would be invited to be on a trip where we would be showing our board members the projects we had underway, as well as discussing the market projections and financial implications of the decision they had made two years earlier to make those investments. No one seemed to know the answer to that obvious question, except that Kenefick had been invited by L. W. Menk to make the trip.

Just as the participants were starting to board the train, I was given a note from our chief dispatcher that six cars of a freight train had derailed in the vicinity of Holdrege, Nebraska, about 150 miles west of Lincoln—on the route the directors' special train was to operate. Since the derailment was on the neighboring Colorado Division, I immediately passed that information on to my boss, and he, in turn, advised John Hertog and Tom Lamphier of the problem. Although it appeared from the message that the main track would be open by the time the special train would reach the location of the derailment, it was necessary to decide at that minute whether to rely on that estimate or to, instead, run the train on the line from Lincoln toward Ravenna and Alliance. Kenefick overheard the conversation among BN's senior officers and said the best thing for us would be to run the train on BN to the connection we had with the UP at Grand Island (ninety-six miles) and then on the UP to Denver. A few board members heard Kenefick's suggestion and expressed their interest in that unexpected opportunity to run over the UP. Hearing those comments seemed to be enough for our officers to decide to take the chance of getting through the derailment site without delay.

Years later, after he had retired, I asked Bob Downing why Kenefick was invited to make that trip, especially in view of the tension between BN and the C&NW over the government having mandated we grant access to a competitor, which had ties to the UP. Downing said Menk never discussed it, nor was

gained in some of the challenges we undertook through the ICC and the courts.

For these reasons, some BN executives were not overly disturbed when the C&NW made its move to gain access to the three mines north of the "boundary" we had established at Coal Creek Junction. Getting the C&NW's check for the additional $27 million for one-half ownership of an additional eleven miles of line might be worth more than being able to continue hauling all of the coal produced by those mines. The combined amount of $98.2 million from the C&NW became available to the BN holding company for additional investment in non-rail businesses. Indeed, the railroad was a "cash cow" operation in the harvest mode.

C&NW REAPS THE BENEFITS OF ENTRY INTO THE POWDER RIVER BASIN

Together, the C&NW and UP were successful in pulling business away from BN at lower rates. Once the C&NW

he challenged by any board member, to his knowledge. I have never come across anyone on either BN or UP who knew with any certainty or had any insight on why that happened. My theory is that Menk figured the C&NW would never be able to come up with the financing for 50 percent ownership of the joint line. Menk did not want an unknown—possibly non-railroad organization—to be formed to take the place of the C&NW as the one-half owner of the new joint line we were building. He might have thought it would be far better if the UP—a strong, well-managed railroad company—would move in instead of another company that might be a more adversarial and difficult organization for us to deal with. Once Kenefick would see the extent of work we had underway and heard our projections for growth, there would be no doubt he would energize the UP organization to find a way to get access to the new coal mines in the Powder River Basin.

Further, I believe there was concern over BN's ability to handle coal tonnage that might be far in excess of what we thought were realistic tonnage projections through about 1980. Having a strong, capable organization such as UP as a "partner" (more so than a competitor)—to haul a large part of such additional tonnage and keep BN from becoming overwhelmed by the amount of capital required to handle all of the coal produced—might have been part of Menk's thinking.

Whether this theory is valid or not, it has been good for BN that the C&NW and UP built up enough capacity to handle about half of the tonnage produced. By the early 2000s the mines in the Powder River Basin increased their annual production to about 400 million tons, far in excess of the projections made in the mid-1970s. Handling even 200 million gross tons annually brought some of BN's coal corridors with two-main-track capability to the limits of their capacity. A density of 200 million gross tons per mile per year is far greater than that of any other freight-railroad operation in the world. It might have been too much of a challenge for any railroad to handle twice that amount entirely on its own lines. A three- or four-track main line would have been needed on hundreds of miles, together with a much larger fleet of cars and locomotives, more maintenance facilities, and many other types of support facilities to build enough capacity into BN's network to handle all of the 400 million tons of coal. While it was difficult to stomach the ICC's decision that forced us to accept competition, perhaps we were fortunate in the long run that is what happened.

gained access to the mines in the Powder River Basin, it did not take long for it to build up the tonnage it was handling. Its first victory was getting a contract to move 10 million tons per year to plants operated by Arkansas Power and Light. In early 1984 BN projected that 45 million tons could be diverted to the C&NW–UP by the late 1980s.

By 1993 they were handling more than half the tonnage produced by mines on the joint line. With its higher wage costs (due to not yet having an agreement allowing it to operate with fewer crew members), BN had trouble competing with the lower rates its competition could offer. By having to cut its rates to hold on to as much tonnage as possible, BN's profit margin on coal began to suffer. In time, even the UP began to complain about the low profitability of this line of business. Some years later, Jack Koraleski, UP's executive vice president–Marketing and Sales, stated there was so little profit left in the coal business that, given the choice of running a grain train or coal train over UP's capacity-constrained railroad, he'd choose the grain.[1] Matt Rose, president and CEO of BNSF, expressed a similar view of the coal business at a conference of financial analysts in 2006: "Our coal business is the least profitable business we have on this railroad today."[2]

The C&NW bragged that it had an operating ratio of only about 55 percent in its WRPI operation. The operating ratio for the UP system was consistently a few points lower than BN's starting in 1987, three years after the C&NW started to serve the mines. Having the cost advantage in these years gave the UP and C&NW enough flexibility to undercut BN's rates. What happened clearly was a validation of Richard Grayson's oft-repeated statement that under deregulation, we had to be the low-cost carrier. This new competition further reduced the overall return BN was getting on the huge investment it had made. By having a lower level of density on its coal routes when the C&NW entered the picture, the economies of scale on BN's coal routes deteriorated somewhat. This greatly accelerated the urgency to get a crew-consist agreement in place that would put BN on a cost-competitive basis.

The bottom line was that the electric-power industry and its customers were benefiting greatly from a very low-cost source of energy. Strong competition between two large, well-managed railroad companies in a deregulated environment, and having the ability to become increasingly more efficient, had carried the power companies through the challenge they faced in finding an alternative source of fuel to put them in compliance with the Clean Air Act.

To their credit, BN and UP aggressively took on the challenge to make the coal business more profitable than it was when aggressive rate cutting became a way of life. Numerous examples of such initiatives could be cited, but those that produced the broadest gains were the following:

- Increase in the number of tons per car by conversion from steel to aluminum car bodies
- Increase in train length from a range of 100 to 105 cars in the mid-1980s to as many as 150 cars
- Continuation of programs to extend two-main-track

territory, thereby reducing train delay for meeting trains on single-track segments, and to increase line capacity
- Acquisition of locomotives with higher tractive effort, thereby reducing the number of units required
- Reducing fuel consumption by improving train operating procedures and by acquiring more efficient locomotives
- Improving operating techniques to reduce the cycle times of unit trains, thereby reducing the number of train sets in service

Overall, the transition to having a second railroad operating its trains on the joint line to serve several customers worked out quite well. By the time of the entry of the C&NW, the people of BN's Alliance Division had developed an efficient, well-organized operation on the newly opened joint line. Under the contract for ownership of that line, BN performed all roadway maintenance (including the signal system) and dispatched the line. Future capital expenditures would be paid for on a 50-50 basis. The C&NW furnished motive power and operating crews for its trains. We were pleased the C&NW was able to provide good-quality locomotives, mainly the SD40-2, SD50 and C30-7 models, for its trains. Given its history, there had been some fear the C&NW would have only old, unreliable power available. Many of the locomotive units used on the C&NW's trains were owned by the UP and used in "run through" service on the lines of both railroads.

To its credit, the C&NW had trained its operating personnel quite well with respect to train handling, the BN's rules and special instructions, and the physical characteristics of the new line. The C&NW's managers and employees worked cooperatively and professionally with BN's dispatchers and managers, and with the crews operating BN's trains. The ability to work well together was a credit to the division and region-level employees and managers of both companies, in the case of BN, primarily Division Superintendent Joe Yeager and his staff: Cal Evans, Jerry Doughmann, Gene Mamer, and Ken Wilkowski. There were no apparent efforts on the part of the C&NW's people to openly ridicule or demean BN as a vanquished or conquered empire, nor did BN people try to give the C&NW a hard time as it moved in to serve the mines and take business away from BN. The situation at

---

1   Fred W. Frailey, "Powder River Stories," *Trains*, April 2010.
2   Tom Murray, "Where's That Coal Train Going?" *Trains*, April 2010.

the time of the transition could have turned messy and combative between employees and the supervisors of both companies, but it did not.

BN AND ITS PEOPLE IMPACTED ADVERSELY

The downside of the ICC's decision to force BN to allow the C&NW to operate on the "crown jewel" BN had created through its courage and ingenuity was, of course, its loss of business and having to accept a lower level of return than it deserved from the investment it had made. In addition, BN had to furlough about thirty crew members at Guernsey who had worked on trains operated into the mines. About the same number of BN operating personnel had to be cut off at Gillette, Edgemont, and Alliance as well.

Those who lost their jobs were a mixture of people who had taken the risk of moving out to a forlorn area of the country to stake their futures with BN, as well as many others from the local area who had been fortunate to get a job that gave them a much higher standard of living than they would likely to have had with any other employer. BN employees were well aware that their company had done everything possible to keep the C&NW from gaining access to mines on the joint line, or at least, to delay that from happening for as long as possible. In time, attrition took its toll, and those who stayed in the area again had the opportunity to work for BN. Some transferred from train service to vacant positions in other crafts such as train dispatching or track maintenance.

The amount of tonnage that BN was moving out of the Powder River Basin did not recover to the pre–C&NW level until 1994, ten years later. Even with new mines continuing to open, and with the higher demand for low-sulfur coal, it took a long time for BN to build its base of business back to where it had been. For several years after the C&NW started to serve mines on the joint line, the C&NW had a significant cost advantage over BN through the agreement it had made with its unions to operate its trains with only a three-person crew. It took until 1992 for BN to be able to make a similar agreement with the employees in train service who worked on lines still under the agreements between the former CB&Q (Burlington) Railroad and the United Transportation Union, which required four-person crews.

Table 43.1.

COMPARISON OF TONNAGE HANDLED BY BN AND C&NW FROM MINES JOINTLY SERVED, 1984–1996 (MILLIONS OF TONS)

Year	BN	C&NW
1984	72.6	2.4*
1985	64.7	18.6
1986	57.0	23.4
1987	58.0	32.5
1988	68.0	37.3
1989	72.1	42.6
1990	72.6	49.0
1991	71.3	58.4
1992	69.7	57.2
1993	70.6	73.9
1994	70.7	86.7
1995	87.8	103.5
1996	94.3	109.9

* C&NW's first train was loaded on August 16, 1984

Source: "A C&NW Coal Line Study" by Mike Lenzen, *NorthWestern Lines*, Summer 1997

Note that BN handled 73.6 million tons in 1984, the year in which C&NW began to serve mines on the Joint Line. C&NW was successful in drawing business from BN, and obtaining a large part of the increased tonnage produced in later years. It was not until ten years later that BN again handled the tonnage it had hauled before the C&NW's incursion.

Note also that in 1995, BN merged with Santa Fe and C&NW with Union Pacific.

The C&NW bragged that it had an operating ratio in the low 50s on the lines operated by its WRPI subsidiary. See Table 43.1.

With the addition of a high volume of profitable business, the C&NW advanced upward from being a marginal company teetering on the brink of bankruptcy, without enough resources to adequately maintain its property and equipment. It also benefited greatly from the large

Table 43.2.
COMPARISON OF OPERATING RATIOS

Year	UP	BN
1980	86.5	92.1
1981	86.1	91.9
1982	89.9	93.6
1983	89.3	82.3
1984	89.9	78.7
1985		80.7
1986	81.8	97.2
1987	83.1	85.1
1988	81.4	85.3
1989	82.5	85.5
1990	82.2	87.2
1991	98.0	105.3
1992	80.7	87.3
1993	80.4	86.1
1994	79.2	83.4
1995	78.7	92.3

Source: *Moody's Transportation Manual*

Table 43.3.
C&NW RAILWAY OPERATING INCOME AND
OPERATING RATIO, 1980–1990

Year	Operating Income	Operating Ratio
1980	74.8	92.0
1981	74.3	92.4
1982	d 11.9	101.5
1983	58.2	93.2
1984	62.1	92.9
1985	38.4	95.7
1986	85.6	90.9
1987	100.0	89.8
1988	116.9	88.2
1989	120.2	87.3
1990	150.6	84.3

d indicates deficit

Source: *Moody's Transportation Manual*

volume of interchange business it built up in those years as the UP's preferred connection for interchange at the Missouri River. Even with hauls of only about five hundred miles (between Chicago and Fremont, Nebraska), the C&NW built its density up to a respectable and efficient level of 20 million gross tons per mile.

Also during those years, the C&NW was able to acquire the Rock Island's main line route between St. Paul and Kansas City, which was the most direct route of any of the several railroads that competed in that corridor. By then, the C&NW had the resources it needed to upgrade that line as well. The provisions in the Staggers Act governing the abandonment or sale of unprofitable branch lines gave the C&NW a long-awaited and badly needed opportunity to rid itself of many miles of underperforming assets. Together, these events and initiatives finally made the C&NW a satisfactory merger candidate for the UP. It was consummated in 1995.

With the rapid and impressive growth in Powder River Basin–originated coal moving on the UP, it went all-out in typical UP fashion on projects needed to expand its capacity and make the kinds of improvements to enhance its operating capability. A second main track was built on the entire 162-mile North Platte branch line. A large support yard and staging area for unit coal trains was added to the large complex of maintenance and operating facilities in the yard at North Platte. A third main track of 108 miles was built on UP's main line between North Platte and Gibbon, Nebraska, the junction of UP's line to Fremont and Council Bluffs and to Kansas City. From Gibbon to Topeka (222 miles), a second main track was built, together with several major line changes and line relocations to reduce curvature and build by-pass routes to take the railway line out of two communities, Hastings and Marysville.

While the work done by UP to build high capacity and

efficiency into its operation was impressive, it was not as big a challenge as the task BN had faced in having to upgrade and expand capacity on virtually all of the main-line segments of the former Burlington and on a large part of the lines of its two subsidiaries—the C&S and FW&D railways—to coal-train standards, and to get all of it done in only six years. For both BN and the C&NW–UP teams, what was accomplished in those years was impressive. The main line corridors of both companies became the highest-tonnage and highest-density freight lines operated anywhere in the world, past or present. The ability to build their track and train-control systems to the standard required, and then run the operation efficiently, is perhaps the best example of what has made the freight railroads of North America the envy of the world. And, it should be noted that all of the work done by BN and UP was financed with private capital with no government aid.

CHAPTER 44

UNION PACIFIC'S DRIVE
TO INCREASE CAPACITY

We Can Handle It.

A SLOGAN ON THE SIDE OF UNION PACIFIC BOXCARS IN THE 1970S

UNION PACIFIC ALSO INVESTED LARGE sums from 1995 to 2000 in a program it called "Project Yellow III" to build the capacity needed to handle the coal that it and the C&NW obtained under contract. UP spent $855 million for a combination of many additional miles of main track, constructing a separate yard for inspection of coal trains and for making car repairs at North Platte, building by-pass lines around some communities, and purchasing short-line mileage to create an alternate route in Kansas. In addition, the UP had to increase its fleet of motive power.

In gearing up to handle the new business, the UP did not have to upgrade or expand capacity on as many routes as BN had to, over a wide geographical area. UP also had the advantage of having track on its coal routes that had been built and maintained at a very high standard throughout its history. The UP's situation was in sharp contrast with that of BN, which had to start moving heavy tonnages over hundreds of miles of track with low capacity and maintained only for a light density of traffic. UP also has had the advantage of consistently strong earning power from the time it was headed by E. H. Harriman in the 1890s to the present day. The UP maintained a much lower operating ratio than BN during the 1970s and early 1980s, when BN had to go all out in making heavy expenditures that were chargeable to operating expense.

UP had the advantage of low grades on its coal routes. The only grade of any consequence was a one percent grade (ascending eastbound in the loaded direction) of seven miles, a short distance east of Shawnee Junction, on the line segment upgraded by the C&NW. East of South Morrill, Nebraska, UP's line was on a river grade with maximum curvature of only 2 degrees. In contrast, BN had the disadvantage of a grade of 1.55 percent on its route leading out of the Powder River Basin, over Crawford Hill.

When the C&NW–UP combination began moving coal in 1984, it was in a much better position on operating and maintenance costs than BN had been in its early years of moving coal. The C&NW had finished upgrading and constructing the track it needed to reach its connection with the UP. While the UP had not yet expanded the capacity of its branch line between O'Fallon and South Morrill to the connection with its main line just west of North Platte, the line was in good condition. The C&NW had to upgrade only forty-five miles of track, compared to the thousands of miles that BN had to upgrade. Once the coal trains reached North Platte, they moved on a high-quality roadway for the rest of their trip on the UP. This gave the UP and the C&NW an advantage right from the start of having little if any delay to trains for track-maintenance work, slow orders, or line congestion, as BN had to endure for ten years. As a result the C&NW–UP had an immediate cost advantage over BN, which they exploited successfully by being able to offer lower rates.

Overall, the UP was in a much stronger position to make the level of investment required to compete with

BN in pricing, and in getting its capacity expanded at a rapid rate. Although the UP's merger partner, the C&NW Railway, struggled for several years to come up with $76 million to acquire 50 percent ownership in the Coal Creek–Shawnee Junction line built by BN, having to assist the C&NW in making an expenditure of that magnitude was not as challenging for a company as strong as the UP.

Following is a listing of some of the major capital projects carried out by the UP to increase its capacity and overall operating capability, both before and after its merger with the C&NW:

- Rebuilding of 45 miles of the C&NW's "Cowboy" branch line and construction of 6 miles of new line between its connection with the BN–C&NW joint line (at Shawnee Junction) and Crandall.
- Construction of 56 miles of new line to connect with the line built by the C&NW from Crandall to connect with the UP at Joyce.
- Investment of $27 million to acquire 50 percent ownership of 11 additional miles of BN-owned track (between Coal Creek Junction and Caballo Junction) to serve mines north of the original joint-line purchase.
- Construction of 165 miles of second main track on UP's line between O'Fallons (junction with UP main line west of North Platte) and South Morrill, including CTC.
- Construction of a third main track for 108 miles between North Platte and Gibbon, Nebraska (junction of UP lines to Chicago and Kansas City).
- Construction of a new sub-yard at North Platte to facilitate the inspection of coal trains, adding "spares" to fill out the consists of empty trains, the changeout of locomotive units scheduled for maintenance, and for light in-train repairs of coal cars.
- Construction of a second main track on the entire line of 222 miles between Gibbon and Topeka.
- Acquired a 107-mile short-line railroad in Kansas to be made part of an alternate route between Gibbon and Topeka.
- Construction of 9.2 miles of track to reach the Gerald Gentleman power plant near Sutherland, Nebraska, to provide the customer with and alternate carrier to BN.

- Construction of a five-mile by-pass line around Hastings, Nebraska, including an overhead crossing of the BN's Lincoln–Denver line. Two no. 30 turnouts with movable point frogs allowing a train speed of 60 miles per hour were installed as part of the project. Allowing trains to run that fast in crossing over between main tracks or when moving to a diverging route is a strong indication of UP's resolve to build more capacity into this high density route.
- Having the new line in service eliminated a slow, single-track bottleneck route through Hastings. The UP also built a major line change around Marysville, Kansas.
- UP made revisions at four locations on its Gibbon–Topeka line that eliminated four "diamond" crossings of BN branch lines and to Rock Island crossings. At each location BN was the "junior" road and responsible for the cost of maintaining the crossing frogs. Over the years the UP became exasperated with the difficulty the BN section crews and welders on these branch lines had in maintaining these crossings to prevent temporary speed restrictions of 25 or even 10 miles per hour. UP made agreements with BN to make small line changes that would allow the crossing frogs to be replaced by two switches to enable BN trains to cross its line. UP installed the two switches at its expense to finally get rid of the problems of enduring slow train speeds on its high-density lines. Of course, this was a very good deal for BN as well, to be rid of the expense and pressure from UP in maintaining these crossings. The point of mentioning these smaller projects is to show the resolve the UP had to remove impediments and build a line of railroad that would allow it to perform at a very high level.

It was impressive to see how rapidly the UP was able to get large projects completed. The coal business increased far beyond projections made as recently as the mid-1980s with the passage of the Clean Air Act of 1990, which further limited emissions from power plants. It became necessary for BN and UP to jointly finance the $100 million needed for construction of a third main track over the entire distance of the 103-mile line they jointly owned in the Powder River Basin. For fourteen miles, a fourth main track was built as well, in order to provide enough capacity to move up to 400 million tons of coal per year on the joint line. With that much track capacity, maintenance crews are able to take portions of one main track at a time out of service to perform inspections and maintenance work as necessary.

In later years, after its acquisition of the C&NW, the UP restored double track in some segments of the line just east of the Missouri River that the C&NW had reduced to single track in order to lower maintenance expense and in anticipation of future needs for rail, tie and ballast replacement. Near Boone, Iowa, UP built a new double-track bridge, 2,685 feet long and 190 feet high, over the Des Moines River. The bridge was designed to allow two trains to run over the bridge simultaneously at 70 miles per hour. Over the years, speed restrictions had been placed on the old bridge, and only one train at a time was allowed to cross. These restrictions had become so severe that the UP found it necessary to invest $43 million in a new bridge. Having the resources to undertake projects of this magnitude by both UP and BNSF demonstrates the importance of having strong railroad companies in position to generate and respond to major new marketing opportunities such as the movement of low-sulfur coal presented in the Powder River Basin, as well as for moving millions of containers for international business, unit trains of grain to foreign markets, and most recently, train loads of oil being extracted in western North Dakota. The mines, the electric-power companies, and the consumers of electricity in much of the United States have been very fortunate to have two strong, capable railroad companies serving them with reliable, low-cost transportation.

CHAPTER 45

ENERGY TRANSPORTATION SYSTEM, INC. (ETSI)

A PROPOSED COAL-SLURRY PIPELINE

Absent the threat of cream-skimming slurry pipelines,
the railroads can do the big coal transportation job ahead while
simultaneously expanding and improving their over-all sevice to the nation.

TESTIMONY OF L. W. MENK BEFORE THE SENATE COMMITTEE
ON PUBLIC WORKS, WASHINGTON, D.C., JUNE 11, 1975

• *a serious threat to BN's investment in additional capacity* •
• *aggressive pursuit of right eminent domain to cross*
its competitor's right-of-way •

DURING THE 1970S, THE PROPOSED construction of coal-slurry pipelines became a serious threat to the investment BN was making to handle the forecasted tonnages of coal from the Powder River Basin. If one or more such pipelines were built, it would greatly reduce the density of rail traffic and destroy the economies of scale inherent in the rail business. It was not an exaggeration to say coal pipelines could ruin the future anticipated for BN, its employees, and shareholders.

Coal slurry was to be a mixture of pulverized coal and water, pumped through pipelines to large electric-power plants. The plan for construction of a pipeline was produced by a new company organized as Energy Transportation System, Inc. (ETSI). It was jointly owned by ARCO (owner of the Black Thunder Mine), Bechtel (an engineering, construction, and consulting company), and the Arkansas Power and Light Company. Among

ETSI's board members were such notables as George Schultz and Casper Weinberger, who had held cabinet positions in the Nixon administration.

To reach its identified markets, ETSI needed to have the power of eminent domain to build across the rights-of-way owned by BN and other railroad companies. BN organized an all-out assault against legislation at the federal level that would require it to allow ETSI to cross under its tracks. BN worked aggressively to get support from agricultural, environmental, and conservancy organizations that opposed the diversion of enormous quantities of water to the pipeline. Among the organizations that joined forces with BN were Ducks Unlimited, Pheasants Forever, and the Izaak Walton League, as well as farm organizations and the residents of railroad towns. Employees were enlisted to write their U.S. congressmen and senators to oppose granting eminent domain or the use of vast quantities of water in the arid areas where the coal was mined. Obviously, the rail unions also had a big stake in this fight, and they put pressure on their elected representatives to oppose construction of the pipeline.

Early in the game, ETSI had approached BN to become a partner in the project. This happened before the C&NW Railway announced its intent to serve mines in the Powder River Basin. BN looked seriously at allowing coal-slurry pipelines to be built on its right-of-way. Bob Downing, president of BN at that time, provided Roger Grant, author of a history of the C&NW, the following account of BN's determination to not join the ETSI venture:

> We declined their offer because we knew enough about the economics of rail vs. pipeline that the unit coal train was actually more economic. For financing very little equity was to be provided by ETSI and it was probable that the debt would have to be guaranteed by BN and the utility consumers since ETSI itself did not have much financial strength.[1]

ETSI proposed to transport water hundreds of miles from a reservoir on the Missouri River in central South Dakota to the head of the pipeline in the Powder River Basin. At the destination of the coal slurry, the wastewater would have to be dealt with in some way. One proposal was to build a "twin" pipeline to move it back to Wyoming and use it again. States that were dependent on water from the Missouri River opposed giving up their water, and from that, refused to grant eminent domain to cross property owned by railroads, farmers, ranchers, and countless other private land owners. At that point, ETSI took its case to the federal government. During the

battle between the railroads and ETSI, ARCO decided to withdraw from the project.

Because the economics of the project were so unfavorable, it collapsed under its own weight. To recoup the large investment that ETSI and its investors had made in trying to get support for their project, they filed suit against the railroad companies that had united in opposition to it. They accused the railroads of applying restraint of trade. Rather than risk an adverse judgment from a jury in Texas, the railroad companies (except for the Santa Fe) decided to negotiate settlements. It cost the BN $150 million. The C&NW paid $15 million, the UP, $60 million, and the KCS, $82 million. The jury in a federal court in Texas issued a verdict of $320 million against the Santa Fe.

In addition to settling with ETSI, BN agreed to pay $58 million to Houston Light and Power Company (HL&P) and to "adjust" the rates in the remaining twelve years of its contract with HL&P. The Texas utility had sued BN for alleged conspiracy with other railroads to block construction of a coal-slurry pipeline to serve its power plant.

The fight was very costly to BN and the other railroads in terms of the large amounts paid to settle the case, as well as in the tremendous amount of management time and energy that went into developing support for BN's opposition to construction of coal-slurry pipelines. It was a battle that had to be won for the sake of investors, employees, and anyone else whose future was tied to the success of the rail industry. Without having the large concentration of coal tonnage on the tracks of much of its system, BN would not have been able to develop into the highly successful company it became by the early 1980s. The return on its $2 billion investment in upgrading the railroad and expanding its capacity would have been much lower.

---

1   Letter from R. W. Downing to H. Roger Grant, November 9, 1995, in H. Roger Grant, *The North Western: A History of the Chicago and North Western Railway System*, 231.

CHAPTER 46

COAL RATES

CONFRONTATION WITH OUR BIGGEST CUSTOMERS

*BN's coal rates are the lowest of any railroad in the nation and barely
let the company recover the cost of hauling the coal.*
BN NEWS, OCTOBER 1980

*San Antonio officials say they have enjoyed a savings of
more than $100 million over the cost of gas in just three years.*
R. M. BRESSLER, SPEAKING TO THE
WESTERN COAL TRANSPORTATION ASSOCIATION, SEPTEMBER 1980

*• early establishment of "missionary rates" to attract incremental tonnage,
to move on lines not maintained to a standard high enough for a sustained movement
of heavy-tonnage trains • resistance in later years when higher rates
had to be set in order to get an adequate return on investments
made to transform the coal routes to a higher standard •*

A GREAT DEAL OF ANXIETY DEVELOPED IN Burlington Northern during its struggle to establish rates for coal transportation at a level high enough to yield a satisfactory return on its invested capital. The problem began to develop in the early 1970s, well before the passage of the Staggers Rail Act of 1980, which relieved railway companies of a great part of the restrictions they had in negotiating rates and entering into contracts with customers. The transition into this new era of freedom from the requirement for ICC approval of rates was difficult for the railroad industry and the electric-power industry, both of which also had been heavily regulated for most of their history. The power companies were accustomed to a world in which the rates they charged their customers were regulated. They also were guaranteed a specific rate of return based on their operating costs and on the amount of capital invested in their generating plants and the infrastructure they built for distribution of the power they generated.

The rail industry did not operate under such a guarantee, either before or after it was deregulated. One might have thought the power companies—perhaps better than any other industry—would have understood the stress BN was under to make a reasonable return on the large amount of capital it had invested to move coal to their generating plants. Instead, the power companies decided to fight for rates on coal shipments below the compensatory level the ICC and the courts approved,

277

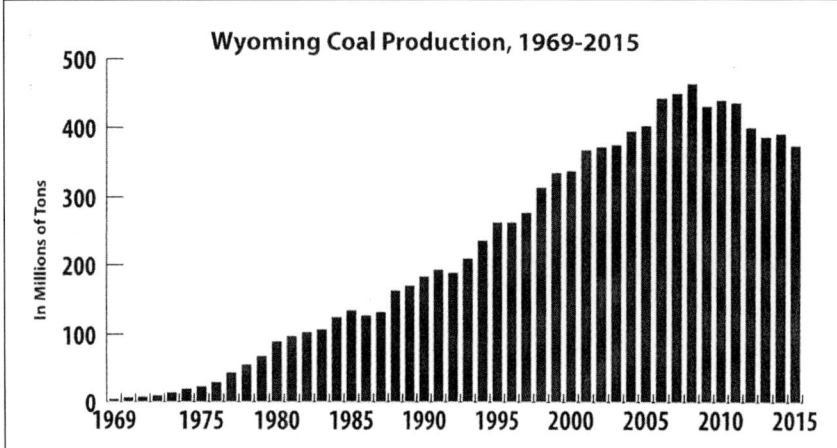

Wyoming Coal Production, 1969-2015

Figure 46.1. The tonnages shown above do not include mines of the Powder River Basin that are located in Montana. The annual tonnage from those mines ranged from about 40 million tons to 45 million tons in the peak years of production in both Wyoming and Montana. WYOMING COAL ASSOCIATION.

rather than determine those rates by negotiating across the table with BN's pricing officers. They battled BN through the ICC and the courts, in hopes that the rates set through those challenges might be lower than what they could have negotiated. If successful, they thought it would help ease the complaints and the pressure they might receive from their customers for otherwise higher rates for electricity. It would also reduce the amount of pressure they would have to put on their own regulators for higher rates needed to cover the price they had to pay for delivered coal.

The problem between BN and its customers in the electric-power industry largely came about when BN notified those customers of the need to substantially increase rates that had been agreed upon a few years earlier when they started to burn low-sulfur coal mined in the Powder River Basin. At that time, this was entirely new business for BN's predecessors, who looked upon it as incremental business. The rates were set high enough to cover only the incremental costs of moving the coal: the "above the rail" costs, with limited recognition of the routine costs of track, bridge, and signal maintenance, but well below the full cost of the movement. No major capital investment in track capacity or track upgrading had yet been made for these new movements of coal. The increase in tonnage was believed at that time to not be great enough to cause any significant accelerated deterioration or wear and tear on the track structure. The tonnage of coal to be moved was modest, amounting to only two or three trains per week for a given movement. All in all, it was

thought that this new business posed no challenges of consequence to the cost of maintaining the track, nor would any major capital programs be required.

LEGACY OF THE EARLY "MISSIONARY RATES"
The new rates set for these new movements of low-sulfur coal were sometimes called "missionary rates"—rates that were appropriate for getting into a new line of business and, hopefully, that might encourage other power companies to consider the use of that type of coal from mines located on BN. At that time, no one could foresee the hundreds of miles of track that would have to be upgraded, or that large new maintenance facilities would have to be built for a greatly expanded fleet of locomotives and cars. Nor was it seen that the capacity of the network would have to be expanded with additional sidings, second main track, and CTC, projects that would require investments of hundreds of millions of dollars.

Even for those specific new movements of coal, there was not much knowledge or concern at that time for the amount of wear and tear that would soon occur on track that had not been maintained or upgraded to a high main-line standard. To get a reasonable return on investments of such magnitude to upgrade the track, the early missionary-level rates could not be maintained; rates had to be raised to much higher levels. When BN brought this issue to the attention of customers who were receiving coal under the low rates, the responses BN received were highly unfavorable.

One rate in particular serves as a benchmark or point of reference for the arguments of the power companies that claimed the rates later proposed or set by BN for coal movements to other power plants were excessive, unreasonable, or unfair, when compared to that rate. This benchmark rate was the one established for the movement of coal from the Big Sky Mine near Colstrip in southeastern Montana to a power plant at Cohasset, Minnesota, located about eighty miles northwest of Duluth, and operated by the Minnesota Power and Light (MP&L) Company. Coal for the plant at Cohasset moved on the main line of the former NP to Fargo, then north on the Great Northern to Grand Forks, and then east to Cohasset. Since the NP's main line was maintained to a very high standard, the addition of about three coal trains per week caused no significant issue of capacity or concern about track maintenance in the short term.

However, on the 181 miles of the GN line between Grand Forks and Cohasset, there were 155 miles of 90-pound conventional bolted rail. The rail, tie, and ballast condition had been adequate for the amount of tonnage being run over that line at that time. The business moving on that line was mainly grain, destined for Duluth and Superior, most of which was still being moved in 50-ton capacity box cars. Conversion to 100-ton capacity covered hopper cars had just begun. With the amount of tonnage handled, the 90-pound rail might have been satisfactory for many years to come.

However, not long after the coal began moving on the 90-pound rail, the defect rate began to increase. That and other maintenance problems with the 90-pound rail soon made it necessary to start a program to replace it with heavier rail over the next few years. Using second-hand welded rail in the 1970s would cost in excess of $10 million. Even on the former NP main line, the heavy tonnage began to take its toll on the track, requiring the installation of thousands of additional rail anchors to restrain the rail from "running," due to the forces generated in the movement of loaded trains, especially when braking. None of these costs were anticipated when the movement of coal started on these line segments.

In making plans for construction of a new power plant at Cohasset, MP&L and the BN's predecessor roads reached what was termed "an understanding" on a basic rate to be applied for the movement of coal, together with

Table 46.1.

REPRESENTATIVE COAL RATES IN EFFECT IN JULY 1977 (RATE PER TON-MILE)

Origin Mine	Destination	Rate*
Colstrip, Mont.	Becker, Minn.	$ .80
Colstrip, Mont.	Cohasset, Minn.	.75
Colstrip, Mont.	Superior, Wisc.	1.24
Belle Ayr, Wyo.	Armarillo, Tex.	1.53
Belle Ayr, Wyo.	Elmendorf, Tex.	1.50
Belle Ayr, Wyo.	Flint Creek, Ark.	1.55
Cordero, Wyo.	Elmendorf, Tex.	1.49

Author's note: Until passage of the Staggers Act in 1980, deregulating railroad pricing, railroads were required to file all rates with the ICC.

* For shipments in unit trains

Source: BN Coal Tarriffs 9 and 12, quoted in Jerry Fruin and Robert Crnkovich, "Western Coal Transportation Rates for Minnesota Users," Department of Agriculture and Applied Economics, University of Minnesota, February 1978, revised April 1978.

a built-in escalation formula for future rate increases. The agreement also contained a "gross inequity" clause for rate changes if deemed necessary. When BN found it necessary to increase the rate to a level higher than allowed under the escalation formula, it invoked the gross inequity clause. When MP&L refused to accept the higher rate, BN published the higher tariff rate with the ICC.

The ICC rejected the proposed rate as unreasonable and stated that the rate agreement between MP&L and BN constituted a contract. BN argued that contract rates were not allowed at the time (in 1979, before the Staggers Act became effective) and therefore, there was no contract to be enforced. Both parties knew at the time that such contracts were not yet authorized by law, and in fact, the informal, non-binding agreement did not even specify the tonnage to be moved within a particular time. Under the escalation formula, the rate would have increased from $6.59 to $10.03. The rate proposed by BN was $11.03.

Since BN failed to obtain approval from the ICC for a rate high enough to provide a satisfactory return on the capital it needed to invest for this movement, BN filed an appeal with the U.S. Court of Appeals, 6th Circuit. In November 1981, the court upheld the decision of the ICC, based on the fact that the dispute on rates was pending at the time the Staggers Rail Act was enacted into law.[1] In time, MP&L and BN negotiated a long-term contract to cover rates and service, as allowed under the Staggers Act.

A total of fifteen of BN's utility customers made the same contention as MP&L—that the "letters of understanding" BN sent to its customers in the 1970s with proposed rates constituted a contract. On that basis, the argument was that these rates could not be increased. In a case with Kansas Power and Light Company, a federal district court ruled that the utility did not have a contract with BN, as contested. With that decision in hand, BN was free to take action to raise its rates. Thus a precedent for BN to raise rates to an acceptable level of profitability was established. From that point on, BN sought to negotiate long-term contracts and avoid litigation, unless negotiations failed or a customer refused to negotiate.

The basis for establishing a higher rate for the Cohasset move was straightforward and therefore easy to identify and quantify. However, as more and more coal movements began elsewhere on the BN system, the issue of adequacy of rates became far more serious for BN. Millions of dollars had to be invested to expand the capacity and upgrade the track over long distances. In 1974 BN committed to an investment of $2 billion, which included $100 million for a new line of 120 miles to serve mines to be opened in Wyoming, together with a much larger fleet of locomotives and a large new facility to maintain them. In addition, inflation was rampant in the years of the highest rate of growth in the coal business.

In a speech Richard Bressler gave at a meeting of the Western Coal Transportation Association in September 1980, he related that BN's railroad company had earned $41 million before tax in 1979, for a return of less than 4 percent on the investment BN had made in its coal-hauling business in recent times. Clearly, action had to be taken to improve the rate of return, both through rate increases and by running the railroad more efficiently,

now that most of the projects for expansion of capacity and operating capability had been completed.

LENGTHY DISPUTE WITH SAN ANTONIO'S ELECTRIC-POWER UTILITY

Of all the coal-rate cases disputed between BN and its customers, the one that may have been given the most publicity and dragged on the longest was the case involving City Public Service, the city-owned electric-power utility in San Antonio, Texas. The movement of coal to San Antonio was a 1,250-mile haul for BN, including 801 miles on two BN subsidiaries, the Colorado and Southern and the Fort Worth and Denver railways, both of which had been maintained only for light-density service. Once the heavy movement of coal to San Antonio and other power plants in Texas and Oklahoma started on those lines, heavy upgrading and expansion of capacity had to be undertaken on the lines of the C&S and FW&D. Those projects were on top of all the work already underway on BN's Alliance, Nebraska, and Yellowstone divisions, and on the Chicago Region in those years.

Initially, BN set the rate for the San Antonio move at $11.94 per ton. Various expert groups that were called in to analyze BN's costs determined that the rate BN needed to recover its full costs would be between $17.57 and $20.08 per ton. When the power company went to the ICC for relief from a rate it considered unreasonable, BN proposed a rate of $18.23 in the application it made to the ICC.

L. W. Menk, BN's chairman, presented the following argument in favor of BN's request in a statement quoted in part in the July 1978 issue of BN *News*:

> The utility company realizes it cannot defeat our rate request on economic and legal grounds and is applying political pressures on the ICC.... Even at the reasonable freight rates we are asking, electricity customers will get a better break if their utility company uses coal for fuel instead of continuing to rely on high-priced intrastate gas and oil.... The present rate ($11.94 per ton) is forcing us to subsidize San Antonio because we cannot even meet our costs.

Menk went on to say that BN's rate of return on equity in 1977 was only 2.25 percent on its railroad operations

---

1   664 F. 2d 568.

and 4.2 percent overall, compared to the 13 percent rate of return on equity by San Antonio's power company.[2]

Further, the rate paid for electricity in San Antonio had been made higher than necessary because of the city's decision to place a tax of 14 percent on the gross revenue of its power company. This practice amounted to a subsidy of $30 million to the city's budget, intended to help keep property taxes low. Elimination of the subsidy would have allowed the power company to reduce the rates it charged its customers.

The issue of rates for coal to be delivered to San Antonio's city-owned power company developed into a long ordeal, with no resolution or agreement on rates being made until 1983, seven years after BN first quoted a rate. Following is a chronology of the disputed rate:

- 1971   Decision by the City of San Antonio to substitute coal for natural gas for the generation of electricity.
- 1974   San Antonio entered into long-term contracts for coal with two suppliers in the Powder River Basin. Negotiations were conducted with BN for rates to transport the coal. BN and Southern Pacific quoted a rate of $7.90 per ton (SP handled the move between Fort Worth and San Antonio).
- 1975   BN and SP raised the rate to $11.90; San Antonio filed a complaint with the ICC.
- October 1976   The ICC established a temporary rate of $10.93 to allow the movement of coal to commence, pending further review.
- January 1977   BN and SP petitioned the ICC to increase the rate, alleging the current was "below a maximum reasonable rate" when compared to similar movements.
- 1978   The ICC set the maximum rate level at $16.12. Both railroad companies and San Antonio were dissatisfied with that decision and petitioned for reconsideration.
- January 1979   ICC set a new maximum rate of $17.23. BN and SP filed tariffs at this rate.
- June 1980   The U.S. Court of Appeals ruled that both the $17.23 and the $16.12 rates were "arbitrary and capricious [and] without defensible rationale."

The case was remanded to the ICC. San Antonio reduced its payments to the $10.93 rate set in 1976.

- November 1980   BN and SP filed a tariff requiring San Antonio to prepay at the $17.23 rate before service would be provided. The ICC suspended that tariff, stating that the decision of the court precluded any rate except the $10.93 rate set in 1976. San Antonio unilaterally reduced its payment of freight charges due to the $10.93 rate. BN and SP then filed a tariff requiring prepayment at the $17.23 rate before coal would be moved to San Antonio. ICC suspended that tariff. Both parties carried on the controversy through other forums, including the Texas District Court, which ruled in San Antonio's favor.
- April 1981   The ICC vacated the order that had set the $10.93 rate, stating it would next determine what the rate "should have been" when it was set in 1976, and for the period when the higher rates were in effect. BN appealed to the 5th Circuit Court, arguing that only the ICC, and not the Texas District Court, had jurisdiction to enjoin BN and SP from collecting their filed tariff rate.
- May 1981   The court denied the request of San Antonio for a stay on the tariff rate of $17.23. As a result San Antonio finally had to pay under that rate.

The U.S. Supreme Court ruled that San Antonio was required to pay at the tariff rate between June 24, 1980, and May 1, 1981, the time during which San Antonio had been paying only at the rate of $10.93 per ton. Even after the Supreme Court ruling, San Antonio argued before the D.C. Circuit Court that there still were unresolved issues remaining. However, the circuit court declined to hear any further argument for reimbursement of $24 million of the allegedly "unreasonable" charges between December 1978 and July 1980. During the time San Antonio failed to pay the tariff rate, BN and SP lost $19.833 million in handling three million tons of coal for the city. In December 1983, the federal district court awarded that sum in damages to the railroads. The court of appeals ruled against San Antonio and ordered the payment of interest in addition to the $19.833 million reimbursement.

Thus ended the long saga of seven years (1976–1983) to establish a rate for transporting coal to the power plant at San Antonio. BN and the City of San Antonio

---

2   "BN Chairman Accuses Texas Utility Firm of Using Political Pressures in Coal Haul Rates," BN News, July 1978.

spent large sums to advocate and defend their positions. This case also consumed a great deal of management time. The uncertainty of the outcome of this dispute (and similar—but not as lengthy or controversial—disputes with several other utilities) raised questions on the part of some investors, financial analysts, and BN board members on whether it would be possible for BN to make a satisfactory return on the enormous investment it was making to handle the coal to locations scattered throughout much of the system.

There were many confrontations over rates between BN and a large number of other electric-power companies it served. From the early to mid-1980s, most of these disputes had to be resolved through the courts. Once these cases were settled, a basis could be established for setting rates on a business basis in the future, rather than through the courts or intervention by a government regulatory agency. Once the C&NW (and later, the Union Pacific, upon merger with the C&NW) obtained access to mines in the Powder River Basin, the strong competition that resulted was enough to overcome the assumption that since BN had a monopoly on the business of hauling coal, there was a chance its rates could be challenged successfully through the legal or regulatory process of appeal. Instead, negotiations became centered around the question as to which of the two railroads, BN or UP, could offer the best rate and service plan to the customer.

POWERFUL MESSAGES
FROM BN'S NEW CEO

Richard Bressler came to BN in May 1980, right in the middle of controversies over rates with several of the utilities, some of which were BN's highest revenue customers. Bressler wasted no time in expressing his views on the coal business and the way BN had been working through the issues with its customers. Kaufman's *Leaders Count* gives an account of Bressler's assessment of the situation we were in with the electric-power industry. His appraisal of the relationship was severe. When the Law Department showed him a list of the litigation in progress, Bressler reacted: "These are all our best customers. They're all suing us. What's going on here? How can you run a business in which you're in court with all

your biggest customers? . . . I'm going to go out and visit these customers."[3]

After holding such meetings, Bressler reported these customers told him they had never met an executive from BN until then, except perhaps a lawyer or marketing person. Norman Lorentzsen takes issue with that contention, recalling that he and Ivan Ethington (senior vice president–Marketing in the late 1970s) visited most if not all of BN's coal customers more than once to update them on the projects underway to expand capacity to handle the rapidly expanding coal business.

In a speech Bressler gave to members of the Western Coal Transportation Association, he clearly stated where BN stood after investing large amounts in the coal business, and the approach he would be taking on the issues the power companies had with BN. Following are excerpts from Bressler's speech:

> I look at an array of tariffs and figure out that relatively little of that $41 million (BN's income before tax in 1979) came from hauling coal. . . .
>
> There's another aspect to this coal rate problem. And that is the massive litigation—litigation that is demanding our attention and costing you and us millions of dollars. . . .
>
> . . . we at BN will be careful about future investments in coal-hauling capacity—at least until the picture is clear. . . . we have invested more than our cash flow for many years. Those decisions were based on the premise that the business opportunity would ultimately reward this aggressive investment program. I will not hesitate to continue such a program if I believe that the ultimate return is there. . . .
>
> A second conclusion I've drawn is a real need to improve communications with our customers. To do that, I've embarked on a program to visit all our major customers and shippers in the coal-hauling business. . . .
>
> Our present studies show that the vast majority of our customers are not dis-advantaged by burning Powder River Basin coal and having Burlington Northern transport it to them, compared with their alternate fuel sources. San Antonio officials, for example, say they have enjoyed a savings of more than $100 million over the cost of gas in just three years.

---

3   Kaufman, *Leaders Count,* 217.

However strong or lacking the relationships may have been between the senior level of BN management and those customers, BN faced tremendous challenges in making the coal segment of its business more profitable. BN had to continue to move ahead with the large projects it had underway in upgrading the railroad's track, expanding its capacity, improving its operation performance, and reducing the number and severity of train accidents caused mainly by deficiencies in the condition of the track. These and other initiatives had to be intensified to reduce the cost of moving the coal. Regardless of accusations of managerial ineptness or dereliction, a lack of vision, or being out of touch with the market in the past, the "top line" (revenue) for the coal business had to be enhanced through litigation, negotiations, and any other means that would get the rates on coal raised to the level where we could produce a satisfactory return on invested capital. No amount of charm, business acumen, or relationship building was going to accomplish that in the short term. For many of the rates in dispute, we would have to continue to slug it out in the courts and in the halls of the ICC.

Although the deregulation of pricing in the Staggers Act gave the rail industry the kind of opportunity it had needed for many years, its provisions were not set up to deal with a situation that had become as disorderly, contentious, and adversarial as the situation BN had in dealing with the electric-power industry. There could be no hope for withdrawal from litigation at least until decisions were handed down by the courts on the pending cases. After that, BN would have the opportunity to get set up to handle rate matters with each coal customer on a business basis in the future. It took several more years to resolve all of the "old" cases through the courts, but it got done. The people of BN owed a great deal to the lawyers on its staff and several outside legal firms, who successfully progressed those difficult rate cases to solution.

In the hope of facilitating its process for review and evaluation of contested coal rates, the Surface Transportation Board (STB) issued a set of procedures called "Coal Rates Guidelines" (1 ICC 2d 520) in 1985. In the STB's announcement of the "improved" process for ruling on large rail-rate cases made on October 30, 2006, it advised that under its guidelines, "captive shippers should not be required to pay more than is necessary for the carriers involved to earn adequate revenues. Nor should they pay more than is necessary for efficient service. And captive shippers should not bear the cost of any facilities or services from which they derive no benefit."

In a fact sheet that followed the announcement, the STB further advised that most captive rail shippers could seek relief under the concept it called the "stand-alone cost." In this process, the cost of building and operating a new "stand alone" railroad to handle the particular movement in dispute would be determined. Under that constraint, the rate at issue could not be higher than what the stand-alone railroad would need to charge to cover all of its costs, including a reasonable return on investment. The STB intended this new process to serve as the means necessary for regulatory protection for captive shippers against unreasonably high rates. Also, it would insure that railroad companies still would be able to earn an adequate level of revenue, and from that, a reasonable return on the capital they invested to handle that particular segment of their business.

As a final note on this topic, it is interesting to note that on a nationwide basis, rail rates on coal delivered to electric-power companies declined by 42 percent from 1984 to 2001. The utilities and their customers have benefited greatly from the efficiency of rail transportation, together with the railroads' success in reducing their costs and their concern for the health and needs of the power companies. One would be hard-pressed to find any case where an electric-power company has reduced its rates for electricity by any magnitude, either during or since those years, and certainly not by 42 percent.[4]

---

4   U.S. Energy Information Administration, "Coal Transportation Issues," 2007.

CHAPTER 47

THE FUTURE FOR COAL

*We invested heavily, and now the capacity and operations of the Powder River Basin lines
are very, very impressive. . . . Less than 10 years later, I don't anticipate we'll see that
level of coal volume [in 2006, when BN moved 287 million tons] again.
That leaves us with millions of dollars in investment in what will
eventually be stranded assets. . . . It is the fastest-changing story in railroading.*
STATEMENT OF MATTHEW K. ROSE, BNSF EXECUTIVE CHAIRMAN,
AT THE U.S. ENERGY INFORMATION ADMINISTRATION'S ANNUAL CONFERENCE
IN WASHINGTON, D.C., QUOTED IN
"POWDER RIVER BASIN COAL IS FAST BECOMING A 'STRANDED ASSET,'"
BY TOM SANZILLO, *HIGH COUNTRY NEWS*, AUGUST 5, 2015

• *additional* EPA *requirements to reduce emissions from power plants* •
decline in coal tonnage likely to be gradual, over many years • *the number
of sets of unit train equipment in service remains at a near-record level in 2015* •

IN THE MID-1970S, WHEN BN MADE THE decision to go all out and seize the opportunity to move an unprecedented amount of coal over a large part of its network, there were a number of business analysts who advised top management that this new demand for low-sulfur coal would have a life of about twenty years, to approximately the end of the twentieth century. By then, the boom would be over and BN would have to revert to its historic dependence on agricultural and forest products, it was said.

In November 1976 I recall making a one-on-one hi-rail inspection trip east of Lincoln, Nebraska, with a senior-level officer to look over some of the trackwork projects we had completed that year. He told me that shortly before I would reach my retirement age in 2004, one of my last responsibilities would be to remove much, if not all, of the second main track, new sidings, and support facilities we had completed in the last few years—because by that time, the coal boom would be over, and electric-power generation would have converted to nuclear power and hydrogen as their sources of energy. Anything twenty years out seemed like forever to me as a division superintendent at age 37, so I passed it off as no more than an interesting piece of advice or speculation.

Instead of dismantling the infrastructure rebuilt in those years, it has been encouraging to note that BNSF continued to build more capacity in its coal corridors, with such major projects as creating a second bridge over the Missouri River at Plattsmouth and adding a second main track to fill in nearly all of the single-track segments that remained between Lincoln and Alliance. Commitments as large as these major projects demonstrated the confidence BNSF had in the demand for coal from the Powder River Basin for years to come.

Nuclear power has, of course, fallen out of favor because of fear of serious accidents that might result in the release of dangerous amounts of radiation. Technology for the use of hydrogen has not yet been developed. However, the availability of unlimited quantities of low-cost natural gas has moved in and already replaced some of the coal used for power generation.

In 1976 BN carried 33.4 million tons of coal from mines in Wyoming and Montana. Over the next several years, our tonnage continued to increase rapidly, although our growth was set back when the government-sponsored competition from C&NW started in 1984. It took until 1989 to gain enough tonnage from new mines and additional demand for Powder River coal to build its tonnage back to the level it moved in 1984.

What really set off the next big wave of increased coal business in the early 1990s was the passage of the Clean Air Act amendments of 1990, which introduced a cap-and-trade system to control sulfur dioxide emissions from electric-power plants. Increased burning of low-sulfur coal saved the utilities the expense of either buying allowances to continue burning higher-sulfur coal or installing and operating pollution-control equipment. The new regime for "tradable" permits made low-sulfur coal even more preferable than it was before implementation of the new provisions in the act of 1990.[1] More power companies over an even larger part of the country had to shift to low-sulfur coal, which quickly resulted in even heavier demands on BN and UP to expand the capacity of their coal network even more.

EARLY PROJECTIONS WERE FAR EXCEEDED

The projections for 1980 that we had worked under in the 1970s were challenging and impressive, but those made for the early 1990s and beyond required another quantum leap in capacity and operating and maintenance capability. The gross tonnage on the lines serving the Power River Basin became by far the highest of

any railroad in world history. It was a real credit to the operating and maintenance people to be able to dispatch and manage an operation of this scale.

THE PEAK WAS REACHED

The tonnage produced by the mines reached 496 million in 2008. Parts of the joint BN–UP line were moving coal at the rate of 445 million gross tons per year. A deep recession set in shortly after this peak was reached, causing lower demand for electric power for a few years. But the drop in tonnage was more the result of the newly formed capability of energy producers to extract enormous quantities of natural gas at a very low cost. Gaining this capability coincided with increased pressure to reduce emissions of carbon dioxide gas, which was the one contaminant from burning coal that had not yet been addressed by EPA regulations. From 2008 to 2012, the production of coal decreased by 13 percent. That decrease brought the tons of coal produced down to about 400 million tons, about the same amount as in 2004. Anyone who was part of the operation in 2004 would surely agree that they had a very busy and challenging operation to run—it was not drying up.

There are a number of uncertainties yet to unfold that will govern how much, if any, of the loss of coal business experienced so far may be recovered. One factor is whether the process of hydraulic fracturing is contaminating underground water supplies used for drinking and irrigation. Another is the wild fluctuation of the price of gas from short-term changes in the demand from heavy industry, and on the chance of increase in the overseas demand for natural gas produced in the United States. In the past two years, the price of gas has increased enough from its lowest level to make coal less costly. Once the price of gas rises above four dollars per million BTU, coal becomes less expensive than gas. New technology is being applied to a power plant under construction in Mississippi that makes it possible to capture 65 percent of the carbon produced. The carbon dioxide gas would be moved by pipeline for use in enhanced oil recovery at aging oil fields. Getting this much of the carbon removed from emissions might bring the level of emissions down to about the same level as natural gas. However, because

---

1   Meghan R. Busse and Nathanial O. Keohane, "Market Effects of Environmental Regulation: Coal, Railroads and the 1990 Clean Air Act," Center for the Study of Energy Markets, University of California, September 2004.

of the high cost of this technology at this stage of development, it is uncertain whether it will be economically feasible to use it on other installations.

All of these factors make it unlikely that any new power plants will be designed and built to burn coal. From most recent reports, it appears the amount of coal moved out of the Powder River Basin will continue to decline sharply in the next few years. Matt Rose, BNSF's executive chairman, gave a pessimistic outlook for coal shipments at a conference sponsored by the U.S. Energy Information Agency (EIA) in the summer of 2015: "I don't anticipate that we'll see that level of coal volume [287 million tons in 2006] again. That leaves us with millions of dollars in investment in what will eventually be stranded assets."[2] Rose was referring to the year in which BNSF started another program of heavy investment to increase line capacity on some of its major coal routes.

Union Pacific's chief financial officer, Rob Knight, advised an investor conference in June 2017 that about 30 percent of the nation's electricity generation currently comes from coal, which is where UP expects coal's share to stay over the long term. Until a few years ago, about one-half of our electricity was generated at coal-fired power plants. Knight also stated that UP's coal loadings were up 22 percent in the second quarter of 2017 compared to the same period a year ago. In the April 21 edition of the *Omaha World Herald,* it was reported that the price of natural gas had increased from $1.92 per MBTU to as much as $3.60, causing some electric companies to revert to coal.[3]

Surprisingly, a report on the future of coal in *Trains* magazine in March 2016 contained a table of tonnages projected by the EIA through 2040 showing an increase from 392.1 million tons mined in 2015 to 429.5 million tons in 2040. Clearly, these numbers show no decrease in tonnage that would leave BNSF and UP with unused capacity. With this sharp difference between BNSF's outlook and that of the U.S. Department of Energy (DOE), some reconciliation will have to be made in the long-range plans of BNSF, UP, DOE, and the electric-power industry. It should also be recognized that even with one-half of the coal tonnage moved in recent years, BNSF would still be handling upwards of 100 million gross tons on some of its lines. Substantial funds for track maintenance and asset renewal on coal routes would still have to be provided for in the company's maintenance-of-way budgets.

EXPORTS MARKETS—QUESTIONABLE

There is some optimism that export markets for Powder River coal will offset some or all of the decline. In spite of the high cost of moving unit trains of coal over the mountain grades, and on a circuitous route to the north Pacific coast, this segment of business has grown. Three new facilities for transfer coal from rail to ships may be built, possibly at Longview, Cherry Point near Bellingham, and at Vancouver, Washington. Export coal is now being handled at the rate of two trains per day through Roberts Bank, British Columbia.

If the export business continues to grow, it may be necessary for BNSF to invest in additional capacity. Accomplishing that on the mountain passes will be very expensive. It may not be feasible to expand capacity to any extent on the low-grade route along the Columbia River. One possibility would be to replace the track that was removed from the line of the former SP&S between Spokane and Pasco in 1984. A non-capital solution to some of the problem of capacity with today's volume of business has been the use of directional running of trains and crews between Auburn and Pasco. Making this change was an ambitious undertaking, but indications are that it is working satisfactorily.

It is not surprising that strong objections have been raised to even the rather small number of coal trains moving today through the pristine, scenic areas of Montana, Idaho, and the Pacific Northwest. Those concerns may expand into pressure on the governments of the United States and the countries wanting to buy Powder River coal not to allow coal to be exported and burned, polluting the air elsewhere in the world. The argument would be that the United States should not supply fuel that is dirty and a detriment to the health of people living elsewhere. Rather than coal, we would offer to export our natural gas to those countries.

2   "Rose: Coal Lines Could Become 'Stranded Assets,'" trains .com, August 10, 2015.

3   Bill Stephens, "Union Pacific sees coal traffic holding steady," *Trans Industry Newsletter,* June 8, 2017.

At the time when the prospects for the long-term movement of coal were dimming somewhat, BNSF suddenly found itself with yet another marvelous opportunity to gear up to handle large quantities of another form of energy, in this case oil produced by hydraulic fracturing in areas where it had been uneconomic to produce oil on a vast scale. The application of "fracking" has involved BNSF mainly in the northwestern part of North Dakota and in eastern Montana. The challenge to BNSF was reminiscent of the 1970s in the Powder River Basin, in which large-scale investment in capacity expansion became necessary within a short period of time. It is another nice problem to have.

PART VII

THE PEOPLE
OF THE RAILROAD

CHAPTER 48

HOW AND WHY THEY CAME

WHO WERE THE MEN AND WOMEN WHO took a chance and moved out to remote areas of Nebraska, Wyoming, and southeastern Montana in the 1970s, to work in an industry that had been written off by many as having no future, was on the way to bankruptcy or being nationalized, and had no real economic value to the country, and was about to be relegated to museum status? Some had left railroad companies that were at or nearing financial failure, but still, they wanted to work as railroaders. They believed the newly formed BN, the nation's largest railroad, would be able to make it, with the amazing opportunity it had to move large tonnages of coal.

There were recent college graduates who saw a future for the rail industry, once regulatory restrictions were relaxed enough to give railroads a chance to compete in the marketplace for transportation service. Others saw a chance to be able to move, settle down, and raise families outside of expensive, high-crime, crowded urban areas. Once they learned that BN was hiring people for jobs that paid good wages in sparsely populated, low-cost areas, they were ready to go.

Many people who had their roots in these remote areas had chosen not to move to large cities with more job opportunities. They saw railroad employment as the key to having an income high enough to allow them to continue the rural lifestyle they had grown up under. They were glad to give up the jobs they had as ranch hands, barbers, store owners, teachers, or government workers, to get the higher wages the railroad paid.

The people who came to BN had a mixture of talents and life and work experiences, and a range of dreams or hopes for their futures. Through this diversity in backgrounds, they contributed to a fabric that would make the expanding operation on BN a success for shippers, investors, employees, and the general community the railroad served.

They did so in the wide range of technical functions that a railroad must have at the field level—in the maintenance of track, bridges, signals and communications systems, with machinists and electricians maintaining locomotives and cars; in the management and direction of the flow of trains; and in the on-board operation of trains. Not all of the people BN hired could adapt to the kind of culture and work life that is basic on a railroad, and some soon decided to move on. But most of those hired learned to accept the railroad way of life and work and made it a career of thirty years or more. In fact, they thrived on it. The company and its leaders are grateful to the thousands who heard and accepted the call, and had enough confidence in BN's future to "sign on" and go to work learning the trade they were hired for. All of those who were hired and "stuck it out" and developed the skills their jobs required make up the "hall of fame" of a railroad company that was transformed into a real story of success in U.S. economic history.

WHAT THEY CONTRIBUTED

The people who were new to Burlington Northern successfully developed and applied their talents, strengths, and dedication to what became a "cause": to build and upgrade thousands of miles of railroad to make it capable of moving the heaviest tonnages ever run on a railroad. Because of their spirit and the contributions they made, the foundation for BN's success was built in a few years. We should remember those railroad workers of the 1970s, their supervisors, BN's operating and maintenance officers, department heads, senior-level executives, and board members who made it possible through their support. This chapter and a later chapter entitled "A Tribute to the Sons of Martha" acknowledge employees and leaders at all levels who made it possible to develop the franchise BN was blessed in having.

This chapter also gives credit to a representative group of the professional railroaders who were well "on board" in their careers before the start of the coal boom. Some of them had requested assignments on the coal corridors, "where the action was," in those times. Others were directed by senior management to relocate and take promotions or reassignments to positions on which their talents, experience, and job knowledge were needed more than anywhere else on the railroad. They became coaches and mentors to less experienced officers, supervisors, foremen, and the newly hired employees.

Together they built the foundation of competency that was vital to improving safety and the quality of work being performed by the large number of inexperienced people we were hiring, training, and promoting. Railroad professionals working at the division and region level on the coal corridors built a railroad that could provide quality service and produce the rate of return that was expected for the capital we had invested in plant and equipment. Together with support from BN's technical trainers, our first-line supervisors and mid-level managers developed a work force and a corps of officers who could perform at the standards required of those wanting to have a long-term career in the business of railroading.

CHAPTER 49

THE PEOPLE WHO CAME TO BN

• W. T. (BILL) REILLY •

*Mr. Reilly has done a very good job in taking a group off
the street and making railroad people out of them.*
DAVID P. LYSAKER, LOCAL CHAIRMAN,
BROTHERHOOD OF LOCOMOTIVES ENGINEERS

• eight years on perhaps the toughest job on BN in the 1970s •

BILL REILLY CAME TO THE ALLIANCE DIVISION as trainmaster in 1974. He was promoted to assistant superintendent–Transportation in 1976 and held that position for eight years. In that position, Bill was the lead officer for train and yard operations almost from the start of the heavy upgrading programs and the rapid, month-to-month increase in the number of coal trains operated. He soon established himself as the anchor of experience and knowledge for operations on the division. A large number of operating officers who worked for Bill came and left in those years through promotions, lateral transfers to other territories or other departments, and resignations and terminations.

Bill has some startling recollections of his early days in Alliance. On his first day on the job, a crew called on duty at Alliance was headed up by a conductor who had been promoted only three months earlier, and the two brakemen were "brand new," making their first trip. The engineer had worked as a conductor, but he was making his first trip since he was qualified to work as a locomotive engineer. This level of experience was not unusual for many crews called in those early years. It was always a

relief to the officers when there was even one crew member who had worked as a brakeman somewhere else for at least a short time before coming to work on the Alliance Division.

The new hires came from all over the country. Some were ranch hands from the Nebraska-Wyoming area or had left the jobs they had in local businesses. Some had been teachers. For a while there were no barbers left in Alliance because all of them had gone to work for BN. Bill recalls the problem that came out of hiring ten or eleven members of a motorcycle gang that was passing through. Together, they bought a house to live in. Two of them turned out to be satisfactory employees, but the others were soon terminated due to a host of problems they chose not to overcome. One of the two men he kept on the payroll was promoted to trainmaster a few years later.

Some new hires for train service had college degrees and had even taught at the college level. Some of them stayed with the railroad only until a "better" job outside the railroad opened up, where they thought they might make greater use of their education. Bill recalls hiring a sixty-two-year-old Baptist preacher from the South. Another new hire had served as an artillery officer in the U.S. Army. He was promoted to assistant trainmaster about two years after he was hired. Many of those hired stayed on and made a career out of the job they took with BN about forty years earlier.

Hiring so many local people caused tension and resentment on the part of some of the townspeople, because of the much higher wages the railroad paid compared to local businesses. The large number of people moving into Alliance, Edgemont, and Gillette to work for BN put pressure on the schools and local government agencies that provided essential community services. Because these were rather small communities, housing soon became a major problem. Employees shared apartments and motel rooms; some pitched tents in city parks or slept in their cars or pickups, often for a few weeks until suitable housing became available. Many local residents began to rent sleeping rooms in their homes to take advantage of the opportunity for additional income. Due to the lack of housing available for sale at the time of his promotion and transfer, Bill was unable to relocate his family from Livingston to Alliance for six months.

Norman Lorentzsen recalls, "At the terminals which had limited housing, we established programs with contractors to build nice homes, initially not to exceed $50,000. If the homes were not sold within a specified time after completion, we agreed to pay interest and eventually acquire the homes. That never happened since the homes were sold as fast as they were built. A second contractor started another development with home is in the $25,000 to $35,000 price range, and these homes also were sold as rapidly as they were completed."[1]

By and large, new employees who moved into the area managed to tough it out and did not give up on their new jobs out of frustration in learning the work, getting used to the railroad way of life, or dealing with short-term problems in finding housing. They knew they had good jobs and enjoyed living out in the wide open spaces of western Nebraska and Wyoming. For many, it was a big relief to know they could bring their families out to places that had a low crime rate, a healthy small-town atmosphere, and good schools.

Bill recalls the difficulty in having enough skilled dispatchers to cope with such a high volume of trains on a single-track railroad, in which trains were moved under timetable and train-order authority. CTC was being installed on all subdivisions, but it was not in service on the entire division until 1980. For some of the older dispatchers who had grown up dispatching lines with no more than about six trains per day, the "new order" was tough to handle. BN was fortunate to have several experienced dispatchers come to Alliance from other railroads. All were well qualified, and once they learned their new territory, they were able to handle the heavy workload. BN's school for training new dispatchers attracted many capable young people for those very important jobs. Some of them soon became candidates for promotion to officer positions.

Even the old yard at Alliance was a challenge to operate in those times. The yard had only ten tracks and none of them were long enough to hold a unit coal train of 100 to 110 cars. Hence, each train had to double over on arrival. Even though these were unit trains, they had to be yarded for inspection before going to a mine for their next load, then have the locomotives inspected and fueled or possibly changed out for maintenance and "spares" switched into the train if any bad order cars were found in the inspection. By the late 1970s, the yard tracks were extended and many additional tracks were built to make Alliance the hub for the coal operation.

Bill had to break in the new supervisors as they were brought into the division and still keep the operation rolling until they got their feet wet, which could take a few months for those with little or no prior railroad experience. During that time the division superintendent and the regional assistant vice president–Operations evaluated the newly assigned officers to see if they had the potential to develop well enough to handle the challenges of their new position. Bill served under four division superintendents in his eight years on the Alliance Division. With the revolving-door approach used in the training and development of first-line supervisors—not to mention their burn-out rate—Bill was indispensable to the operation. The company could not risk losing his experience and capability. He was the anchor who could be depended on to keep things going through very difficult periods of congestion and the shortage of

---

1   N. M. Lorentzsen, letter to author.

such resources as track capacity, crews, and locomotives. Somehow, through all of the tension and pressure he was under, Bill did not burn out.

A number of the junior officers Bill supervised rose to high-level positions in the company in the 1980s and '90s. Much of their success came from the experience and development they got while working under Bill and a few other stalwart officers under him who knew how to hold things together during those years of unequalled challenge in the rail industry, at least in modern times.

The job of assistant superintendent–Transportation on the Alliance Division in those years ranked as among the toughest on BN, including such other notable operating assignments in Cicero, the Twin Cities, Kansas City, and Spokane.

In 1982 Bill was promoted to terminal superintendent at Alliance. He was replaced on the Alliance Division by Cal Evans, another solid, seasoned veteran of the tough years on the Alliance Division. Bill took over as terminal superintendent in Fort Worth in 1986.

· C. J. (JAKE) GREELING ·

· *made the decision to take a step backward and head west for a new career* ·

JAKE GREELING IS A GOOD EXAMPLE OF SEVERAL established railroad officers who saw the opportunity to become "part of something big," when he decided to leave his position on another railroad and seek opportunities on BN. Jake held the position of district assistant superintendent on the Illinois Terminal Railroad (IT), a terminal and line-haul railroad that operated between St. Louis and Springfield, Illinois. In 1974 Jake decided to give up his position on the IT to try his fortune on BN's Alliance Division. In an interview a few years ago for the newsletter of today's Powder River Division, Jake recalled what caused him to make his move: "I took a step backward because railroads in the East were starting to fail. It was worrisome. I wanted to stay in the railroad industry, and I knew the coal in the Powder River Basin would be booming."

Jake hoped to be hired for an officer position right away, but at that time, BN was still trying to work off the excess number of managers it had following the 1970 merger. Also, this was just before the "coal boom" really began. Instead of giving up on BN, Jake decided to take a job as a brakeman at Sheridan, Wyoming, and show BN that he had "the right stuff" to be an officer when the opportunity came. He stuck with his decision and, less than a year later, was selected for the locomotive engineer training program. Eleven months later, Jake was promoted to road foreman of Engines, his first officer position on BN. He was soon promoted again, to trainmaster at Alliance, and then to assistant superintendent

at Gillette in 1980. Altogether, Jake worked on the Alliance Division for nine years. He later advanced to higher level positions in the Operating Department at other locations and finally, to director of Interline Services at Fort Worth in 1987.

Jake was among the early corps of officers who were expected to get increasing numbers of trains over the many miles of track riddled with slow orders and where much trackwork was in progress. Most of the division was operated under train-order authority, which was a complex system when used in a high-density operation, and especially demanding for the many new hires they had in those years. It was a real challenge for division officers to be responsible for coaching newly promoted conductors and engineers on such issues as the complexities in the rules governing the superiority of trains; the required spacing of following trains in non-signaled territory per Rule 91; giving proper flag protection to the rear of a train; and numerous slow orders and Form Y train orders issued to protect track gangs working on the main track. The only part of the division with a signal system in service at that time was the 238-mile line between Alliance and Ravenna, and 26 miles with an automatic-block system over Crawford Hill. CTC was being installed on all lines that had no signal system in service, but until it was extended, the complex, old, traditional timetable-train-order system was the only way trains could be moved.

Officers such as Jake were the prime source of knowledge for new employees as to application of all the rules and procedures they had been taught in the training program, once they started to run trains. It was difficult to hire and train people fast enough to handle the ever-growing number of trains being run. Officers in the field bore a tremendous responsibility for the safety of the operation and for getting trains over the road to meet the loading schedules at the mines and then to their destinations. In addition to having inexperienced people running trains, there were many newly promoted dispatchers managing the operation on congested lines, together with newly hired operators copying and delivering train orders in the field.

Many of the foremen running track gangs also were new to their jobs. They were not always as familiar and confident as they should have been in enforcing the rules for setting up the signs protecting their work limits and "talking" the train crews through the Form Y train orders issued to protect the gangs while work was underway. It fell to the road foremen and trainmasters to follow up and coach the foremen of the gangs on blending the need to get trains over the road with getting the trackwork done. On many days, there could be several work trains out on each subdivision unloading rail, ties, or ballast. All of that work had to coordinated with the maintenance supervisors, dispatchers, conductors, and engineers of the work trains, and foremen of the gangs. BN was fortunate to have a few officers such as Jake who were experienced and able to manage through these challenges. Jake summed it up: "The whole division had to be rebuilt to take the stress and burden of the coal freight."

The trackwork got done, the new CTC was put in service, the track was made safer, train speeds were increased, and on-line congestion was lessened. Fewer train crews might be needed, but only for a short time, since business continued to increase. It took about six years of very hard work to get the railroad up to the standard needed for what soon became the highest-traffic-density lines (in terms of gross tonnage per mile, per year) of anywhere in the world. It took the skill and dedication of a team—field managers, including the operating supervisors, track maintenance supervisors, signal supervisors, dispatchers, and professional trainers in all of the specialties—to bring the quality of the operation up to the capability needed to get the job done.

Jake and that team had a major role in coordinating the train operation with the heavy construction work involved in building the new 127-mile Gillette–Orin line. Train crews and dispatchers had to be trained and readied for the opening of the new line, which was to provide greatly improved operating capability. BN was to gain a shorter, more efficient route for handling much of its increasing coal business. Jake recalls the first train run on the new connection to the Orin line: "It was a great sight to see that long unit train coming around the connection at Bridger Junction. It was a special event, a historic moment."

The years from 1974 through the opening of the Orin line in 1979 were the toughest to manage through. Jake and many of his associates regard those years as among their most rewarding and productive in their development as seasoned, highly competent railroad operating officers.

• C. P. (CAL) EVANS •

• promoted from conductor ranks after twenty-four years of service •
• had the ability to handle train crews and the experience and
leadership capability needed to shape new hires into the railroad mold •

CAL EVANS HAD TWENTY-FOUR YEARS OF service as a brakeman and conductor on the Galesburg seniority district of the Chicago Division when he decided to pursue an officer position through the "This Way Up" program in 1977. After passing all the tests and scrutiny given to candidates for advancement from the ranks to assignment as a junior-level operating position, Cal was named assistant trainmaster in Tacoma. He soon established his capability as a manager and was transferred and promoted to trainmaster on the Alliance Division.

Bill Greenwood was brought in as division superintendent at Alliance in 1979, and found that he needed to bring more officers to the division who already had strong knowledge of operations, plus the strength and leadership capability needed to deal with the inexperience and lack of proficiency among the large number of new hires in train and engine service. Cal Evans was among those who met such criteria—he was out of the same mold as the other experienced "pillars of strength" who came to the division in those times, among them Bill Reilly, Jake Greeling, Don Maze, and Tom Lynch.

At the time Cal came to the Alliance Division, the average age of employees working as conductors, engineers, and brakemen and in yard switching was about 24. Much of the time, the only employees on train crews who were qualified to take the lead were the engineers. The training they had received in the Engineer's Training Program made them more knowledgeable and proficient in any aspect of the operation than many of the conductors who had worked as brakemen for only a year or so before being promoted to conductor.

Cal and Adrian Hertog, an expert trainer and rules instructor, set up classes to further raise the proficiency of the newly promoted conductors. They also found it necessary to set up a more concerted effort to train newly hired brakemen. Up to then, new hires were given only a few hours of classroom and field training before they were placed on the extra board. The field training amounted to only a few "student trips" on a train. At that time, train crews still consisted of a conductor, two brakemen, and an engineer. As the proficiency of crew members was enhanced, they became qualified to help new employees practice and further develop the skills they had been taught in their abbreviated classroom training program. With three "regular" employees with a new brakeman on a crew, there was ample time for them to work with the new hire and still get the train over the road, or for yard crews to finish the work they were assigned. However, if all of the employees on the crew had only minimal experience, a student brakeman might not be able to learn much from them.

The second part of the effort to raise the overall quality of employees being put to work in train and yard service was to make a closer evaluation of them at the time they applied for work. Out of desperation to get enough people hired quickly to meet the growing need for more crews, it seemed that nearly anyone who could pass the physical examination and the quick background check could be hired. Cal intervened in the hiring process by conducting one-on-one interviews with the applicants. With the years of experience he'd had in working with new hires at Galesburg, Cal had good success in evaluating which candidates had the right general attitude and personality makeup needed for railroad work. Cal called it "looking for the cut of the eye": a blend of a candidate's intelligence, body language, reflexes, and general attitude toward life and work. From this initial screening, the rate of turnover among new hires was reduced, and fewer of those hired had problems in adjustment to the demands of railroad life.

Cal found that nearly all of the employees who'd had prior railroad experience worked out well. They were glad to be able to stay in the rail industry and found it refreshing to work for a railroad company that had the resources needed to improve the track and take full advantage of the unique opportunity for growth they found on BN. And, because they already knew railroad work, they were a big help in on-the-job training of new hires, and soon became qualified for promotion to conductor or to enter the locomotive engineer training program.

Unfortunately, a number of the younger employees got in trouble through the use of drugs and alcohol. With their level of income, they often were targets for drug dealers. Peer pressure entered into the problem, even for some who had been rehabilitated through the counseling program, and evaluated by the counselors as good sobriety risks. During the time he was assigned at Guernsey, Jake Greeling recalls that in post-accident testing, about 80 percent of the employees tested positive.

Cal brought local law-enforcement officers and union officers into meetings he set up with employees who had gone through the company's rehab program, to give even stronger emphasis to the fact that the use of drugs and alcohol seriously jeopardized the safety of the railroad operation and public safety on the streets and highways, as well as had a serious effect on the safety and well being of their families. These employees were warned that if they were found in violation of the company's rule on the use or possession of drugs or alcohol while on duty, they would have no second chance. In the end, it took the threat of the 50 percent random testing program instituted under federal law to clear alcohol and drugs from the railroad work scene. Because of the nature of railroad operations, employees cannot be under the influence of anything that will impair them from performing at their best.

In cases of rules violations of various types, whether assessing responsibility for accidents, misconduct, or failing to be available for work when called, operating officers were required to conduct an inordinate number of formal investigations (hearings) in those years. Cal recalls that two days of every week were set aside for holding investigations. On those two days, each of three officers designated that week to handle formal discipline would conduct as many as six investigations. Discipline, ranging from a reprimand and some coaching to suspensions without pay for a specified number of days or dismissal, would follow. Some employees learned faster than others from progressive discipline. As might be expected, some employees were making so much money that being given time off from work did not particularly concern them, especially if their "offense" was missing a call, and only a censure or suspension of five days or so would result. Some found discipline a way to get some relief from having to work seven days every week.

Most of the new hires who made it through two years of service matured enough to accept railroading as a way of life, and settle down to become good, responsible members of the work force and the community. But for many of them, getting to that level was a struggle. Certainly for their supervisors, having to coach, mentor, and discipline such a large number of people who were new to the kind of behavior, skills, and rules that compliance required in railroad work was a tough burden. Cal likened those times to the gold rush days, with people coming to Nebraska and Wyoming from all over the country to work on the railroad and get rich. Having to handle this heavy burden of personnel issues consumed a great deal of the supervisors' time and took them away from direct supervision of the train and terminal operations that were vital to getting the coal moved.

The number of crews required in chain gang (pool) service doubled from 1976 to 1977 alone. Every additional set of unit train equipment put in service required the equivalent of four more people to be hired and trained. Cal credits Bill Greenwood with giving him the authority to establish a quality program to train new brakemen,

increase officer staffing to the level needed for effective supervision of the rapidly increasing work force, and deal with the increasing complexity of the operation. In 1979, when Cal came to the Alliance Division, the staff consisted only of two trainmasters and two road foremen of Engines at Alliance, a trainmaster at Gillette, and an assistant superintendent for Transportation.

The operating timetable for the Powder River Division for 2015 lists twenty-one trainmasters based at Alliance alone. They work on twelve-hour shifts for four consecutive days, followed by three days off. In the 1970s, the officers had no days off and, on most days, worked close to sixteen to eighteen hours. These officers and their bosses, the assistant superintendent and the division superintendent would work a full twelve-hour day, go home for dinner, and then return to work for another four to six hours. When a derailment occurred, they would head out to the scene and stay there straight through, often without having time for rest, until the line was opened.

For eight to twelve hours on most days, each subdivision was shut down for track-maintenance work. That meant the dispatchers had to organize the operation well enough to get all of the trains moved during the evening and night hours. They usually had several miles of slow orders to contend with—on track that had been disturbed for surface correction, re-ballasting or the replacement of ties on that day, or because track rehabilitation had not yet been started in locations where the track was not safe for trains to operate at the standard speed of 40 miles per hour for loaded coal trains and 50 for empty trains or general merchandise. Between Alliance and Ravenna, there was an outdated slow-response CTC system in place. On the rest of the division, trains had to be dispatched by timetable and train orders. With the low level of experience on most of the train crews, authorizing trains to operate with a "fist-full" of meet orders, "right over" orders, slow orders, and Form Y orders protecting trackwork in progress, plus with second-class train schedules to contend with, there was a major challenge to operate safely and efficiently. The dedication to duty of the dispatchers, the experienced railroaders out there running trains, and the division operating officers was impressive. The challenge of working in the environment of the 1970s should go down in history as one of the most remarkable accomplishments in BN's first decade.

• M. L. (MIKE) HOLSTEEN •

*• a high level of commitment, dedication, and sacrifice,
still a big part of the railroad way of life •
• "for every one of these failures we found out about,
there probably were several others that we never became aware of"[2] •*

MIKE HOLSTEEN HELD THE JOB OF TRAIN-master at Alliance in the division's most difficult years, 1976 through 1979. There also were three road foremen of Engines and an assistant superintendent–Transportation assigned at Alliance in those years, but Mike was the primary contact for the large number of brakemen being hired and trained every week—for the many questions and concerns they had about the challenges of the work life they had just entered. At that time only a minimum amount of training was given before new employees made their first "pay trip." Most of the conductors did not have enough experience to make them very good trainers of the new brakemen assigned to them.

Mike was responsible for the forty to fifty pool crews based at Alliance that worked in three directions, to Ravenna, Edgemont, and Northport. When he first came to the Alliance Division, Mike was the only trainmaster at Alliance for a year as a half, when two trainmaster positions were finally added, plus one at Edgemont, and a combination trainmaster–Road Foreman position at Gillette. Until then, all of the first-line supervisors except Mike were road foremen, who were primarily responsible for the work of the locomotive engineers.

In those early years of expansion on the Alliance Division, Mike and Bill Reilly may have had the toughest jobs of any operating officers on BN. The same could be said for the officers responsible for the heavy load of maintenance-of-way work underway on the division. (I do not wish to imply that all operating positions at other locations on BN were easy or light duty in those years. That certainly was not the case, especially for the operating officers who worked in the Spokane and Lincoln areas, or nearly anywhere on the Chicago Region.)

After starting with BN as a management trainee, Mike worked for two years on the COMPASS implementation team, followed by assignments as assistant trainmaster at La Crosse and Trainmaster at Cicero. In 1976 he was transferred to Alliance and, three tough years later, was promoted to assistant superintendent–Administration at Portland. Mike soon advanced from there to terminal superintendent at Portland and Minneapolis. He then became the division superintendent at Springfield, followed by director–Transportation at Denver and general manager on the northern lines territory. Among the positions he held on BN in later years were assistant vice president–Intermodal Automotive Operations and vice president–Operations for the former Santa Fe lines after the BN–Santa Fe merger. In 1997 Mike was hired by CSX as general manager of its Florence (South Carolina) Division. All in all, Mike had an excellent career.

When Mike and his family began to look for housing in Alliance in 1976, they found only three houses of any size or price on the market anywhere in town. One of the three houses had just been built, and the other two were old and not in good condition. By that time, some new apartment buildings were under construction, but many of BN's new hires still had to live in small towns up to thirty or forty miles from Alliance.

The workload was very heavy for all supervisors, with the number of pool crews having to be expanded frequently to handle the rapid growth in the coal business. A large number of brakemen were being hired from midwestern and eastern railroads that were either in or nearing bankruptcy, losing business as a result, and having to cut their forces. Since these new hires were experienced railroad employees, they needed only a minimum of training. After a few familiarization trips over the territory, they could start to work.

Unfortunately, some of the new brakemen hired from other railroads were later found to be out of work for disciplinary reasons. Since BN policy did not allow its supervisors to make a reference check with previous employers at the time the new applicants were interviewed, there was a risk in hiring people who claimed to have work experience in train or yard service. The "real" character and skill level of a number of these newly hired employees

_____

2 M.L. (Mike) Holsteen, letter to author.

sometimes was not revealed until they had been working for BN for a while. Some had to be terminated when it came to light that they had falsified their employment application or because of unsatisfactory performance on the job. Overall, however, BN was fortunate that so many experienced railroaders who had either lost their jobs, or expected they would be cut off due to the declining fortunes of their company, wanted to stay in the rail industry, even if that meant relocation to a far different kind of place than they had ever lived or worked.

Most of the new hires with no previous railroad experience were from the western part of Nebraska or eastern Wyoming. Many of them had parents, relatives, or friends who worked for BN or its predecessor, the CB&Q, and decided it was time to "get off the farm" and get a job that would pay much more than any other job they could get within a few hundred miles. As was true with new hires anywhere, some had difficulty adjusting to a work life with very irregular starting times that required them to work on nights, weekends, and holidays. Many of them were satisfied with the earnings they could make in working only three or four days per week, and they resisted the company's requirement that they had to take calls for work on a seven-day basis, and often with only a short time off duty between trips.

The turnover among them was high. At most places on a railroad, that kind of problem is manageable, but on the Alliance Division at that time, it was far more serious because there were so many new hires coming on board at one time. Mike recalled that a class for between twenty and forty new hires had to be conducted every week to keep up with the growth in business, together with the turnover among new hires, the number of employees being taken out of service for disciplinary reasons, and resignations of those who concluded that "railroading is not for me." With all of these problems, it was difficult to get the extra board and the number of pool crews built up quickly enough to the level needed to handle the business.

Mike recalls several instances that revealed the extent of experience of some of the crews that had to be put together. Often, there were not enough experienced brakemen, conductors, or engineers who had to be called for work right at the time, to have even one employee on the crew who knew the work well and could give direction to the new employees. Mike provided an interesting account of how this lack of experience affected the ability of switch crews to get the work done in the yard at Alliance: "On nights and on weekends it was not unusual for yard crews to have only a few months' experience for both the Foreman and the Switchmen. In some cases the Engineer would also have only a few months' experience. This would literally result in a yard crew switching only a handful of cars in an eight hour shift or not completing the switching of a handful of bad orders out of a hopper train."

Mike cited an extreme example of what that inexperience and lack of judgment and basic "railroad sense" could lead to:

Another example of lack of experience of crews involved a local train 368 that did local work from Edgemont to Alliance. The first crew expired on the hours of service at Belmont at the top of Crawford Hill. The relief crew was called out of Alliance to taxi to Belmont and bring the train to Alliance. The crew was given the information on the train which consisted of three small locomotives and 101 cars. The crew talked on the taxi trip to Belmont about the possibility of having to double into Marsland because of the tonnage of the train and the locomotive consist. At one point after the crew had been enroute for some time the Dispatcher came on the radio and asked the Engineer where he was located. He responded that they were coming into Hemingford. The Dispatcher responded that he had the Conductor on the phone at Belmont and the rear of the train had not left Belmont. It was determined that the original crew that was relieved at Belmont had cut the crossing at the east end of Belmont and this relief crew left Belmont with only three cars and had traveled approximately 30 miles to Hemingford and was oblivious to having only three cars. There were numerous failures that caused this. The scary thing is that for every one of these failures that we found out about there probably were several others that we never became aware of.

Through the 1970s, heavy-maintenance work was in progress on the entire division. On most days, and even on weekends, the track was taken out of service for twelve hours to give the gangs an uninterrupted period to get heavy work done such as relaying rail, replacing ties, ballast sledding, and undercutting. An eight-hour workday was planned for such gangs, but fairly often, that stretched into twelve hours before all machinery was moved clear of the main track, and the day's work was "closed up" with the unloading of enough ballast, followed by lining and tamping, to put the track back in service. That left twelve hours to the dispatchers to run a day's worth of trains on a line riddled with slow orders and to deal with the queues of several trains that had been held during the day and would have to be moved before the track would have to be taken out of service for the next day's work.

An interdivisional run of 238 miles had been set up on the Alliance–Ravenna line in the early 1970s. Although CTC had been in service on that subdivision since 1951, it was installed with power switches at only one end of the sidings and with a spring switch at the other end. At the time this CTC was installed, the density of traffic was very light, so there had not been a problem in getting trains over the road until the number of trains operated on that line increased dramatically in the mid-1970s. The CTC system's capacity was so low that it could take up to twenty minutes for signal and switch indications to come from the field to the dispatcher. Having this kind of delay impaired the ability of dispatchers to move trains, causing train delay and greatly reduced the capacity of the line. To overcome this problem it was necessary to install microwave links to carry the code and convert all of the spring switches to power switches.

Mike recalls that getting a train over this entire subdivision with two crews was considered a success. With priority being given to loaded trains, three or four crews often were required to move an empty train from Ravenna to Alliance. It took until 1970 to complete the construction of the 10-10 configuration (10 miles of single track followed by 10 miles of two main tracks), an enhanced CTC system, and a major reduction in the number of miles under slow orders to build capacity up to the level needed for an efficient operation.[3]

With the lack of experience on many train crews, and because a thorough understanding and respect for the rules of railroading not yet developed in many newly hired employees, a number of incidents occurred that could have led to accidents and injuries. Unfortunately, some accidents did occur, but the number of "near misses" far exceeded actual occurrences. Getting real professionalism instilled in all of the train and engine crews was a tremendous challenge. Over the next several years, this was accomplished, and the Alliance Division moved from a position at or near the bottom of the division safety standings to a much more respectable level. Its successor division, the Powder River Division, has consistently been one of the safest on today's BNSF Railway.

In making spot checks while out on the railroad and while conducting efficiency tests, Mike would note some cases of a failure to provide flag protection to the rear of a train in non-ABS territory when it was stopped or could be overtaken by a following train. Failure of the brakeman and conductor to get off the locomotive or caboose to make a walking inspection of the train while it was stopped was another fairly common fault. At that time, there still were two brakemen on every road crew, and since nearly every train they worked was a unit train, they would often have little or no on-the-ground work to do on an entire trip. Keeping employees alert and attentive during the monotony of a long, slow trip with no switching to be done en route could be difficult. It was easy for complacency to set in.

Until the second trainmaster position was established at Alliance, there was no officer who could be directed to handle Mike's responsibilities for a weekend to allow him some time off. About thirteen months into his tour of duty, Mike was finally able to schedule a two-week vacation. Three days into his vacation, he was called back to work. He did not get his vacation time in until four months later. It was not until 1979 that the division officer staffing was increased, commensurate with the ever-increasing tonnage being moved and the number of operating employees being added to the payroll. The number of gross tons being moved between Alliance and Ravenna shown in table 38.2 is representative of how rapidly the workload increased from year to year.

Mike certainly was not the first or only operating supervisor on the railroad to have had a job with a heavy workload that would continue without relief for a long time. To handle a very demanding job under such conditions required a high level of commitment, dedication, and sacrifice on the part of railroad supervisors and their families. In those times, and yet today, those requirements still are a big part of the railroad way of life for field operating officers.

_____

3   M.L. (Mike) Holsteen, letter to author.

· THOMAS H. (TOM) LYNCH ·

Treat the employees right, but don't be their local chairman.
They will respect you and will get the job done.
ADVICE FROM TOM'S FIRST AND TOUGHEST BOSS,
NICK CARTER, TERMINAL SUPERINTENDENT AT DENVER

TOM LYNCH WORKED ON THE ALLIANCE Division in the early years of the coal boom, from 1973 to 1975, when the large track-upgrading programs were just getting underway. The relentless movement of heavy coal trains was taking its toll on the track, requiring many miles of slow orders to be placed. Track-caused derailments were on the increase, and the hiring of large numbers of new brakemen and trackworkers was in full swing. It was a time of pressure and demand on division officers to somehow overcome unprecedented pressure to move more and more trains on many miles of marginal quality track—and get it done with many inexperienced train-service employees holding responsibility for safety under a complex system for dispatching trains.

All of this required field-level supervisors who had high energy, could tolerate frustrations, and had enough ability and experience to overcome the hour-by-hour deluge of problems that hit them. Tom was one of a group of field officers on the Alliance and Lincoln divisions who really earned their stripes in those years before many company officers at higher levels grasped the magnitude of the growing challenges that were developing out on those parts of the railroad. What kept spirits high was that the company was providing a huge amount of resources of new track, bridge and signal material and new locomotives and cars, as well as the support of the senior-level executives who knew what the real railroad was all about in those times.

Having a superior officer such as Wayne Arntzen at the region level to quell the concerns, criticisms, and second-guessing from a few factions outside the division kept field officers from having to spend time on non-essential, make-work inquiries, advice, and direction. The people on the divisions could see they were part of something great, and they knew that with the vast resources being deployed, to keep the trains moving while trackwork was underway, would pay off. It was a far better situation to be in than watching business and earnings decline, along with the general deterioration in the condition of the roadway that many of them had seen in the years before the 1970 merger. As a result, morale and enthusiasm were high, although the pace was exhausting and could be almost overwhelming at times.

This was the atmosphere in which Tom and his peers worked in those years. Tom came to Gillette as a combination trainmaster–road foreman in 1973, after working for three years as an assistant trainmaster in the Denver terminal. Tom recalls there were only two houses up for sale in Gillette. To give some relief to the shortage of housing, BN placed several mobile homes on its vacant property adjacent to the yard. At that time only five or six pool crews were needed at Gillette, Edgemont, and Sheridan. By 1976 the pool at each of those points increased to twenty-five or twenty-six crews. A number of the employees hired for train service left their jobs as ranch hands or members of oil-drilling crews. It was difficult to hire and train employees fast enough to increase the pool as business grew, which meant that there usually were no employees available to staff an extra board to fill vacancies due to vacations or sickness or for employees who needed an occasional day off from the seven-day regimen to take care of personal business.

Tom recalls a Saturday when there was no rested crew available at Edgemont in time to go on duty for a "hot" westbound train. Tom says he "drove the streets" of Edgemont, looking for an engineer to call for the train. He found one in the middle of pouring concrete on the driveway at his house. Tom offered to finish smoothing the wet concrete if the engineer would agree to take the train. The engineer had worked for BN for some time in Wisconsin before deciding to move out to Edgemont, long enough to understand that Tom would be in trouble if the train had to be held. All of the officers who worked in territories of rapid growth in those years could tell stories of the creative things they had to do at times to put crews together to keep the trains moving.

Tom is very complimentary of the commitment of Art Fiedler, manager–Locomotive Training, in those years to get employees qualified at locomotive engineers fast enough to run trains moving the rapidly increasing tonnage of coal: "It was guys like him that through their actions insured that the railroad could handle the coal transportation and helped beat down the ETSI proposed coal slurry pipeline. What would BNSF and UP be today if that pipeline would have been built?"[4]

After a tour of duty as terminal manager and assistant superintendent at Lincoln, Tom was promoted to division superintendent in Fort Worth in 1983. Later, he served as the Labor Relations officer for the Twin Cities Region. Since retiring from BN, Tom has served as chairman of the board of the Siouxland Ethanol plant at Jackson, Nebraska, served by the short line that acquired BNSF's O'Neill line. It has been a very successful venture for the local farmers and other residents who invested in the plant. Together with two other ethanol plants and five large new grain elevators, BNSF found the O'Neill line was generating so much revenue that it bought the line back from the short-line operator who had acquired it several years before.

∘ DONALD L. (DON) MAZE ∘

People will follow you if they sense you have direction and know where you are going.
DON MAZE IN "DIVISION BIDS FAREWELL TO DON MAZE,"
PRIDE OF THE NORTHWEST, MARCH/APRIL 2003, PAGE 2

DON MAZE HAD AN EXCELLENT CAREER WITH BN, advancing to the important line-operating position of general manager of a division. Soon after Don started his railroad career as a brakeman working between Galesburg and Cicero, Illinois, changed to engine service in 1973 as a locomotive fireman at Aurora, Illinois. At that time, this still was the entry-level position in which prospective locomotive engineers were trained. Even after the path was established for dealing with the long-standing issue of the need for a fireman on a diesel-powered locomotive, railroads still were required to use firemen on passenger trains and on 10 percent of the freight trains and yard-switching assignments designated as "must fill" jobs.

Don was soon promoted to locomotive engineer, and then to road foreman of Engines at Seattle in 1977, his first supervisory position. In 1979 Don was offered and accepted promotion to trainmaster at Alliance. It was very helpful to the Alliance Division at that time to have such an experienced operating employee from the "big leagues" of the Chicago Division available to work with new hires on lines handling a higher and higher density of traffic. Don also had the advantage of having worked under the complex CB&Q schedule of agreements that was also in effect on the Alliance Division, so he did not have to master the operating rules as a new officer, on top of the agreements that governed employee compensation and working conditions, to become an effective leader at the first level of management.

Don recalls his biggest challenge as having to train and guide "all the new hires in train service. This was a very serious challenge for the safety side as we were still operating with train orders. We continually had problems with crews blowing [failing to comply with the requirements of train orders and the temporary signs placed to restrict the movement of trains through areas in which trackwork was in progress] 'track flags.' We had extensive testing to correct this. I spent a large amount of time riding trains and educating crews about what was going on and what were their responsibilities."[5]

The serious problem of Rule G violations that had to be dealt with immediately and effectively. Rule G was the provision in the operating rules that the use or possession of drugs or alcohol (later expanded to include controlled substances) while on duty or subject to duty was prohibited. The frequency of violations of this rule required Don to conduct two or three formal investigations (hearings) each week for crews based at Alliance

---

4   Tom Lynch, telephone communication and email to author.

5   Letter to the author, October 14, 2012.

alone. In post-accident testing, Don recalls, about 25 percent of employees tested positive for drugs or alcohol. This created a situation so serious for safety in the operation that it led to such severe methods as random testing for the presence of drugs or alcohol in blood or urine, and the use of "sniffer" dogs to detect the presence of drugs being brought on company property and equipment. It was found that most of the employees charged and disciplined under Rule G were binge drinkers and had not yet become problem drinkers or drug users. Even so, the problem was serious and had to be dealt with severely.

Don remembers the social situations that company officers and their families confronted while living in the small "railroad towns" in such a remote area of the country. There were times when spouses and friends of employees who had been disciplined for rules violations confronted Don's wife and the spouses of other officers with unpleasant, disrespectful comments when they met in stores, schools, and other public facilities. The children of officers reported being harassed and threatened by the children of some employees who for one reason or another disliked the standards their superiors expected them to uphold as professional railroaders.

Some employees and family members resented being disciplined for failure to be available when called for work, especially when they could be given only the minimum of eight hours off duty while at home between trips. Others would become upset if they could not be allowed to get some time off and miss a trip or a shift. The shortage of adequate housing made some railroad employees irritable and frustrated, by having to live in conditions not suitable for rest and privacy, and at the standard they thought they should have as well-paid workers.

Such conditions were tough for both employees and officers to deal with in the early years of rapid growth, and with a large number of people moving into these small communities within a short time. There was great diversity among people in terms of cultural backgrounds, interests, values, and aspirations. To their credit, nearly all of the new entrants were grateful for the opportunity to have good jobs and to establish roots with a good company and in a community. Most of them stuck it out. Many were rewarded with opportunities for advancement or to be trained in new skills.

Don wrote, "For the most part the employees were great and worked hard to make this effort work. They were aware that coal was a big deal for the railroad and their careers, and they wanted to be a part of it. The bad actors were weeded out fairly quickly and the Local Chairmen were old heads that I dealt with and they had the same goal, a successful railroad and worked with us to improve the quality of workers."[6]

Don's next promotion was to assistant superintendent of the Portland–Vancouver terminal, followed by transfer and promotion to assistant superintendent–Transportation at Spokane and soon, to division superintendent of the Cascade and Nebraska divisions. He returned to the Pacific Northwest as general manager at Seattle before closing out his career. Don expressed appreciation for the opportunity he had been given to advance, without having a benefit of a college education. On one of his last days of service before retiring Don reflected on his career: "It's been a great career. . . . I went further on the railroad than I ever thought I would when I started out learning to be a brakeman. I've always been pleased with the railroad."

Don's father, Archie, had worked as a brakeman and conductor on the Aurora Division for thirty-seven years. He served as local chairman of the union for many years and took great pride in Don's accomplishments. In learning Archie had been voted out of office after serving for many years as the representative of train service employees, I asked one of Archie's fellow workers what could have caused employees to vote him out of office. His reply was that after making several trips to visit Don after he was promoted to an officer position, Archie would come back with a different view of things. "He became too much of a company man," I was told. A majority of employees still expected their local chairman to take a tough line with the company in areas of disagreement or contention. They saw Archie moving too close to the company's position on some issues and decided it was time for a change in union leadership.

On Don's last day as general manager of the Nebraska Division, he was given a fine tribute by Lorri Savidge, senior claim representative: "It's been a great privilege and pleasure working with you. You were a tremendous asset to this division and the railroad. We all thank you for making us part of your team."[7] Indeed, it was a fine tribute to a highly respected and capable line-operating officer.

---

6 Letter to the author, October 14, 2012.
7 "Maze Heads Northwest," *Nebraska Division Whistle Post*, December 2001/January 2002, p. 1.

· WAYNE L. ARNTZEN ·

We come to work here because we want to work and not just to earn a pay check.

A COMMENT HEARD HUNDREDS OF TIMES ON THE RAILROAD IN THE 1970S

· *leadership in the time and place of unprecedented challenge and opportunity* ·

A GREAT MANY PEOPLE DESERVE CREDIT FOR THE success and rewards BN reaped from the investments it made to haul coal out of the Powder River Basin. Several have already been mentioned in this book, and a Hall of Fame should include hundreds of dedicated, highly capable people who made it possible. Of those who were involved at the field level, Wayne Arntzen would stand out as the leader who best manifested the qualities of leadership, job knowledge, and professionalism that got the job done on schedule and as planned. Wayne served as assistant vice president on the Denver Region from 1975 to 1977, and as regional vice president from then until his decision in 1984 to leave the company.

BN and many of its people owe a great part of their success to Wayne's ability to raise people's sights and inspire them to perform at the top level of their capability, develop new skills, and take pride in their work. He was successful in helping the operating divisions and the region's engineering and mechanical units perform at the high level of proficiency that was necessary. At the same time he was successfully leading and developing this team, Wayne kept learning and improving his own capabilities. He worked well with technical experts who held positions at the system level, suppliers, and others who were developing better materials, methods, and machinery for getting the work done. Wayne was able to recognize that he had some subordinates who knew more about some aspects of the work than he did, or had better capability in certain areas of management. He would listen to them and very often applied their advice on handling a problem or a project rather than relying strictly on the plan he initially had in mind.

I was fortunate to work for Wayne in three positions: first, as assistant division superintendent of the Chicago Division at the time of the 1970 merger, a few years later as division superintendent of the Rocky Mountain Division, and then as superintendent of the Lincoln Division (later renamed the Nebraska Division) in the heaviest years of rebuilding that part of BN. I consider the time I spent in working for Wayne as among the best of times in my career development.

Wayne had no formal education beyond the high-school level. He started his career with the Burlington as an operator at stations on the Fox River line in Illinois. He was soon promoted to a junior officer position. Before long he was on a "fast track" and advanced: assistant superintendent at Omaha, assistant general superintendent of Transportation, assistant general manager, assistant to the vice president–Operations, division superintendent of the Chicago and Twin Cities Divisions, assistant vice president of the Billings and Denver regions, and vice president of the Denver Region. The company and its people were fortunate to have an operating officer of Wayne's capability—both to lead a major part of BN's program to upgrade and expand the capacity of the railroad, and to build a team of region and division officers who could handle the challenge and who were able to develop and carry out the plans needed through the stages of growth in tonnage that were anticipated, year to year.

Wayne had the confidence of the senior level of management, and those at the middle and junior levels as well. He was actively involved in what was being accomplished on all of the Denver Region, but especially on the lines of the Alliance and Lincoln divisions, where a very large part of the upgrading work was required. Not an expert on track engineering or maintenance, he had learned how to manage those functions very well. Most of the officers he brought into the Denver Region in those years were young, energetic, and eager to have a part in the unique and awesome challenge and opportunity BN had in those years.

Wayne decided to leave BN in 1984 at age 54. His decision was very unfortunate for the company and its people. He should have worked for at least another six years. Unfortunately, Wayne seemed to be viewed as working contrary to the agenda of a new team that was designated to lead the Operating Department when we merged with the Frisco in 1981. Part of the problem was that some members of that team had little or no understanding or appreciation of what it took to manage a growing business such as we had with the movement of low sulfur coal. They did not seem to appreciate—nor care to learn much about—what work had already been done, and what remained to be done and the plans for how it should be done. Admittedly, the new team was good at managing road and terminal operations and reducing the cost of moving "general freight," but managing the ever-growing coal operation required a far different set of skills and resources. Equally, if not more, important was the vast difference in leadership styles of Wayne and some of the leaders of the new team. More is covered on this issue in the chapter on the Frisco merger in my

second book. Rather than trying to continue weathering this hostile atmosphere, Wayne decided to leave.

I would like to have had a chance to visit with Wayne at some length after he retired to document his experiences and recollections on what was done under his leadership in building the railroad up in the 1970s. However, Wayne was not interested in talking about the past or staying in touch with people of the railroad, even with those of us who had worked closely with him for a long time. I received two cordial, heartfelt letters from Wayne not long before he died.

His departure in 1984 was a big loss for BN. After battling the U.S. Supreme Court, James J. Hill remarked, "I've made my mark on the surface of this earth and they can't wipe it out." Wayne left his mark on BN through his successful efforts and those of hundreds of people on the Denver Region whose heart, soul, and skills in running the railroad made BN successful. For BN's successful effort in moving the high tonnages of coal out of the Powder River Basin, I don't believe there is any one person who should be given more credit than Wayne Arntzen.

· GEORGE K. LAMPHIER ·

The most knowledgeable track man we have.

. . .

*There's nothing else I'd have rather done . . .
you can see track built, upgraded. It's a good feeling.*
"GEORGE LAMPHIER: SUPER GANDY,"
BN NEWS, APRIL 1980, PP. 6 AND 7[8]

AMONG THE MOST ADMIRED AND RESPECTED people on BN in the 1970s and early '80s was George Lamphier, director of Maintenance Programs for the system Engineering Department. Highly knowledgeable on all aspects of roadway maintenance work, George had "learned" the entire railroad within a short time following the date of the merger. George may have had a higher level of trust and credibility with all levels of management than any other officer in any department. His judgment and "track record" in decision making for the

allocation of maintenance dollars, for priority of work that needed to be done, and for evaluating the quality of work performed in the field was above reproach. George always was "the last word" on track-maintenance matters. He was never challenged or second-guessed, even by the maintenance experts who held high-level positions somewhere in the company. Through all of this he became known as BN's "super gandy."

Mike Martin, who served as superintendent on the Yellowstone, Nebraska, and Colorado divisions wrote,

> George Lamphier had a critical job. . . . Top management counted on George and had full confidence in him to know

---

8   This tribute could have been made by hundreds of BN people who worked with George.

and understand every milepost on the railroad, at least, on the coal routes where the situation was deteriorating and became desperate. But what about the field officers? . . . Could George Lamphier, a roadmaster from the former Northern Pacific, get the respect of maintenance-of-way officers who were from the GN, the Burlington and the SP&S? The answer to that question was an absolute "yes."[9]

George spent at least half of his time out on the railroad. The reports he prepared after each trip he made over a subdivision were informative, objective and were used to form both short-term improvement projects (i.e., work to be done immediately) and major work to be programmed in the future. George did not miss anything. On his hi-rail inspection trips, he coached and instructed local supervisors and foremen on how to make good use of the resources of material, machinery, and manpower they were provided. It was important to George to see that conditions were improving from one inspection trip to the next.

Instead of dreading and trying to avoid an inspection trip with a senior maintenance officer, the track people and the division superintendents looked forward to George coming to their territory. Knowing they would always be able to learn something from him, they looked forward to a chance to convince him they needed an additional tamper for a few weeks, a few hundred (or thousand) more ties, additional rail anchors, or other resources of some kind. George's response to such appeals was to evaluate how effectively they were using the resources they had already been provided. Those who were showing good results were able to prevail upon George for more material or machinery needed to correct an immediate problem. George would make notes of specific improvement work that would have to be programmed for the next maintenance season. Those whose track quality was not improving were given polite but firm, clear evaluations on what was expected of them. At the end of an inspection trip, George would indicate to us that he would be reporting his findings to the region maintenance officers or to someone on his staff, and expect them to follow up as to progress made in correcting any deficiencies, and in moving approved projects

toward completion. The reports George prepared on the subdivisions on which the heaviest improvement work was underway were read by senior operating officers and even by Bob Downing, Norman Lorentzsen, and Tom Lamphier when they held the office of president of the railroad.

The track people saw George as "one of their own." They knew he played no politics, he had no "in groups or out groups," and was fully "merged," i.e., not biased toward one or the other of the roads that made up BN. George would give recognition to those who were making progress in getting the work done and who ran a safe, efficient, and quality maintenance operation. Even with the heavy demands put upon George to allocate the resources needed to upgrade the main lines handling heavy tonnages of coal and competitive merchandise, he still was able to allocate a reasonable amount for basic maintenance work on branch lines and in most yards each year. He also saw to it that the non-coal routes on the northern tier and in the Pacific Northwest were not being neglected in favor of the coal routes.

George may have been the only person who got over all of the 23,000 miles of road BN owned in the 1970s. He made walking inspections of much of the yard trackage, even in the major terminals. He managed to get over all main-line subdivisions twice a year, and more often on the parts of the Denver and Chicago regions that had the heaviest work programs.

George had the ability to recognize spots where trouble was developing below the surface, such as when a condition was not yet at the threshold where it became obvious and a slow order would be needed. He would point out such problems while they were still small, before they would warrant taking labor, money, and material off a programmed job that would have to be set back and moved to a problem area in which such work was not programmed. Such capability headed off the need to place slow orders and affect service and avoided the need for a heavy influx of resources sometime later, when conditions had deteriorated beyond the "routine" stage, as well as the host of internal criticism and questioning that would likely follow.

Overall, George played a big part in the improvement of track throughout the BN system. A great many people benefited from his tutelage.

---

9   Mike Martin, "The Yellowstone Division Story: From 1973 through 1980," *The* BN *Expediter* (October 2011): 7.

· D. R. (DON) ROGERS AND ROY V. BRAWNER ·

· *held in high regard as experts in their field* ·
· *served as councilors, mentors, teachers, technical experts, evaluators, and expeditors* ·
· *their evaluations and statements on track conditions held above*
reproach by senior executives and division officers ·

UNDER THE DIVISION STRUCTURE FOR management of operating and maintenance activities, the primary accountability for getting the work done safely, to standard, and within the budget lay with the superintendents of the fifteen operating divisions. The superintendents reported to the assistant vice president– Operations at the region level. The Engineering staff of the AVP–O included one and sometimes two experts on trackwork, who served as the "eyes" of the AVP–O as to how well the work on the track was progressing and to oversee the preparation of the annual work program. George Lamphier kept a finger on the pulse of the regions and divisions to keep the chief engineer and the vice president–Operations at headquarters updated on the quality of the track and the progress being made on the year's work programs. It was very important that these track experts could relate well to senior executives, whether their careers did or did not have some firsthand experience in track maintenance earlier in their careers.

During the years of heaviest program work, there were two regional engineering officers whose knowledge, skills, and leadership capability contributed greatly to the success we had in completing the programmed work: Don Rogers from the Denver Region, and Roy Brawner from the Chicago Region. Don and Roy had earned their stripes as foremen, general foremen of track gangs, roadmasters, and as the heads of roadway maintenance at the division level. In addition to reporting to the AVP–O and the vice president of the region as to the progress being made, they provided "heads up" advice to the division superintendents regarding problems they observed or situations that would become problems unless some corrective action was taken. At the same time, they consistently recognized successful efforts make by the division forces in correcting problems and in the successful completion of program work. They were out on the railroad almost every day of the maintenance season.

Don and Roy held jobs in which they could have been seen as a threat, a stooge, or no more than a pipeline that passed gossip—or even misinformation—to the region heads, who might then overreact to small problems, and from that, issue criticism or evaluations that would undermine the morale of a division's supervisors. But by their serving constructively as coaches, mentors, and facilitators in getting the work done, such toxic relationships never developed on the Denver and Chicago regions. A spirit of trust, openness, and respect between the regions and divisions was developed and maintained very well in those years of challenge, stress, and pressure to get all of the large work programs and projects completed on schedule and at the level of quality that was expected. Everyone was pulling in the same direction, on the team, and there was only one team: there were no in-groups or out-groups among the maintenance supervisors, the superintendents, or their superiors at the region and system level. In those respects, it was an ideal culture for us to work in.

Following is a list of activities and responsibilities that Don and Roy undertook in facilitating the work in those years:

- On-the-job training and coaching on safe work practices and on getting all work done to the standards specified in the instructions issued by the System Engineering staff.
- Evaluation and development of candidates for promotion to supervisory positions, e.g., roadmasters, general foremen, and track inspectors.
- Insuring that machines coming out of shops after repair or that had recently been acquired were in place in time to allow gangs to start and continue work on schedule.
- Inspect work completed and underway as to the quality performed and what corrective action, if any, was needed to finish the job.

- Preparation of annual and five-year programs for heavy maintenance and upgrading.
- Planning ahead on both routine and program-level work needed to head off problems.
- Development of short-term and ad hoc programs needed to address particular local needs for additional training, and when new types of material, machinery, or standards were being introduced.
- Helping to insure that the required types and quantities of material and tools were available at the time required.
- Having a high-energy, affirmative approach to the job and the ability to work with diverse groups of people, to tolerate frustration, and to regroup and get everyone on board to overcome problems, whether caused by human failure or an act of God.
- Able to celebrate success by giving credit to those who did the work, and get everyone ready to move on to the next challenge.
- Above all, get everyone on board to demonstrate and live out their commitment to safety.

Don and Roy had similar types of experience in the early days of their careers. Don first worked on a track gang on the NP in Idaho. When he was cut at the end of the work season and was not called back early in the spring, he hired out on the GN and rose through their ranks up to the 1970 merger. Don ran the 75-man steel (rail laying) gang on Lines West of the GN for several years before being promoted to District Roadmaster.

A few months after the 1970 merger, Don was promoted to assistant superintendent–Roadway Maintenance on the Yellowstone Division. He was among the first maintenance officers selected to begin the program of "integration." To his credit, Don was able to earn the respect of his superiors at the region level and the supervisors who reported to him, all of whom had their roots on the NP. They leaned heavily in favor of NP maintenance practices and were beholden to the engineering officers of the NP whom they had worked under for thirty years or more. Don was promoted to engineer–Track on the Denver Region in 1975, at the beginning of the all-out effort to upgrade and expand capacity on the Lincoln

and Alliance divisions. In 1982, Don was promoted to engineer–Maintenance for the Seattle Region.

In his early years on the CB&Q, Roy ran steel gangs and was promoted to roadmaster and general foreman. At the time of the merger in 1970, Roy was promoted to engineer–Maintenance on the Chicago Region, where he immediately had to deal with the problems of inadequate track maintenance that had built up over several years. This problem became even more critical when the heavy-unit trains of coal began moving over nearly every subdivision on the region. Roy was in that high-pressure position through all of the 1970s and into the '80s when massive track-upgrading projects were undertaken on virtually every main line segment on the region.

George Lamphier thought highly of both Don and Roy, regarding both of them well qualified for promotion. George decided to retire shortly after the Frisco merger. As happens in any merger, the succession plans for management personnel are disrupted when there are new people who must be slotted into the plan. In time, however, we were able to bring Don into the headquarters Engineering staff as assistant chief engineer–Maintenance. Later, Roy was also brought in to headquarters, and he and Don split the system when they were given a set of responsibilities similar to those George had handled so well in BN's first twelve years.

The track men on any railroad are seen as "the salt of the earth." They are hard workers and remain highly committed to their tasks and responsibilities through any of the challenges that come to them, through weather, heavy workloads, budgets that may be too tight, and the never-ending pressure to reduce their forces and increase productivity. Their work has evolved from having to rely on labor and hand tools and a few small machines on gangs to highly mechanized, as well as technologically advanced systems for track inspection and maintaining track to handle tonnages that were not envisioned by anyone at the time BN was formed. BN was well blessed with a fine cadre of people at all levels of its engineering and maintenance work in its track, bridge, and signal departments. Don and Roy stand out as excellent representatives of that tradition.

· EDWARD L. (ED) BAUER, JR. ·

· on a very fast-track career, to head the Mechanical Department at age 39 ·
· BN's push to adopt AC technology for heavy-duty freight locomotives ·
· made his mark in getting the new shop at Alliance
up and running with hundreds of new hires ·

ED (EDWARD L.) BAUER WAS PROMOTED FROM general foreman of the locomotive maintenance shop at Northtown to assistant master Mechanic at Alliance in 1978. Only six months later (September 1978), Ed was promoted to the new position of shop superintendent for the new $42 million combined car and locomotive shop that was nearing completion at Alliance. Ed was age 31 at the time. He reported directly to Al Stranek, the head of the Denver Region Mechanical Department, rather than to the division superintendent. The position of assistant superintendent–Mechanical for the division (held by Fred Albert) was maintained to manage mechanical activities outside the Alliance terminal facility. The new facility was fully opened in the fall of 1979.

Ed and his team had two years after his appointment to get a work force of 550 additional people hired and trained to maintain six hundred locomotive units and to inspect and make running repairs as needed on the thousands of coal cars passing through Alliance every day. Only about 150 people were employed in the old roundhouse and car-repair facility at the time. Most of the people hired to staff the new facility came from farms and ranches in western Nebraska. Only a few of the new employees came to BN as qualified machinists, electricians, or pipefitters, either from another railroad or an industry that employed people with such skills.

The Mechanical Department at Alliance had to set up training on a 24/7 basis for several months, using old converted passenger cars and box cars for classrooms. With agreement from the unions, the period upgrading apprentices to journeymen was reduced from the standard of two years to one. The methods used for teaching and "on the job" experience had to be greatly accelerated to get enough qualified people ready for work in the new shop. To add to the challenge of establishing a skilled, professional workforce in such a short time, the average age of supervisors assigned to the new shop was twenty-seven years.

To assist in training new employees, and to upgrade locomotives with high failure rates, Ed set up a program for the Alliance shops to rebuild two locomotive units each month. The diesel engine was stripped down to the engine block. All major parts were removed and replaced with reconditioned components. Each rebuilt locomotive was moved outside the shop building for two days of rigorous testing before it was put back in service. Having new employees involved in the rebuild program accelerated the learning process.

In addition to managing the intense task of getting the large new shop complex ready for service, Ed had to contend with day-to-day problems occurring out on the division. Even with all of the track upgrading that had been done, and the training given to operating employees to improve rules compliance in the previous several years, there still was an average of one yard or road derailment about every other day that had to be dealt with. Ed recalls one stretch of eleven days in which there were eleven derailments with damage of some extent to three hundred cars and twelve locomotives. Having to spend so much time on derailments was a distraction from getting the new shop operational.

Ed recalls an especially bad accident in 1979 in which eleven of twenty locomotives on an empty coal train derailed in a washout on Crawford Hill. In a flash flood, a raging river washed the embankment of a high fill out around the culverts, creating a deep hole. Since the welded rail did not break, continuity was maintained in the signal system, which prevented indication on a block signal of the loss of integrity in the track structure. The washout was very deep, making recovery of the locomotive units very difficult. After spending three and a half days in pulling the units out of the hole, only four of the seven units had been recovered.

Caterpillar tractors from two wrecking companies were used, plus several Cats owned by mining companies, which were as desperate as BN to get the line back in service. Since the new line being built between Donkey Creek and Orin had not been completed, all coal trains still had to be routed over Crawford Hill and through Alliance. There was no alternate route available in case of an accident. The economics of keeping the line out of service any longer to recover the locomotives forced the decision to leave three units in the deep hole so the track could be rebuilt. Having nine units in the locomotive consist on that particular train was the result of an imbalance of motive power. Generally, no more than seven units were placed on a given train, in case an accident that serious might occur. In this case, two units were to be set out at Crawford for helper service.

Ed also recalls the operating and maintenance problems the Alliance Division had during the severe winter of 1979. A number of trains got stuck in snow drifts, stranding the train crews. Some had to be rescued by employees sent out with snowmobiles. On one occasion, the crew was stuck on the train for so long that the stove in the caboose ran out of oil. The crew ripped the wood off the walls of the caboose and burned it to keep from freezing.

Under Ed's leadership and with guidance from an excellent team of mechanical managers and employees, the new shop was made ready to take on the maintenance of a large part of BN's ever increasing fleet of locomotives acquired for coal service. Having a fleet large enough to handle the business was of prime importance, but it was also necessary that the quality of maintenance would improve to the level needed to drive down the rate of in-service failures affecting the operation of what had already become the highest-density rail line (in terms of gross tons per mile) in the world.

Ed was involved in the development and oversight of a program called for improving locomotive reliability called SEARCH (System Evaluator and Reliability Checker) that enabled BN to predict the maintenance a locomotive unit needed, or the timing for replacement of components in its diesel engine or electrical system. Before long, SEARCH evolved into the capability for remote monitoring of locomotive "health" while a unit was in service, pulling a train. In the early 1970s, when the SEARCH program was developed, the SD45-type units were the newest in service. Designated units were fitted with a wiring "harness" that went between sensors added at several critical points on the locomotive. Since at that time there was no on-board computer to store fault indications, SEARCH relied on the sensors and the actual operation of the locomotive's systems to determine the fault. When the wiring harness was plugged into a terminal at the maintenance facility at Havre or Seattle, a computer-generated diagnostic sheet was printed out to aid in finding and fixing locomotive faults.

Carl Stendahl Jr., then an officer in the air-brake section of the Mechanical Department, described the SEARCH system: "It is like the system on all of today's automobiles where you plug in an OBD reader to determine faults logged on the automobile's computer and the technician uses the codes to determine what is needed to fix the car. . . . When SEARCH first came out it was truly innovative, there was nothing else like it."

The Dash-2 electrical system was built into EMD's next generation of units, plus troubleshooting diodes and SEARCH, the output of BN's SEARCH generated a great deal of interest from the rail industry in self-diagnosis capability and resulted in EMD's decision to offer it in the new locomotives they began to produce in 1974. Not too many years later, EMD and General Electric equipped their new locomotives with on-board micro-processors that would reveal trouble spots in both the mechanical and electrical functions of a locomotive. SEARCH was the forerunner and catalyst for those advanced systems, an expensive one to build and operate.

At the outset of the new program, the range of days between failures was only from twenty-five to thirty-five days. Through application of the Demming program quality process, the technical training center at the Johnson County Community College in Overland Park, Kansas, and MMC (Mechanical Maintenance Control Systems) for root cause analysis, BN increased the mean time between failures to an average of seventy-five to eighty-five days. This level of improvement was achieved even before micro-processors could be placed on locomotives to carry out a self-diagnosis of failures. The fleet of SD40-2 units (all owned or leased by BN) evolved to a level of performance superior to the new SD60 units owned by Oakway and maintained by EMD.

Ed is proud of what was accomplished in those years to improve the safety performance of employees in the Mechanical Department and in all of BN. By giving close attention to compliance with safe work practices and through improved training programs and assistance of safety experts from the Du Pont Company, the rate of injuries was eventually driven down to a frequency ratio of less than 1.0, about one-tenth the number of injuries that occurred in the mid-1970s, when large numbers of new hires were being brought into the rapidly expanding operation. Ed also attributes much of this success in safety to having employees fully involved in the process of managing the safety process, by helping them become committed to making safety a way of life both on and off the job, and by establishing accountability at every level. This kind of leadership brought the Alliance shop to a much improved standing among BN's largest mechanical facilities.

The success Ed had as superintendent of the new shop in Alliance put him in good position for rapid advancement. In an interview conducted by *Railway Age* magazine, a group of senior managers in the Mechanical Department headquarters office in St. Paul early in 1979 expressed that all of the office-based assignments Ed had handled in his first eight years at BN were a long way from what he would have to handle as the shop superintendent in Alliance—a field job with high accountability and no regular hours and which involved interaction with the workforce. Gus Welty wrote, "Frankly . . . , they didn't know quite what would happen when he became Superintendent in April, [but they] maybe resting a bit easier these days: Bauer moves easily among old heads and new hires; he seems clearly in control, and the shop is producing—despite having an extraordinarily high percentage of apprentices among its locomotive and car shop forces."[10]

---

10   Gus Welty, reprint of "BN's New Alliance Shop: Big, Capable, Expandable," *Railway Age*, August 3, 1979.

After serving at Alliance for two and a half years, Ed took over as director–Mechanical for the Twin Cities Region and, only a year later, as director on the Denver Region. In 1986, at age 39, Ed was named vice president–Mechanical for the entire BN system. His responsibility included all system car and locomotive matters, including system heavy-repair facilities. In addition to the mechanical function, Ed was given responsibility for freight-car management and car distribution. Early in his career, Ed was seen by many as a likely candidate for high-level positions, and he soon met those expectations. As head of BN's Mechanical Department, Ed led the development of a number of new methods, procedures, and technologies that made BN a leader among the railroads of North America. One of the highlight projects was the "Power by the Hour" concept for locomotive ownership, maintenance, and performance that Ed developed jointly with BN's Finance and Transportations departments (refer to chapter 20 in *Transformation of a Railroad Company*).

Another was the application of AC technology to locomotive traction motors, in which BN took the lead in inducing EMD to design locomotives in which it would be available. Ed wrote, "EMD and GE were pushed very hard [by BN] to develop AC technology because of BN's competition in the Powder River Basin with the Union Pacific. GE felt the rail industry was not ready for AC, but EMD agreed to build two prototype AC units for development. We tested these units for about a year before we placed the order for the first 350 new alternating currency (AC) units. History was made and BN continued its coal success." These and other impressive projects and initiatives under the direction of Ed Bauer and other leaders and employees in BN's Mechanical Department are covered in some detail in chapter 42, "Mechanical Department Upgrades."

· H. (HANK) BLITZ ·

· a mechanical officer of the strength, intelligence, and leadership capability
needed to keep the rapidly increasing fleet of cars and locomotives up and running ·

IN ADDITION TO THE MAINTENANCE-OF-WAY AND transportation functions, there were "heroes" in the Mechanical Department in those years, among the managers and employees who had to maintain and service more and more locomotive units as the fleet was expanded for coal service. From 1972 through 1980, 1,405 new high-horsepower, high-tractive units were acquired. From BN's formation in 1970, the fleet of owned and leased locomotives increased from 2,128 units to 2,925 units by the end of 1979.[11] At that time BN policy required that each unit be assigned to a specific shop for routine, scheduled maintenance.

Until the new locomotive maintenance shop was opened at Alliance in January 1978, the "coal power" had to be added to the work load that had been established for the existing shops at Lincoln, Northtown, Denver, and Glendive. There was no capability available at the old roundhouse at Alliance to handle much maintenance work other than such basic tasks as adjusting brakes, changing headlights, fueling, and cleaning cabs and windows.

The shop at Lincoln was located in a major coal corridor. Because it was a modern, well-equipped shop with capacity to maintain 410 units, much of the burden fell on it to handle the ever increasing load until the new shop at Alliance (with a capacity of 600 units) was opened. In 1975 Hank Blitz was transferred from Alliance to the Nebraska Division as assistant superintendent–Mechanical. Blitz was a strong, effective leader. His "track record" over his nearly forty years of service showed he had the ability to get maximum production out of a large locomotive-maintenance and -servicing facility such as we had at Lincoln. He was seen as a no-nonsense officer who would not settle for excuses for poor performance in safety, quality of work, or keeping the out-of-service ratio for Lincoln's assigned units at a low level.

As a strong, silent type, he had a strong presence and did not need to resort to a demeaning, threatening style of leadership that some managers will adopt in running a big operation under stressful conditions. Blitz knew how to develop and channel people's talents in a positive, constructive direction. He was able to remold or salvage some supervisors and employees who at times were going off the deep end under the pressure the Lincoln shop was under. Rather than dismissing or demoting them, Blitz knew how to build up their capabilities so they could become more effective managers. Blitz was a good team player. His goal was to provide the Transportation people with locomotives at the time they were needed for service. He did not run his operation strictly for achieving the Mechanical Department's goals on its budget or specific productivity measures. At the same time, Hank expected the Transportation people would get Lincoln's assigned units back to the shop by the date they were scheduled for maintenance, so he could maintain efficiency and order in the shop's workflow.

Blitz showed broad shoulders in dealing with suggestions and complaints from the Terminal and Division Safety committees, as well as from individual employees and union representatives. Even with seemingly frivolous or low-priority items brought to the Mechanical Department for handling, Blitz's attitude was, "If it's important to you for safety, we'll get after it."

Blitz also knew how to work around "the system." For example, when we needed modifications, repairs, or improvements made on company service cars—bunk cars, kitchen cars, or shower cars used on track gangs—and we could not wait a year or more for authority to have the work done at one of the large, system car shops, Blitz would get it done in the division's car-repair facility at Lincoln. He would not insist on having an AFE or other formal authority before doing the work. He was able to get that kind of work done without blowing the budget or otherwise causing someone in the headquarters office

---

11   Net fleet size, after retirements, sale, or other disposition of older units taken out of service as new units were put in service. Burlington Northern annual reports, 1971 through 1980.

of the Mechanical Department to overreact and demand the division be taken to task for not following formal procedures.

In return, our building-maintenance and track people would not attempt to delay or avoid responding to Blitz's requests for correction of a condition that was hindering productivity or was an annoyance or distraction to Mechanical employees. Having maintenance officers of the strength and professional capability of Hank Blitz was of great value to the company in meeting the challenges we had before the capacity of the new locomotive shop at Alliance was available. Certainly, not all the credit for getting through those times should go to Hank Blitz. The managers of the locomotive shops at other locations also had to pick up the slack in those years, and did so very admirably.

⋅ DAVID L. (DAVE) HOWLAND ⋅

⋅ new train-crew personnel needed mentors and teachers more than a supervisor ⋅
*⋅ a few employees deliberately ran through a switch just so they could get suspended
for fifteen days and go back to the Midwest to see their wives and kids ⋅*

DAVE MOVED INTO THE ALLIANCE DIVISION IN 1984, not long after two very serious accidents had occurred, one at Wiggins, Colorado, and the other near Newcastle, Wyoming. He was directed to get into the problem of some train crews being found under the influence of drugs or alcohol while on duty or subject to duty. These serious accidents had put on high alert the entire BN system, but especially on the Alliance Division where both accidents had occurred. There was a great deal of concern about how the large number of new, inexperienced employees—some of whom were "following the crowd"—were spending their time while off duty between trips, rather than being as concerned as they should have been about their jobs or their safety to comply with the company's rules pertaining to their fitness for duty.

Dave was assigned the position of trainmaster at Guernsey, an away-from-home terminal for train crews. He was one of only two operating officers for the territory between Guernsey and Alliance, which had a heavy flow of traffic following the opening of the new line between Gillette and Orin. He also had responsibility for operations on the south end of the new line. Dave came to the Alliance Division with prior experience as a trainmaster, yardmaster, clerk, and switchman at Galesburg, a large terminal. He had gained the background needed to handle personnel issues while working for some time under the direction of several seasoned, competent operating officers. The Alliance Division was in need of managers with such ability, rather than having to place as field supervisors more corporate management trainees, who were fresh out of college and lacking much, if any, supervisory experience.

Dave found the situation at Guernsey quite distressing: "When I got there, 90 percent of the urine tests required under regulations for post-accident testing came back 'hot' for drugs or alcohol, and crews were running through switches just so they could get suspended for 15 days and go back to the Midwest to see their wives and kids. Two years later, 99 percent came back clean and we had established regular vacation times again."[12] The possibility of disciplinary action had not been a deterrent sufficient for some employees to comply with the rule forbidding the use or possession of drugs or alcohol.

The danger facing BN was that the "red flags" revealed by the proverbial "accident pyramid" were being played out. Many safety experts forewarn that for every 1,000 unsafe acts that lay at the base of the pyramid, one catastrophic injury or accident is likely to occur. Hence, any organization concerned about safety must concentrate on reducing the number of unsafe acts, whether they are "near misses" or minor incidents caused by careless acts, taking short cuts, a lack of training or supervisory oversight, defective equipment, bad attitudes, or distractions. With the high rate of positive readings in post-accident testing, having a high frequency rate for injuries, and a number of minor

---

12   Dave Howland, email to author.

train accidents, BN was moving up the pyramid and in increased danger of a serious accident occurring. Hence, the company needed to focus more intently on what was happening at the base of the accident pyramid.

Dave found the quality of employees with prior experience that BN had hired to be quite capable and reliable. Too often, that was not the case with some of the employees who had been hired from the surrounding area, unfamiliar with the regimens of railroad life and what the railroad had to expect of employees working in a very safety-sensitive environment.

Because of the turnover of employees and continued rapid growth in business, the company had not been able to allow employees to lay off from work to return to distant locations for family time—they didn't have an agreement to establish permanent seniority at the new location, and they were unable to obtain housing at their new work location. As the operation improved, and the cycle time on unit train equipment was reduced, fewer crews were required. That took enough pressure off the crew board so a few employees at a time could be given reasonable time off from work for personal business or to go home for short visits. Also, it reduced the need for as many crews to be called to work at their home terminal after only eight hours off duty.

Dave found that the newly hired employees needed a mentor or trainer to supplement what they had been taught while in the training program for new brakemen. They needed on-the-job instruction as they began to put into practice the information and training they had been given. Dave found it was not enough for an operating officer to think of himself or herself as the supervisor of the new hires. That meant he or she had to spend a great deal of time teaching: instilling the principles on "how to railroad," to get the work done safely and be able to better handle such routine instances as having to start a heavy train on a grade without breaking a knuckle or pulling a drawbar. Other lessons included how to understand and comply with the complex sets of train orders that were being issued involving train meets and temporary speed restrictions, and for the protection of track maintenance work in progress. Officers had to be ready to field countless other questions on in the code of operating rules and the rules that governed train handling and air brakes. During his time at Guernsey, Dave was expected to work on the same 24/7 basis, with no provision for any scheduled time away from work, as were all other field operating officers worked in those times.

Crews often had little time off duty between trips other than the minimum of eight hours for rest. When they were fortunate to have a little more time off duty between trips, there often was nothing for them to do to pass the time in the small towns where they tied up, other than to watch TV, play cards, or socialize at a local establishment where they were welcome—sometimes a tavern or bar, unfortunately. The company attempted to meet the need for something "constructive" for employees to do while waiting to go back on duty, by providing video games, pool tables, and tubes for rafting on the river that flowed through Guernsey, and by helping the owner of a restaurant stay open twenty-four hours a day. To help keep crews informed of their position on the crew board while off duty, BN started to put the line-ups on the local-area cable TV network. That gave employees better information for planning their rest and to schedule time for activities important to their families.

During the time Dave was assigned at Guernsey, BN undertook a major project to reduce curvature, "daylight" a tunnel (remove the rock and earth above the track so it is not longer a tunnel), extend several sidings, and generally upgrade the line through the Wendover Canyon east of Guernsey. Trains had been restricted to only 10 miles per hour for several miles due to the heavy curvature and the number of derailments occurring on old jointed rail. It was one of the last major roadway-improvement projects to be undertaken on the Alliance Division. The work included one of the first installations of concrete ties on BN. As soon as the work was completed, the cycle time for coal trains routed on that line was greatly reduced and improved alignment eliminated the problem of track-caused derailments.

Dave was promoted to assistant superintendent at Alliance in 1986 and filled in as acting terminal superintendent for six months at one point. Later, he served as the director of customer service at Denver. From there he moved into the Intermodal Business Unit and advanced to positions with increased responsibility, up to director of Intermodal Operations for the BN system. Later, Dave left BN to take positions with C. H. Robinson Worldwide, Schneider National, and APL Logistics. He had a very successful career, serving as APL's vice president for Land Transportation Services until his retirement in 2014.

· THE HATZENBUHLERS ·
EDWARD L. (ED) AND DAVID (DAVE)

· *a very fine father-and-son team of experienced locomotive engineers* ·

EVERY EXPERIENCED RAILROADER WHO DECIDED to pull up stakes and move to Alliance, Gillette, Edgemont, or Guernsey in the 1970s or '80s would have an interesting and meaningful story to tell about that transition. Many of them had lost their jobs in the rail industry due to bankruptcies, as in the case of those who had worked for the Rock Island. Some were cut off when the Milwaukee discontinued operations on its West Coast extension, i.e., west of the Minnesota–South Dakota state line. Others who were working on the bankrupt lines that later made up Conrail knew their jobs were in jeopardy. They were experienced railroaders who wished to stay in the rail industry rather than try to get jobs in manufacturing, construction, or administrative work of some type.

Some who were interested in working "out west" may have already had jobs on viable railroad companies, but wanted to move their families to an area with less crime, a lower cost of living, better schools, and overall a better quality of life. There were outdoorsmen who wished to live near an abundance of wild game and opportunity for outdoor recreation.

Together with their families, they decided to pull up stakes and move to the railroad communities in Nebraska or Wyoming and start over. For some, moving to smaller towns and in a somewhat remote part of the country, far from relatives and friends, would be a big adjustment, but overall, they saw going to work for BN as a wonderful opportunity to have a job with good pay and benefits, and with a railroad company that had a good future.

Even though they had to start work with a new seniority date, they knew they would be better off in the long run than they would if they continued to try to work for railroads that were struggling to keep operating and losing business from not being able to compete in the marketplace for transportation service. Also, they knew that with the high turnover rate among new employees hired without prior railroad service, they might move up the seniority roster rather quickly. Continuation of the

rapid growth of the coal business meant that BN would have to keep hiring for the foreseeable future, meaning there would soon be many new hires below them.

Ed and Dave Hatzenbuhler, father and son, were among those who decided to terminate railroad employment in the Midwest and head out to Alliance to hire out. Both had been locomotive engineers with the Milwaukee Road. A few years earlier, Ed had held the position of assistant superintendent at La Crosse. In a program of reorganization and downsizing, the Milwaukee abolished his position and transferred him to an assignment in the Milwaukee's terminal in Bensenville, a suburb of Chicago. Instead of taking that new position, Ed decided to exercise his seniority as a locomotive engineer and bump in on a job at Mitchell, South Dakota. Ed's son, Dave, was working as a locomotive engineer for the Milwaukee at Ottumwa, Iowa.

This was in 1979, when the Milwaukee was in bankruptcy, with a future that was uncertain at best. Where Dave was working, and generally throughout the Milwaukee system, the track had deteriorated badly due to the Milwaukee's lack of money for tie replacement, relaying old rail, remedying the ballast section, and other basic maintenance work that is necessary to keep a railroad safe and efficient.

Ed and Dave began talking about the opportunity to go to work for BN at locations in or adjacent to the new mining areas in the Powder River Basin. They decided to head for Alliance to look things over and determine if they should leave the Milwaukee and apply for work with BN. In reflecting on the visit Dave and his family made to Alliance to evaluate the community and decide whether they should make this bold move, Dave tells of his wife's reaction as they drove through the sand hills of western Nebraska. It was dark as they neared Alliance, and she observed there were very few farm lights or other signs of life for miles and miles. Just then, as they crested a small rise on the road, the floodlights at the diesel shop and the yard were shining brightly and could be seen for miles. That sight helped give them the encouragement they needed to keep going and to pursue employment with BN.

Dave recalls the morning when he and his dad got in the line of somewhere between 100 and 150 people who were applying for work on the Alliance Division on that day. Keep in mind that the hiring of large numbers of people in this part of the railroad had been going on without letup since the early 1970s. While they were waiting in line to be interviewed by a personnel manager, a clerk handed out a form for the prospective employees to fill out while they waited in line. The same person asked the applicants if any of them had worked as a locomotive engineer. Those who raised their hands were brought to the head of the line and were interviewed within a few minutes. By 10:30 A.M. the Hatzenbuhlers were lined up for physical exams, rules, exams and then familiarization trips on the territory they would be running on. Dave and his dad often joked that the senior Hatzenbuhler had a difference of five minutes in their position on the seniority roster, with Ed being ahead.

By that time, BN had set up an accelerated sixty-day program for qualifying foreign-line engineers. There were seven engineers in the class in which Ed and Dave were placed. For the first forty-five days, they made familiarization trips over the division, followed by three weeks of training in St. Paul on BN's prescribed methods for train handling and general operating practices. Because of the urgency to get qualified engineers ready for service, the foreign-line engineers were trained and qualified ahead of the student engineers who had no prior experience as engineers. This decision caused some consternation among the student engineers who felt they had been run around. As a result, BN did not set up any more special, accelerated courses for engineers with prior experience. They had to take their training and placement on the seniority roster according to their dates of hiring, and were not allowed to move ahead of the students who had no prior experience. However, the applicants who had experience working on "foreign" railroads were given only one chance to pass the series of exams given in the training program; the student Engineers were given three chances.

When they were lined up for the qualifying trips, generally three trips over each crew district, Dave noted he was not scheduled for any trips on the 238-mile interdivisional run between Alliance and Ravenna. He was told that since the most senior engineers held pool turns in that territory, it was likely he would not be called for any trips on that line for a long time. However, it turned out that for his first paid trip as a qualified engineer, he was called to run a coal train from Alliance to Ravenna. The conductor asked Dave if he thought he could make the trip, even though he had never been over this subdivision. Dave told him he thought he could, although the trip would have to be slow. With the guidance of the conductor as to the locations of speed restrictions and other operating challenges en route, together with reference to the timetable, special instructions and grade profile, they set out and made the trip safely. Dave and many other employees and managers have mentioned how often the grand total of seniority of a four-person crew was less than two years. Good training and the conscientious attention to safety by managers and employees in all crafts got BN through those years with remarkably few accidents. Unfortunately, some of those accidents were extremely serious and resulted in fatalities and major damage to locomotives, cars, and track.

Dave recalled the challenges new employees faced in finding housing in those times. He and his family did not move to Alliance until he had passed all of the exams, to be sure he was found to be fully qualified, was placed on the engineer's extra board, and had started to make "pay trips." At this time, they two daughters and moved into a two-bedroom apartment. Soon, they built a house. For many other employees who moved to Alliance, no housing of a permanent nature was available. Some lived in their cars for several days until something opened up. A group of seven former Milwaukee employees rented a two-bedroom apartment. Since they were working long days and irregular hours, all seven were never in the apartment at one time. Their families stayed behind until they could find suitable housing. Construction of new houses and apartments was underway, but far too slow to meet the demand.

Dave has interesting recollections of the work ethic that was typical of many new hires in those years. Employees who had prior railroad experience knew what the work was all about and had been committed to it as a career for a long time. But for many of the new hires, the adjustment to railroad work and life was difficult. Unless they came from a railroad family, they had never experienced anything like a 24/7 requirement for availability to go to work—and the effect that had on one's ability to get rest or to have much time for a social or family life, or even much certainty as to when they might have time off duty. As a result, the turnover rate was high. To help ingrain the new people into the railroad way of life, employees with experience imparted their knowledge and ideas to the new people on what it took to somehow mesh personal interests with the demands of their jobs. From their prior service, they were in position to explain how things work—how all of the safety rules, special instructions and operating practices had to be blended together to get trains over the road and to switch cars in the yard. Dave recalled how many of the new people found what they had been taught in classes more credible and understandable when they heard it explained and clarified out on the job by experienced employees.

After working for only about a year as an engineer, Ed was offered the supervisory position of road foreman of Engines at Alliance. He was age 50 at the time. Ed accepted the offer and soon made a very positive impact on the operation and on the performance of employees in engine service. He also coached the many young, inexperienced operating officers on the division in those years. A short time later, Ed was promoted to trainmaster, with responsibility for the heavy coal corridor between Edgemont and Ravenna. Ten years later, he retired as manager of Operating Practices at Alliance, with responsibility for the on-the-job portion of training and evaluation of new engineers. Ed retired in 1991 after twelve years of service on BN. His many friends and acquaintances were saddened by his death in March 2012. Many fine tributes on Ed's work and his character were placed on the BN Alumni website. Among the most notable recollections of Ed's work were the following:

From W. E. Greenwood, division superintendent at Alliance:

> It was practically impossible to get experienced supervisors to come to Alliance in those days, so when I heard we had a former Milwaukee operating manager running an engine on the Alliance Division, it didn't take me long to interview him, become quickly impressed, and offer him a job as a Road Forman. He took it and had an immediate impact on bringing some maturity into an otherwise

inexperienced work force where it would not be unusual to have five crew members where the total seniority between them could be less than three years, and the oldest one would be 21! This of course led to unbelievable safety, discipline, and rules compliance issues.[13]

From Greg Mangieri, trainmaster:

> Ed was the top line of a very short list of management immigrants from struggling railroads throughout the country who were absolutely critical to our efforts to keep the division afloat while we struggled with training and educating entire high school graduating classes in Nebraska, Wyoming, and South Dakota to operate a railroad. Ed was critical in the education effort of not only these young employees but also myself and my young entry level counterpart management level people.[14]

From Dave Howland, trainmaster, to Dave Hatzenbuhler: "I was lost when I got to the Alliance Division and not sure where to start, but your Dad was as steady as ever and always gave good advice. Seemed like whenever it felt like a bottomless pit, Ed would show up and want to go to the canyon to do some testing (and a little fishing!). Things always went better after that."[15]

After two years of service as a locomotive engineer, Dave Hatzenbuhler was promoted to road foreman of Engines at Centralia, Illinois. That was the start of a quick series of promotions through the chairs of the Operating Department, and on to assistant vice president–Corporate Development. In 1995 Dave left BN to set up a new company, Main Line Management, to provide consulting service on various types of transportation issues. During his time with BN, Dave was able to complete the requirements for a B.A. degree, with concentrations in History and Political Science.

Dave was representative of many of the new hires in those years whose excellent job performance got the attention of BN managers, from which they moved into leadership positions. Ed was among the seasoned veterans of railroad service who were hired as craft employees,

---

13   W. E. Greenwood, conversations and exchange of messages with author.

14   Greg Mangieri, conversations and exchange of messages with author.

15   Dave Howland, conversations and exchange of messages with author.

but qualified and ready to move into a supervisory position as soon as they had opportunities to demonstrate their capability to a division officer. By hiring with BN, they had positioned themselves to be in the right place at the right time for advancement. By getting on the right team, they succeeded in having careers at a level

they likely would not have had anywhere else in the rail industry. Having the courage and drive to leave the past behind and start their careers over paid off handsomely for these managers, for the hundreds of employees BN hired in those years, and for their new company.

⋄ A. H. (ART) FIEDLER ⋄

You, as a Locomotive Engineer…

⋄ *he developed a generation of proud, professional engineers* ⋄

AT THE TIME OF ART FIEDLER'S DEATH IN 2011, a number of BN operating officers who had been trained by Art as locomotive engineers gave him very high marks for the professionalism he instilled in them and the hundreds of other employees he trained for engine service. With business growing rapidly in the 1970s and into the '80s, getting employees trained and qualified as locomotive engineers as quickly as possible was vital.

Without having enough engineers in service, the coal simply could not be moved out of the Powder River Basin. The old routine of candidates working as locomotive firemen for anywhere from five to ten years to learn by osmosis before being examined and evaluated for promotion could no longer be relied on, since the position of fireman had been largely eliminated in the 1960s. Also, it took far too long to get engineers trained under the historic, informal system.

When BN set up an engineer training program in 1973, Art was brought in as an assistant manager. He was moved up to head the program as manager in the next year. Previously, Art had worked as the road foreman of Engines at Fargo, having started his railroad career as a fireman on the Tacoma Division of the NP in 1945. Jack Christensen, a fellow road foreman, wrote, "he acquired almost legendary status among the Engineer trainees due to his motivating instruction and emphasis on the importance of the job. The phrase 'You, as a Locomotive Engineer' would introduce many of his presentations on the duties and skills connected to locomotive operation."[16]

After a series of interviews and evaluations, trainees spent a minimum of thirty days in on-the-job training in the territory where they would be working after they became qualified as locomotive engineers. Next, they were given three weeks of classroom instruction at the training center in St. Paul, covering air-brake equipment, the rules and practices for train handling, instruction on the operating rules, and technical material covering the mechanical and electrical functions on a locomotive. In those years, training simulators had not yet been developed. The only training tools available for classroom instruction were displays showing the operation of air brakes on cars and locomotives.

Following the classroom instruction, the trainees returned to their home division for a period of sixty to ninety days under a designated locomotive engineer instructor. If recommended for promotion by a qualifying officer on their home division, trainees were given a series of final examinations. Upon passing these examinations, trainees were certified and promoted to locomotive engineer. The program was highly effective, turning out hundreds of highly competent, professional railroad people for a position that was critical to the company's ability to run a safe, efficient operation, and to be able to move the rapidly increasing volume of business throughout the BN system.

In addition to the training given for the operation of locomotives and proper train handling, trainees going to the Alliance Division and parts of the Nebraska and Hannibal divisions had to be qualified to operate trains in non-CTC territory. Operations on such lines were

---

16   Jack Christensen, comp., "Arthur Fiedler, 1927–2011," *The Mainstreeter* (Summer 2012).

governed by a complex set of operating rules, together with train orders and special instructions for operating on a given territory. Conductors and engineers working those territories had to be highly proficient on those rules and able to handle complex train orders issued by dispatchers for train meets, "wait" orders, "right over" orders, slow orders, and the types of orders issued to protect the work limits set up for track maintenance. They had to be able to handle heavy trains of 14,000 tons or more, on grades of as much as 1.5 percent, with helper engines in one location on the Alliance Division.

Locomotive engineers were entrusted to handle millions of dollars of equipment safely on lines with a high density of traffic. The newly trained engineers often were the most experienced member of a crew, which put the burden of responsibility on them for the trip and for giving direction to new conductors and brakemen, even though conductors were designated to be in charge of a train. They were thrust into the "chain gang" way of life, with irregular hours and often with limited time at home between trips. It was a real credit to Art Fiedler and the engineers who did the classroom and field training that the operation went so well in those years.

A number of fine tributes to Art were written by BN operating officers who were trained for engine service in the classes conducted by Art:

From Jake Greeling:

All of us who went through the Locomotive Engineer training program in St. Paul could never forget this man. I can still hear him say, "You as a Locomotive Engineer." He certainly was a champion for the railroad, one of the most dedicated people I ever knew.

Tom Lynch recounted:

I was Road Foreman–Trainmaster in Gillette at the time the Orin line was under construction. . . . Art spent a lot of time in Gillette working with Locotrol and helping the train Engineers coming off the street and from other railroads across the country. He was always available by phone when he went back home for questions that came up.

Dennis Prewett remembered:

Art made a whole generation of Locomotive Engineers understand that they are professional railroaders. His early influence on me in the basement of the GOB in St. Paul had a lot to do with my career goals. Good training, and lots of it, is the backbone of a company's ability to perform.

Dennis Prewett added that Art was a great teacher who emphasized repeatedly that as engineers, they would be making an important contribution to the success of BN. He was skillful in getting across the attributes and character of a professional person. Art told his students that as engineers, they would be earning more than even the superintendent of schools in the areas where many of them would be working. Dennis recalled he felt more pride in becoming qualified as an engineer under Art's instruction than he did when he was awarded his Bachelor of Arts degree. Art sought to restore engineers as the "aristocrats of labor," as they had been recognized by the general public earlier in railroad history.

Joe Donovan, who worked as a locomotive engineer in the Twin Cities terminal and in the chain gang between Minneapolis, Staples, and Dilworth, remembers Art's willingness to listen. Even after his promotion to general road foreman for the system, Art would take time to call individual engineers and discuss specific operating problems and get their input as to the best train-handling practices to be employed in particular situations. Joe remembers that Art was very knowledgeable on all aspects of train braking, and that he could always be trusted to stand by instructions he gave and support the engineers, once he agreed that their actions in a controversial situation were in line with the operating rules and best practices.

Carl Stendahl, BN's director of Car and Locomotive Air Brakes and a close associate of Art for many years, wrote, "Art is one of the only men I have ever known who actually, 100 percent, walked the walk when it came to acting on his beliefs. Art, without fail, treated everyone with respect, and he expected everyone else to do the same."[17]

In later years, Art was promoted to general superintendent–Air Brakes and general road foreman, the position designated on a railroad as the number-one authority on the system for all technical matters in involved in the operation of trains and engines.

17   Jack Christensen, comp., "Arthur Fiedler, 1927–2011," *The Mainstreeter* (Summer 2012): 26.

• ELDON FICKE •

The first couple of nights when trains passed,
it sounded like they were coming right inside the bunk car.

ELDON FICKE IS REPRESENTATIVE OF THE LARGE number of young people who were hired for trackwork in the 1970s and either have retired or soon will be eligible for retirement. They have accumulated a great deal of experience, mentoring skills, and the ability to apply good judgment in carrying out a wide variety of maintenance work. They developed the ability to spot and correct problems in the track before they become serious situations affecting the operation. Because so many of these capable people are reaching retirement within only a few years, the railroad has a great challenge in training and developing competent replacements.

Eldon started his career as a grinder operator on a four-man welding gang on the Lincoln Division at Aurora, Nebraska. At the time Eldon came to work, the transition was being made from the use of oxygen and acetylene to electric welding and wire-feed technology. The gang was assigned to grind rail ends on lines laid only with jointed (conventional bolted) rail, weld up chipped joints, and repair rail end batter.

Eldon described the away-from-home living conditions at that time:

> My first year on the railroad, I stayed in . . . a 40-foot box car, with one end being a tool car filled with gasoline, oxygen and acetylene tanks. The other end is where I stayed. I was fortunate as I had that 20-foot space to myself, and the other three members of the gang each had their own bunk cars which were converted passenger cars. A waycar [caboose] stove provided the heat, and I walked to the depot to get water. The first couple of nights when trains passed, it sounded like they were coming right inside the bunk car, but as time went on, I got used to the noise and was able to sleep through the night. The main track was about 25 feet from where our bunk cars were parked.[18]

Several years later, BN agreed to provide lodging in facilities close to the location where gangs worked. When remote areas where not enough lodging was available to house large gangs, bunk cars still had to be provided, or the company would pay a lodging allowance to employees who would then have to "improvise" by using their campers, tents, or trailers, or work out other arrangements with local residents. The first agreement for company-provided lodging required two employees to share a room. It was not until the contract negotiated in 2013 that each employee was provided a single room.

When the large work programs were started in the mid-1970s, Eldon moved to the pre-plating gang and later to gangs that were building extensions to sidings or additional double track. Eldon recalls the skills and competencies of the supervisors he worked under in those years, and the character of some of his co-workers:

> As time went on more and more new, young employees were added with all the new construction and upkeep. We realized that the work was hard, but didn't know any better until you look back on it after forty years. The old roadmasters at the time, Dale Cross, Perry Schneider, Rich Barton, Vic Jensen and Merle Fivecoat are just some of those I worked with early on. Again, we didn't realize at the time how much knowledge these guys had, until now when we are in their shoes, trying to replace all the people who were hired in the '70s and early '80s. And remember, at that time some of the new hires were not exactly stellar employees, with the hippy generation going on.
>
> Sometimes we thought they [the old Roadmasters] were kind of old fashioned in their ways, but at the same time we didn't realize they were doing a lot of things they hadn't had to deal with earlier in their careers, like building new double track, working with ribbon rail, and the fast advancement of technology in machines. . . . They were learning on the go but still taught us the basics of good railroading.[19]

---

18 Eldon Ficke, conversations and exchange of messages with author.

19 Eldon Ficke, conversations and exchange of messages with author.

In 1979 Eldon became an assistant foreman on a steel gang (rail-laying gang) on the Alliance Division. There were two 100-man steel gangs working between Sidney and Northport, one gang working behind the other. Nearly all of the work on a steel gang has been mechanized, but back in the 1970s, there still was enough work that had to be done by hand to require thirty laborers in addition to a force of seventy machine operators, mechanics, foremen, and assistant foremen. Eldon also served as a surfacing-gang foreman and welding foreman. His first management (exempt) position was assistant roadmaster on a rail gang. Later, he was promoted to roadmaster at Bridgeport, the territory where BN first installed concrete ties. At that time the highly mechanized P811 machine that lays concrete ties and replaces rail in the same operation had not yet been built and put into service. Instead, the concrete ties had to be placed by using a Mannix sled. Eldon later served as roadmaster on several lines on today's Nebraska Division. Currently, he works as the division's supervisor of Engineering Support at Lincoln.

Eldon recognizes the great progress that has been made to improve the safe work performance of track crews on the forty years he has been working. He attributes it to "the pride and good work ethic of the Nebraska Division employees, and the enforcement and compliance with rules. . . . But for sure, the Nebraska Division is a safe place to work, year in and year out."

Eldon's father, Larry, served as assistant superintendent–Roadway Maintenance of the Nebraska Division through the years of heavy upgrading work that was done over the entire division from the time of the 1970 merger and into the early 1980s. A large number of young engineering and maintenance officers were developed under Larry's tutelage, among them Dave Hestermann, who is BNSF's assistant vice rresident–Chief Engineer. Currently, Dave is overseeing large track capacity-expansion projects in North Dakota, the same type of work he handled as a young gang foreman trained by Larry Ficke in the 1970s.

· DONALD V. (DON) ZEISS ·

. . . as quick to push his bosses as his contractors if things aren't going like he thinks they should.

MANY OF THE LARGE CONSTRUCTION PROJECTS completed in the 1970s and early '80s were managed by Don Zeiss, assistant director of Construction Projects. In overseeing the construction of the new hump yard at North Kansas City in the late 1960s, Don established his capability for getting a large project completed on schedule, within budget, and at the standards specified in the company's AFE (Authority for Expenditure) and per the specifications in numerous contracts with suppliers and builders. Construction or rebuilding of a large yard was especially complex, due to the broad range of technology to be installed and in the types of work to be performed by a large number of specialized contractors. All of the work had to be carefully coordinated to avoid conflicts and delays in order to keep the projects on schedule.

When it was necessary to keep an existing yard in at least partial operation during construction, the task became even more complex. Such was the case at Northtown in Minneapolis when a new hump yard was built on the site of a large, old flat switching yard. A large part of the switching capability of the old yard had to be given up in stages for four years while the new yard was being built. The implications of getting even a few new tracks completed and turned over to the terminal managers on schedule were great, since the Twin Cities terminal was put under a lot of stress in the role it had in implementing the consolidated operating plan for the new company.

Gordon Mott served as assistant superintendent in the early years of construction of the Northtown yard. He was one of the operating officers who worked closely with Don as more and more of the old yard had to be turned over to him for renewal and reconfiguration of the old tracks, and the construction of many new tracks and various support facilities. Gordon speaks favorably of the relationship he and other terminal operating officers had with Don in those years:

> I think we worked well together in bringing that project to completion, largely because it was a case where two reasonable people had a common understanding of the objective. Don had a reputation of being over-bearing . . . and certainly he tested the "new kid" when I first showed up as assistant terminal superintendent. I had an advantage in having some understanding of the project from having been regional industrial engineer, but mainly, I spent enough time in his construction trailer to understand what he was needing to accomplish. Don was an extremely good planner and his plans for the next week of what tracks would come out and what he would give back, and when, were rarely wrong which certainly promoted a feeling of trust. At the same time he knew that I would do everything possible to give him the tracks that he needed and therefore when I said that I just flat couldn't do it he rarely challenged me. Mainly, he never lied. He was also very good at doing the occasional extra thing to help us out when he could. . . . It doesn't surprise me at all that he accomplished what he did.[20]

In addition to a new hump yard and two control towers, the Northtown project included a new locomotive-maintenance shop with the capacity for maintaining four hundred units, a fueling facility, a large car-repair shop, and a maintenance facility for WFE (Western Fruit Express), a subsidiary in charge of maintaining refrigerator cars.

Don proved to be a master in overseeing the work on such projects. Over the next ten years, he went on to head the construction of several other large projects:

- A $70 million taconite storage and lake-vessel loading facility at Superior, Wisconsin
- A new mechanical complex at Alliance, Nebraska, including a large locomotive maintenance shop with capacity for eight hundred units, a locomotive-servicing facility, a running maintenance facility for car repair, and a new division office building
- A shop building at Mandan, North Dakota, equipped for car repair, and with provision for locomotive maintenance in the future

For each project, Don moved his residence and set up an on-site office for the duration of the construction and start-up. In 1980, while the project at Mandan was underway, BN News identified Don as "BN's renowned ramrod." By that time Don had forty years of service in the Engineering Department.[21]

Bruce Anderson, the head of BN's Engineering Department, expressed to BN News that Don was "as quick to push his bosses as his contractors if things aren't going like he thinks they should. He's just as hard on us in getting what he wants as he is on the people out on the job with him. Whether it's getting people, materials, plans, whatever, he hounds us until he gets them. Rank doesn't mean a thing to him."[22] There were some who said Don was a cross between General George Patton and John Wayne.

A one-time subordinate, J. H. Robinson, wrote, "It's true that he will step on a few toes while working on a project, but you can be assured this is because said toes were interfering with progress . . . [Don] accomplished more than was expected of him, and he doesn't care who has the pleasure of putting on the final coat of paint."[23] Don was proud that he never stayed around for the ribbon-cutting ceremony at any of the projects he managed. He had already left to get set up on his next assignment.

In addition to being a capable engineer, Don Zeiss had strong leadership and administrative skills. He knew how to coordinate the work of the contractors to insure they did not get in each other's way. To be able to do this required a quick mind and the ability to make adjustments to keep the project on schedule. Getting these major projects completed on schedule was vital to BN in achieving the objectives in the profit plan, and for gaining the capabilities it needed to improve its service and to handle its rapid growth in business. Don played a major part in reaching these objectives. The edifices he built will have very long lives, and to a great extent, they define what the company is and what it stands for. Through it all, Don maintained that first, he was a railroad man, and secondly, an engineer.

---

20 Email message, Gordon Mott to the author, September 27, 2013.

21 "Construction at Mandan and Sioux City Helps BN Handle Rail Traffic," BN News, August/September 1980, page 6.

22 "Zestful, Zealous Don Zeiss Zooms in on Problems as Alliance Project Construction Director," BN News, December, 1978, page 11.

23 Letter to Editor, BN News, January, 1979, page 22.

· A. B. (ALTON) CROSS, JR. ·

"Nobody would hire me. For three years,
I heard nine hundred ways for employers to say, 'No.'"

THE AVERAGE AGE OF NEW HIRES ON THE Alliance and Nebraska divisions and throughout the Chicago Region was low, with a large number in their early to mid-twenties. Alton Cross was a notable exception. It was unusual to hire a person of age 58 for any position on a railroad and especially as a switchman-brakeman. But if the applicant was in good physical condition, there was no reason not to. With relatives in railroad service and some prior railroad experience himself, he knew what he was getting into in terms of the physical work activity, irregular hours, etc.

During the time Bill Reilly served as terminal superintendent in Alliance, Alton was the foreman of the third trick switch crew assigned to the south end of the new yard. Bill made it a practice to visit with all of the switch crews on duty every day, including those who worked the night shift. He was impressed by the encouragement he observed Alton giving his crew members, most of whom lacked experience in switching, and the respect they showed for him. Alton was soon recognized for the people skills he had developed as a major in the Army Reserve and a minister in the Baptist church. Cal Evans says he thought of Alton as the piece of wood you would place in an uncovered pail of water in a caboose to keep the water from slopping out of the pail. Alton had a calming, stabilizing influence on the turbulent, Wild West atmosphere of those times.

By performing his duties well as a new hire, Alton came to the attention of his supervisors. He was encouraged to apply for advancement to a supervisory position through BN's "This Way Up" program. At age 59, Alton was named assistant trainmaster at Guernsey, Wyoming. In a short time, he was promoted to trainmaster at Alliance. Alton's success is a good example of BN's willingness to not pass up good candidates because of age or other factors not directly related to the job.

In an article in the February 1982 issue of BN *News,* Alton related the difficulty he had in getting hired for any kind of job after serving as a minister for twenty years. "Nobody would hire me. For three years I heard 900 ways for employers to say 'No.'" Having had some prior experience in railroad work, Alton contacted BN's agent in Houston as to opportunities for work where the coal was moving heavily. The agent referred Alton to BN's personnel officer in Alliance. It turned out to be a good break for all concerned.

· DENNIS PREWETT ·

"Hundreds of people worked just as hard as I did....
I am really proud that I got to work with such a great bunch of people."

DENNIS PREWETT BEGAN SERVICE AS A BRAKE-man on the Alliance Division in 1975. After graduating from New Mexico Tech with a B.A. degree in Biology, he heard of opportunities for employment with BN through a fellow student who was working at Alliance. The possibility of working for a large, substantial company with a big operation appealed to Dennis, although he had no particular interest in railroading at that time.

Dennis had gone on a year-long motorcycle trip to see the country. He recalls,

My motorcycle had quit running so I went to Alliance on a Greyhound bus. I had $50 and a few clothes. I worked for a building contractor for a few months until the railroad hired me. My friend was willing to let me rent half a duplex for $35 a month. BN was not hiring right at the time I arrived in Alliance, but I found work with a house builder for the short-term period. Every day that I could I went down to the railroad depot and asked the Train-masters when they planned to hire me. I made sure they knew my name. In May 1975, Jake Greeling, trainmaster, put me first among a group of new hires. I always appreciated getting that job. The officers of the Alliance Division always treated me well. They worked very long hours and were professional in every way.

After only two weeks of service as a brakeman and switchman, Dennis was made the foreman of a switching crew on the night shift in the yard at Alliance. The two helpers assigned on his crew had even less experience. Dennis recalls that first night as a foreman:

> We took the lists of cars to be switched and pulled out the tracks and put the cars where the list showed them to go. It wasn't pretty, but I don't remember tearing anything up. We got the work done. There was very poor lighting in the yard, but we didn't know any better. I don't remember anything about the engineer except that he swore at us and called us "snot-nosed kids." The yardmaster yelled at us on the speakers that were along the switching lead that we were the rottenest help that he'd ever had. We probably were.

A fair number of the new employees who came to BN in those years had prior railroad experience. Dennis recalls many of the best of the new hires had been cut off by the Rock Island when it went out of business.

In a short time, Dennis was moved to the position of hostler at the roundhouse as preparation for entering the locomotive engineer training program. He recalls how quickly he developed proficiency in servicing locomotive units. He and a helper were able to fuel and sand seventy units in an eight-hour shift. They also were required to assemble the units they serviced into consists of three or four units each, in the old, outdated servicing facility. (The new locomotive shop and servicing facility were still under construction.)

Dennis was promoted to locomotive engineer on August 1977. He recalls how crews were sometimes transported by helicopter from Alliance out to relieve road crews that had "died" on the twelve-hour law. This happened when the 112-mile line between Alliance and Edgemont became severely congested due to the increasing number of trains being run, and before the track was upgraded and CTC was put in service. Dennis recalls that it sometimes took more than two crews to move trains over that subdivision. There were cases when a relief crew was able to advance a train by only one station. On two occasions, Dennis and his crew did not "turn a wheel" in the twelve hours they were on duty.

In those times, before trains could be dispatched under CTC rules, trains were operated under train order and timetable authority, a complex system of rules that traditionally had taken two or more years for most new employees to master. Many train dispatchers were inexperienced as well. The train crews were required to understand and comply with numerous train orders issued on each trip, including "right over, wait and meet" orders, additional orders received en route that superseded or annulled their orders, and slow orders, plus orders governing the movement through areas where track gangs were working. To add to the complexity, there were second-class train schedules in effect, requiring train crews to know how to apply the rules involving the superiority of trains by class and direction.

Many stories could be told that reveal the stress that local supervisors and their crews were under in those times. In one instance, Dennis was ordered to run a train at night on the new line being built from Donkey Creek toward Orin. When he told his supervisor he had never been over that line, nor had the conductor, the supervisor told him that since he had a locomotive with a headlight and a timetable with the listing of sidings, he had to make the trip or be charged with insubordination. It was a case of desperation and poor judgment made under pressure. Dennis and his crew made the trip safely.

Dennis remembers life on the railroad in those years was "like a battle." He worked for ten months without having a day off. Even while off duty at their home terminal between trips, crews hardly ever had more than the minimum of eight hours off duty required by law. They were expected to tie up in 11 hours and 59 minutes, and not work the full 12 hours authorized by law, so they would have to be given only eight hours off, and not the 10 hours required if they had worked to the limit of 12 hours. Dennis added, "My experience was not unique. Hundreds of people worked just as hard as I did. A lot of us have some good stories. I am really proud that I got to work with such a great bunch of people."[24]

---

24 Dennis Prewett, conversations and exchange of messages with author.

After working as an engineer for just over two years, Dennis accepted promotion to road foreman of Engines at Greybull. A year later, he was named the trainmaster–road foreman on the Camas Prairie, a railroad of 256 miles in Idaho, jointly owned by BN and UP. In a short time, Dennis was promoted to manager, the position that designated him as the chief executive of the Camas Prairie. He returned to BN as director of Costs and Budgets in 1985, followed by appointment as terminal superintendent in Spokane, a stint in the Marketing Department as director of Customer Service in the Consumer Products unit, back to the Operating Department as chief of staff for the Northern Region, assistant superintendent and division superintendent at Fort Worth, director of Unit Train Operations, and finally, director of Safety in 1996. Dennis has continued his railroad career as president of the TNW Corporation, a holding company for three short-line railroads in Texas, owned by the Murchison family.

In recalling the time he worked on the Alliance Division in the boom years of the late 1970s, Dennis says, "It was a big deal for all of us who had so little experience. It seemed like we were having a big role in such an important enterprise. I found railroading to be a good job. I have never been bored. I never really learned any job before getting [promoted to] another."

· ROBERT E. (BOB) PELAVA ·

An experienced engineer transfers to Sheridan and,
before long, is elected general chairman

AN INTERESTING STORY OF A JOB TRANSFER WAS that of Bob Pelava, a locomotive engineer at Minneapolis. I got to know Bob in the early 1970s while I worked as terminal superintendent in the Twin Cities terminal. One day, Bob came in to tell me he had decided to transfer and go to work as an engineer at Sheridan, Wyoming. Bob had about twenty-five years of service and also was serving as local chairman for the engineers represented by the Brotherhood of Locomotive Engineers. I thought at first he had just made up a story to see how I would react. After all, who would give up twenty-plus years of seniority and in effect start over in chain-gang service at a new place? But in a few minutes of conversation, Bob convinced me he was serious. He said that since he was an avid big-game hunter, he knew Sheridan would be a good location for him to live. And, with the shortage of engineers out there, he knew he would soon be one of the senior engineers at Sheridan (or anywhere else on the Alliance Division) and would be able to hold a turn in the pool as soon as he made his qualifying trips over the territory.

At that time, engineers did not have system-wide seniority, so Bob had to begin work at Sheridan with a new seniority date, no different than a new employee hired with no prior railroad service. There were times when the number of engineers on the crew board at Sheridan had to be cut, due to a seasonal decline in business or when fewer older engineers were on vacation. To keep working, junior employees such as Bob had to place on the board at Edgemont, about 120 miles east of Sheridan, until conditions changed enough to again require more engineers at Sheridan. In 1990 an agreement was made by Labor Relations officers and the general chairman of the Engineers for system seniority, meaning that employees could take their hiring date with them to any location on BN.

Bob ended the conversation by saying that I could be sure he would be actively involved soon after he started work out there. Not surprisingly, it wasn't long before I heard he had been elected local chairman. Not too much later, we met at a joint labor-management meeting in St. Paul. He reminded me of his pledge that he would soon have his name in lights again. Bob had been elected general chairman to represent the Engineers of the entire BN system. In the mid-1980s, we sat across the table from each other at a number of meetings that were part of an effort to find common ground for moving the company into the future.

Soon after Bob started to work in Sheridan and could vouch that indeed Sheridan and the Gillette–Edgemont area were good places to live and work, several more engineers from the Twin Cities moved out there. Most of them were interested in raising families outside of an urban environment, where where they thought the pace of the operation would be slower than in a major terminal such as the Twin Cities. Getting engineers with that kind of experience in service out on the coal lines was a great help to BN in moving the rapidly growing amount of business coming from the mines in Wyoming and Montana.

· THE SOGN PACK ·

A father and three sons, all railroaders

DOUGLAS (DOUG) SOGN HAD WORKED AS A brakeman for the Milwaukee Road at Sioux City for two years when he decided to join BN at Edgemont. Soon, after he came on board, he was promoted to conductor. Doug's three sons, all railroaders on BN, were with him on the last trip he made before retiring after thirty-five years of work on BN. In a tribute to Doug and his sons, the employee newsletter of the Powder River Division quoted Doug: "It was great to have the entire Sogn crew at the train crew [on his last trip]. It meant a lot to me."

Similar stories could be told about other families who benefited from the opportunities BN provided them for good careers in the rail industry for many years, and into the next generation. Two of the Sogn boys are locomotive engineers and one has been promoted to terminal trainmaster. "I am proud to see they have good careers and great opportunities ahead of them," Doug said as he closed out his career.[25]

· SHEILA HUSS ·

"… probably the best career opportunity for a woman, or for that matter, a man."

SHEILA HUSS FIRST WORKED FOR BN AS A TRAIN order operator, yard clerk, and crew caller. In March 1976 she qualified for the training program for locomotive engineers. In an interview with Bob Wiedrich of the *Chicago Tribune* early in 1977, Sheila said she anticipated promotion to engineer in a few months.

She provided interesting comments on the life and work of a woman working a job that had been limited to men throughout railroad history:

It's a tremendous opportunity for me. It's probably the best career opportunity in the United States for a woman or, for that matter, for a man. If I had a college degree, I wouldn't do half as well. We've got college graduates—teachers and so on—already working on the railroad as Brakemen. … The opportunities for advancement are great and so is the work.

Sheila is a third-generation railroad employee. "It's not a job. It's a career. It's a way of life. It's something that gets in your blood. It's like joining a big family." Wiedrich wrote that Sheila's words "sounded somewhat like a management's sales pitch. But they were not. They were sincere. You can tell it. Her enthusiasm is infectious."[26]

---

25   "Final Run for Sogn Patriarch," *Powder River Reflection,* third edition, 2011.

26   Bob Wiedrich, "She's a Railroader—Simple as That," *Chicago Tribune,* re-published in the *Alliance Times-Herald,* March 5, 1977.

• BARB (BARBARA) SCHAFER •

"If the wheels were turning, I was happy to be going somewhere."

"IT WAS NEVER THE SAME EVERY DAY—DIFFER-ent times, trains and schedules. It was quite a challenge. I felt I had to prove myself on a man's job. I gave 110 percent and tried not to worry about the small stuff. I just tried to do my job and be pleasant about it. I did not dwell on things. I tried to always keep a positive attitude."

Barb Schafer retired after thirty years of service. Leaving her job as a unit secretary at a hospital in Omaha, she began work as a brakeman at Alliance in 1979. The next year, she was promoted to conductor and, two years later, to locomotive engineer.

Having grown up on a ranch, Barb said she liked to work outside. She said the railroad was "a good fit," since she had never liked working conventional 8-to-5 jobs. "If the wheels were turning I was happy to be going somewhere. I like working with machinery. . . . I've always worked odd hours."

Her advice to fellow railroaders is "to maintain a positive attitude, be careful, ask questions if they don't understand, remain aware of their surroundings at all times and make sure they have a clear understanding of their role in the task at hand. With a moment of inattention it can be all over."

Barb is representative of many women who were successful in mastering the work on railroad jobs that had been held exclusively by men until the 1970s. Breaking through that barrier is a real credit to them and to the supervisors and fellow workers at BN who gave their support to this major change in railroad life and culture.

• KATHY AND STEVE STRAIGHT •

"We women had to work twice as hard to be half as good
as the men even if we superseded some of them in ability."
"I wouldn't change a thing, thirty-six years later."

STEVE AND KATHY BEGAN WORKING ON THE Alliance Division in the mid-1970s, Kathy as a train order operator and Steve as a clerk. Less than a year after she was hired, Kathy was promoted to dispatcher and, a few years later, to assistant trainmaster and trainmaster. When Kathy retired in 2012, she held the position of director of Administration at the headquarters office of the Powder River Division at Gillette. She had a good career, as evidenced by accolades from co-workers such as: "Although I asked her the same questions repeatedly, she treated each question like it was the first time I had asked. . . . She treated each . . . request like it was the most important thing on her daily to-do list. . . . She always went the extra mile to help others, despite daily challenges."[28]

Kathy first worked for the C&NW Railway as telegrapher clerk at Casper, Wyoming. She recalls being given no training on how to copy and repeat the train orders issued by dispatchers—she had to learn by listening to the interaction between the dispatchers in Chadron and Sioux Falls and other operators in the transmission of orders on the phone line that was dedicated to the dispatchers. In a fairly short time, she was able to pick up the terminology and routine that was required by rule in the issuance of orders and their delivery to the crews of passing trains. Her father worked as a conductor on the

---

27   "Dedication and Determination," *Powder River Reflection,* October 2009.

28   Patty Whitlock, quoted in "Straight Bids Division Farewell," *Powder River Reflection,* 4th ed., 2012.

C&NW. She had hoped to be accepted into the locomotive engineer training program, but when she was not given that opportunity, she decided to go to work as an operator on BN.

After observing her work as an operator for less than a year, the division officers encouraged Kathy to train and qualify as a dispatcher at Alliance. Upon completion of the training program, the agreement with the dispatchers' union provided the company sixty days to evaluate the work of a new dispatcher and decide whether the trainee met the standards or if he or she should be terminated. Kathy was disqualified during the period allowed for qualification, a disappointing setback.

When Kathy learned that candidates for dispatcher positions were being trained in the office in McCook, Nebraska, she took this second opportunity to become qualified. This time, she was successful. Before long, she was able to place on a dispatching position at Alliance. Her capability for a supervisory position was recognized, and in a short time she was promoted to assistant trainmaster. A few months later she was promoted to trainmaster, followed by promotion to director of Administration for the Powder River Division. When the Powder River and the Colorado divisions were consolidated, Kathy was set back to a terminal trainmaster position and within a few months, was named trainmaster for the Butte Subdivision, a challenging territory that included operations over Crawford Hill. In time, the director's job opened up, and Kathy was appointed to it again.

Kathy was the first woman to become qualified as a dispatcher at Alliance. Other women had entered the training program for dispatchers, but they had failed to qualify or remain employed as dispatchers. She recalls the resentment in those years over women working railroad jobs such as dispatching that had been held exclusively by men until then. It took some real courage and perseverance for her and other women to overcome that stigma by proving they could do the work, and finally become accepted as members of the railroad family. Kathy recalled, "We women had to work twice as hard to be half as good as the men even if we superseded some of them in ability." In looking back at those early years, Kathy believes that having to work through times of adversity drove her to develop stronger capability and potential for an operating officer position.

Nearly all of the employees in the field (i.e., train crews and track gangs) that the dispatchers worked with also were young and had not yet become imbibed in the old railroad culture, and most of them were able to work well with the women who were new dispatchers and first line supervisors. Kathy recalled that, as new railroaders who were trying to become more and more proficient on their jobs, the younger employees would look out for each other. They developed a trust and confidence among them that prevailed through their careers. It took longer for some of the male supervisors and dispatchers to develop a similar rapport with women working on non-traditional jobs than it did for most of the employees who worked out in the field.

Kathy recalls that in dispatching the 238-mile line between Alliance and Ravenna, more time was used in lining up the crews called to relieve those that had tied up under the hours of service law than in actually directing the movement of trains out on the line. On the job that dispatched the territory between Edgemont, Gillette and Reno Junction (on the new Orin line), it was common for a dispatcher on the day shift to have to issue as many as one hundred "track and time" authorities to track gangs, track inspectors, and signal maintainers, which put a terrific load on the dispatcher. The responsibility for authorizing and protecting those employees was in addition to the primary responsibility of a dispatcher, that of moving trains across his or her territory.

The workload on what became the heaviest tonnage railroad lines in the world soon required the assignment of two dispatchers to each dispatching district. One "moved the trains" and provided the authority to maintenance employees that they needed to "take the track" for their work. The second one handled the "paperwork": managing the information that had to be placed on the train sheet and updating records on the status of crews at work and relief crews, and of those who had run out of time under the twelve-hour law.

On November 6, 1979, the day the first coal train ran on the newly completed Orin line, Kathy was called to work as an operator for the midnight–8:00 A.M. shift at Glendo, Wyoming, a small station on the Canyon Subdivision between Bridger Junction and Guernsey. She was the first employee to open and go to work at the small old tool shed that had just been remodeled enough to house an operator assigned there to copy train orders for the large number of trains that were to be routed via Guernsey instead of going "the long way around" through Alliance. The depot was equipped with only a desk, a chair, and a dispatcher's phone—there was no commercial phone available to make contact with the outside world, if needed. The position at Glendo was one of eight that had to be established between Reno Junction and Northport, since CTC was not yet in service. On the thirtieth anniversary celebration of the opening of the Orin line, Kathy recalled, "I have seen from dark territory and hooping up handwritten train orders to trains, to quad track [four main tracks] and signals. It is truly amazing what we've seen in the past three decades."[29]

Steve started his career as a clerk at Alliance and soon transferred to train service as a brakeman. He completed the training program for locomotive engineers in 1975 and worked in engine service until his retirement in 2009. Shortly before retiring Steve was named "Engineer of the Year" for the Powder River Division. The

division deems this award as "prestigious professional recognition." In accepting this award, Steve said, "The competition is tough. There are many great people on this division who are good at what they do. I wouldn't want to have to pick."[29]

In recalling his thirty-five years of service, Steve's advice to fellow railroaders was: "Anyone can make a mistake at any time. Don't think it won't happen to you, because about the time you think that way something will blind-side you." The company has been fortunate to have many employees of the character and caliber as the Straights for upwards of thirty and forty years. Through the challenging years of the 1970s, when hundreds of people were being hired and the railroad was in a state of tension in trying to upgrade its roadway and move more and more trains, these employees and the company stuck together and made it work. As with the Straights, most of them worked through all of those years with no personal injuries. Upon his retirement in 2009, Steve said "I wouldn't change a thing, thirty-five years later."

· MICHAEL (MIKE) TURLEY ·

On the day he was hired, Mike was given an hour and a half
to buy work clothes and boots and return to the office for his first pay trip.

MIKE TURLEY WAS HIRED AS A BRAKEMAN AT Alliance in 1976, when the early days of growth and opportunity were in full swing. Only ten months later, he got the chance to go into "engine service" as a hostler at the roundhouse. (A hostler is an employee designated to move locomotive units within servicing and maintenance areas, or to or from the locations where trains arrive or depart in a terminal complex.) On that job, he established himself well enough to qualify for prerequisite training as a student locomotive engineer. After only thirty days, he had progressed enough to be ready to go to St. Paul for three weeks of training, both in the classroom and the state-of-the-art (Mike calls it "rudimentary") simulator. That was part one of the accelerated

program BN established to qualify new hires and employees who qualified for the opportunity to transfer from their craft, first to work as a brakeman and switchman, and later, to the engineer's training program.

By choice, Mike began part two of the program as a student engineer at Sterling, Colorado. He made rapid progress toward becoming qualified as an engineer. After making only forty-three of the seventy-six student trips required in part two, Mike was evaluated as fully qualified to become a full-fledged locomotive engineer. For three months, he was the youngest engineer on the Alliance seniority roster. Because BN's greatest need for engineers was at Alliance, Mike was "forced" to mark up on the extra board at Alliance rather than at Sterling, his

---

29   "Memorable Moment in Time," *Powder River Reflection*, December 2009.

30   "Steve Straight of Alliance," *Powder River Reflection*, October 2009.

home. Most of the time, he worked between Alliance and Edgemont. In the summer of 1979, he was able to hold a regular job for several weeks on a work train (used for unloading material for track gangs) based at Hyannis, a station fifty-nine miles east of Alliance. In the fall of 1979, three years later, Mike had enough seniority to work on the engineer's extra board at Sterling, where he preferred to live and work.

In coming to work for BN, Mike had the advantage of a few months of service with the Union Pacific at Denver. Mike qualifies that "experience" by noting that he was furloughed most of the time. He actually worked for only about five months of the two and a half years he was with the UP. On his website, Mike wrote, "It was common practice for railroads to hire too many people, only to furlough them when winter came and freight slowed down." During the time he was furloughed, Mike did roofing for a friend who owned a housing construction company and enrolled at a trade school to become qualified as an aircraft mechanic. He then worked for Denver Beechcraft and a Chevrolet dealer as an automotive electrician and a service dispatcher. When Mike learned that BN was hiring for train service at Alliance, he decided to go back into railroading.

On the same day he was hired at Alliance as a Brakeman, Mike recalls the trainmaster gave him a purchase order that allowed him to go to a store in Alliance and buy work clothes and boots. He was given an hour and a half to get that done and report back at the office for his first pay trip. He also bought a house trailer for a place to sleep—no houses or apartments were available. For the next eight or nine months, Mike was consistently called for work at both his home terminal and away from home terminal after having only the mandatory minimum of eight hours off duty.

Soon after Mike was placed on the engineer's extra board at Alliance, he was called to run a loaded coal train of 14,000 gross tons to Ravenna, an interdivisional run of 238 miles. Up to that time Mike had never worked on that line, even as a brakeman. He was told to "go anyway," since the head brakeman who would be with him on the engine had been over that line a few times. They made it safely, although Mike said he held the speed way down to help ensure no mistakes were made in complying with speed restrictions, signal indications, and all of the requirements contained in the "fist full" of train orders he was given. Such was life on the Alliance Division in those early years of the coal boom. It is apparent that Mike's supervisors had a great deal of confidence that he could apply good judgment and already had enough experience to handle a train under the circumstances he was thrown into that day.

In September 1985 Mike and his wife decided it was time for them to leave BN and move back to the Denver area, where they lived before he started work with BN. They found they preferred living in the Denver area to towns the size of Sterling and Alliance. Mike went back into aircraft maintenance, this time with Continental Airlines. In 1988 he was contacted by Montana Rail Link (MRL), which had just started to operate, to see if he would come to work again as a locomotive engineer. When Mike learned from MRL that he and his wife could live in Billings, a city to their liking, he took the job. Mike put in fourteen years with MRL, retiring in 2012.

In thinking back over the "boom" years he put in with BN, Mike says, "It was great to have been a part of it." BN was fortunate to have had the service of so many young, enthusiastic employees who could learn fast and adapt easily to the fast-paced, ever-changing operations of those times. It was not a place for those who wanted set routines and predictability in their work or a lack of challenges on every trip. There were hundreds of railroaders such as Mike who made things go under difficult circumstancs, especially in the early years of the boom.[31]

---

31   To learn more about Mike Turley's work experience on BN and MRL, refer to his website, http://bearstoc.com/MRL/hist/career.html.

· BOB BYE ·

After about fifteen minutes on the use of the new distributed power technology,
Bob was told by the road foreman to "take the train"

BOB BYE MADE AN INTERESTING AND SURPRIS-
ing career change, going from a career of twenty years
in ministry for the Presbyterian church to hiring out as
a brakeman at Edgemont. Bob was soon promoted to
conductor and, four years later, to Locomotive Engineer.
He worked mainly between Edgemont and Gillette and
on the Donkey Creek–Orin line.

By the time Bob started railroad service in 1984(?),
the heaviest part of the track improvement and capacity
expansion projects had been completed, including the
installation of CTC and construction of many miles of
second main track. Bob experienced the introduction
of distributed power (DP)[32] to unit coal train service
that began in October 1998. He operated the first test
train with distributed power, which was loaded at the
Cordero Mine to 19,800 tons. It was far heavier than the
trains of about 14,000 tons that had been the norm since
the start of unit train service in the Powder River Basin
in the early to mid-1970s. Bob recalls he received only
about fifteen minutes of instruction on the use of DP by
the local road foreman of Engines, who then told him to
"take the train." He was given no training on either the
simulator or in a classroom session, as was soon set up
for all engineers who would be handling trains set up
with DP. It is obvious the road foreman had a great deal
of confidence in Bob's ability to safely handle a train with
this new type of power configuration and with very little
instruction or practice.

The train was powered with three locomotive units
on the head end and one DP unit on the rear of the train.
Bob managed to handle it over the undulating terrain
between Edgemont and Donkey Creek without it breaking in two or having any other difficulty. The use of DP
power was implemented very quickly over the entire
division.

When asked if he had much opportunity to use his
ministerial skills in his years in the life and culture of
the railroad, Bob replied that indeed, he often served
as a counselor, friend, and advisor to fellow employees
who were facing such challenges as relationships with
family members, difficult financial problems, or issues
that arose from working in the irregular, unpredictable
conditions that are part of working in chain-gang service
or on the extra board.

When Bob first started to work in the pool that handled
trains between Edgemont and the mines along the new
Donkey Creek–Orin line, crews ended their tour of duty
when they reached the mine where their train was to be
loaded. A crew from Gillette designated as the "loading
crew" would then man the train while it was being loaded
and run it back to Donkey Creek. Those crews were made
up of a conductor, engineer, and brakeman, and they
were paid at "local" rates, a rate somewhat higher than
crews were paid for handling a "through" train. This was
a costly operation, and BN offered a proposal to eliminate
the brakeman's position on the loading crew and pay the
conductor and engineer on a loading crew a flat rate of
two hundred miles, or two basic days, for each tour of
duty. The Gillette-based crews were to be given exclusive
rights to handle trains as they were being loaded.

Employees based at Gillette voted down this proposal due to the loss of jobs and their anticipation of
lower earnings, and a "sick out" of several days ensued.
Employees from Edgemont did not join in this protest.
Contract railroad companies were then engaged by the
mining companies to handle the trains while they were
being loaded and on the property and tracks owned by
the mining companies. The one-person crews provided
by the contractors were paid at a lower rate than BNSF
paid its employees. The contract crews also were trained

---

32 Distributed power enables remote control of unmanned
    locomotives placed at the rear of a train or within a train
    consist by the engineer in the lead locomotive unit. The pur-
    pose is to reduce stress that may exceed the tensile strength
    of couplers or drawbars on heavy tonnage trains moving
    on grades, and for better control of train slack action on
    undulating territory.

to make the mechanical inspections and air-brake tests required before trains moved onto BN–owned trackage. This move reduced BNSF's costs and resulted in a loss of jobs for employees based at Gillette. It was an example of efforts undertaken to reduce costs and maintain margins on competitive, low-rated business.

Overall, Bob is glad he made the radical change from the ministry to running trains on the world's heaviest tonnage railroad line. He felt rewarded through the camaraderie and teamwork that developed in a group of people who worked together in the rather unique and challenging culture of a railroad. Today, Bob is an active member and a director of the Great Northern Railway Historical Society. He has done considerable research on the rear-end collision of the two sections of the Great Northern's Empire Builder that occurred only a few weeks after the end of World War II at Michigan, North Dakota, resulting in thirty-four fatalities. Many of those killed were military personnel.

Two years ago, Bob helped organize a gathering of a large number of descendants of those killed, sixty years later, at Michigan. Through his ministry and pastoral skills and his railroad background, he has been able to better inform many of the descendants of the circumstances of the accident. By sharing of thoughts and remembrances of those killed, many of those present found some closure and consolation. A large stone monument has been placed in a park at Michigan, near the scene of the accident. Bob's interest in the people who had family members killed in this accident arose from his recollections of the years he spent as a youngster growing up in nearby Grand Forks, and of the service his grandfather had as a locomotive engineer on the GN in those years.

CHAPTER 50

SIGNAL SYSTEMS AND THE PEOPLE WHO BUILT THEM

*• signal and train-control systems to enhance
safety, increase capacity, and raise the quality of service provided •
train crews bet their lives on what the signals tell them hundreds of times on every trip •*

On the major coal corridor of 829 miles between Lincoln and Huntley, only the 238 miles between Ravenna and Alliance were operated under CTC rules in the mid-1970 when large tonnages of coal began to move. A modified version of CTC with only limited operating capability was in service on that subdivision. Between Lincoln and Ravenna (126 miles), an automatic block signal (ABS) system was in place, and while that was important for the safety of the operation, it did not add much to the capacity of the railroad. Also, ABS was in service for 12.5 miles in the area of Crawford Hill, to enhance safety on that heavy-grade territory.

There was great urgency to get CTC installed as quickly as possible between Alliance and Gillette and between Lincoln and Ravenna, to increase the capacity and safety of the operation. CTC had to be installed on the new line of 127 miles being built between Donkey Creek and Bridger Junction. Next in priority for installation of CTC were the lines between Bridger Junction, Northport, and Alliance; between Northport, Sterling, and Brush, Colorado; and then, the line of 232 miles between Gillette and Huntley. On the Nebraska Division, in addition to the Lincoln–Ravenna line, 112 miles of CTC had to be in service on the line between Lincoln and Napier that would become the designated, most direct route for trains going to Kansas City and beyond, as soon as the new bridge at Rulo was completed.

While the installation of 1.112 miles of new CTC was underway on the Lincoln, Alliance, and Yellowstone divisions, it was vital that the old, "poor man" CTC between Alliance and Ravenna be upgraded. The number of trains being operated far exceeded the capability the old CTC was designed to handle. With the old technology of the pre–solid state CTC machine, and the limited capacity of the code line between the CTC machine and the signals and switches in the field, there often was a delay of about twenty minutes for the signal and switch indications to come back to the dispatcher in response to the commands the dispatcher had just sent out to the field. This lag in the transmission of indications severely hampered the dispatchers and restricted the capacity of the railroad.

It would be an understatement to say that getting all of this work done within five years put a very heavy work load on BN's Signal Department, no less imposing than the burden the track-maintenance forces had in building many miles of additional track capacity and upgrading hundreds of miles of track. To accomplish this, the Signal Department had to hire and train large numbers of people to do the basic work of laying cable, erecting signal masts, powering siding and crossover switches, and conducting exhaustive testing before even short segments of new CTC could be cut over and put in service. Before such field work could be started, design and planning for each segment of CTC had to be completed by signal engineers at the system and region level.

At the same time, new signal maintainers had to be trained and qualified to maintain the new CTC, once it was placed in service. These positions required technical expertise in electronics and circuitry and on the operating rules, to insure that the signal system they were charged with maintaining would function properly. Since train crews bet their lives every day on proper functioning of the signals, there was absolutely no room for error. Such stringent requirements put a heavy responsibility on those charged with selecting, training and qualifying candidates to work as Maintainers.

Chuck McCormick recalls the challenge of getting new employees trained well enough so they could begin work. It was similar to those faced by the track-maintenance and construction people in those years.[1] A new system-training center had just been opened in St. Paul, which was very good news to the signal engineers working on field installations. All new employees were supposed to go through the program, but with the rate at which people were being hired, the training center could not take them in fast enough to meet the needs in the field. As a result, courses had to be set up in the field and held evenings in the gang's bunk cars or in a motel meeting room. In the years of the heaviest work load, there were from sixteen to eighteen gangs at work on the Alliance Division alone. Each gang had ten or twelve employees.

The new workers came primarily from ranches and farms in Wyoming and Nebraska. Once hired, most of them made a career in the Signal Department, and many became qualified to work as signal supervisors. Many of the new employees were promoted from the gangs to signal maintainer, a safety-sensitive position, within only a year after they were hired. This became necessary when no qualified maintainers would bid on newly created positions in remote areas. This lack of experience put additional tension into the job of a signal supervisor.

The job of a signal maintainer is particularly unique, as described by Pete Swendsrud, a signal inspector: "For the most part, we are working alone, and there isn't anyone else to watch out for you."[2] The article continued: "Because they are lone wolves and often called out in less than ideal conditions, many of the maintainers have developed a meticulous nature when it comes to safety." Bill Kreutzer, signal supervisor on the Powder River Division, added, "It takes a certain self-discipline as a lone worker, in the middle of the night, to avoid taking shortcuts and follow proper procedures consistently." A signal maintainer, Dustin Jackson, recalled his early days on the job: "Nobody expects you to know everything. There's a lot of leadership and help. They test you to see if you're willing to learn."[3]

The jobs of signal maintainer, signal inspector, and supervisor are largely unheralded in the rail industry, in spite of the demands put on those employees and managers for perfection and the need to maintain the train-control system to such quality that failures that may affect safety or cause delays to trains do not occur. The safety of the operation and the ability to run trains on schedule are highly dependent on their dedication, job knowledge, and commitment to work as a team with dispatchers, conductors, engineers, and the track-maintenance forces.

Also in the years of rapid hiring and advancement, women began to move into non-traditional positions in the railroad work force, including work on signal gangs, and later, as signal maintainers and managerial positions in that department. Kathy Pettry had earned a degree in Horticulture. She graduated from the twenty-fifth class in BN's Signal Apprentice Training Program in 1982. After working on construction crews for three years over a wide area, Kathy became a signal maintainer on the Donkey Creek–Orin line. In BN News, January 1983, Kathy reported on her progress: "As a maintainer, I'm on call 24 hours a day. . . . It's interesting work and I really like the people I work with. I think the job was harder at first because it seemed I had to prove—as a woman—I could do the job. It's gotten much easier, though, as time goes on and people realize that I can get the work done." Chuck McCormick rated Kathy as

---

1   Chuck McCormick held the position of region signal engineer at Denver in the early 1980s, preceded by assignments on the Nebraska Division at Lincoln and on the headquarters engineering staff. He served as director of Signals for BNSF's Chicago–Seattle corridor for six years, 1995–2001.

2   "Signal Groups Mark Ten Years, Zero Injuries," *Northern Light*, BNSF newsletter for employees on the Twin Cities Division.

3   "Jackson Finds New Home in Newcastle," *Powder River Reflection*.

an outstanding maintainer who worked in what was a trying time for many.

In addition to new construction on the Denver Region, the Signal Department had a major role in the construction of the new hump yards at Northtown, Galesburg, and North Kansas City. A major system-wide project was undertaken in the early 1980s to eliminate the open-wire pole lines that carried signal and communication lines for the signal system and voice and data communication.

Getting the communication system on to microwave and fibre optics and to electro-coded track circuits eliminated the exposure to ice, wind, falling trees, and other vegetation that impaired the reliability and consistency that are vital in those systems. When completed in 1974, bn's microwave network became the largest privately owned and operated system of its kind in the world. It represented an investment of $22.3 million for a 4,502-mile system in ten states.

CHAPTER 51

A TRIBUTE TO THE SONS OF MARTHA

Now as they went on their way, he entered a certain village where a woman named Martha welcomed him into her home. She had a sister named Mary, who sat at the Lord's feet and listened to what he was saying. But Martha was distracted by her many tasks; so she came to him and asked, "Lord, do you not care that my sister has left me to do all the work by myself? Tell her then to help me." But the Lord answered her, "Martha, Martha, you are worried and distracted by many things; there is need of only one thing. Mary has chosen the better part, which will not be taken away from her."

THE HOLY BIBLE, NEW REVISED STANDARD VERSION, LUKE 10:38–42

• *to those who have the accountability for results and for getting the work done safely, on schedule and within budget* • *we do not want to forget the names of railroad workers, first line supervisors, and the mid-level operating and maintenance officers whose skills, experience, and professional capability transformed Burlington Northern into a high performance, growth company. Together, many of them accomplished more for the company and developed their talents to a higher level than many thought they would ever reach.* •

SEVERAL YEARS AGO, I CAME ACROSS A monument in Churchill, Manitoba, with a tribute to the workers who built the railroad line north to Hudson Bay, containing a quote based on Rudyard Kipling's poem, "The Sons of Martha." Kipling wrote, "It is their [the sons of Martha] care that the gear engages; it is their care that the switches lock. It is their care that the wheels run truly." For the people of a railroad company, Kipling's writings can be considered to cover those who run the trains and yards, those who maintain the roadway and equipment, and the professional engineers who design and build it. They are the uncelebrated people who run the trains and switch cars in yards. Their work meets the basic and essential needs of the railroad.

Kipling believed the sons of Mary (possibly defined on a railroad as the support departments at headquarters) are dependent on the employees out in the field who are with Martha, getting the work done safely, efficiently, and up to standard. Certainly there were many "Sons of Martha" working in territories and in specialties not directly involved with upgrading the railroad for the coal boom. They continued to handle BN's non-coal business that was termed by some leaders as "our real bread and butter." A series of tributes could also be written about their work lives and what they have contributed to BN's success.

The "Sons of Martha" is based on the Biblical passage in Luke 10:38–42 in which Martha asks Jesus to order her sister, Mary, to give her some help in the kitchen, rather than avoiding such work by sitting in another part of the house and listening to Jesus as he teaches. Jesus rebukes Martha, saying that Mary had chosen "that good part"—she recognized a higher calling than to do the work that would help make Jesus and the other guests comfortable.

On a railroad, one could say the sons of Martha make life comfortable and achievable for those who work in the headquarters office and design and plan the services need to bring the revenue in, plus deal with a host of government regulators, consultants, and the suppliers of materials and services. Kipling would designate them the well-blessed sons of Mary. At times, the sons of Martha may be uncelebrated, taken for granted, or otherwise unappreciated. But in the years that BN rebuilt and expanded a large part of its system, they were recognized as the group upon which the company was depending to get the new and upgraded material placed in the track, and get the operation shaped up so a higher return on invested capital would be realized, and shippers would be getting the service they demanded.

In closing this section on the heavy work done to upgrade and expand the capacity on so much of the BN network, I thought it would be well to give credit to those considered "the real doers": the line managers and the track, bridge, and signal workers who were held responsible for getting the work done on the major coal corridors on which the track had to be upgraded in the 1970s and early 1980s.

In making this tribute, I will begin with the work of the division superintendents. They were at the highest level where an operating officer is still close to the real work of a railroad. They were generally known and recognized by the work force as the person "who really is in charge of what is going on." On BN, the superintendent had responsibility for all of the major functions of the Operating Department: Transportation, Maintenance of Way, and the Maintenance of Equipment. He was expected to get the leaders and workers in all three functions to work together as a team. He made the decisions on whether the needs and priorities of one function should be put ahead of one or both of the other functions, should a conflict or tension arise. Usually the superintendent was not an expert in all of the three functions, but he had to have the capability to manage the budget, set goals and standards, achieve satisfactory performance in safety, deal with personnel matters, and provide general leadership in each function.

The superintendent and his staff of assistant superintendents were expected to provide the support, leadership, and direction needed for development of first-line supervisors and employees holding skilled positions, among them the foremen of track gangs, roadmasters, machine operators, electricians, machinists, dispatchers, clerks, B&B (bridge and building), signal technicians, crew callers, conductors, and engineers whose functions were vital in getting the work done. The superintendent was expected to recognize differences in talent, experience level, strengths, and weaknesses in the work force, both employees and their managers.

This required him to be able to compensate for weaknesses in the organization and to jump into the fray when necessary to support the effort. He needed to be able to apply coaching skills when problems arose, but not to criticize excessively. The superintendent needed to help those who lacked experience or had made a bad call to understand how to do it better the next time. At the same time, the ability to apply good judgment and take a firm stand were needed in handling cases of rules violations and poor job performance, especially when that either did or could have affected safety in the operation. It was important for the superintendent to recognize success when his people achieved production goals, made efficiency improvements and the completion of major projects.

The divisions were expected to "run the show," and the region and system officers were to serve as coordinators, advisors, and evaluators of results. In the end, the divisions had the primary accountability for getting the work done safely, on schedule, and to standard.

THE SUPERINTENDENTS[1]
Alliance Division

On the Alliance Division, the work had to be done under traffic, and heavy unit trains had to be run on an infrastructure that was totally inadequate for the task at hand. The division superintendent had the task of organizing the work and getting the resources lined up to complete the many large projects authorized to upgrade the track and expand the capacity of the railroad. Employment had to be increased rapidly in all three of the operating and maintenance functions, presenting challenges in recruiting, housing, training, and discipline and often having a low ratio of supervisors to employees.

Eventually, the superintendent was able to convince top management to allow an increase in the number of field supervisors commensurate with the greatly increased number of employees in train, engine, and yard service, a solution to a host of problems that was long overdue. A number of experienced employees were convinced to take promotion to supervisors, to give some balance to the large number of inexperienced management trainees who had been brought in to what was then the most demanding territory on BN. With the completion of major projects, the division was able to greatly reduce the cycle times for unit coal trains and virtually eliminate derailments caused by deficiencies or weaknesses in the track structure.

---

1   This section contains a listing of the operating and maintenance officer positions that reported directly to the Superintendents of the divisions on the Denver and Chicago regions, and the Yellowstone Division of the Billings Region. Virtually all of the heavy maintenance and capital projects needed to upgrade the track and expand the capacity of the railroad were carried out on those divisions. While there was a heavy movement of coal on each of the three divisions of the Twin Cities Region as well, their track was in good shape. It did not require heavy upgrading, except for the line between Sioux City and Willmar on the Minnesota Division, and relay of the 90-pound rail between Grand Forks and a power plant at Cohasset, Minnesota, with heavier rail.

Nebraska Division
(named the Lincoln Division prior to 1976)

The superintendent and his staff got the resources and plans in place and organized the work to get heavy trackwork projects started on schedule. They undertook the hiring of large numbers of employees and had the challenge of getting trackwork done in the face of having to run several priority merchandise trains in daylight hours, in conflict with the work expected of the track gangs.

The division managed major trackwork projects on each of the six main-line subdivisions radiating from Lincoln in time to handle the large year-to-year growth in tonnage. Several major expansion projects were carried out, among them construction of several miles of second main track, building siding extensions, conversion of some lines from dispatching trains under train order authority to CTC, constructing line changes, upgrading bridges, and having a large number of gangs upgrading branch lines enough to handle 100-ton capacity cars used for grain and fertilizer.

Colorado Division

The Colorado Division was formed in 1976 as a new division to reduce the load on the Alliance Division, and also to manage operations and maintenance on the lines of the C&S Railway. The division had the challenge of upgrading the 230-mile C&S line south of Pueblo from the standard for a secondary line to a heavy-duty main line, including major line changes and a complete relay of the rail. Operation of an increasing number of Texas and Oklahoma–bound unit coal trains on the 114-mile joint line of the Santa Fe and Rio Grande railroads used by the C&S between Denver and Pueblo became increasingly complex as business increased. Also in those years, the C&S yards in the Denver terminal were consolidated with BN's main yard, which included switching for the Santa Fe's business by contract.

Fort Worth Division

A new division headquartered at Fort Worth was formed when the FW&D Railway was merged into BN. Major upgrading work started by the FW&D in the mid-1970s had to be continued to bring the quality of 454 miles of track between Texline (the north end of FW&D ownership) and Fort Worth up to the standard needed for a heavy tonnage railroad.

Ottumwa Division

The division superintendent provided oversight and direction on major track upgrading and embankment-stabilization projects on all of the 262 miles of double-track and 51 miles of single-track railroad on the main line across the entire state of Iowa and into Galesburg. A generally poor tie and ballast condition had to be overcome, and the work had to be done on schedule in the face of daytime Amtrak and priority-merchandise trains. The old two-hump yard at Galesburg had to be kept in full operation despite its generally poor condition and the outmoded hump-control technology that impaired productivity and the flow of traffic through the yard.

Hannibal Division

The division's forces had to rebuild 578 miles of main track on all of its main-line subdivisions (except for the 272 miles between Bushnell and Kansas City that were not designated to be a corridor for coal trains) from secondary line status to heavy-duty main-line standards. The greatest challenge was to handle the heavy tonnage that started to move on the 296-mile north-south line through Illinois between Bushnell and the Ohio River at Metropolis (except for trackage rights of sixteen miles on the Missouri Pacific Railroad) well before the upgrading was completed.

Chicago Division

The division superintendents had to carry out heavy track work on the division that was considered by most people who knew the railroad as the "hottest" of any on the system in terms of the number of priority trains scheduled and the complexity of the operation. Heavy upgrading work was necessary not only for coal, but for the growing intermodal and general merchandise business, as well as the suburban passenger service between Chicago and Aurora. All of the work had to be done under heavy traffic. The division had to restore

authorized speeds of 60 miles per hour for freight trains and 79 miles per hour for passenger trains that had been reduced to 40 on many miles of its two-main-track and double-track lines between Aurora and Galesburg and on the northern corridor line between Savanna and St. Croix near the Wisconsin-Minnesota state line.

Yellowstone Division

The division superintendent was challenged to upgrade 232 miles of mainline between Gillette and Huntley, Montana, from secondary main-line status to the standard needed for heavy unit trains. The division's 419 miles of former NP main line (Huntley–Mandan) was handling far more tonnage than anticipated in the merger plan. Even on a line that had been well-maintained leading up to the 1970 merger, heavy coal tonnage took its toll. A large program was required to install additional rail anchors and adjust the superelevation on the multitude of curves on parts of the line. Jointed rail had to be replaced with continuous welded rail, starting with the curves. Also, new lines were constructed to serve new mines at Belle Ayr, Decker, and Colstrip.

ASSISTANT SUPERINTENDENTS— ROADWAY MAINTENANCE

One of the toughest jobs on the railroad in those years was that of the assistant superintendent–Roadway Maintenance, who was responsible for the maintenance of track, bridges, and buildings and for all new track and bridge construction work (except for large bridges and buildings when the work was contracted out). This was the level where "the buck stopped" in getting the work done. These maintenance officers were held responsible for track-caused derailments, for the quality of work done by the gangs, and for on-the-job safety on gangs often staffed with large numbers of inexperienced employees, including machine operators and even the foremen of the gangs. The assistant superintendents were expected by the division superintendent and the regional and system maintenance-of-way officers to get all work done on schedule and within budget. No excuses were accepted for failures or deficiencies, since they were right at the point of action and had direct authority over all work programmed.

Region and system engineering staff

Although the assistant superintendents held the primary accountability for all roadway maintenance work, there were several system and region engineering officers whose role was very instrumental as well. At the system level, George Lamphier, director–Maintenance Planning for the entire BN system, was known throughout the railroad as the "super gandy." George deserves a place of prominence in the annals of BN history. George was highly knowledgeable, very competent, and an excellent teacher. Track-maintenance personnel across the system knew they could count on George for support and an objective critique of their work. His decisions and evaluations always were accepted by other track "experts," even those at the highest level of management.

At the region level, the position of maintenance engineer and engineer–Track served as a coach and developer of roadmasters, gang foremen, and track inspectors. They were a valuable resource in areas where a heavy work load was underway and where it was vital to improve productivity and the quality of work performed.

ASSISTANT SUPERINTENDENTS– MECHANICAL

The biggest challenge for the assistant superintendent–Mechanical on the divisions of the Denver and Chicago regions was in getting the required maintenance work done on the expanding fleet of motive power for coal service. Until the large new shop at Alliance was completed in 1978 and the new work force was hired and trained, other shops had to take on the large burden of work that could not be handled in the old roundhouse in Alliance. The heaviest of the work load fell mainly to the shops at Lincoln and Denver. Considerable unscheduled maintenance work also had to be handled "as needed" at the old roundhouse in Galesburg and the shop in Cicero known as the Clyde Diesel Shop. The capacity of the facilities at Galesburg and Cicero had been greatly reduced in the late 1960s when the former CB&Q centralized its locomotive maintenance in the new shop built in Lincoln.

Also adding to the work load of the division mechanical forces in those years was the large number of derailments. Although most derailments in those years were track-caused, mechanical forces had to be sent out from car-repair shops in the large terminals to re-rail the cars, make the repairs needed to return the cars to service, or prepare them for movement to a system car shop for heavy repairs.

Due to the backlog of older locomotive units in need of classified (heavy) repairs in those years, some major re-wiring work had to be set up in Galesburg and Cicero on a temporary basis, further adding to the work load at facilities that were no longer set up for such work. More of that kind of work became necessary as the failure rate increased on older and less reliable units. The fueling and servicing of locomotives needed for the increased number of trains being operated also put an additional burden on facilities not designed for that kind of work load. Until new and expanded fueling facilities were built at Alliance, Lincoln, Denver, Mandan, and Guernsey, the old facilities at many locations had to be called upon to support the effort to keep trains moving through intermediate points with a minimum of delay.

ASSISTANT SUPERINTENDENTS– TRANSPORTATION

These positions held responsibility for the train operation, including dispatching. The incumbents had to spend much of their time with the track-maintenance people in setting the amount of time that track gangs could take the track out of service each day. Every effort had to be made to give these gangs between eight and twelve hours of uninterrupted time. During most of the maintenance season (April through November), there were some main-line subdivisions on which all trains needed to be run at night for several weeks. However, it was not possible to shut the railroad down for that many consecutive hours on lines on which daytime priority merchandise, intermodal, and Amtrak trains were scheduled.

Since CTC was not yet in service on many lines that were being upgraded, both work trains and regular trains still had to be run under the timetable-train order system. With only about twelve hours available to dispatchers to run upwards of twenty trains on single track, it was a challenge to get all of the trains run through the areas of track work at night and by the time the track would have to be taken out of service again for work the next morning. Adding to the difficulty of getting trains moved was the extent of slow orders in effect where work was in progress. It was important to keep the lines of communication open between dispatchers, conductors, and engineers as to what was going on, to help keep their morale up during times of disruption, delays to trains, and day-to-day changes that had to be made in the train operating plan. The assistant superintendents and their subordinate officers had to be available to answer questions and listen to input and questions from employees on the requirements of the large numbers of slow orders, Form Y orders (those issued to protect the gangs while at work), "wait," "run late," and "meet" orders issued by dispatchers. Train operating employees had to be cautioned on the need to be especially alert in areas where track, bridge, and signal material had been placed along the track for work in progress, and where rough, uneven ground on walkways in yards and along tracks out on the road could not be avoided.

Having a similar magnitude and types of challenges also underway on adjacent subdivisions often made it difficult to avoid losing an unacceptable amount of running time for a given "through" train. The work had to be scheduled to avoid having the same train hit with delays several times over its entire run. Managing the movement of unit ballast trains and trains carrying continuous welded rail and other types of track material that needed to be unloaded in time to keep gangs on schedule was another challenge. Work trains often had to be run even when there was a shortage of crews and motive power to handle the revenue business, and often on weekends.

The assistant superintendents–Transportation had to be available on a 24-hour basis to help keep things going. They had to work through many points of tension with the Marketing Department, System Transportation officers, and the supervisors in charge of track work. In those years, the number of trains increased greatly on many main-line segments, making it necessary to hire and train large numbers of new brakemen and getting many new locomotive engineers trained and qualified.

The disruption to the regular flow of traffic made it difficult to keep the wage expense for train and engine crews within the limits of the budget. The same was true for the portion of car hire expense based on the time a car was in service on BN. Extra locomotive units had to be put in service at considerable expense to handle the large number of work trains being run. Operating officers were pleased that so much emphasis was being placed on improving the track and expanding the railroad's capacity, but in the short term while the work was underway, it often was difficult to provide consistent service, keep congestion from building up, and keep crews and motive power in balance.

In the early years of upgrading single main-track lines on the Denver and Chicago regions, there were few sidings in condition strong enough to handle the loaded coal trains. Having to require priority merchandise trains to take the siding and get delayed when meeting a loaded coal train of 100-ton capacity cars was unpopular with officers and employees who had been brought up to go all out to keep the "hot" trains on time. On the double-track and two-main-track territory across Iowa, Illinois, and on the line along the Mississippi River in Wisconsin, we had to take one main track out of service at several locations every day for track work. Except where a few short segments of two-main-track CTC were in service, careful planning was required in running trains "against the current" of traffic, since both tracks were signaled for operation in only one direction, i.e., with the current of traffic. For reasons of safety, train speeds had to be reduced and gangs were required to stop working while a train was passing a gang, which added to train delay and congestion.

All of these factors required a real "heads up" operation on the part of everyone involved, from the division's operating officers to the train crews, dispatchers, and foremen of the gangs. The working environment was unforgiving if any mistakes or oversights occurred. The responsibility for safety rested with all levels of management, but primarily with the division officers. A failure that resulted in an accident could have a bearing on an officer's career. This was an occupational hazard in being one of the "Sons of Martha," and part of the culture and norms on a railroad.

TERMINAL SUPERINTENDENTS

The superintendents in charge of running the large terminals built for the switching and makeup of general merchandise trains had to adjust their operations to handle unit trains as well. Tracks designated for yarding inbound trains with general freight also had to be used for an increasing number of unit trains that had to be moved off the main track and into a yard for a mechanical inspection, to be held for crew and power changes, to allow higher priority trains to pass, or during the hours a curfew for track maintenance was in effect. In some yards, tracks that had to be used to hold a coal train had not yet been upgraded to the standard needed to safely handle a complete train of 100-ton loaded cars. Any defective cars identified in inspections of the unit trains had to be repaired on the spot if possible. If that could not be done, the "bad order" cars had to be switched out of the trains and moved to maintenance facilities for repairs. This requirement took yard-switching crews away from their regular work, thereby increasing costs and possibly delaying "regular" freight moving on a schedule.

A unit train might have to occupy a yard track for two or more hours. At Kansas City, a great deal of the officers' and yardmasters' time was needed to coordinate the interchange of coal trains with the connecting lines, mainly the Missouri Pacific and Kansas City Southern. With the large amount of track work underway out on the line of road each year for several years in succession, there were times the yards had to hold several trains for the "curfews" set up for heavy maintenance work. Unit trains of ballast sometimes had to be held in a yard for one or two days for scheduled unloading. Empty ballast cars would come back into a yard "piece meal," requiring the yard to hold those cars long enough to assemble a unit train before it could be run back to a ballast loading facility. At Alliance, enough "spare" cars of each ownership and type of coal car had to be held on several tracks to fill the trains to the standard consist whenever bad-order cars were found during inspections. The yard had to be expanded to handle such work effectively and to keep train delay to a minimum.

The regular, scheduled flow of trains in and out of terminals often was disrupted by curfews and congestion out on the line of road. This could cause big surges in the flow of traffic, making it difficult to maintain service commitments and operate at the budgeted or planned number of yard switching crews. The plan for having motive power available per the service plan could be thrown off base, causing additional congestion and difficulty in moving trains out on schedule, or through the yard as fast as was needed to keep it fluid. All in all, a great deal of give and take was required on the part of the managers of terminals as well as those who supervised the line of road. Selfish, insular notions of what was best, or even fair or reasonable for a given terminal, often had to be set aside "for the greater good." Getting all of these factors considered and blended together into a plan required careful and frequent coordination among all of the divisions and terminals, the mechanical facilities involved, and the roadway maintenance forces, together with the System Transportation office in St. Paul.

THE FIRST-LINE SUPERVISORS

The first-line supervisors[2] who reported to the assistant superintendents had much the same accountabilities and pressures as their bosses. They deserve just as much credit for what they contributed in time and talent to "the cause" in those years of transforming so much of BN to a high density, "big time" operation, handling more tonnage than any other railroad had ever handled. In time, it is hoped that a book or essay devoted solely to the work done to upgrade the railroad in those years might be written, and include the names of all of those supervisors.

It is important to recognize that in the years when such priority and attention were given to the Denver and Chicago regions, there still was plenty of activity to manage on the Twin Cities, Billings, and Seattle regions of BN. Impressive growth was underway in the intermodal business, and the first double-stack container trains and solid trains of import automobiles were being run on the northern corridor. The amount of grain moving in unit trains for overseas markets was far beyond what had been forecasted even a few years earlier. Even on

---

2 Includes primarily the positions of roadmaster, B&B supervisors, signal supervisors, trainmasters, road foremen of Engines, general foremen, and mechanical supervisors. Some of those titles could vary slightly from one division or facility to another.

some lines that were not coal corridors, the amount of business being handled was in excess of what had been projected in the merger plan for 1970.

The work done by the operating and marketing people on those regions and by officers at headquarters kept the revenue and operating income high enough to maintain profitability and dividend payments in the years when so much money was being invested in the infrastructure to handle the coal efficiently and economically.

Most of the Sons of Martha generally worked on a seven-day basis, with twelve hours or more on duty on most days, depending on the demands of the job. Their off-duty time could be interrupted for derailments, employee injuries, questions about priorities, or handling anything that interfered with the flow of traffic. If things were going reasonably well, an effort was made on some divisions to allow first-line supervisors to have every other weekend off (unless a fellow officer or subordinate who would normally cover for him on such weekends was on vacation), or until some type of disruption to the operation would occur. If that happened, they would likely be called back to work. In many terminals, where the weekend days usually had the heaviest volume of business, any time off for supervisors had to be given on the lighter days, at the middle of the week.

Some of the toughest challenges came in managing the work schedule and placement of train operating employees in "irregular" or pool service who also were subject to call on a 24/7 basis. Signal maintainers and section foremen also were required to be available for calls to handle disruptions on a 24-hour basis.

A strong camaraderie developed among the officers, and they had great satisfaction in seeing success evolve from their efforts to overcome challenges and make the operation run well. As time moved on, the company got better at scheduling time off from work that field officers could count on having. Being able to plan for family activities, rest, and renewal improved to the quality of work life they had in the railroad environment.

Support, guidance and direction came down to the division superintendents from the four levels of management above the division officers. Nearly all of BN's operating and maintenance officers at the region and system level had worked up "through the chairs" at the division level. They understood and had not forgotten the need for hard work and accountability that was required of the people in the "front lines" of the railroad. However, it was at the division level that the battles had to be fought to get the work done on schedule, at the standard required, and with support for the safety of employees who performed the physical labor out on the track and in the shops, and for those who ran the trains and yards.

CHAPTER 52

SAFETY

THE FOUNDATION FOR SUCCESS

• safety: the foundation block that reveals a company's culture, quality of leadership,
professionalism, attitudes, and whether it cares about its people •
• a transportation company's standing is built largely on its
safety record and the quality of its service
• safety must be embedded across all business processes
across the company • a solid commitment to safety must run
through all levels of management and all departments of the organization •

BASIC TENETS FOR LEADERSHIP IN SAFETY
There is no single measure of greater importance to an organization than its performance in safety. None can claim it is successful, by any measure, until it is running a safe operation. There is no stronger indication of the character of a company than how well it respects and supports the needs of its people for safety. Safety is the foundation for success in all facets of a company's business. Once that foundation block is set, a company has a far better chance to become successful in its service, financial results, level of morale of its people, and reputation among its peers in the rail industry and with the general public. It is more important to its success than productivity, achieving success in the marketplace, or increasing profitability. Safety is of particular importance in railroading because the environment in which a railroad operates is unforgiving. With a mistake there may be no second chance.

Management is responsible for safety. Having an operation that is safe means that management from the top down is seen as serious about safety and committed to it as a way of life in the company. It is "how we do things here." Safety must be embedded in all business processes. If a railroad company is not performing well in safety, it tells you there are deficiencies in its leadership. When top management demonstrates it is serious about safety and cares about the well being of its people, the entire organization takes that to heart and decides to make things safer. Employee involvement in the process of enhancing safety is essential for success. Once employees get on board, they will become active and assertive and begin to take responsibility for their safety.

To get to that level, management must first accept that it is responsible for safety in the company and "stay the course." It must be seen as the number one priority and everyone must "walk the talk." The right resources must be provided, among them training on safe work practices and getting the unions to be part of the process, together with good tools, equipment, and facilities. There must be a resolve to deal affirmatively with concerns for safety and the expectations that employees have for safety in their work place. With such encouragement, employees will make their personal commitment to safety, and peer pressure will cause their co-workers to do the same.

BN'S RECORD IN SAFETY

The North American rail industry has made great strides over its history in improving safety, for both employees and in the reduction of train accidents, especially in the past twenty years. Railroad companies, large and small, have an amazing record of improvement. Since performance in safety is such a strong indicator of the quality of leadership in a company and the professionalism of its work force and of its culture, it is necessary and appropriate to examine Burlington Northern's performance in safety. Table 52.1 shows the frequency ratio for employee injuries on BN from 1970 through 1995, when BN merged with the Santa Fe Railway. The frequency ratio is the number of reportable injuries to employees per 200,000 employee-hours worked.

Under FRA standards, an injury is reportable if the employee loses time from work and /or the employee receives medical treatment for the injury. The ratio is equivalent to the number of employees out of one hundred who were injured during the year. It should be noted that some changes have been made over the years in the interpretation or definition of exactly what constitutes reportability. However, none of these changes would be considered great enough to cause any significant difference in reportability, or in the calculation of the frequency ratio for the years shown.

As a benchmark for evaluating BN's safety performance over its history, it should be noted that the frequency ratio on today's Class I railroads (including BNSF) has been in the range of 1.0 to 2.0 for the past several years. The numbers clearly show that safety has improved greatly on all railroad companies. It is obvious from the numbers in Table 52.1 that BN got off to a rough start, with ratios in excess of 12 in its first two years. In its third year (1972), the ratio was cut in half, indicating the company took some strong steps in setting safety as its top priority.

As safety improved on BN in the those early years, so did its service, its financial performance, and the general quality of its operation. That amount of improvement still was not enough to put BN at the level of the best railroads in safety for those years, but at least it was a respectable showing. Two years later, in 1975, safety performance deteriorated, and by 1976 the frequency ratio of 12.86 set BN back to the level it had been at shortly

Table 52.1.
FREQUENCY RATIO*
FOR EMPLOYEE INJURIES

Year	Ratio
1970	12.53
1971	12.17
1972	6.46
1973	6.81
1974	8.58
1975	9.84
1976	12.86
1977	13.10
1978	13.90
1979	12.54
1980	8.51
1981	4.66
1982	3.97
1983	3.53
1984	3.79
1985	5.90
1986	6.13
1987	8.45
1988	8.44
1989	8.96
1990	11.68
1991	13.51
1992	12.31
1993	6.98
1994	3.85

* Defined as the number of reportable injuries per 200,000 employee-hours worked.

after the merger of 1970. It remained at that level for four years, through 1979.

It is interesting to reflect on the environment in which BN's people were expected to operate and maintain the railroad from the time of the 1970 merger and into the early 1980s. First, we had the challenge of integrating the operations of three large, far-flung railroad companies plus the smaller SP&S. Overall, it went well, but it was certainly not without some tension, uncertainty, and reluctance on the part of some managers and employees, particularly at some of the large common points. Some employees were adversely affected due to issues of seniority, changes in working conditions, or having to relocate. These factors were major distractions that could have had an effect on employees' ability to concentrate and work safely.

The main factor that caused the deterioration in safety starting in 1975 was the number of injuries to employees hired in large numbers within a very short time to work on track gangs in Wyoming, Nebraska, Iowa, Illinois, and Missouri. Having hired hundreds of new people for these gangs, together with having many new, inexperienced gang foremen and other first-line supervisors, was the basis of the problem. In those years, the newly hired trackworkers were given only a little training before they started work. Doing basic trackwork with hand tools and handling heavy pieces of track material resulted in many injuries. As mentioned at the outset of this chapter, work on the railroad was unforgiving, and even a small mistake could result in an injury. Also, because the work was heavy and dirty, in all kinds of weather, and often in rough terrain (causing bad footing), there was a high turnover on these gangs. New people were constantly being brought into these gangs to replace new employees who would leave after working only a short time. Many new employees found they did not like having to do hard physical labor outside in all kinds of weather, or that we expected new employees to quickly learn the work and become productive. A number of those hired were satisfied with the earnings of only two or three days' work per week and did not want to work for the balance of the week. This outlook made them unreliable, and they had to be terminated. The turnover rate required us to keep bringing more new hires into work on the gangs.

In addition to hiring large numbers for the track gangs in those years, we hired many people to work as brakemen and switchmen. After only two or three days of orientation to the work, they made their first pay trip. The conductor of the train they were called for may have had only a few months of service, and often did not have the confidence or experience needed to train new brakemen and switchmen on the job. Because such a large number of new brakemen and switchmen often were being hired within a span of only a few weeks, it was difficult for trainmasters, safety supervisors, and trainers to follow through with all of the new hires, as to their progress as they made their first pay trips. Minor injuries to new employees from slips, trips, and falls, and in such tasks as applying hand brakes and lining switches had a negative impact on the safety of the Alliance and Nebraska divisions in particular. Also, the ratio of first-line operations supervisors to employees in train and engine service was far too low at many locations, due to the rapid growth in employment and failure to authorize additional supervisory positions soon enough.

Many of the new employees hired for train service went to work in territories in which trains still operated under the system of timetable and train order authority. With the high density of traffic on those lines, having to dispatch and operate trains without the benefit of CTC or even an automatic block-signal system was a challenge even for experienced employees.

By 1980 much of the heavy work in upgrading track and in building track to expand capacity had been completed. Employees not fit for railroad work had been weeded out. Much better training programs had been developed, for both new hires and existing employees. The newer first-line supervisors had gained experience, making them more capable in carrying out their responsibilities as leaders, coaches and knowledgeable railroad officers. BN's overall performance in safety began to improve.

It should be noted that the poor performance in employee safety in the mid- to late 1970s was on the six divisions of the Denver and Chicago regions. On the other nine divisions of the BN system, safety did not deteriorate. Frequency ratios continue to run in the range of two to six on those divisions in each year.

CHAPTER 53

EMPLOYEES' STATEMENTS ON SAFETY

*I've seen it go from a local line in serious disrepair to an ultra-modern
state-of-the-art segment—the busiest stretch of railroad in the world.*

BILL CLARK, TRAINMASTER/ROAD FOREMAN
QUOTED IN *BN NEWS*, SEPTEMBER–OCTOBER 1986

THIS CHAPTER CONTAINS REFLECTIONS OF BN people who were employed on the divisions that were charged with gearing up to handle the enormous work load in upgrading the railroad and moving a heavy volume of traffic while that work was underway. It is interesting and rewarding to hear from those who made their careers on BN reflect on "where we've been and where we are going."

Many employees have taken the opportunity to evaluate the career choices they made when they hired out in the 1970s, the skills they developed as railroaders, and the pride and satisfaction they developed in their profession, their company, and the relationships they built. They see that a good foundation was built for those who follow them: a safety culture and a company with a good future. Overall, they reveal how they made BN a better place to work than it was when they started their railroad careers.

Especially on safety, it is rewarding to hear about the excellent performance of today's BNSF Railway. A railroad's safety performance is the strongest and most important indicator of how well its people have done and what kind of shape the railroad is in today, at the time they retired after some twenty-five to forty years of service. Some divisions and work units are achieving frequency ratios of less than one, which no one who

worked in the 1970s thought would even be possible. It means that having a goal of zero injuries for large groups of railroad workers over a sustained period is not unrealistic. It has been proved that achieving safety and having high rates of production are not in conflict. When large numbers of new people were being hired for the massive track- and signal-upgrading programs, and many new brakemen were being hired, there was no model or system available for training or work observation on a large scale that would instill safe work methods, correct bad work habits, or otherwise assist employees in taking care of themselves and their fellow workers out on the job. Instead, training was done out on the job through osmosis and informal interaction with what few experienced employees were working at many places in those years.

It was not uncommon to have frequency ratios in excess of ten on the divisions with the heaviest work load. This meant that over 10 percent of the employees sustained a reportable injury within a year, a level that had to be reduced. By the late 1970s, managers and employees at the division and shop level were coming up with more structured and organized ways to use the skills and job knowledge of what was often few experienced employees in a work unit to help turn the situation around. At the same time, managers at all levels came to realize they needed training to better manage safety as a work process, and to develop the leadership skills needed to effectively build a safety culture—a way of life—as our top priority. The management of basic railroad work had to be more than "get 'er done" and having to tolerate injuries in the process of trying to maximize production.

For the most part, management at the senior level had not experienced a safety challenge of that depth and breadth in their formative years, and hence, had to be convinced that safety would not come about only by taking tough disciplinary action or simply ordering managers on the front line "to do better." The support of top management for training and providing the resources needed to correct potentially hazardous workplace conditions and better trained people on how to work safely in the railroad environment was vital. It was necessary that managers at all levels better demonstrate their personal commitment to safety with consistency and determination. We had to learn how to win the hearts and minds of the people.

Those kinds of attitudes toward safety were developed over several years into an enduring commitment throughout the company to the well-being of employees and managers. The results have been outstanding. Several employees and managers who were hired in the early days of the coal boom and are still working and have shared their thoughts and lessons learned as they reflect on the vast improvements made in safety over the years as they were about to begin their final trip or work their last shift:

> I am thankful to have had such a good career that provided good benefits and retirement. Think safety every day so you can return home to your family. Do it by the book. The rules are implemented for your safety. —*Gary Hudson, locomotive engineer, who retired with nearly forty years of injury-free service*

> Don't assume anything. It will get you in trouble. Always be willing to learn and apply what you've learned. Treat people how you want to be treated. —*Wayne Green, locomotive engineer*

> I always told them [while serving as an instructor for conductors] to keep their head on a swivel. If you look and listen, you will work safely. It has been a good life. It's hard to see it coming to an end. —*Willie Marks, conductor*

> The company makes it as safe as they can for us. Just stick with it, and you can have a great career. I loved telling people what I did and having them see the finished project. I'm proud to say we had no injuries during my time as roadmaster. —*James Mashek, surfacing gang foreman, and former roadmaster*

> Do not take shortcuts. Listen to what your co-workers have to say, especially those with experience. That is the best way to learn. —*Jim McElwain, locomotive engineer*

> There is no room for error in the type of work we do, but it can be done with proper training and equipment. The main thing is being committed to working safely. Our goal is to provide the safest, most efficient, professional electrical crews out there. —*Jim Hartman, electrical supervisor*

> The No. 1 battle out here is fighting complacency. We must always keep that in the back of our minds. Communication is key to fighting complacency. —*Blane Morse, conductor, as crews on the Butte Subdivision (territory that includes the operation over Crawford Hill) reached two years of injury-free service*

> Our safety accomplishments are a testament to Alex Adam, Lyle Horton, and Ken Willey [operations supervisors]. They are always there to help, guide and inform employees, whether in the field or otherwise. They take the time to know each employee. This is a source of pride. Safety becomes a culture. New employees follow the lead of experienced employees. —*Will Hall, locomotive engineer (another employee working on the Butte Subdivision)*

> The trick is to learn the most basic way to get things done from the people you work with. When I look back, the people who did things the simplest way were often the people who were never injured. That's because they did what they were supposed to do well. —*Henry Benjamin, conductor*

> I am proud of my day-in and day-out dedication to safety. This makes me realize that I do make a difference to the safe movement of trains. —*Fenton Johnson, tamper operator*

> We want people who want to be here, not those that think they have to be. Maintain a positive, wanna-do attitude and a safety-conscious mentality. Things will always

come around. It is never as bad as it seems. You make your environment what it is. —*Brian Chatten, division engineer*

The people and camaraderie you find on the railroad can't be found anyplace else. We take safety seriously and therefore truly watch out for one another like a brother's keeper. —*Ray Bennett, water service foreman (In an interview, Bennett advised new employees to pay attention to and take advantage of the experienced employees' wisdom and knowledge to avoid making the same mistakes they did.)*

There is no such thing as a small accident on the railroad. —*Doug Sogn, conductor*[1]

Statements such as those listed above are representative of the attitudes that have led to the vast change in employee safety performance that developed from the mid-1970s and to the present day. A clear indication of how far we have come can be seen in Roger Nelson's statements about the struggles that were faced in the early days to improve the safety and overall work performance in all departments of the division and region organizations. Roger served as a line-operating officer and progressed to the level of vice president of Safety in the 1980s. Roger gave an evaluation of the prevailing safety culture of the early days: "I think the prevailing attitude amongst both the management (at all levels, not on just the divisions) and the employees was just to 'keep from drowning' and 'get 'er done' without regard to good safety practices and processes. Production at any cost."

To bring about a quantum improvement in safety, Roger recalled how BN appointed him to lead a system task force in the early 1990s to determine the best practices the divisions should employ to improve safety. As part of that process a team of safety professionals from the Du Pont Corporation was brought in to apply the systems of leadership and management in safety that had been applied successfully on other railroads and in many other industries. Together with the leadership of retired Air Force general John Chain, a profound improvement in safety came about in the next two years.

The preceding statements are strong and meaningful. Clearly, they are from the heart and represent the depth and strength these managers and employees developed

as railroad professionals in the span of twenty-five to forty years. They are from the large groups of young, spirited, adventurous people who came to BN's version of the Wild West to get a job, learn railroad work, and become an important part of a challenge that no other railroad had faced since the transcontinental lines were built in the 1880s.

It is amazing how well things turned out on that part of the railroad, when we think back on the times when we were hiring people off the street, with limited opportunity or capability to screen or evaluate their potential and willingness to contribute positively to the efforts of the day. Turnover was high, and there were some whose attitudes and job performance was so low that they had to be worked out of the company through progressive discipline. There were some who came to the realization that their views on what a career should be and what the railroad company stood for were so incompatible, that they had to agree with us that it was time for them to move on. It is heartening these days to read and hear favorable comments from so many who stuck it out in the early days and became solid "railroad citizens," as role models, mentors, and participants in activities set up to improve safety and the quality of work done in the operation and maintenance functions.

From a personal standpoint, a number of thoughts come to mind as to what many of us went through in those early days and how fortunate we were to be in at the right place at the right time to have those experiences. Part of the effort toward improving morale and building the base needed to improve safety was to get the point across to employees that the company's goals and strategies would benefit of all of its constituencies—owners, investors, customers, employees, and the parts of the general community we touched in one way or another.

We were trying to show concern for the well-being of our people by bringing in large machinery to help clean up the property by picking up years of accumulation of scrap material that was unsightly and, at times, formed tripping hazards. We went all out to clean up the trash and spillage in yards and facilities that endangered safety or at least made it an unpleasant place to work. We rebuilt yard tracks that were in poor condition and improved walkways between and along tracks. We remodeled and upgraded lunch and locker rooms, painted buildings,

---

1   The above comments from employees and managers are contained in various issues of *Powder River Reflections,* a newsletter issued to employees and their families.

cleared brush, cut the weeds, and applied chemicals to suppress dust on roads going through or adjacent to working areas in yards. We rebuilt hundreds of grade crossings that were in bad shape and a legitimate source of complaint. Training programs were being set up for both new and experienced employees in all specialties. As a result of such efforts, we made the railroad look like we were in business, we had a future—we were not running a junk yard. Most importantly, we were making the railroad a safer place to work.

Having a much better-looking (and safer) property, together with the hundreds of new locomotives we were putting in service and the thousands of miles of track we upgraded, was helping employees establish pride in their company. They saw we had the authority, the ability, and the resolve to correct the backlog of conditions of concern to employees, and that BN was committed to building a strong railroad for the future. This really hit home in the lean years in the 1970s when dividends and executive pay were reduced but yet, the management did not defer or cancel upgrading projects in order to raise or maintain BN's net income in the short term. Employees also learned that the senior management in place in those years consisted of people who knew what railroading was about, and that they spent time out on the railroad to see first hand what was going on. By establishing confidence in these and many other ways, more and more employees were getting on board with the program and the number of "attitude-caused" injuries began to decline.

By 1980 the number of train and switching accidents caused by defective track or equipment also showed a downward trend. The safety record of the Chicago and Denver regions was no longer dragging down the record of safety for the entire railroad as it had in earlier years. This was the beginning of a period of remarkable year-to-year improvement on all measures of performance in safety, which continues to this day.

CHAPTER 54

THE COAL BOOM'S EFFECT ON COMMUNITIES

· a return of the "Wild West" · a time of opportunity
mixed with tension for BN's new hires and their families
· a social challenge for both the new hires BN brought in and the non-rail residents ·

DURING THE 1970S, HUNDREDS OF PEOPLE moved into the Alliance–Edgemont–Gillette areas to take advantage of new opportunities to work for BN and the mining companies. Most of BN's new employees were hired for train service and to work on the track maintenance and construction gangs. In addition, about six hundred people were hired at Alliance to work as machinists, electricians, and laborers in the large shop built for maintaining the growing fleet of locomotives, and for running repairs on the cars that made up the unit coal trains.

The population of Alliance grew from 6,862 in 1970 to 9,920 in 1980. BN's employment in Alliance rose from less than 200 in 1970 to 900 by the fall of 1977, and to about 2,200 by the mid-1980s. Between 1970 and 1979, 1,169 new homes were built in Alliance. Under Norman Lorentzsen's direction, BN facilitated the construction of fifty new houses that could be acquired for less than $50,000. These houses sold as fast as they were built.

In the Gillette area, the mining companies and their contractors were hiring literally thousands of people in those years to construct production and loading facilities at the new mines. As the mines opened, BN brought people into Gillette to operate trains to and from the mines. Several large track gangs were set up to replace ties and rail, upgrade the ballast section, and construct new or extended sidings and many miles of second main track. Contractors brought people in for construction of the new 127-mile line between Donkey Creek and Orin and tracks to for serving the mines. Gillette grew from a population of only about 8,000 in the late 1960s to 20,000 by 1980, and 35,000 in 1985.

This rapid growth in population put a real strain on such community services as schools, streets, health care, law enforcement, and fire protection. Some schools were forced to hold double sessions until they could be expanded. Local residents had to pay higher property taxes and assessments to fund the construction and staffing of additional facilities for community services. In the absence of zoning regulations and with a general lack of preparation for such a boom, "man-camps," barracks, mobile homes, and other forms of temporary lodging were hastily put together. Newly arriving workers who could not find even such temporary housing sometimes used motel rooms in eight-hour shifts, pitched tents in public areas, or slept in their cars, pickups, and campers for a few nights until they could find somewhere more suitable "to hang their hat."

In addition to lodging, there were some problems of anti-social behavior in such a freewheeling atmosphere. With a lack of "normal" entertainment and recreational opportunity, community-provided social services could not keep up with expanding problems of having new arrivals from diverse backgrounds and behavioral patterns suddenly thrown together in a fairly small community.

This was the tough environment into which BN brought the workers it needed in Gillette, Edgemont, Alliance, or smaller towns within a reasonable driving distance of their new jobs. It was a challenge for newly hired or transferred employees to have to move in and go to work under these conditions, but even more challenging for them to obtain housing suitable for their families. Needless to say, there were BN people who had some difficulty in adjusting to the conditions that came with their new jobs. Living in a boomtown culture was far different from life in an established suburb in a large metropolitan area, or a quiet small town elsewhere on the BN system. Suddenly, they had to move in and begin to live among people with different kinds of backgrounds, some of whom could be typed as "rough necks" or "rowdies." School-age children in particular often had a tough adjustment to make.

In Wyoming, the State used severance taxes paid by the mining companies for assistance to local communities such as Gillette where the pressure was so great. The general problems faced by the community service agencies are described very well in *A New History of Wyoming* by Phil Roberts:

> Older residents, living off of Social Security or other fixed income, watched property prices spiral upward and, with the rising assessments, the tax bills on their already-paid-for homes. Small business operators, accustomed to hiring employees at the minimum wage suddenly faced competition for laborers from the much better paying mines.... Some [teachers] left education altogether and took jobs in mining or became land developers. High schools reported increasing incidents of students dropping out before graduation in order to take jobs paying $40 to $50 per hour in the mines.... Town councils were overwhelmed with demands for water and sewer taps for housing developments.... The State helped with funds for law enforcement and provided an increasingly greater proportion of public assistance, once the province of the cities and counties.... Law enforcement officers, accustomed to dealing with traffic matters or an occasional shop-lifting complaint, were besieged with calls in involving assaults and more serious felonies.... A University of Wyoming sociologist ... referred to the social impact of boom-town living as leading to the "Gillette syndrome" [defined as shorthand for the dark side of energy development] ... the toll on humans having to

live in isolated towns with few civic amenities, in overcrowded substandard housing, and with scant options for entertainment beyond gambling, drinking and vice.[1]

A paper entitled "Social Consequences of Boom Growth in Wyoming," written by Eldean V. Kohrs in 1974, describes how the "Old West" of a hundred years ago was revived in Wyoming in those years. He characterizes the concerns, tensions, and inadequacies of "old timers" that had to be dealt with even in modern times. Classrooms were set up in trailers similar to the housing in which some of the children lived, although the city of Gillette responded quickly by constructing new schools that soon became highly rated for both academics and athletics. Kohrs found that the new entrants into life in the Gillette area "forsook the pursuit of higher needs in order to cope with the [more basic] needs for adequate water, sanitation and social survival."[2]

It was interesting to learn that three doctors from the New York and New Jersey area decided to move to Gillette in 1975 to help meet the need for health care for the growing population. One of the doctors was a general practitioner, one a general surgeon, and her husband, an anesthesiologist. They had to set up their practice in a double-wide trailer while awaiting completion of a new clinic building. In the announcement about the coming of these three doctors, the hospital administrator also advised that three more doctors would soon be coming to Gillette. He appealed to inactive nurses living in the area to come back to work. If these nurses would work just one day each week, "it would sure help out," he added.[3] An example of the difficulty in holding onto professional people was the decision of some teachers to take high-paying jobs at the mines or with the railroad. The city had problems in attracting replacements due to the shortage and high cost of housing.

To their credit, most of BN's new employees and their families stuck it out. Employees learned their work, and spouses and children adapted to a new life. The quality and professionalism of the railroad operation

---

1　　Phil Roberts, *A New History of Wyoming*, chapter 19.

2　　Eldean Kohrs, "Social Consequences of Boom Growth in Wyoming," paper presented at the Rocky Mountain Association of the Advancement of Science Meeting, April 24–26, 1974, Laramie, Wyoming.

3　　*Gillette* (Wyoming) *News-Record*.

improved, and in time, the territory of the Alliance Division achieved first place in safety. Gillette has evolved into a respectable city of about 40,000 residents, with good shopping and business areas, new and expanded schools, a modernized infrastructure, new health-care facilities, and good opportunities for recreation.

In a few years, things settled down in the communities that had experienced the turmoil and pains of rapid growth. New and better housing became available, and some housing was freed up as construction projects were completed at the mines and on the railroad. Large numbers of workers holding temporary jobs then moved on. Many railroad supervisors who demonstrated capability to manage the "big time" operation in the Powder River Basin were promoted and transferred to jobs in other parts of the BN system. That in turn created vacancies and new opportunities for other aspiring managers to move in and have a part in managing the unique operating and maintenance challenges BN had on its Alliance Division.

CHAPTER 55

BECOMING A
GREAT PLACE TO WORK

*• for those who recognized and seized the opportunity to help
upgrade and expand BN's capacity to handle unprecedented growth
in what a railroad does best—to move heavy tonnages on a high-density railroad •
for railroaders who'd maintained their faith that the rail industry would have a
good future if it could be deregulated. • for many railroaders who had
lost their jobs on other railroads due to bankruptcies, restructuring,
or their employer's inability to compete in the
changing market for rail services •*

BN BECAME A GREAT PLACE TO WORK FOR many people in the 1970s when its leaders decided to accept the challenge to expand its capacity to handle the large tonnage of coal that would be mined in the Powder River Basin. In addition to expanding the capacity of a large part of the railway, it would be necessary to upgrade the quality of the track over hundreds of miles in order to reach the many electric-power plants that would have to convert to low-sulphur coal as their source of energy. Since there were many power plants on the lines of connecting railroads in addition to those on the lines we owned, we had to work closely with several other railroad companies and the plants they served to ensure the coal could be moved efficiently and at a price satisfactory to both the railways and their customers.

The rapid growth in the coal business provided unprecedented opportunity for careers in the rail industry for thousands of people. It was in stark contrast to what had happened a few years earlier to the people of the Penn Central when it went bankrupt and lost large amounts of business due to deterioration of its roadway and equipment, and its ability to provide reliable service. Not long after this debacle, the Rock Island and Milwaukee had to declare bankruptcy. Basic maintenance was being deferred on a number of other midwestern and eastern railroads as well. In contrast, BN was growing and was willing and able to invest the large amount needed to carry out the projects we determined necessary to meet its challenges and opportunities.

BN's managers had to develop new sets of skills to manage business growth and expansion of capacity in contrast to years of cost-cutting and reducing the asset base, in an effort to maintain a satisfactory level of earnings. Employees had to be trained in the operation of new types of track-maintenance machinery and in methods for maintaining track to the standard of quality needed. Most of the newly promoted foremen of track gangs had to be trained on new standards and work methods to ensure the work was completed at the standard of quality needed to handle the ever-increasing tonnage of coal being handled. They also had to be trained in supervisory skills. The gang foremen and first-line supervisors (primarily roadmasters) of roadway maintenance were expected to improve their supervisory skills so they could better manage large numbers of inexperienced workers. Because the work of maintaining and building track required hard physical labor, long hours in all kinds of weather, and being away from home nearly every night for weeks at a time, we had a high level of turnover on some gangs. Getting the work done on schedule and at the standards necessary for safety of train operations, and preventing injuries to gang members was a big challenge in those years.

In those years, we were fortunate to have senior executives who were experienced and knowledgeable on railroad work at the basic level. They understood and supported the efforts of supervisors in the field to get the work done. The executives demanded the work be done up to standard and never stood in the way of acquiring the best available machines, tools, and material. These senior officers took time to get out on the railroad to see the work in progress and to evaluate the quality of work completed. They showed respect for the junior level supervisors and division officers who were held responsible for safety, quality, and getting the work done on schedule and within budget.

The experts in engineering and maintenance who were sent out from the system and region offices did not come out to simply find fault and write negative reports to senior executives. Instead, they came out to assist in getting the work done, to help in planning the work coming up ahead of us, and to lend their expertise in every possible way. Because the involvement of these experts was so vital in the success we had in getting the work done in those early years, it is important they be named, even at the risk of unintentionally leaving someone out. At the system level, the expertise and methods used by Bruce Anderson, Dick Brohaugh, and George Lamphier in the process of oversight and training were exemplary. On the Denver Region, Don Rogers and Ingvold Lyngby provided much valuable assistance, as did Don Merrill, Roy Brawner, and Rex Duryea on the Chicago Region.

Having the division-type organization structure in place in those years, rather than the more common departmental organization structure, was very helpful in the effort to get the work done in a coordinated, team-based way. Under the division concept, the division superintendent had responsibility for all three of the functions of the Operating Department—roadway maintenance, transportation, and mechanical (the maintenance of cars and locomotives). Having these responsibilities under one officer at the division level was vital in establishing a team effort for getting the roadway projects done, coordinating the train and yard operations with those projects, and helping ensure the Mechanical Department could complete the maintenance and servicing of locomotives at the time that trains had to be run through the areas where trackwork was underway, and to have locomotives and ballast cars ready for work train service when scheduled.

Having agreement and understanding at the top level of the Operating Department and the executive level that responsibility for all three functions would remain decentralized at the division level was very helpful. If the departmental type of organization had been in place, we would not have been as successful in getting the work done. There would have been at least a tendency for walls, chimneys, or stovepipes to be built up between departments, and less communication among the experts in each function; more direction would have had to come "top down," and it would have been necessary to sort out priorities and tactical at the region or system level rather than at the level where the work was being done.

For everyone, the work was hard and demanding but very rewarding. Good performance by gangs and particular individuals was recognized by the "revered" experts in roadway maintenance who were named earlier in this chapter. Quality work was not just taken for granted. The best foremen and junior-level supervisors were given opportunities for rapid promotion and additional training. Everyone had the opportunity to learn, advance, and develop the skills they needed to meet the standards for their work. High standards of personal and management conduct were established and adhered to at all levels in the company. By their conduct and the attitudes they displayed toward the work of the railroad, senior level executives were seen to "walk the talk."

To a great extent, BN moved away from the historic militaristic culture that had long been in place in much of the rail industry. Management by fear, threats and intimidation was not an acceptable way for BN people to give directions or to motivate people. Instead, the people of the railroad were shown respect as individuals and as a team, as well as for the work they carried out. Corrective action was applied in a constructive, affirmative way. However, there was no tolerance of unethical conduct or unsafe, slipshod work. As a result, a great team was built. Many junior-level supervisors developed skills and achieved results they had no idea they even had in them, much less that they would ever develop. They learned how to get results. They moved onto good careers and achieved self-satisfaction in a life that at times was hard, but good.

In those years, BN was a company relatively free of corporate gamesmanship. It was not dominated by internal politics. There was only one team, and everyone was on the team. All of the work we were doing was seen as a BN effort, for the benefit of the electric-power industry, BN's shareholders, and its employees. We were united toward achieving the objective of getting the work done. It is amazing that so many BN workers found success in an industry that many had written off as nearly dead, no longer needed, and not a place for people to invest their talents and education. Instead, those who came into BN had jobs in which they would have pride, challenge, and opportunity. It was a place of work they could be proud to show family members, former classmates, and neighbors with pride.

There was a general feeling of relief that we had the resources to overcome shortfalls of the past that had resulted in some deterioration of the track and maintenance facilities on some parts of the railroad. We were pleased to have a management that insisted the property be cleaned up as work progressed. That included basic "housekeeping," meaning that scrap track, bridge, and signal material was picked up, that material kept on hand for maintenance was properly organized and stacked, that safe walkways were maintained in yards, buildings were painted, weeds and brush were cut, drainage was restored in yards and out on the road, and a host of other basic tasks. By getting these tasks done, we knew we were running an up-and-coming business, not a junk yard, or working for a company about to go out of business. The property of the railroad became a credit to the communities it passed through. Employees could take pride in the company for not only having a job, but also because they were working for a progressive organization.

A number of good programs for employee development were established in BN's first ten years. One was named "This Way Up," for employees to become qualified for promotion to management positions. This program

provided a source of talent in addition to the Corporate Management Training Program for which recent college graduates were recruited each year. It gave employees from the ranks the opportunity to advance and to obtain formal training and education. The Employee Assistance Program provided counseling and remedial programs for cases of alcohol and drug abuse, and for employees in need of assistance in working through financial and family problems.

By 1980 the heaviest part of the work of upgrading the railroad and expanding its capacity was completed. The challenge had been met. There still were some important projects to be completed and, of course, there was the need to continue giving proper and fully adequate attention to maintaining the roadway, equipment, and train-control systems to a high standard. We could not rest on our laurels. A high level of competency, precision, alertness, and avoidance of complacency would be needed. The railroad would not take care of itself. A strong budget for routine maintenance would be needed over the long term.

We were at the threshold of being able to run the railroad at a much lower operating ratio and with a reduced level of capital investment. In the next few years, we would be seeing the benefits from the large capital investments we had made and the increased amount of operating expense put into upgrading the track. The financial community would find that, indeed, the operating ratio would go down, thereby increasing operating income, and the anticipated return on invested capital would be realized.

It was at this time that the BN Board decided to hire a new CEO who they thought would have the ability to substantially increase the earnings from BN's Resource Division. The board also was anxious to improve the financial performance of the railroad, following many years of heavy capital investment and spending far more for maintenance of roadway and equipment than other large rail systems at that time. In hopes of better managing both the railroad and the resources, the board hired Richard Bressler, an executive with the Atlantic Richfield Company whose background was in finance and planning. Also at this time, BN and the Frisco Railroad had made application to the ICC for authority to merge. In advance of receiving authority to consummate this merger, the board hired the Frisco's president and CEO, Richard Grayson, who had been planning to retire at the time. Grayson agreed to stay on to head the railroad, oversee implementation of the merger, and provide the kind of leadership needed to improve the overall performance of the railroad.

Soon after coming to BN, Bressler proclaimed that our earnings on the massive investment made to move the coal were unsatisfactory. I recall the day on which the C&NW finally paid us the $76.3 million for its 50 percent ownership of the joint line between Gillette and Orin, and a conversation involving a few senior executives in which some believed we were far better off with the $76.3 million in hand than if we had been able to retain all of the coal business for ourselves. Try telling any of this to the people of today's BNSF, or any of the BN people "who were there" in the 1970s and early 1980s when we built up the railroad through countless challenges and showed the world that BN was up to the task of moving the coal. We believed in and carried out the plan that would get the job done.

It was a "frontier type" experience for hundreds of people who moved out to Alliance, Gillette, Guernsey, Lincoln, and many small hamlets in between those places to start a new life. They moved in from other railroads and many locations throughout the United States. The overall "coal project" provided jobs for the hundreds who otherwise might never have had such a good job (or have yet today).

We built and maintained a network of rail lines to a standard and quality of work as good or better than that of any freight railway in the world. We acquired and maintained excellent equipment and a well-trained, committed group of people who get the job done, day in and day out, in the field and in the Network Operations Center at headquarters. We now have a strong partnership with the electric-power companies and with the operators of the mines. We are moving an unprecedented amount of tonnage, with hundreds of trains of nearly 20,000 gross tons moving over a high percentage of the rail lines in North America. Those who got the job done in the early days of this endeavor and who keep it running today should have a great deal of pride and satisfaction on what they accomplished as professional railway people.

It is good to see that BNSF continues to invest capital in the coal business so it can be run even more efficiently. A recent example is the extension of a second main track through Grand Island, Nebraska, up to the viaduct that crosses the Union Pacific's main line. Another recently completed project that added to the capacity of the railroad and improve operations is the construction of a second bridge over the Missouri River at Plattsmouth, Nebraska. From 2014 to 2016, several miles of second main track and additional sidings were built in North Dakota to add the capacity needed to handle the rapid increase in crude oil shipments. Also in 2014, BNSF announced plans to build a second bridge of over 4,000 feet over Lake Pend Oreille to add capacity on the corridor between Spokane and Sandpoint. Construction of a second main track of 3.6 miles between Graf and the east end of the yard at North La Crosse has been completed. This addition will provide relief to a bottleneck of long standing.

PART VIII

IT'S BEEN
A GREAT ADVENTURE

IT'S BEEN A GREAT ADVENTURE

*Our policy in these 10 years has been to build and develop our assets to improve
Burlington Northern's growth in the future and to enhance
the value of the owner's investment.*

…

*In 10 years, Burlington Northern's consolidated
net income has grown at an average annual rate of 22 percent.*

LETTER TO SHAREHOLDERS, *BURLINGTON NORTHERN ANNUAL REPORT FOR 1979*

*Most men who have really lived have had in some shape their great adventure.
This railway is mine.*

JAMES J. HILL'S LETTER TO THE STOCKHOLDERS
ON RETIRING FROM THE CHAIRMANSHIP OF THE BOARD OF DIRECTORS, JULY 1, 1912

THROUGH A COMBINATION OF GOOD PLANning, leadership, and execution of its strategies, BN developed into a strong business enterprise. The new company was blessed with a strong franchise and a cadre of dedicated, highly committed people in all of the departments and technical specialties that make up a railroad organization. All in all, it was a great adventure for thousands of people in a company that fulfilled the vision James J. Hill had to create a consolidated rail system that would be stronger and far more efficient than any of the four component railroads would be if they remained independent.

BN was blessed with talented people who were able to formulate the plans and strategies needed to take advantage of new opportunities that came its way, among them the electric-power industry's need for vast tonnages of low-sulfur coal, gaining authority to merge the four "Hill Roads," and the desire investors demonstrated by assisting the company in financing $2 billion of projects needed to expand BN's capacity to provide efficient transportation service for new and developing business opportunities.

The unprecedented and unexpected opportunity to move hundreds of millions of tons of low-sulfur coal over a large part of the company's network transformed BN from the rail industry's years of retrenchment into a fast track of growth that strengthened its franchise for the long term. The investments made through the 1970s began to pay off by 1980 with increased earnings. A single completed project would not always produce great benefit by itself, but as more and more projects to expand capacity and upgrade the property were completed, the synergy or linkage among them made the operation more fluid and efficient. Very importantly, the revenue projections were being fulfilled, although there still was disappointment in the level of rates we could charge on several coal movements.

During these years, the investments BN made in its non-rail transportation and resource-based companies began to pay off as well. BNAFI, the air-freight forwarder subsidiary, grew from a start-up company to become the nation's second-largest domestic forwarder, and one of the top five in international business. Plum Creek Lumber Company advanced to a major player in forest-products manufacturing and expanded its land holdings to become the second-largest private timberland owner in the United States. Exploration and production levels were greatly expanded in the company's oil and gas holdings, ranking it just behind the six major worldwide oil companies.

The landmarks or mileposts that set the course or had the greatest influence on BN's progress in its first ten to fifteen years were the following:

- Approval by the ICC and the U.S. Supreme Court of the application to merge the two northern lines and the Burlington Railroad, and to lease the SP&S Railway.
- Having the strength and ability to survive in the early to mid-1970s in the face of the failure of two neighboring railroads, the Milwaukee Road and Rock Island, and while a number of other railroads struggled to survive as independent, privately owned companies.
- Creation of quasi-governmental units to operate intercity passenger trains and commuter service to relieve railroads of operating losses, and having to make the capital expenditures necessary to continue providing passenger service. At last, it was recognized that rail passenger service was a social obligation and that it was no longer justified to require a privately owned company to provide it without financial assistance.
- The requirements of the Clean Air Act of 1970 for the electric-power industry to reduce emissions, which led to demand for the transportation of heavy tonnages of coal over a large part of BN's network.
- Investment of $2 billion to expand BN's capacity and upgrade the track, bridges, and train-control system on a substantial part of its network.
- The vast improvement in the quality, strength, and capacity of the network that built a strong foundation needed to promote growth in several lines of business, in addition to coal.

- Building and strengthening the company's culture in its first ten years by requiring consistency in adherence to the standards of performance, ethics values, goals, and mission it had established.
- Provided job opportunities for thousands of new hires, with some of the highest levels of wages and benefits offered anywhere in American industry.
- Developed a reputation for BN as "a great place to work" for those whose attitudes and values fit into the culture the company was developing.

In its heaviest years of rebuilding and expanding (1974–1978), BN contributed a great deal to society:

- Paid taxes of $941 million, including payroll, property, and federal and state income taxes.
- Paid its shareholders $105 million in dividends.
- Repaid more than $613 million in debts; borrowed an additional $683 million to invest further in roadway, equipment and support facilities.
- Made wage outlays, including fringe benefits, of $5.1 billion

The decisions BN made in its capital spending from 1974 through 1978, the years of the heaviest expenditures, enabled it to rebuild and expand capacity on a large part of its rail network. Examples of that massive investment were the construction of a new line of 127 miles to reach mines that would soon be opening; construction of 419 additional miles of new second main track, mine spurs, and siding extensions; laying 5,203 miles of new replacement rail; installing over 14 million ties; and installing CTC on 681 miles. The horsepower of the locomotive fleet was increased by 24 percent, and $47 million was invested in new locomotive and car-maintenance shops in Alliance, Nebraska. In the same period, revenue ton-miles increased from 81.3 billion to 116.3 billion.

All of BN's constituencies benefited from the success it had in building the railroad into a sound business with a good outlook for the long term. For its people, it provided the base needed to create and develop more business opportunities and enhance careers by offering technical training and opening opportunities for advancement. For shippers, BN was able to provide low-cost, dependable service that helped them open new markets for goods and commodities they produced. The general public benefited from the low-cost transportation service that BN provided for consumer goods, food products,

Table Conclusion.1.
THE FIRST DECADE:
A VERY SUCCESSFUL EFFORT—
BURLINGTON NORTHERN AND CONSOLIDATED
SUBSIDIARIES, 1976–1980

	1976	1977	1978	1979	1980
(millions of dollars except per share amounts)					
Operating income	150	145	192	256	420
Net income	73	77	114	176	223
Earnings per share	2.85	2.87	4.26	6.55	7.65
Operating revenue and sales	1,887	2,094	2,515	3,225	3,954
Revenue ton-miles* (billions)	89.4	98.5	111.3	135.0	155.3
Contribution to pre-tax income:					
Transportation**	81.2	67.0	94.6	145.6	299.4
Non-transportation	68.4	77.8	97.4	112.0	120.4

The numbers shown above are clear evidence of the progress BN had made by 1980. The increase in business handled (revenue ton-miles) was impressive. It caused BN to be looked upon as a growth company. The year-to-year improvement in financial performance also was impressive. The challenges of BN's early years had been met and by 1980 the company was well-positioned to handle continued growth and continue to improve the return from the large capital investments made in the 1970s.

* Tonnage handled by rail
**Includes BN's railway company, its air-freight forwarder, and
 truck line
Source: Burlington Northern Annual Report for 1980.

and the coal used in the generation of electricity. The general citizenry experienced cleaner air from the use of the low-emission coal transported long distances to power plants.

The unions representing BN workers benefited from the success they had in negotiating lifetime job protection and guaranteed wages. Even when railroad employment decreased, job positions on BN had greater benefits, more security, and better working conditions in a company that had a strong future. Through modernization and the investment BN made in new technology and in generally upgrading its property and equipment, communities could look upon BN as a progressive, good corporate citizen in which they and BN's employees could take pride.

By the close of its first decade, BN had accomplished a great deal, but much remained to be done to achieve the rate of return that was expected from the large amount of capital invested in the railroad. Also, pressure was mounting to accelerate the development of BN's vast holdings of land, forests, and oil and gas reserves. To speed that process along, the Board of Directors decided in 1980 to bring Richard Bressler in as its new chief executive officer to head BNI, with responsibility for both its Transportation and Resources divisions. At about the same time, Richard Grayson was brought in from the Frisco Railroad to head the railroad company. These changes in leadership brought about a total transformation of the company, with major changes in culture, goals, priorities and methods of operation. This era in BNI history is covered in my second book, *Transformation of a Railroad Company: Burlington Northern, 1980–1995.*

RECOLLECTIONS, INSIGHTS, AND KNOWLEDGE TO BE SHARED

BY NORMAN M. LORENTZSEN

AUTHOR'S NOTE: EARLY IN 2011, I WAS *fortunate to have several meetings and exchanges of correspondence with Norman Lorentzsen on specific matters he felt were significant in the three and one-half years he served as president and CEO of BNI. Norman was interested in having an opportunity to have his recollections and viewpoints documented, especially since such opportunity had not been afforded him in recent writings on BNI history. With the many challenges, projects, and opportunities that occurred in BN in its first decade, it is vital to have documentation made available for research in the future and the chance to properly record insights provided by the person who was in a position of leadership from the company's founding and into its tenth year.*

Norman emphasized the improvement in earnings over the first ten years, which was reflected in the price of BN stock. The stock was split two for one in February 1980. The investment community showed confidence in BN, even after the decision to omit the dividend payment on common stock in the last half of 1975.

This chapter contains excerpts from communications Norman provided on a variety of subjects of importance to BN in those early years. The explanations and interpretation he provided are helpful in understanding the rationale and the context in which decisions were made. Norman was very frank and open in revealing his feelings and his evaluations of how some of these matters were viewed by some board *members and other senior managers at the time. Some of his comments are responses to various earlier writings on BN history with which he either disagreed or wished to lend his interpretation. Other comments are his responses to questions or comments I have presented to him on events and challenges of those years.*

FINANCING IN THE EARLY YEARS

The first two years were greatly impacted by the Penn Central collapse, and secondly, the extremely poor tie condition on the CB&Q. It was the intention after merger to bring a stock issue public to provide funds for necessary projects to implement the physical track changes. The Penn Central collapse delayed raising the required capital for over a year.

TIE CONDITION ON CB&Q LINES

The normal tie-replacement program for lines of the former GN and NP had ties distributed on line at merger time but Bruce Anderson, assistant vice president–Engineering, found it necessary to make emergency tie replacements on the former CB&Q. For the next three years, tie replacements on both the GN and NP lines were reduced to provide ties needed to restore reasonably safe conditions on the former CB&Q. We were told that payment of the CB&Q dividend to the NP and GN before merger had been the first priority, and expenses including tie replacements were severely reduced accordingly.

Author's note: For several years before the 1970 merger, the Burlington cut back on track-maintenance expenditures in order to produce earnings high enough to sustain the amount of dividends payable to the GN and NP, who together held about 98 percent of its stock. This happened while betterment accounting was mandated for railroad companies, which meant that a very large amount of the money spent even for major track maintenance projects such as tie and rail replacement was charged to operating expense. The betterment system allowed railroads to "manage" earnings by adjusting the amount spent for trackwork. By the time of the merger, the tie condition on much of the Burlington, even on its main lines, was marginal. In the fall of 1972, the tie condition on the two-main-track territory of 125 miles between Aurora and Galesburg, Illinois, had reached the point where train speeds had to be reduced to 40 miles per hour. Since this line segment was one of the most important on BN's Chicago Region, having to take this kind of action was of great concern at all levels of management. Quick action was taken to send a fully mechanized, highly productive tie gang from the Montana Division to supplement the Chicago Region's gangs, to put as many ties in the track as possible before freeze-up. It was necessary to pick up ties that had already been unloaded for a tie replacement program on parts of the main line across Montana and ship them to the Aurora–Galesburg line. Some of the necessary work had to be carried over until the next spring. I do not remember the average number of ties that had to be replaced to restore track speeds of 79 miles per hour for passenger trains and 60 for freight trains, but I expect it was about 1,000 per mile, or about 250,000 ties for the entire 250 track miles.

With the rapid buildup of unit coal train traffic that occurred on the Chicago and Denver regions in the 1970s, heavy tie renewals, ballast replacement, and rail replacement became necessary on virtually every main-line subdivision (and even on some branch lines) on the former Burlington. Fortunately, BN was able to finance these heavy expenditures long enough for the work of upgrading the railroad to be completed. The result was a vast improvement in the company's track condition by 1980.

THE PLACE OF WILLIAM J. QUINN[1]

Mr. Quinn was to be made vice chairman of BN at the time of merger. He had a meeting with a fairly large number of BN people at which he spoke about his feelings. He explained that BN, with him as vice chairman, created a very top-heavy management group, more than was needed. He felt the prudent thing to do was to return to the Milwaukee Road as president. Apparently, the Milwaukee Road kept the opportunity available for him.

DECISION TO FORM AN AIR-FREIGHT FORWARDING COMPANY[2]

Prior to merger, Jim Nankivell proposed starting an air-freight forwarding company. This idea was put on hold due to the pending merger. Within the first year of the merger, this idea was alive. Preliminary work as to stations to serve and on obtaining federal approval was underway. It was thought that in the excess employee group, there was someone who could run and manage such a company. The concept and what should be done became my responsibility. A review of the available people convinced me that we should go outside. We found a man named Larry Rodberg, a vice president for an airfreight firm in Seattle. Nankivell and I arranged to meet him. It was concluded we needed a second executive, and Rodberg recommended George Ryan from the same air-freight company. I had Ralph Merklin interview him with favorable results. We employed both of them.

There was opposition from several air-freight forwarders to our application for federal approval, which we finally received. At my request, BN had provided $1 million for operating expense and $1.5 million for capital. Within the first year, we went from a ten-city operation to multiple stations. We were in the black after our fifth quarter. We kept expanding. Within the third year, we were international, serving the Far East, Europe, all over.

---

1   Author's note: In 1966 Mr. Quinn, president of the Milwaukee Road, replaced L. W. Menk as president of the Burlington when Menk moved to the Northern Pacific.

2   Author's note: This subsidiary company was named Burlington Northern Air Freight, Inc. See chapter 34 in *Transformation of a Railroad Company* for additional details on its development, financial performance, and sale.

We began to charter 747s out of Japan, as well as smaller airplanes as the volume required. . . . We were #3 in the rankings of handling air freight and continuing to gain on our competition. Bressler sold BNAFI to Brink's and used that money to buy El Paso, which turned out to be a less than desired. Rodberg left BNAFI since Brinks wanted him to move the headquarters from California to I believe, Chicago. If you look at FedEx and UPS today, one can't help but ask where BNAFI, with the leadership we had, would be today. El Paso is still far from being a shining light. . . . We paid BNI back the full operating loan and began paying dividends to BNI. In its ninth year of operation, BNAFI revenue was $355 million with operating income of $25.3 million. Based on both revenue and tons handled, BNAFI was the nation's number two air-freight forwarder. By the tenth year, these dividends to BNI were becoming a significant part of BNI's earnings.

Bob Wilson [a BN board member] kept saying that we could not take care of the demand for both locomotives for BN and planes for BNAFI. I had Rodberg explain that our need for purchasing our own planes was probably another three to five years in the future. By the time we would have to buy planes for BNAFI, their earnings would be adequate to buy their own planes.

BNI was a transportation and resource company. Air freight was certainly a part of what the company should be. It was a business that was growing by leaps and bounds—witness Federal Express, Costco, and others. We had come from zero to being in the upper four or five companies and gaining. Bressler thought he could do better with El Paso, which was a "dud" then and failed to make improvement. Selling BNAFI and the sale of the NP line between Sandpoint and Laurel were not good decisions.

PROJECTS UNDERTAKEN
IN ANTICIPATION OF MERGER

For the new connection at Casselton, the actual track changes were made before M-Day. However, the signal changes were completed after M-Day. For the North-town yard, not much additional land was needed, but the property was acquired before M-Day. At Hauser, Idaho, the first track built was a siding of about two miles in

length. Construction started while I was superintendent of the Idaho Division. The terminal at Yardley [Spokane] had a grade crossing where traffic was discouraged but had to be kept available. A train of over one hundred cars had to contend with this crossing. Between the end of double track, about six miles east, the next siding was at Cocolalla and of limited capacity. To facilitate eastbound train movements, here was Hauser with the normal 400 feet of property (200 foot right-of-way) but additional land was available. Along siding at Hauser was a great help. Secondly, I had strong feelings about getting out of the Yardley facility and making Hauser a terminal. Additional property would provide room for a fueling station for both eastbound and westbound trains. Locomotives would remain on the train. I believe the additional property was purchased well before M-Day. There was substantial opposition by the public in the area over an extended period of time, and many changes were made to overcome the opposition.[3]

POSSIBILITY OF MERGER
WITH THE SANTE FE

I knew John Reed, chairman, and Larry Cena, president. From time to time we had many discussions. On one occasion, Larry said we ought to consider merging the two roads. I told him that was an excellent idea, but first, at BNI we had to get our stock up to a reasonable level. I told him that at that time, if we merged we would be giving away far too much. I had breakfast with John Reed many times. He was a fine man, as was Larry Cena.

---

3   Author's note: The new hump yard proposed for construction at Hauser in the merger plan was never built except for six tracks of about 7,000 feet in length, built in the early 1970s and used to "stage" unit grain trains and, at times, to handle overflow traffic from Yardley, the main yard in Spokane. In 1982 we made the decision not to undertake construction of a yard at Hauser with full humping capability. By that time, so much business had been converted to movement in intermodal equipment or in unit trains that there was no need for an additional hump yard at the western part of the system.

FINANCING OF THE DECKER AND COLSTRIP LINES

One of the earliest mines to open was the one owned by Peter Kiewit, which involved construction of about fourteen miles of track. In order for BNI to save cash, Kiewit agreed to accept BNI stock with a value equivalent to the cost of building the grade and construction of the track. Our stock was moving up and for Peter Kiewitt, it was a very good deal. They also paid for the coal cars involved. At Colstrip, we did the track construction with our own funds.

COMPASS (COMPLETE OPERATING MANAGEMENT INFORMATION SYSTEM)

Frank Coyne was in charge of getting a management information system installed on the Southern Pacific. He was brought into BNI to move a system forward that could be used by the merged company. As operating vice president, it became my responsibility to get one system into place as promptly as possible. But each company thought their system was it. . . . The CB&Q and GN systems were obsolete. . . . We had all three make presentations to the user group, but each one kept supporting their own units. We finally came up with the idea of a questionnaire that could be *Yes* or *No* on about 100 items. . . . Our next meeting took about ten minutes. . . . A tabulation of the results favored the NP system [COMPASS] which was nearly identical to the TOPS system of the Southern Pacific.[4]

COAL RATES TO POWER PLANT AT COHASSET

The unit coal train move to Cohasset was set up with a very low rate. Our initial handling took a big bite out of the revenue. As the coal movements developed, the electric-power industry always looked at the Cohasset rates and wondered why they couldn't get the same low rate. The chairman and president of Minnesota Power and Light, the owner of the Cohasset plant, and our sales people worked together to get this rate set. The matter of

---

4　Author's note: For more on the implementation of COMPASS, see chapter 22.

interchange of the Cohasset trains between the NP and GN at Fargo was very inefficient. Trains had to go east to Dilworth, and then return west to Fargo for the GN to take the train and move it to Cohasset. After merger, we constructed trackage to permit coal trains to move directly to the former GN line en route to Cohasset.

COAL-SLURRY PIPELINE

The proposal to build a slurry line to move coal from a mine to a Midwest utility or to a water terminal for movement to southern utilities created much incorrect data, as well as constantly stating the railroad could not meet the needs of the utilities. . . . The pipeline people claimed their costs would be lower, that they could be a more assured supplier, and other issues like coal dust from open cars, etc. I was the only rail representative at four of the conferences where all parties—utilities, coal producers, pipeline people, and other interested suppliers were brought together. . . . At the height of these debates, the state of Wyoming disclosed the study made in 1975 with slurry costs of 750 BTU's per ton mile vs. rail costs were 300 BTU's. . . . I discovered very soon that the ideal spot to be a speaker was after ETSI [the company set up to build a slurry pipeline] or any other competitor had made their presentation. We developed the slogan, "Water is the Life Blood of the West." ETSI at one time considered bringing the water back to the origin point via a second pipeline. The economics of such a proposal were not very good. John Kenefick of the Union Pacific was at the last conference, and it was the first time UP had been involved.

THE GENERAL OFFICE BUILDING

Prior to merger, the NP remodeled its one-half of the GN–NP building. It included replacing windows, air conditioning, and many smaller building changes. I do not recall the exact cost, but it was in the range of $6 million. The GN portion was done later, after the merger. . . . Earlier writings say there was no passage between the GN and NP buildings prior to merger—NOT SO. There were two passages, on floor ten and floor two. Both required keys.

A SHOWCASE MERGER

Modern Railroads in October 1973 had an eight-page section entitled "Showcase Merger." Since much of the success of BN depended on the Operating Department performance, both Bob Downing and I were very much involved as well as the regional officers. The original study made by the Merger Committee contemplated starting all priority train schedules on day one. Doing so was part of the Penn Central failure. In discussing this with Ivan Ethington and John Robson, we decided not to do that, but rather, get all of the priority schedules going within the first week. It was a good decision.

MOVING TO SEATTLE

After trying to get Governor [Rudy] Perpich [of Minnesota] to lower personal income taxes for several years, I had a study made (as did Lou Lehr of the 3M Company) about potential savings. The results indicated that being in Seattle produced a very good return. A good example of Minnesota taxes was [Ralph] Merklin on getting his first paycheck in Minnesota [after being transferred from Seattle]. Even with a salary increase of over $10,000 per year, Merklin's net pay was less than it was when he worked in Seattle. 3M went to Austin, Texas.

DECISION TO CUT THE PAY OF EXEMPT PERSONNEL AND OMIT THE DIVIDENDS IN THE THIRD AND FORTH QUARTERS OF 1975

The result of the difficulty in raising capital following bankruptcy of the Penn Central, and the need to increase the tie-replacement program on lines of the former CB&Q, included a decision to reduce the pay of senior executives by 10 percent and 5 percent for middle-management personnel. It also involved a reduction in the dividend declared on stock of 50 percent. Managing our cash during the next year was of major importance and dictated where and how we would spend our dollars. These decisions permitted us to proceed with the highest-priority physical changes that were planned for the merged company. Very careful attention was necessary to maintain adequate funds. We had close coordination with the Finance Department. By 1976 improvement in our financial condition, plus the capital raised, resulted in completion of the remaining merger items. The dividend was brought back to the original payment and salaries for both senior and middle management were restored in 1976.

DECISION TO RETIRE

It was in early 1979 (February or March) that I began thinking about retirement. While I enjoyed the work, the constant travel and the absence from home and family resulted in our decision to seek a change. I spoke to Mr. Menk and indicated my desire, including staying as an officer for one more year in some position without day-to-day responsibility, but not as CEO. The result was that we got a search firm. The board appointed a committee of five people, myself, Menk, Don Dayton, and two more. Menk suggested I stay on the board and also be chairman of the Executive Committee. This was approved by the board. I was with Menk when we met Bressler. We never had a formal meeting of the five-person committee. I believe Menk, having gone through the Norton Simon struggle, wanted to get something done without more struggle. Bressler was "an outsider"—an oil executive—and would be acceptable. Secondly, Menk wanted the Frisco merger completed.

As to additional candidates, I knew the man who was chairman of Green Giant at that time. He was well recommended and recognized for his work. Menk said I should pursue him as to whether he had any interest in BNI. I did that, and he was interested. But he also told me that another large national firm also was interested in him. He needed a commitment before we could give one to him. He took the other job and went to a much larger corporation.[5]

---

5   Author's note: Norman's decision to retire was the threshold of a new era for BN. Under his leadership and that of John Budd, Lou Menk, and Bob Downing, the company had been formed through the merger of four railroads. Up to then, it was the biggest merger in the history of the rail industry, and was judged as highly successful. Those leaders took on the challenge of the largest expansion of capacity and upgrading programs and growth in business experienced on any railroad since the days of building the large transcontinental rail systems. Even with the large expenditures associated with those challenges and opportunities, the company's financial performance *(continued, next page)*

At that time, the BN board decided it was time to take a much different course for the future, by recruiting a CEO from outside the rail industry. The new CEO's set of goals and priorities and his assessment of the strengths and weaknesses of the company's lines of business would take us on

showed year-to-year improvement. Many good candidates had been identified and developed as leaders for the future. The time was right to begin to reap the benefits of the large investments made in infrastructure, equipment, and people in the company's first decade.

a much different course. Together with merging with the Frisco Railroad and preparing to operate in a deregulated environment, BN soon was transformed.

For thousands of employees and other stakeholders, the first decade had presented them good jobs with a dynamic, growing company in which they were able to develop their talents and skills. In those years, no other railroad company or other basic industry offered such opportunities. For that, we are grateful to Norman Lorentzsen and other leaders of BN in its early years.

GLOSSARY

AUTHOR'S NOTE

Contains terminology for both rail and non-rail enterprises owned by Burlington Northern.

ABS (AUTOMATIC BLOCK SYSTEM)   Defined in the Consolidated Code of Operating Rules, edition of 1967, as "A series of consecutive blocks governed by block signals, cab signals, or both, actuated by a train or engine, or by certain conditions affecting the use of a block." Although the operation of trains is protected under ABS in a series of wayside signals, trains must still be given movement authority through train orders issued by a dispatcher and by schedules contained in the operating timetable for a specific territory. An ABS system may indicate a broken rail, and if connected to a slide-detector fence, protection to trains from rocks or other material that may foul the track.

AC LOCOMOTIVE   Equipped with AC traction motors, thereby not affected by the limitations of the maximum continuous-current ratings or the short time in operating ratings that locomotives equipped with DC traction motors have in service.

AC TRACTION MOTOR   First applied to North American railroad freight locomotives in the late 1980s. Has advantage over DC motors by having a higher adhesion rating in the range of 40 percent compared to 20 to 25 percent rating with DC motors, and having no brushes or commutator to maintain. AC refers to alternating current, in which electric current reverses its direction at regular intervals.

ADHESION COEFFICIENT   The percent of the weight on the driving wheels of a locomotive that is available for traction. It may be as low as 10 percent with a wet rail condition, to 40 percent on dry, sanded rail, or with AC traction motors.

AFE (AUTHORITY FOR EXPENDITURE)   A business form used in the process for review and authorization of projects requiring a capital expenditure. It may include some work that has to be charged to operating expense. May be required for construction, purchase of service or physical change to be made to an existing facility, acquisition or rebuilding of locomotives and cars, or retirement of an asset.

AGREEMENT   A contract between the railroad company and the union or unions representing employees of one or more crafts, as to rates of pay, working conditions, handling pay claims and grievances, the scope of work, and many other issues.

ALIGNMENT   The location of a line of railroad in terms of its tangents and curvature. It is not defined in terms of the grades on which it is built.

ALLOCATIONS (NO. OF)   A provision in merger implementing agreements for train operating crews on some districts of the former GN and NP that allowed employees to "follow their work" when trains they had previously manned were diverted to a newly designated "preferred route" on another former railroad. An example was the allocation of a specified number of GN crews from the Minneapolis–Willmar–Breckenridge line into the Minneapolis–Staples–Dilworth chain-gang pool that operated trains on the newly designated preferred route of the former NP. This provision kept crews from losing a large part of their work when "their" trains were to be diverted to the line of the "other" railroad.

ANODE   Copper that was partly refined was cast into anodes for further refining by electrolysis, in the form of slabs. Anodes were shipped by rail from the smelter in Butte, Montana, to a refinery in Black Eagle, Montana, for further refining.

ASSIGNED CREW   Employees on assigned train crews report for work according to a schedule or other routine as to days of work, days off, and the time to report for duty.

BAD ORDER   A car designated by an inspector as needing repairs before further movement is made.

BALLAST   Generally, a crushed hard rock such as granite, trap rock, quartzite, or dolomite placed on the roadbed to help distribute the load imposed by trains, to provide stability to the track structure and good drainage to help maintain good surface and alignment.

BALLAST CLEANING   Ballast fouled with mud or other contaminants is excavated from the track and screened to separate any remaining "good" ballast from the mud, and then the cleaned ballast is returned to the track. New ballast will be unloaded to replace the fouled ballast that has been removed.

BENTONITE   A soft, plastic, porous rock composed of clay minerals and silica, bentonite is greasy and absorbs large quantities of water, accompanied by an increase in volume of about eight times. Its presence causes great difficulty in track maintenance and quickly contaminates even high-quality ballast.

BLOCKS   Groups of cars coupled together for movement to the same destination, junction, or set-out point.

A train may be made up with several blocks to reduce the amount of switching or rehandling enroute or at the train's destination. In contrast, an unblocked train will have to be switched at a distant point for delivery to industries, for interchange to another railway, or to connect with another train.

BOARD FOOT   The unit of measurement for lumber. Calculated by multiplying thickness (inches) times width (inches) times length (feet) divided by 12.

BRAKEMAN   Works under the direction of the train conductor in switching, inspecting cars, making air-brake tests, and coupling and uncoupling cars.

BRIDGE TRAFFIC   See OVERHEAD TRAFFIC.

BTU (BRITISH THERMAL UNIT)   The amount of heat required to raise the temperature of one pound of water by 1 degree Fahrenheit.

BUFF ACTION   Caused by compression forces on couplers and draft gear when slack runs in on cars in a train. Also see DRAFT, the forces "opposite" buff.

CHAIN GANG   A group of locomotive engineers, conductors, and brakemen set up to operate freight trains on a "first-in, first-out" basis, with unscheduled hours for going on duty and generally with no assigned days off. Also known as the "pool" on many railroads. Origin of the term "chain gang" is in the nature of the work, with employees not having much freedom to really get away from the job, due to irregularity of hours worked and considerable unpredictability (and short notice) as to when they will be called to go on duty.

CLASSIFICATION   Switching cars in a yard according to destination or other category for movement out of the yard on a train or a local industry switching assignment. Groups of cars with a common destination or other status may be called blocks.

CLASS I RAILROAD   A railroad with annual revenue in excess of $250 million (in 1991). A regional railroad has annual revenue between $40 million and $250 million, and at least 350 miles of track. Short-line railroads have revenue of less than $40 million.

CLEAR-CUTTING A forestry and logging practice in which most or all trees in an area are uniformly cut down. Removes all trees, regardless of size, in an area in one operation. Used by foresters to create certain types of forest ecosystems and to promote select species to grow in an even-age stand.

CODE LINE A two-wire line running between a CTC control machine and the signals and powdered switches in the field operated by a train dispatcher or control operator. Control codes are sent via the code line to corresponding signals and switches. To avoid the danger of a code line being damaged or torn down by weather or vandalism, as well as the cost of maintaining a pole line with open wire, railroads either buried the code line or installed electri-coding by sending electrical pulses through the rails. (Source: Michael J. Burgett, "The Engineering Basics of CTC," www.ctcparts.com, and "Pole Line to Be Eliminated," BNSF *Whistle Post,* Nebraska Division newsletter, December 2001/January 2002.)

COMPASS (COMPLETE OPERATING MOVEMENT PROCESSING AND SERVICE SYSTEM) Installed by BN in the early 1970s (see chapter 22 for details on what this system was designed to deliver). Among many other capabilities, COMPASS was a system of quality control over freight-car movement on the entire BN system.

COMPENSATION (OF GRADIENT) The easing of gradient on curves to compensate for the additional resistance caused by curvature.

COMPOUND CURVE A continuous curve with two or more curves of different degrees connected without a section of tangent track between those curves.

CONCENTRATOR Centrifugal concentrators rotate to allow heavier and more dense minerals to flow to the outer surface of a bowl, which are then removed for further upgrade.

CONDUCTOR Employee in charge of a train and responsible for giving directions to other employees on the train, the locomotive engineer, and Brakemen.

CONSIST A list of cars and locomotive units in the order in which they stand in a train. Also contains such information as destination of each car, whether the car is loaded or empty, and the commodity it contains.

CONTAINER A large "box" generally built in lengths of 20, 40, or 53 feet, used in common for shipments to be made in one or more modes of transportation, rail, truck, or marine.

CREW CALLER Employee who maintains and inputs data as to the status of train crew members, and initiates calls to them when they are needed for duty or when they are displaced from their position by exercise of seniority by a senior employee.

COMMITMENT A pledge or promise, declaration of an opinion, position, or belief.

CONDUCTOR Employee in charge of a train.

CONTROLLED SIDING A siding within the limits of CTC territory on which a signal indication authorizes use of the siding.

CTC Centralized Traffic Control. A system for governing the movement of trains in which a dispatcher or a control operator under the direction of a dispatcher is authorized by signal indication as opposed to written or verbal authorization. Usually incorporates the remote control of switches with the control of signal indications. (Refer to the railroad's General Code of Operating Rules for a more strict, "official" definition.)

CURRENT OF TRAFFIC The provision for movement of trains in one direction on a main track as specified in the operating rules.

CWR Continuous welded rail. Conventional 39-foot or 80-foot rails welded end to end, thereby eliminating the need for rail joints and their high cost of maintenance.

DEADHEADING   Transporting a train crew by train, auto, bus, or other means to a location where their work will begin, or to go off duty.

DIED ON THE ROAD   When a train crew has run out of working time under the Hours of Service Law and must be relieved from duty.

DIRECT CURRENT (DC)   Electric current that flows in one direction only.

DIRECTIONAL RUNNING   The use of each of two somewhat parallel lines for trains moving in one direction only between two common points to gain the capability of a double-track line. With this designation, line capacity is increased and the delays for trains meeting at sidings on a single-track line are eliminated. An example that worked well was between Spokane and Pasco, in which westbound trains ran on the former SP&S line and eastbound trains on the former NP line.

DISPATCHER   Employee responsible for directing and authorizing the movement of trains on main tracks.

DISTRIBUTED POWER   A locomotive consist set up for remote-control operation in conjunction with locomotive units at the head end of the train. The units operating as distributed power may be placed at the rear of a train or cut into the train at the location that will allow the most efficient train handling for the engineer.

DIVISION SUPERINTENDENT   The Operating Department officer in charge of a designated territory referred to as a division. Responsible for all train and terminal operations; on some railways, may also have responsibility for maintenance-of-way and the mechanical (maintenance of equipment) functions. Reports to a general manager.

DRAFT   Tension (as opposed to buff or compression) created in couplers when train slack is "out."

DYNAMIC BRAKING   A means of braking by using the locomotive's traction motors as power generators, thereby controlling and reducing train speed. The power

generated from the traction motors is dissipated through dynamic braking resistors.

ELECTRIC FLASH BUTT WELD   A weld of rail ends made by electric welding. Can be made either in the field or in a stationary rail-welding plant.

EMBANKMENT   A bank of soil or rock built up from the natural ground surface to bring the roadbed up to the desired grade line.

END-OF-TRAIN TELEMETRY DEVICE   Mounted on the trailing coupler of the last car in a train to provide air-brake pressure and indication of motion at the rear of the train (whether moving or stopped) to a device mounted in the cab of the locomotive.

ESCARPMENT   A steep slope formed as an effect of faulting or erosion, separating two relatively level areas of differing elevations.

EXTRA BOARD   A group of conductors, locomotive engineers, or brakemen set up to fill vacancies created by assigned and "chain gang" employees who are unavailable for work due to vacations, illness, or personal business. Extra-board employees are also used to make up an "extra" crew if no chain gang or assigned crews are available to man a particular train or yard switching crew.

FALSEWORK   A temporary structure built during construction of a bridge to hold components or the permanent structure in place until construction has advanced enough to support itself. In the case of the new bridge built over the Missouri River at Rulo, Nebraska, falsework was built adjacent to the piers of the old bridge that would be kept in service for the new bridge. The new truss spans were placed on the falsework by barge cranes and held on the falsework until the contractor was ready to remove the old spans. The new spans were then moved from the falsework onto the piers that had just been cleared. The track was out of service for only forty-eight hours to get the new spans in place.

FIELD WELD   A weld made out on the track rather than in a stationery welding plant by a thermite process or electric welding to eliminate joints between rails.

FINAL TERMINAL DELAY (FTYD)   A penalty payment made to crews when kept on duty in excess of thirty minutes from their time of arrival at the switch or other location designated at the entrance to the yard until relieved from duty. The exact conditions under which this payment was due varied somewhat among railways. This payment was eliminated when the trip rate bases of pay was adopted.

FIXED BASE OPERATOR   Provides such aeronautical services as fueling, general servicing and general maintenance, parking, hangar storage, aircraft rental, and flight instruction.

FREIGHT FORWARDER   A company established to consolidate less-than-carload freight shipments into carload lots.

FREQUENCY RATIO   Number of employee reportable injuries per 200,000 employee hours worked. As a rule of thumb, it is also the number of injuries per 100 employees.

FROG   A steel component in a switch (turnout) placed at the intersection of the two running rails to permit wheels and flanges to move from either rail to cross the other rail, depending on whether the switch points are lined for a "through" (straight) movement or to enable the movement to diverge onto another track. (See also *Railway Ages Comprehensive Railroad Dictionary* or Christopher F. Schulte, *The Dictionary of Railway Track Terms.)*

GED (GENERAL EQUIVALENCY DIPLOMA)   A document given to a person who did not complete high school but passed tests on four subjects that show high school–level academic skills.

GENERAL CHAIRMAN   Employee elected to represent the interests of employees of a given craft system-wide in contract negotiations and in the appeal of claims, grievances, and disciplinary action issued at a local level, to system officers of the company.

GENERAL MANAGER   A middle-level management position in the Operating Department reporting to the vice president–Operations. Generally there are from two to four division superintendents reporting to a general manager.

GRADE (PERCENT OF)   The rise or fall of track over a distance of 100 feet. Example: A rise of one foot in 100 feet is a grade of one percent. Generally, grades of 1.0 percent or less are considered light grades, heavy grades between 1.0 and 2.0 percent, and mountain grades 2.0 percent or greater.

GRADE RESISTANCE   The resistance to movement on a grade due to gravity, 20 pounds per ton of train weight for each percent of grade.

GROSS TON-MILES   The sum of ton-miles handled in a train, made up of the weight of the freight handled plus the empty weight of the cars. It includes the weight of empty cars in the train but does not include the weight of the locomotive.

HELPER   A locomotive used to assist a train in ascending a grade when the "primary" locomotive assigned to a train will not produce the amount of tractive effort needed to avoid stalling on a grade, or if the tonnage would cause excessive strain on the drawbars of cars at or near the head end of the train.

HI-RAIL   A highway vehicle equipped with retractable flanged wheels that can be placed on a track for inspections or to carry material or machinery needed to assist in maintenance of track, bridges, or signals.

HOLDING COMPANY   A business structure allowing a parent firm to control other firms, often outside one's core industry. A holding company does not produce goods or services itself, but enables the ownership to control a number of different companies with benefits from sharing of operating losses and ease of divestiture.

HORIZONTAL DRILLING   Directional drilling of an oil well in a horizontal direction rather than vertical. It penetrates a greater length of the oil reservoir and thereby offers significant production improvement over a vertical well. (Source: Summary of Schlumberger Oilfield Glossary, https://www.glossary.oilfield.slb.com/.)

HOURS OF SERVICE   A federal law and regulation in place primarily to restrict train crews and yard crews from working in excess of twelve hours on a given tour of duty. Also regulates the amount of time that dispatchers and signal maintainers may work. The full provisions of this law are much more complex than indicated in this summary, but the underlying principle is to limit the number of consecutive hours of work and to provide for a minimum number of hours of rest before going back on duty.

HOSTLER   Employee from either engine service or locomotive maintenance who moves locomotives to and from fueling facilities, shop tracks, the locomotive washer, and couples units into consists for outbound trains or switching service.

HUMP YARD   A classification yard in which cars are pushed by an engine up an incline ("hump") from which cars roll by gravity when uncoupled to various tracks according to destination, outbound train or other designation. These cars will then be pulled from the opposite end of the classification tracks and moved into tracks that are designated for blocks of cars to be assembled into outbound trains.

HYDRAULIC FRACTURING   Creating fractures in rocks and rock formation by injecting a measure of sand and water into the cracks underground to force to open further and allow more oil and gas to flow out of the formation. (Source: https://www.investopedia.com/terms/f/fracking.asp.)

INTERCHANGE   The movement (transfer) of cars from one railroad company to another, usually carried out at a common point. May also be handled through a third party or intermediate railroad such as a terminal switching company.

INTERMODAL   The movement of freight shipments by a combination of modes including rail, truck, and/or ship, usually in 20-, 40-, or 53-foot containers.

JOINTED RAIL   Also known as conventional bolted rail or CB rail. The laying of conventional 39-foot rails end to end, staggered, connected by a steel bar (called a joint bar or angle bar) that fits against the rail and usually secured by four or six track bolts, thereby holding rails in position.

LEVERAGED LEASE   A lease agreement partially financed by the lessor through a third-party financial institution. The lending company holds title to the leased asset. (Source: Investopedia.com/dictionary.)

LOCAL CHAIRMAN   Employee elected to represent employees of a given craft at a local level in the appeal of time claims, handle grievances, and represent employees at formal investigations (hearings) conducted by the railroad company to determine the facts and responsibility in alleged or possible violations of rules.

LOCOMOTIVE TYPES   Types of diesel-electric locomotives produced by the Electro Motive Division of General Motors referred to in the text.

FT   Freight locomotives equipped for multiple-unit operation from a single control. Most often, used in two-, three-, or four-unit consists. Rated at 1350 horsepower per unit.

F3, F7   F-type unit upgraded to 1500 horsepower.

F9   F-type unit upgraded to 1750 horsepower.

GP9   Locomotives designated "general purpose" unit, initially built for road switching and local freight service. Soon evolved to use in multiple in "through" freight service. Horsepower evolved from GP5 unit with 1350 horsepower to GP60 with 3800 horsepower.

SD   Locomotives designated "special duty," initially intended for heavy drag-type freight service, e.g., to serve mining operations or other operations that will benefit from having six traction motors instead of four as with the F and GP types. Having a higher weight on the driving axles produces higher tractive effort. With the electric current spread over six traction motors, a SD-type locomotive can be operated at a lower minimum continuous speed without danger of overheating the traction motors. In time, the SD-type unit evolved as the mainstay of motive power for heavy-unit train service as well as intermodal and "general merchandise" type trains on most large railroads.

LOW-SULFUR COAL   Contains one percent or less sulfur by weight. For air quality standards, contains 0.6 pounds or less sulfur per million BTU.

MAINTENANCE OF EQUIPMENT   The function in the Operating Department responsible for the maintenance of cars and locomotives.

MAINTENANCE-OF-WAY AND STRUCTURES   The function in the Operating Department for the maintenance of roadway assets, i.e., track, bridges, buildings, and signals.

MANNED HELPER   A locomotive consist used to assist heavy trains in getting over grades, controlled by a locomotive engineer.

MASTER LIMITED PARTNERSHIP   A type of limited partnership that is publically traded and does not pay taxes from the cash distributions made from its cash flow. Those distributions are taxed when the MLP is sold. Often used by forestry and oil and gas companies due to their having a higher cash flow than taxable income due to high depreciation and other tax deductions. (Source: Investopedia.com/dictionary.)

MILLION GROSS TON-MILES   The number of tons of traffic (in millions, including the weight of cars and locomotives plus freight handled) that have passed over a given section of line on an annual basis.

MOVABLE-POINT FROG   A frog design with points that will be moved in the same position as the points on a switch. See also FROG.

MU   Multiple-unit operation in which two or more locomotive units are coupled to allow control from a single unit.

NON-ABS   Non-ABS territory is sometimes referred to as "dark territory," meaning there are no wayside signals in place to protect trains from danger of collision and to warn of unauthorized occupancy of the track by a train or engine, or of switches improperly lined against the movement of a train or to indicate the possibility of a broken rail. In non-ABS territory, crews must protect against following trains by flag protection, unless a manual block system or track warrant control is in place to maintain spacing between trains moving in the same direction.

OPERATOR   A position established to issue instructions on the movement of trains per instructions initiated by a dispatcher. In some cases, an operator may operate a machine controlling the position of signals governing train movements, again, under the direction of a dispatcher. The position of operator has largely disappeared due to advances in technology and simplification of rules that allow a dispatcher to communicate instructions directly to the crews of trains rather than by train orders copied and delivered by an operator.

OPERATING DEPARTMENT   The department of a railway responsible for the operation of trains and terminals, maintenance of track, bridges, buildings, and signals, as well as cars and locomotives.

OPERATING RATIO   The ratio of operating expenses to operating revenue.

OUT-OF-FACE   Extended or continuous maintenance work done on a track segment without having to skip some locations not included in the work program. This type of program for tie renewal, replacing rail or track surfacing, is in contrast with spot maintenance work, which covers only short segments of work to be done at a number of disconnected locations.

OVERHEAD TRAFFIC   Shipments that are routed on part of the "trip" on an intermediate or "bridge" railroad, i.e., on a railroad that neither originates or terminates those shipments. Overhead or bridge traffic often is very competitive, hence, subject to diversion to another railroad.

PERMANENT BRIDGE   Constructed of steel and/or concrete. Programs to replace bridges originally constructed of timber were started by most railways in the 1880s, at least on main lines.

POSITIVE TRAIN CONTROL (PTC)   A safety overlay system for train control designed to prevent collisions with other on-track equipment, trains, or cars; to prevent trains from exceeding authorized speeds; and to protect roadway-maintenance workers by enforcing limits of authority and restrictions pertaining to train movement where CTC is in service.

POWDER RIVER BASIN   A geologic structural basin in southeast Montana and northeast Wyoming about 120 miles east to west and 200 miles north to south. Supplies about 40 percent of the coal used in generating electricity in the United States. Classified as sub-bituminous containing about 8,500 BTU/pound and a low content of sulfur.

PRESIDENTIAL EMERGENCY BOARD (PEB)   A three-member board that may be appointed by the president of the United States under the Railway Labor Act to study and make recommendations for the settlement of contract negotiations between railway companies and unions.

PULL-APARTS   The creation of a gap of excess length between two rails, often due to contraction of rails in extremely cold temperatures. Often, the track bolts in one rail end will be sheared as the rails separate and create the gap.

RAIL CREEP   Also known as running rail, the longitudinal movement of rail occurring under trains or in temperature fluctuations. Controlled by placement of rail anchors.

RAIL GRINDING   Use of a self-propelled, on-track machine to remove irregularities on the surface of rail such as corrugation or to re-profile rail to restore its designed profile following heavy wear from passing trains.

RAIL RUNNING   Also known as "rail creep," longitudinal movement of rail due to fluctuations in temperature and forces on rail caused by trains passing over it. Can be mitigated by applying rail anchors under compression to the base of the rail.

RAIL WEIGHT   The weight of a three-foot section of new rail.

RAILWAY LABOR ACT   A 1926 federal statute recognizing the right of collective bargaining in the railroad and airline industries. It governs such labor-management matters as collective bargaining, mediation, grievance arbitration, and labor protective provisions.

REDUCED CREW CONSIST   Operating a train or yard switching crew with fewer employees than what had been a standard crew size (i.e., a conductor or yard crew foreman and two brakemen or two switchmen). The first phase of transition to a reduced crew was to eliminate one brakeman or switchman, followed by elimination of the remaining brakeman or switchman to a "conductor-only" crew. Starting in the early 1960s, the position of fireman was eliminated, leaving only the engineer to run the locomotive.

REGULATOR   A track-maintenance machine used to shift ballast and shape the ballast section to the standard level required between ties and on the shoulder with a plow and usually a broom mounted on the machine.

REPORTABLE INJURY   Generally, an injury is reportable if the injured employee loses time from work, cannot perform the normal duties of his assignment, or is given medical treatment. Of course, the regulations on reportability are far more complex than this, but these are among the basic factors that determine reportability.

RERAIL   A particular design for rail sections adopted by the American Railroad Engineering Association and widely used on North American railways.

RESISTANCE   Refers to resistance to motion of a train due to grade (at 20 pounds per percent of grade), plus the rolling resistance of cars, and the amount due to wind and curvature.

REVENUE TON-MILES The total of ton-miles moved, excluding the weight of the cars. Also does not include ton-miles generated in the movement of company material, e.g., rail, ballast, or locomotive fuel.

ROAD FOREMAN The supervisor of locomotive engineers, also known as the traveling engineer on some railroads.

ROADWAY The entire track section including the road-bed, bridges, culverts, embankments, cuts, and fills.

RULING GRADIENT The grade that limits the maximum tonnage that can be hauled with a given locomotive consist. The ruling grade may not be the maximum grade if that grade is so short that the entire train is not on it at any time, or if the effect of the maximum grade will be offset by the momentum of the train.

RUN-THROUGH An operation that may apply to a train and/or motive power. A run-through train is designed to move "through" from one railroad to another at a junction point or terminal without having to change the power from the inbound train to power owned by the railroad that will be moving it beyond that point.

RUNNING MAINTENANCE Routine periodic maintenance performed on locomotives, often on a mileage or time basis to maintain service reliability and efficiency in fuel consumption. Overhauls and damage repair are performed in heavy repair shops.

SHOULDER The part of the ballast section between the ends of the ties and the toe of the ballast slope.

SHOULDER BALLAST CLEANER A machine used to remove fouled ballast from the shoulders of the track section and run it through a screening process to remove contaminants and then place the cleaned ballast back in the track. Clean ballast will be unloaded to replace the amount of fouled material no longer suitable.

SIDING An auxiliary and usually parallel track used for meeting and overtaking trains.

SKELETONIZED TRACK Condition of track when ballast has been removed from between the ties, awaiting the unloading and tamping of replacement ballast.

SLACK Unrestrained free movement between cars in a train. May be mitigated on passenger trains by the use of "tight lock" couplers.

SLEDDING Use of a machine to raise track out of fouled ballast, which can then be plowed out of the track. While the track is raised, defective ties may be replaced. After track is lowered onto the grade, new ballast may be unloaded and the track raised again to get the standard number of inches of ballast under the ties. Additional ballast is then filled in between ties and broadened out to form the shoulder beyond the ends of the ties. The track is then surfaced by a power tamper. A ballast regulator and broom are used to shape the shoulders and brush out excess ballast from the tops of ties and out onto the shoulder.

SLOW ORDER A temporary speed restriction generally due to maintenance work in progress or when work needed to correct the deficiency has not yet been started.

SPRING SWITCH A hand throw switch designed with a spring mechanism that will allow movement from a diverging route without having to stop to have an employee re-line the switch to its normal position.

STAGGERS ACT Legislation named for Rep. Harley O. Staggers of West Virginia enacted in 1980. Provided for deregulation of railroads, mainly in the area of pricing. It provided greater range for pricing with less regulatory restraint, independence from collective rate-making procedures, and authority for railroads to negotiate contract rates.

STEEL GANG Organized and equipped to lay new or replacement rail. Generally consists of about 75 employees for a large project, or 35 or fewer if the work is limited to a small quantity, such as replacing only curve-worn rail.

SUB-BALLAST   Granular material used to separate the roadbed (subgrade) from the ballast section.

SUBGRADE   The roadbed that lies below the track, ballast, and subballast.

SUPERELEVATION   Often called "banking" of curves in which the outer or high rail on a curve is raised higher than the inner rail to allow trains to move at a designated speed above what would be required without any superelevation.

SUPERINTENDENT   A middle-level management position based at a field location responsible for operations and safety on a division and/or terminal. On some railroads, the superintendent also holds responsibility for the maintenance of track, bridges, and buildings, and cars and locomotives, in addition to train and switching operations and overall customer service.

SUSTAINED YIELD   The amount of a resource, generally in forested land, that can be harvested without causing depletion. Short-term harvests are limited to allow for longer-term regeneration of the forest from the trees that remain.

SWING-NOSE FROG   A frog designed with both a primary power switch machine and a second, separate switch machine to position the switch point. The swing-nose design eliminates the gap over which wheels pass on a standard frog and allows the installation of frogs that will allow higher speeds on the diverging route to be taken. It also requires less welding and grinding than a conventional frog.

SWITCHTENDER   An employee assigned to hand operate one or more switches in a limited area to keep trains from having to stop or slow down to a crawl to allow the conductor or brakeman to get off the engine (or caboose) to line those switches.

TACONITE   A very hard low-grade iron ore containing about 27 percent iron that must be processed into pellets with iron content of about 65 percent iron for economical use in steel-making. Bentonite clay is used as a binding agent.

TAMPER   An on-track machine used to get ballast placed under the ties to bring the surface (smoothness) and alignment to standard.

TERMINAL   An operating facility consisting of one or more yards or intermodal transfer facilities. May also include equipment and facilities used for transfer of bulk freight from rail cars to storage or to truck service for delivery. Often defined as the end point for runs made by train crews.

THERMITE WELD   Welding of rail ends together with a compound of iron oxide and aluminum powder.

TIE GANG   Organized and equipped to replace ties on a production basis. Generally consists of about forty employees and two sets of machinery.

TONNAGE RATING   Maximum tonnage to be handled by a given consist of locomotive units. It is based on the tractive effort produced by the locomotive, along with grades on a given territory and the "running time" required per the service schedule or operating plan.

TRACKAGE RIGHTS   A contract or agreement for one or more railroads to use the track of an owning railroad. It may provide for a tenant railroad to furnish the crews and motive power and run its own trains, or it may instead pay the owning railroad to move the tenant road's cars in the owner's trains.

TRACTIVE EFFORT   The "pulling power" of a locomotive. For a diesel locomotive, it is determined by the weight on the driving axles and the coefficient of adhesion, which will vary from 0.18 for older types of locomotives to 0.40 for the modern types with AC traction motors. The tonnage rating for a locomotive is based on its tractive effort and the resistance to be encountered due to grade, curvature, and the rolling resistance from the cars, as determined by weight, cross-sectional area, and wind velocity.

TR RAIL   Refers to "torsion resistant," a rail section designed by the Engineering Department of the CB&Q Railroad for new 112- and 129-pound rail.

THROUGH TRAIN   A train operated with no stops or only a minimum number of stops to perform intermediate switching work.

TRAIN BLOCKING   Grouping of cars placed in a train according to destination, intermediate point at which they will be set out of the train, or other particular designation intended to help get a train over the road and to speed up the delivery of cars by reducing the requirement to do as much switching enroute.

TRAIN CREW   Most often, a freight train crew consists of a conductor and engineer. A brakeman may be included on trains designated to perform "local work," e.g., stopping en route to switch cars at customers' facilities or at yards, or when agreements on staffing require a brakeman be included on the crew of some types of trains.

TRAINMASTER   A first-line field supervisor of train operations and/or terminal operations.

TRAIN-MILE RATIO   The number of reportable train and switching accidents per million train-miles operated.

TRUST   Confidence and reliance on the integrity, strength, and ability of a person or organization. Confident expectation and predictability of the action, behavior, and beliefs of a person or organization. A belief that "we can be counted on to do what we say we will do."

TURNOUT   Provides capability for a train or switching operation to move to or from a diverging route.

TWC (TRACK WARRANT CONTROL)   TWC replaced the rules governing operation under ABS and non-ABS territory, and by timetable and train order authority, when the technology for radio communication between dispatchers and train crews for movement authority was greatly improved. TWC is in effect on secondary and branch lines (although not in all cases) on which train traffic is lighter than on primary main lines where CTC is in place. In the Union Pacific's rules for TWC, it is defined as "A method to authorize train movements or protect men or machines on a main track within specified limits in a territory designated by timetable. (See the railroad's General Code of Operating Rules, 2017, UPRR.com.)

UNDERCUTTER   A track machine that removes ballast generally fouled with dirt and other contaminants to allow replacement with clean ballast.

UNIT TRAIN   A train with a fixed consist of cars that is operated intact between origin and destination. The consist is not broken up (i.e., reduced or added to over the entire round trip) except when mechanical defects are found en route, or when some cars are due to be switched out of the train for scheduled maintenance. Unit trains are usually operated per a standard set for the round trip cycle time. The most common applications of the unit train concept are for the movement of bulk commodities such as coal, grain, and fertilizers.

VICE PRESIDENT–OPERATIONS   The senior management position responsible for the movement of trains, operation of yards and terminals, and the maintenance functions.

YARD   A system or network of tracks usually parallel to each other and connected with each other on both ends to a track called a switching lead. A yard is built within defined limits, used for switching and the classification of cars according to destination, outbound train, or interchange connection.

YARD CREW   Employees who classify cars in yard switching facilities according to instructions from a yardmaster or other designated supervisor or employee, and to switch cars to or from customers' facilities. A yard crew consists of a foreman (sometimes called a yard conductor). Depending on local factors, there may be one or two switchmen (sometimes called helpers) on a crew, in addition to the foreman. When remote control equipment is used by the crew, there will be no locomotive engineer assigned to the crew.

YARDMASTER   Employee in charge of supervising a freight classification yard or switching facility. Supervises the work of yard switching crews.

REFERENCES CONSULTED

Abbott, James, "Santa Fe Coal-Slurry Damages Seen Reaching $750 Million," *Traffic World*, March 20, 1989.

Abbott, R.A., "Concrete Ties vs. Wood Ties: The Debate Continues," *Railway Track and Structures*, March 1989.

Abramson, Howard and Lawrence Kaufman, "BN Changing Its Reputation as Labor Basher," *Journal of Commerce*, (date not known, but likely to be in late 1990s).

"Actions Achieve Results: New Shop Improvement Committees at Havelock Build Teamwork," *BN News*, April 1985.

"A. C. Breakthrough on BN?" *Railway Age*, March 1993.

"A. Kahn, Pushed U.S. Airline Deregulation," *New York Times*, December 29, 2010.

Adams, John, MD, "Galesburg, Illinois," *The BN Expediter*, April 2010.

Adams, Mikaila, "ConocoPhillips Acquires Burlington Resources for $35.6 Billion: Set to Become Leading Natural Gas Producer in North America," *Oil and Gas Financial Journal*, February 1, 2006.

Adelsberger, Dr. Helmut, "Developments in High-Speed Turnout Design," paper written for Voest-Alpine GmbH, date not specified.

Agreement Between Montana Rail Link and Its Employees Represented by the Brotherhood of Locomotive Engineers, October 1987 (date note specified).

Agreement Between Order of Railway Conductors and Brakemen, Brotherhood of Railroad Trainmen and Switchmen's Union of North America and Burlington Northern, Inc. www.smartunion386.org/agree/68merger.

Agreement for Protection of Employees Who May Be Represented by the Brotherhood of Railroad Trainmen in Event of Approval of Merger and Related Applications Filed by Burlington Northern Inc. in ICC Finance Dockets Nos. 21478, 21479 and 21480.

Allen, John P., Stephen, Eichler, Clifford M. Goldberg, and Louise E. Kier, *Operation: Red Block*, Institute for Human Resources, Inc., Rockville, Maryland, October 1988.

"Alliance, a Boom Town," *BN News*, January 1972.

"Alliance Division: Coal, Youth, Growth," *BN News*, October 1976.

"Alliance Division's Maintenance Complex Combines Rebuilding and Training," *BN News*, July 1980.

"Alliance Terminal Injury-Free for More Than One Year," *Alliance Times-Herald*, September 17, 1998.

"Alliance Thrives Amid Tough Challenges," *BN News*, July 1980.

American Mining Hall of Fame Awards Presentation and Banquet, December 1, 2007, induction of Dennis R. Washington.

"An Assessment of Development and Production Potential of Federal Coal Leases," remarks of R. M. Bressler before the Western Coal Transportation Association, NTIS Order #PB82-149378, December 1981.

Andersen, E. R., "Analysis of Operations of Railways That Have Substantially Reduced the Cost of Construction and Maintenance of Way Work," Committee on Economics of Railway Construction and Maintenance, American Railway Engineering Association, 1970.

Anderson, B. G., "Burlington's Northern New Gillette–Orin Line," American Railway Engineering Association Proceedings, Bulletin 878.

American Railway Engineering Association, Bulletin 678, "Burlington Northern's New Gillette–Orin Line," address by B. G. Anderson, Assistant Vice President–Engineering.

American Railway Engineering and Maintenance of Way Association. 2003. *Practical Guide to Railway Engineering* 2nd Edition.

"Appeal Court Ruling Says BN Can't Derail Union Contracts," *Associated Press*, November 30, 1988.

Appendix E Memorandum of Agreement Between Order of Railway Conductors and Brakemen, Brotherhood of Railroad Trainmen, Switchmen's Union of North America and Burlington Northern, Inc., 1968.

Application Under Sections 1(18) to 1(20) of the Interstate Commerce Act Relating to Construction and Operation of Certain Extensions of Lines of Railroad.

" The Architects of Change: Darius W. Gaskins, Jr.," *Railway Age*, October 1967.

"ARES: Boldly Going Where No Railroad Has Gone Before!," author's collection, date and name of publication not stated.

ARES for Safety and Service, Burlington Northern Railroad, July 1989.

Arkansas Electric Cooperative Corporation—Petition for

Declaratory Order, Surface Transportation Board Decision, Docket No. FD 35305 March 3, 2011.

Armbruster, Kurt, "John M. Budd, One of America's Great Railroad Men," Great Northern Railway Historical Society, Reference Sheet No. 318, December 2003.

Armstrong, Michael N. "Concrete Tie Experience on the Burlington Northern," *Railway Track and Structures,* August 1988.

Asgenmacher, Robert, "The Train That Brings the Coal," *Minneapolis Tribune Picture,* August 7, 1977.

"At Alliance, Young People Predominate," BN *News,* April 1979.

"Attitude Survey Important Step in Changing Culture," BN *News,* August–September 1983.

Austin, Ed. 2018. *Burlington Northern: Oregon.* Scotch Plains, N.J.: Morning Sun Books, Inc.

Austin, Ed, and Tom Dill. 1996. S.P.&S.: *The Spokane, Portland and Seattle Railway.* Edmonds, Washington: Pacific Fast Mail.

"AVP Allan Boyce Explains Coal Rates, BN's Skyrocketing Costs," BN *News,* February 1979.

Awai, Herman Tokuo Kealaula, "Double-Stack Containers: Changing the Image of Intermodalism." Masters Thesis, Naval Postgraduate School, March 1992.

Babcock, Michael W., "Efficiency and Adjustment: The Impact of Railroad Deregulation," <cato.org/pubs/pas/pa033.html>.

Bach, Ashley, "Plum Creek Timber Makes Huge Move on Northwest Forest Conservation," Washington Forest Protection Association, November 13, 2014.

Bachman, Ben. 1987. "Stampede! A Northwest Classic Is Reborn." *RailNews,* October, 1987.

"Bakken Oil and the BNSF." 2015. *The BN Expediter,* July, 2015.

Baldwin, William, "This Is Deregulation?" *Forbes,* October 27, 1980.

"Banks Will Finance C&NW Coal Line," *Railway Age,* December 28, 1981.

Barnes, David, "Shippers Urge Senate Panel to Reregulate Railroads," *Traffic World,* December 15, 1997.

Bauer, Edward L., Jr., "BN Explores Car, Locomotive Needs," *Modern Railroads,* November 1989.

Baugh, Odin. 2005. *John Frank Stevens: American Trailblazer.* Spokane: The Arthur H. Clark Company.

Bean, Radford, "Roberts Bank and the Railroads," *Railroads Illustrated,* March 2012.

Bechtold, Timothy Matthew. "Now v. Forever: The Conflict Between Business and Forestry in the Management of Plum Creek Timberlands in Montana." University of Montana ScholarWorks, 1992.

"Becoming No. 1: An Interview with Chairman and CEO Walter A. Drexel. "BN *News,* July 1984.

Bell, Wendell A., communication with author, 2014-2018.

Belmont Tunnel (Nebraska) <wikivisually.com/wiki/Belmont Tunnel_(Nebraska)>.

Berger, Jay, "BN Reopens Milwaukee Track," BN *News,* December 1980.

Berman, Phyllis and Roula Khalaf, "Sweet-Talking the Board," *Forbes,* March 15, 1993.

"'Best Ever' Job Protection Agreements Signed," *Burlington Bulletin,* January–February–March 1968.

"Betterment Committee: A Positive Influence," *Burlington Northern St. Cloud News and Views,* April 1985.

"Bill, Wyoming," *Trains,* September 1992.

Birkholz, Paul, with Al Krug, "Burlington Northern Fuel Tenders," <mtnwestrail.com/wyoming/BNft>

"Bizjet Biz at BN," BN *News,* July 1971.

Black, Jo Dee, "Asian Demand Driving Increase in Coal Train Traffic Through Area," *Great Falls Tribune,* March 13, 2011.

"Black Thunder: Facing the Competition on Hauling Powder River Basin Coal with a Firm Self-confidence," BN *News,* November 1985.

Blaha, John, correspondence with the author, various dates in 2013.

Blaszak, Michael W., "Free to Compete," *Trains,* October 2010.

Bleizeffer, Dustin, "Railroads Play Coal Catch-up," *Casper Star-Tribune,* March 18, 2006.

"BNAFI Sold," BN *News,* May, June, 1982.

"BNAFI to Be Sold," BN *News,* April 1982.

"BN Air Freight Climbs to No. 3 Spot in Industry," BN *News,* February/March 1978.

"BN Airmotive Ranks Near Top in National Survey," July 1980.

"BN and Utility Sign 15-Year Pact to Move Powder River Coal," *Traffic World,* November 7, 1983.

"BN Asks Court to Order Mediation in Labor Dispute," *Traffic World,* January 27, 1991.

"BN: A Turnout 'First.'" 1987, *Railway Age,* May 1987.

"The BN At 10: 'The Amazing Thing Was, It Worked," *Railway Age,* February 25, 1980.

"BN Backs Off on Plan to Auction Grain Cars," *Traffic World,* February 14, 1983.

"BN Begins Drug Tests," *Associated Press,* January 18, 1990.

"BN: Big Bucks for C&S [communications and signals]," *Railway Age,* February 1984.

"BN: Big Switch at Cicero," *Railway Age,* January 1983.

"BN Brings Futures Contracts into the Railroad Business," *Traffic World,* January 18, 1988.

"BN Chairman Accuses Texas Utility Firm of Using Political Pressures in Coal Haul Rates," BN *News,* July 1978.

"BN Considers Move of Some St. Paul Staff," *Mgr.,* December 14, 1983.

"BN Consolidates Repair Operations," *Railway Age,"* June 1992.

"BN Denies Role in Milwaukee Fall," *Railway Age,* April 28, 1980.

"BN Dismisses Dispatcher," BN *News,* October 1984.

"BN Elects Bressler of Arco to President and CEO Posts," BN press release, May 19, 1980.

"BN Fights Drug Dog Injunction," *Associated Press,* October 10, 1984.

"BN, Frisco Join Forces," *BN News,* special edition, January 1981.

"BN, Frisco Win the ICC's Blessing," *Railway Age,* April 28, 1980.

"BN, GE Begin Electrification Study," *Railway Age.*

"BN—Getting a Jump on Tomorrow," *Railway Track and Structures,* September 1983.

"BN Increases Safety and Reduces Costs: A Concrete Example," *BN Technical Bulletin,* July 1988.

"BN Installs Swing-Nose Frog in Nebraska." *Railway Age,* June 1986.

"BN Keeps On Track with Locomotive-Borne Lubrication," *BN Technical Bulletin,* May 1988.

"BNL Development Converts Property to Income," *BN News,* March 1979.

"BN, Montana Reach Accord," *BN News,* October 1984.

"BN News Interview: N. M. Lorentzsen," *BN News,* July 1975

"BN News Interview: Richard M. Bressler," *BN News,* March 1981.

"BN News Interview: Richard C. Grayson," *BN News,* February 1981.

"BN Offshoot to Run Hi-Line," *Billings Gazette,* November 28, 1987.

"BN Oil Refinery," *BN News,* October 1975.

"BN Operates Rail System for South Dakota," *BN News,* August/September 1981.

"BN Pays Milwaukee Road $21 Million for 383 Route Miles," *See-Port News,* April 1981.

"BN Plans $50 Million Alliance Expansion," *BN News,* September/October 1977.

"BN Plans to Offer $100 Million in Preferred Stock," *BN News,* June 1977.

"BN Rated No. 1 by Chicago Commuters," *BN News,* March 1981.

"BN 'Rubber Room' Lawsuit to Resume," *Spokesman Review,* December 6, 1988.

"BN Sets New Safety Mark in '83," *BN News,* March 1984.

"BN Sets Up Its Own R&D Program," *Railway Age,* February 8, 1982.

"BN Studies Structure, Location," *BN News,* July 1982.

"BN Switches Off On Electrification."

"BN's First 10 Years: Highlights of the Decade from Past BN News," *BN News,* March 1980.

"BN's Grinstein: Charting a Steady Course," *Railway Age,* January 1991.

"BN's Operations Department Has Moved South," *BN News,* July 1983.

"BN's 'People Trains' Seek Federal Grant," *BN News,* October 1970.

"BN's Plans for Galesburg," *BN News,* April 1983.

"BN Provides a Fast Track for Innovation." 1989. *Railway Track and Structures,* December, 1989.

"BN Puts Line Up for Sale," *The Montana Standard,* November 26, 1986.

BN Railroad, "An Open Letter to BN Railroaders," *Livingston Enterprise,* October 27, 1987.

"BN Rated No. 1 by Chicago Commuters," *BN News,* March 1981.

"BN Realigns Top Operating Staff," *Railway Age,* October 1992.

"BN Rejects ICC Price on Wyoming Joint Line," *Traffic World,* November 15, 1982.

"BN Retracts from Proposed Kansas City–Northern Line," *BN News,* March 1981.

"BN 'Rubber Room' Lawsuit to Resume," *Seattle Times.*

"BN Runs New Expediter in Chicago–Minnesota Lane," *Traffic World,* October 13, 1986.

"BN, SD Celebrate Core System Opening," *BN News,* November 1981.

"BN Seeks Coal Talks with Utilities," *Railway Age,* October 26, 1981.

"BN Sets Price for Powder River Basin Line," *Mgr.* July 21, 1982.

"BN Sets New Safety Mark in '83," *BN News.*

"BN Sets Up Pilot Projects to Test Hub Terminal Concept," *Mgr.* September 23, 1982.

"BN Shops: Cutting Costs—Not Quality," *Railway Age,* October 26, 1982.

"BN Studies Structure, Location," *BN News.*

"BN Suburban Operations Win Transportation Award," *Chicago Region News,* August/September/October 1978.

"BN Sues Engineer, Brakeman Over Wreck," *Livingston Enterprise,* April 25, 1984.

"BN Suggests a BYOM Party," *Railway Age,* October 26, 1981.

"BN 'Swats' Track Maintenance Chores," *Railway Track and Structures,* October 1987.

"BNT Takes First Place—Ten Times," *BN News,* December 1980.

"BN Takes $435 Million Charge Against Operating Income in Quarter," *Traffic World,* July 1, 1991.

BN TEAM Technical Bulletin:

    Acoustic Bearing Detection, March 1989

    BN Increases Safety and Reduces Costs: A Concrete Example, July 1988

    High-Speed Turnouts, September 1990

    In-Place Tie Treatment, October 1991

    Locomotive Analysis and Reporting System (LARS), July 1991

    Rail Grinding, August 1989

    Rehabilitated Welded Rail, January 1991

    Swing Nose Frogs, October 1988

    Track Surface Management, July 1990

"BN Tests Natural Gas as Locomotive Fuel," *BN News,* November 1985.

"BN: The Measure of Good Management," *Railway Age,* February 28, 1972.

"BN Trains Running Over 140 Miles of Milwaukee Road," *See-Port News,* May 1980.

"BN: Up to the Sky," *Progressive Railroading,* 1987.

"BN's Bid to Scrap Merger Conditions Meets Opposition from Competitors," *Traffic World,* September 14, 1981.

"BN's Expediter: A Pioneer in the Intermodal Industry," *BN News,* July–August 1986.

"BN's Plans for Galesburg," *BN News*, April 1983.

"BNSF Bridge 3.8 over Missouri River," AREMA (American Railway Engineering and Maintenance Association), 2018.

"BNSF Completes Triple-Tracking of Joint Line," railwayage.com/breaking_news.shtml#Feature.

"BNSF Employees Celebrate Orin Line's 30-Year Anniversary," <rtands.com/index.php/news/BNsf-employees-celebrate-orin-lines-30-year-anniv>

"BNSF Moves Forward with Plan to Build Second Bridge at Idaho Bottleneck," *Trains Industry Newsletter*, April 19, 2017.

BNSF Railway Statement on STB Coal Dust Decision, bnsf.com/customers/what-can-i-ship/coal/coal-dust.

"BNSF Reopens Stampede Pass," *Railway Age*, January 1997.

"BNSF Strikes Deal to Buy 'Core Lines' from South Dakota," *Trains,* August 2005.

"BNSF: 'That Big Berkshire Hathaway Railroad Deal." 2016. *Bloomberg,"*November 11, 2016.

"BNSF to Move Coal That Will Power New IGCC Plants in Japan," *Progressive Railroading*, January 17, 2018.

Borchersen-Kelo, Sarah. 2014. "Plum Creek Timber Harvesting Opportunities Across Its Portfolio," *Reit News,* October 32, 2014.

"Breezy Point and Beyond," CTC *Board*, April 1991.

Brehm, Kyle, "The Montana Rail Link Story," *Railfan and Railroad,* April 1990.

"Bressler Named BN Chief," BN *News,* July 1980.

"Bressler: Railroad (is) 'Crown Jewel' of BNI," *Mgr.,* April 1982.

Bressler, Richard, Letter to employees, December, 1980.

Bressler, Richard, transcript of speech delivered to the Western Coal Transportation Association, September 10, 1980.

Bressler, Richard M., "Operating Trains, After All, Is a Business," *Trains,* August 1982.

Bressler, R. M., Testimony before a Senate Subcommittee, reported in *Mgr.* April 1982.

Briggeman, Kim, "Towing the Line 20 Years After Montana Rail Link Took Over Track Owned by Burlington Northern Things Have Settled Down," *Missoulian,* October 7, 2007.

"Bring Money, 'BN Tells Would Be Competitors,' " *Traffic World,* October 19, 1981.

Brohaugh, R. G., "The Gellette – Orin Line, One Year Later," American Railway Engineering Association, *Bulletin 683.*

Brotherhood of Locomotive Engineers, Plaintiff—appellee, v. Burlington Northern Railroad Company, Defender-appellant, 838 F.2d 1102 (9th Cir. 1988).

Brotherhood of Locomotive Engineers v. Burlington Northern Railroad Company, Court of Appeals for the Ninth Circuit, Case No. 85-4138.

Brown, Douglas John, "STB Backs BNSF on Coal Dust Issue," *Railway Age,* December 17, 2013.

Brown, Matthew, "King Coal: Despite Concerns About Climate Change, A Batch of New Plants Confirms the Reign of Coal Is Far From Over," *Minneapolis Star-Tribune,* August 20, 2010.

Brownlee, James H., "Protecting an Asset" (rail), *Modern Railroads,* March 1985.

Bryan, Frank W., "Computers in the Cab," *Trains,* June 1990.

Buchsbaum, Lee, "A Dragon They Just Can't Kill," *Coal Age,* July 2005.

Buchsbaum, Lee, "BNSF: We Won't Pay for Dust Fix," *Trains,* December 2007.

Budd, Ralph, "Northern Rail Lines Across the Divide," *Civil Engineering,* April 1940.

Buekett, John, Derek Firth and John R. Surtees, "Concrete Ties in Modern Track,"*Railway Track and Structures,* August 1987.

Buhayar, Noah, "Buffett Said He Paid a Lot. $15 Billion Later, BNSF Is a Cash Machine. 'He Stole It.' " *Bloomberg News,* November 10, 2014.

"Buffett Buying Burlington Northern Railroad," Associated Press, November 3, 2009.

"Buffett: BNSF Buy 'Even Better Than I Expected,' " *Railway Age,* April 2011."

Burgett, Michael J. "The Engineering Basics of CTC," ctcparts.com.

Burke, Jack, "Northern Tier Agreement with UTU 'Overwhelmingly' Approved by Members," *Traffic World,* May 24, 1993.

Burke, Jack, "Burlington Northern, Soo Line Settle Track Rights Case Out of Court," *Traffic World,* February 7, 1994.

Burke, Jack, "Burlington Will Buy General Motors AC Locomotives for $675 Million," *Traffic World,* March 22, 1993.

Burke, Jack, "Management Shake-up at Burlington Northern Ousts Two Veterans; DHL, Frito-Lay Execs In," *Traffic World,* June 6, 1994.

Burke, Jack, "Rings, Radios, Re-enactments Irk BN Workers; Labor Relations Strained Over Series of Issues," *Traffic World,* February 14, 1994.

Burke, Jack, "Is BN Back? Strong Fourth Quarter, Cost Reductions May Signal Blue Skies, Black Ink Ahead," *Traffic World,* February 15, 1993.

"Burlington Begins Study of Whether to Move Corporate Headquarters," *Traffic World,* June 28, 1982.

Burlington, Great Northern, Northern Pacific Merger: ICC Finance Docket 21478-21480.

Burlington Northern Air Freight 1979 Annual Report.

"Burlington Northern: An Elephant by the Tail." *Forbes,* October 30, 1978.

"Burlington Northern Asks Conditions on North Western's Powder River Line," *Traffic World,* December 9, 1985.

Burlington Northern Railroad: Back on Track, Harvard Business School Case 9-396-214 prepared by Carin Knoop, March 20, 1996.

Burlington Northern Railroad Company, Plaintiff—Appellant, v. United Transportation Union, et al., Defendants—Appellees.

Brotherhood of Locomotive Engineers, Plaintiff—Appellee, v. Winona Bridge Railway Company, Defendant—Appellant.

Burlington Northern Railroad Company v. United Transportation Union a Jw Jg We Railway Labor Executives' Association 848 F. 2d 856, decided May 31, 1988.

Burlington Northern Railroad Company v. United Transportation Union, Brotherhood of Locomotive Engineers v. Winona Bridge Railway Company, United States Court of Appeals, Seventh Circuit, Decided November 18, 1988, and 851 F.2d 1056, M.M. Winter, Petitioner, v. Interstate Commerce Commission and United States of America, Respondents.

"Burlington Northern, EMD Ink 2nd Loco Upkeep Deal," *Traffic World,* July 31, 1989.

"Burlington Northern Expands Cicero Yard, Makes It Intermodal Facility," *Traffic World,* January 10, 1983.

"Burlington Northern Given Good Marks on FRA Safety Inspection," *Traffic World,* May 20, 1985.

Burlington Northern, Inc., Annual Reports, 1969–94.

Burlington Northern, Inc. common stock history—BNSF Railway www.bnsf.com/about/bnsf/financial-information/pdf/bn_history.pdf

Burlington Northern, Inc., et al., Petitioners, v. The United States of America and Interstate Commerce Commission, Respondents San Antonio, Texas, Acting by and Through Its City Public Service Board, Invervenor-Respondent. *No. 76 -1899.* United States Court of Appeals, Eighty Circuit. Decided March 29, 1977.

Burlington Northern, Inc., Notice of Annual Meeting of Shareholders to Be Held April 20, 1995, dated March 17, 1995.

Burlington Northern Annual Report to Employees, 1974 and 1978.

Burlington Northern Bankers Tour, September 24–27, 1979 (notebook containing engineering and operating data, maps and transcripts of presentations to be made on an inspection trip).

Burlington Northern Financial Analysts Presentations, February 18, 1976, August, 1979, and August, 1981.

Burlington Northern Institutional Lenders Tour, September 18-21, 1978 (notebook containing engineering and operating data, maps and transcripts of presentations to be made on an inspection trip).

Burlington Northern Company, Appellee, v. United Transportation Union, Appellant, Fred A. Hardin, et al. Appeal from the United States District Court for the Western District of Missouri, *No. 87-2581.* Filed: May 31, 1988.

Burlington Northern Railroad. <en.wikipedia.org/wiki/Talk %3ABurlington-Northern-Railroad/temp>.

"Burlington Northern Railroad Unit Names Gen. Chain to Post," *Wall Street Journal,* December 5, 1990.

"Burlington Northern Sells Truck Subsidiary," *Traffic World,* Auguat 15, 1988.

"Burlington Northern Spared No Expense in Hiring New Chief," *Wall Street Journal,* May 4, 1981.

"Burlington Northern Splits Operations in 2: Non-Transport Units Isolated, to Be Spun Off," *Associated Press,* June 3, 1988.

"Burlington Northern Takes a Look South." 1970. *Railway Age,* May 25, 1970.

"Burlington Northern: The ARES Decision," Center for Digital Economic Research, Sloan School of Business, Working Paper, IS-93-18.

Burlington Northern: The ARES Decision (A) and (B), Harvard Business School Cases 9-191-122 and 9-191-123, February 21, 1991.

"Burlington Northern's Ditching Program," BN *Technical Bulletin,* May 1989.

"Burlington Northern's Fight to Repel Invaders," *Business Week,* November 3, 1980.

"Burlington Northern's Tamping Technology," BN *Technical Bulletin,* February 1989.

"Burlington Resources Shifting Its Focus," *Seattle Times,* March 2, 1993.

"Burlington Resources, Inc. History," <fundinguniverse.com/company-histories/burlington-resources-inc-history/>.

Burlington Resources Inc., Prospectus, Proposed Merger with ConocoPhillips, February 24, 2006.

Burlington Resources Inc., Prospectus, Sale of 20,000,000 Shares of Common Stock, July 7, 1988.

Burlington Resources, Inc., Information Statement Concerning the Distribution of Approximately 87% of the Outstanding Shares of Burlington Resources, Inc., by Burlington Northern, Inc., November 14, 1988.

"Burlington Resources Shifting Its Focus," *Bloomberg Business News,* March 2, 1993.

"Burlington Spinoff About to Happen," *Seattle Times,* December 5, 1991.

Burns, Alan, "Under a Big Sky," CTC *Board,* December 1989.

Burns, David H., "BN Chicago Region in the Early Days 1970–72," BN *Expediter,* January 2014.

Burns, David H. 2017. "Colorado Division Days 1979–81." The BN *Expediter,* April, 2017.

Burns, David H., "Deregulation and the Rise of Intermodalism," October 2012.

Burns, David H. 2013. "M-Day on the Dakota Division," The BN *Expediter,* July, 2013.

Burns, David R., "M/W Cost Components: Part I—Surfacing," *Railway Track and Structures,* April 1987.

Burns, James B. 1998. *Railroad Mergers and the Language of Unification.* Westport, Conn.: Quorum Books.

Burton, R. C., Jr. "Financing a Railroad," September 1979.

Busse, Meghan R. and Nathanial O. Keohane, "Market Effects of Environmental Regulation: Coal, Railroads and the 1990 Clean Air Act," University of California Energy Institute, September 2004.

Button, John J., Statement Before the Federal Railroad Administration on Control of Alcohol and Drugs in Railroad Operations, September 2, 1983.

Byrne, Don, "Rail Union Head Attacks Safety Board for Seeking 'No-Drinking' Regulation," *Traffic World,* May 14, 1984.

"Cabooseless Operation . . . Safer, Money Saver," *Progressive Railroading,* September 1987.

Callahan, Lynn Casey, "Alliance Thrives Amid Tough Challenges," *BN News,* July 1980.

Callahan, Lynn Casey, "BN Airmotive Takes Off As Leader in 'FBO' Business," *BN News,* November 1979.

Callahan, Lynn Casey, "BN Commuters Leave Car Keys at Home, Still Get to Chicago," *BN News,* February 1980.

Callihan, Lynn Casey, "Coal Makes the Difference at Guernsey," *BN News,* May/June 1980.

Callihan, Lynn Casey, "On the Playing Field: Deregulation Gets Railroads Off the Bench and Into the Game," *BN News,* January 1982.

Cama, Timothy, "Toxic Air Pollution Has Dropped Dramatically, EPA Tells Congress," *The Hill,* August 21, 2014.

Camp, Janice, "The Problem Center: Employee Assistance Expands," *BN News,* August/September 1982.

"C&NW and BN Still Don't Agree," *Railway Age,* November 8, 1982.

"C&NW—Going for a Bundle from FRA," *Railway Age,* December 26, 1977.

"C&NW Seeks Access to More BN Traffic," *Railway Age,* February 1985.

"C&S Engineering Closes a Busy Year," *Denver Region News,* December 1980.

"C&S Merger Plan Outlined," *BN News,* November 1981.

"Can Acoustic Detectors Reduce Bearing Failures?" *Railway Age,* October 1988.

Cantrell, Anne, "Q&A with Jerry Grinstein." www.montana.edu/news/15781/jerry-grinstein-montana-state.

Carr, John, *Western Coal: 2005 Compendium of Western Coal Mines, Electric Utilities and the BNSF and UP Trans That Serve Them,* 2006. Duncanville, Texas, Carr Tracks.

"The Carrot Instead of the Stick: One Division's Experiment in Discipline," *BN News,* Fall 1989.

"Cascade Tunnel Ventilation, dslweb.nwnexus.com/tawhite/CASCADE%20TUNNEL

Caughey, Peter, "Railroad Gives Livingston $1 Million Aid Package," *Boseman Daily Chronicle,* May 7, 1986.

Cawthorne, David M., "BN Places Non-Rail Holdings in Separate Corporate Unit," *Traffic World,* June 6, 1988.

Cawthorne, David M., "BN Proposes Major Overhaul in Work Rules, Pay Practices," *Traffic World,* June 22, 1987.

Cawthorne, David M., "BN Rail to Have Far More Debt After Restructuring Finished," *Traffic World,* June 18, 1988.June 18, 1988.

Cawthorne, David M., "BN Seeks Final ICC Approval of Transport Certificates Program," *Traffic World,* May 27, 1991.

Cawthorne, David M., "Burlington Northern Gets Final OK for Freight Car Futures Program," *Traffic World,* February 3, 1992.

Cawthorne, David M., "Future of Winona Bridge Plan Doubtful After Court Ruling," *Traffic World,* June 20, 1988.

Cawthorne, David M., "ICC May Collide with Congress over Line Sale Labor Protection," *Traffic World,* September 14, 1987.

Cawthorne, David M., "U.S. Supreme Court Likely to Review Rail Line Sales," *Traffic World,* June 6, 1988.

Celis, William III, "New Burlington Northern Unit to Offer Shares," *The Wall Street Journal,* June 3, 1988.

Chesser, Al (president of United Transportation Union), *Project Seventies, 1970.*

"Chicago Accounting Dept. Moves to General Office," *BN News,* November 1970.

Chicago, Burlington and Quincey Railroad Co., *Annual Reports, 1960–1968.*

Chicago, Milwaukee, St. Paul and Pacific Railroad Company, Petitioner, v. United States of America and Interstate Commerce Commission, Respondents. 585 F.2d 254, No. 77-1453. October 4, 1978.

"Chicago's People Trains," *BN News,* July 1970.

Christensen, Jack, "Arthur Fiedler, 1927–2011," *The Mainstreeter,* Summer 2012.

Christensen, Kathryn, and Richard D. James, "Volcano's Continuing Eruption Damages Lane, Stifles Commerce in the Northwest," *Wall Street Journal,* May 20, 1980.

"Chronology of the Northern Pacific and Related Land Grant Railroads," compiled by George Draffan, July 24, 2001.

Circular, $60,000,000 Burlington Northern, Inc., Consolidated Mortgage 8.60% Bonds, Series D, Due 1999.

Citizens' Alliance to Save the Southline, letter to U.S. Senator Max Baucus, April 7, 1987.

"Citizens of Alliance Comment on Effects of BN's Expansion Plans," *BN News.*

Clark, Milt, operations in Montana and Washington (Milwaukee Road), correspondence with the author.

Clark, Milt, "The Milwaukee Road's Seattle Connection," *The Milwaukee Roader,* Third Quarter 2009.

"Clean Coal Project in Texas Wins $450mn in Federal Help," *Power Engineering,* September 2009.

Cliff, Daniel F. 1988. "Business People; Burlington Northern Names 2 Executives," *New York Times,* October 21, 1988.

"CNW Earns $43 Million After $27 Million Loss," *Traffic World,* February 16, 1987.

"CNW Unit Buys Interest in BN 10-Mile Coal Line," *Traffic World,* January 5, 1987.

"Coal Boom: Area Set for Another Boom by 1990," *The News-Record* (Gillette, Wyo.), December 14, 1981.

"Coal Car Repair Shop Opens at Laurel," *BN News,* June 1977.

Coal in Montana-Montana State Legislature <leg.mt.gov/content/publications/Environmental/2004deq.../coal_text.pdf>.

"Coal Line Upgrading, Fremont, Nebraska to Shawnee, Wyoming," *North Western Lines,* Summer 2004.

"Coal Makes the Difference at Guernsey," *BN News,* May/June 1980.

"Coal Mining in Powder River," <coaldiver.org/coal-diver/Powder-River?>.

Coal Rates Guidelines, 1 ICC 2nd 520 (1985).

"Coal Shippers Are Pulling Together on Policy for Railroad Freight Rates," *Traffic World,* May 30, 1982.

"Coal Transportation Issues," U.S. Energy Information Administration, www.eia.gov/oiaf/aeo/otheranalysis/cti.

"Coal Use Expected to Increase in U.S. by 2025," *BNSF Today,* January 2, 2004.

"Col. Crown Calls Frisco Good Investment," *Chicago Tribune,* February 19, 1967.

"Collective Bargaining Under the Railway Labor Act," *The Locomotive Engineer,* May 18, 1984.

"Coming of Age," *BN News,* September-October 1986.

"Common Stock Dividends Resumed," *BN Westviews,* January/February 1976.

"Communications Division: Vital to BN Operations," *BN News,* November 1975.

"Commuter Service Moves 46,000 Daily," *BN News,* September 1976.

"Commuter Trains: Passengers Rate BN's Suburban Services Best in Chicago," *BN News,* July 1970.

"Company Meets Taconite Challenge," October 1976.

"Company Omits Dividend," *BN News,* September 1975.

"Company Reduces Dividend," *BN News,* October 1970.

"COMPASS Points the Way," from author's collection; date and name of publication not names.

"Competition Begins in Basin," *BN News,* October 1984.

"Competition Comes to Powder River," *Railway Age,* October 1984.

"The Computer Age Comes to Southern Pacific," *Southern Pacific Bulletin,* December 1967.

"Computerizing Track Warrants," *BN Lakes Division Monthly Report,* no date shown.

"Concrete Tie Turnouts Emerge." *Progressive Railroading,* October 1989.

Conference of BN Executives in March 1983, excerpts quoted in *BN News,* May/June 1983.

The Consolidated Code of Operating Rules, Edition of 1967.

"Construction at Mandan and Sioux City Helps BN Handle Rail Traffic," *BN News,* August/September 1980.

"Construction Update: Getting Ready to Run," Union Pacific *InfoNews,* March 1984.

"Conversation with the CEO," *BN News,* November 1990.

Conway, Chuch, "Powder River Coal," *CTC Board,* April 1991.

Cook, James, "All Alone by the Telephone," *Forbes,* November 16, 1987 (Richard M. Bressler quoted).

Cook, James, "The Turning Point," *Forbes,* September 24, 1984.

Corbin, Steven B., and Harry W. Zanville, "Why Would Anyone Want to Try to Teach a Course About the Rail Industry and How Would They Do It?" *Journal of Transportation Law, Logistics and Policy.*

Corr, O. Casey, "The Chairman: BN's Bressler Is One Tough Boss – And a Puzzle," *Seattle Post-Intelligencer,* March 22, 1988.

"The Cost Effectiveness of America's Freight Railroads," AAR position paper, May 2014.

"Costly Line Change Pays Off for NP," *Railway Age,* December 13, 1965.

"Court Says Railroad Workers Can Engage in Secondary Picketing," *Associated Press,* April 28, 1987.

Coyne, Frank H., remarks at Burlington Northern–held Seminar on Grain Transportation in 1980.

"Crawford Hill Line Change Completed," *Mgr.,* July 21, 1982.

"(1980) Crew Consist Agreement Between Burlington Northern Railroad and the United Transportation Union" utu324.com/93_crew_consist.

Crew Consist Agreement Between Burlington Northern Railroad and Its Employees Represented by United Transportation Union Effective November 1, 1993.

"Crew Consist Agreement Between Burlington Northern Railroad and its Employees Represented by United Transportation Union Effective May 20, 1993.

"Crew Consist Agreement Signed," *BN News,* October 1984.

Cross, B. B., "Alfred E. Kahn: The Life and Death of the Father of De-regulation," Associated Content, 2011.

Cuff, Daniel F. "Burlington Northern Names 2 Executives," *New York Times,* October 21, 1988.

Currie, Earl J. 2007. *James J. Hill's Legacy to Railway Operations.* Self-published.

Currie, Earl J. 2015. *Building Burlington Northern: The Lorentzsen Years, 1970–1980.* Self-published.

Currie, Earl J. 2010. *BN-Frisco: A Tough Merger.* Self-published.

Currie, Earl J. 2010. *Nebraska Division: Challenge and Reward, 1975–1977.* Sefl-published.

Currie, Earl J. 2011. *Robert W. Downing: One of the Best and Brightest.* Self-published.

Currie, Earl J., Statement Before the Federal Railroad Administration on Control of Alcohol and Drug Use in Railroad Operations, August 2, 1984.

"Customer Service: It's Everybody's Job," *BN News,* Spring 1987.

"The Customer: What BN Is Doing to Win Them and Keep Them," *BN News,* November 1983.

"Cutover Seen as Significant Change by Burlington Northern," *The Plattsmouth Journal,* December 2, 1976.

Danneman, Mike, "Coal and Curves," *CTC Board Railroads Illustrated,"* December 2004.

Danneman, Mike and Dave Gayer. 2010. *Montana Rail Link.* Kansas City: White River Productions.

Darius W. Gaskins, Jr. www.zoominfo.com/people/Gaskins_Darius_79175318.asps.

"Dave Crooker, Resources Division Silviculturist, Uses His Knowledge to Help Forest Renew Itself," BN News, November 1978.

Davis, David D ."Evaluation of Load Environment of Flange-bearing Frogs," Railway Track and Structures, July 2013.

DeBoer, David J. 1992. Piggyback and Containers: A History of Rail Intermodal on America's Steel Highway. San Marino, Calif.: Golden West Books.

DeBoer, Dave, "Stacking the Deck," Trains, November 2011.

"Dedication and Determination," Powder River Reflection, October 2009.

"A Definition of Corporate Culture," BN News, March 1985.

Del Grosse, Robert C. 1991. Burlington Northern: Trackside Guide. Bonners Ferry: Great Northern Pacific Publications.

Del Grosso, Robert C. 1994. Burlington Northern: Railroad Giant of the Pacific Northwest. Bonners Ferry: Selkirk Press.

Del Grosso, Robert C. 2014. Burlington Northern Railroad: Historical Review 1970 – 1995. Halifax, Penn.: Withers Publishing.

"Deliveries of Coal from the Powder River Basin: Events and Trends 2005–2007," U.S. Department of Energy, October 2007.

Dennis Washington, Hawassa Online. www.hawassaonline.com/biography_detail.php?id=18&type=entrepreneurs>.

Dennis R. Washington, 2007 Inductee into American Mining Hall of Fame, quoted in "Mining Foundation of the Southwest."

"Depreciation Accounting for Track Structure," BN document, Effective January 1, 1983.

Derailment of Extra 5701 East at Sheridan, Wyoming, March 28, 1971, National Transportation Safety Board Report Number: NTSB-RAR-72-4, April 26, 1972.

"Deregulation: Terms, Questions, Constraints," Burlington Northern Technical Training, 1980.

"Deregulation: What Happen Next?" Railway Age, October 27, 1980.

DeRouin, Edward M., "Land of the Burlingtons, Part II, Building on the Foundation Laid by CB&Q," Passenger Train Journal, January 1989.

"Differential Pricing: Efficient and Necessary," Association of American Railroads Policy and Economics Department, July 2006.

DiMichael, Nicholas J., "Living with Stand-Alone Costs: The OPPD Decision," Traffic World, January 5, 1987.

Ditmeyer, Steven R., "Burlington Northern and Technology," letter to Trains, January 2015.

Ditmeyer, Steven R. and Edward L. Butt, "S&C (Signals and Communications) Projects Focus on Service, Safety," Modern Railroads, November 1989.

"Division Bids Farewell to Don Maze," Pride of the Northwest, March/April 2003.

"Doctors Plan Gillette Practice," The News-Record (Gillette, Wyo.), December 31, 1975.

Dodge, John, "Positive Train Control Ran Successfully in Years-Long Burlington Northern Railroad Trial," Design News, October 21, 2008.

Doerr, Norman M., "Burlington Northern Railroad Merger from a Non-Operating Department Perspective," 2015.

Donaldson, Gordon. 1994. Corporate Restructuring: Managing the Change Process from Within. Boston: Harvard Business School Press.

Dorin, Patrick C., and Robert C. Del Grosso. 1995. Burlington Northern Railroad: Coal Hauler and Coal Country Trackside Guide. Great Northern Pacific Publications.

Downey, Mark, "Crew, Seattle Dispatcher Misunderstood Clearance," Great Falls Tribune, September 3, 1991.

"Downing Recounts Great Northern, Burlington Northern Merger Days," Vingage Rails, Winter 1996.

Downing, Robert W., Annual Luncheon Address, American Railway Engineering Association, Bulletin 648.

Draffan, George, "Annotated Bibliography on Railroad Land Grants," <www.landgrant.org/biblio>.

Draffan, George, "Chronology of the Northern Pacific and Related Land Grant Railroads," Endgameame.org, July 24, 2001.

Drajem, Mark, "EPA Final Rule for Coal Plants Deemed 'Unfortunate' by Industry," Bloomberg, December 19, 2011.

Drexel, Walter A., "TEAMWORK: Management, Labor Must Unite at BN," presentation to company officers, BN News, May/June 1983.

"Drugs: Risky Business," BN News, October 1985.

Dunwiddie, Paul, "Derailed: Auburn's 'Grand Ol' Lady' About To Make It a Widower," Daily Globe News, September 10, 1981.

Dusell, Forester, "A Partnership Report," 1972 (a one-page handout given to passengers on commuter trains).

Eakin, B. Kelly, and Mark E. Meitzen, "Opinion: After 30 Years, Railroad Deregulation Continue to Deliver," <aolnews.com/opinion/article/opinion-after-30-years-railroad-deregulation j-con>.

Eakin, B. Kelly, A. Thomas Bozzo, Mark E. Meitzen and Philip E. Schoech, "Railroad Performance Under the Staggers Act," Regulation, Winter 2010–2011.

"Economists: Rereg Would Imperil Productivity Gains," Railway Age, November 9, 2010.

"EEOC, Burlington Reach $10 Million Settlement in Discrimination Suit," Traffic World, November 21, 1983.

"Effectively Competing: The Product Has Improved," BN News, June 1985.

Eisthen, Bob, "Union Pacific Builds for the Future," CTC Board Railroads Illustrated," May 1989.

Ekey, Robert, "BN Ponders Its Future in Livingston," Billings Gazette, October 27, 1986.

Ekey, Robert, "Railroad Claims Law Will Double Its Taxes," Billings Gazette, October 27, 1985.

Ekey, Robert and Paul J. Holley, "State Off Track, BN Says," *Billings Gazette,* October 27, 1985.

"Electrification: An Answer to Rising Fuel Costs," *BN News,* January 1981.

"Electronic Search Simplifies Diesel Problem Diagnonis," *BN News,* January 1979.

Ellig, Tracy, "Grain Byer Gene Thayer Grows a Future for Bobcat Athletics," *Mountains and Minds Magazine,* Fall 2007. Press release issued by Montana State University, 2007.

"EMD Takes Over Former BN Shop," *Railway Age,"* March 1989.

"Employee Opinion Survey: 2,419 Managers Speak Out," *Mgr.,* April 18, 1984.

"Employee Opinion Survey Report," *BN News,* March 1984.

Employee Protection Agreement Between Burlington Northern Railroad Company and Brotherhood of Locomotive Engineers, July 18, 1985.

"The End of the Line?" (for cabooses), *BN News,* May 1985.

"The End of the Line: Why BN Sheds Unprofitable Lines," *BN News,* July 1983.

"Energy Boom Will Double Gillette's Population by '78," *Rapid City Journal,* May 5, 1974.

"Energy: Frequently Asked Questions," <uprr.customers/energy/sprb/faq.shtml>.

Energy Transportation Systems, Inc. <www.engagingnews.us/select/Energy-Transportation-Systems-Inc.>.

Engineering Division Data Book. Burlington Northern, December 31, 1979.

"Engineer Is Indicted on 16 Counts of Manslaughter in Amtrak Crash," *New York Times,* May 5, 1987.

"Environment and Economics: The Need for Balance," *BN News,* July 1980.

Enzi, Michael B., *Gillette History (The 1970's),* Wy. V-File, October 1, 1976.

"Equal Employment Opportunity: Two Decades of Progress," *BN News,* Fall 1989.

Etheridge, A. Charles, "BNAFI-Chicago Was a Gamble—But It Proved to Be a Sure Thing," *BN News,* December 1978.

Etheridge, A. Charles, "BN Investment in Minerals Good for Country and Company," *BN News,* October 1980.

"ETSI's Coal Slurry Pipeline' Nothing New'" *Gillette News – Record,* June 5, 1975.

"ETSI Pipeline Project v. Burlington Northern, Inc. (U.S. District Court, Eastern District of Texas)" <furth.com/historical>

"The ETSI Settlement Took $108.5 Million . . .," *Burlington Northern Railroad News Update,* February 1989.

Ettorre, John J., "Rails Cautiously Pursue Testing Despite Drug, Alcohol Rule Suits," *Traffic World,* June 16, 1986.

EuDaly, Kevin. 2004. "Coal." *North Western Lines,* Summer 2004.

"The Energy Crisis and Wyoming Coal." Burlington Northern, 1979.

"The ETSI Settlement Took $108.5 Million . . . ," *Burlington Northern Railroad News Update,"* February 1989.

The Facts About Air Quality and Coal-Fired Power Plants. <instituteforenergyresearch.org>.

Farney, Dennis, "Unkindest Cut? Timber Firm Stirs Ire Felling Forests Faster Than They Regenerate," *Wall Street Journal,* June 18, 1990.

"Fast On/Off Equipment Buys Tie Gangs Track Time," *Railway Track and Structures,* October 1989.

Feurer, Keith, "Look Who Came Out on Top," *North Western Lines,* Summer 2004.

Field, David, "Shippers' Agents Debate Linertrain; Opportunity, Threat to Future Seen," *Traffic World,* June 10, 1985.

"First Quarter Random Drug Testing Results Tallied," *Burlington Northern Railroad Inside,"* May 1990.

Forman, Kim, quoted in "1970 News" in <Wikipedia.org/wiki/Talk%3ABurlington_Northern_Railroad/temp>.

Forman, Kim, "BN Timber and Land Division Manages Resources with Eye to Needs of Future Generations," *BN News,* August 1977.

"Final Run for Sogn Patriarch," *Powder River Reflection,* 3rd ed., 2011.

"Fort Worth Move Completed," *BN News,* October 1984.

"Fort Worth Relocation," *Mgr.,* June 20, 1984.

Fox, Wesley, *Powder River Coal and the BN's Denver Division,* Arvada, Colo.: Fox Publications, 1999.

Frailey, Fred W., "A Different Way to Run a Railroad," *Trains,* October 2013.

Frailey, Fred W., "Powder River 'Blowout' Sends BNSF, UP Reeling," *Trans,* September 2005.

Frailey, Fred W. 1989. "Powder River Country: America's Last Rail Frontier," *Trains,* November, 1989.

Frailey, Fred W. 2010. "Powder River Stories: Five Moments That Made a Difference in the Life of North America's Busiest Coal Railroad." *Trains,* April 2010.

Fram, Alan, "Drug, Alcohol Use Widespread in Rail Industry, Accident Crewmen Say," *Associated Press,* February 25, 1988.

Frank Christopherson, Locomotive Engineer, Leader and Volunteer, an oral history conducted by Tom Gannon and Tim Schandel, compiled by Earl J. Currie, December 2011.

"FRA Orders Safety Audit of Burlington Northern," *Traffic World,* August 27, 1984.

Fredrix, Emily, "Officials Sign Documents to Close Sale of Core Rail Line," Associated Press, December 17, 2005.

"Freight Railroads: Industry Health Has Improved, But Concerns About Competition and Capacity Should Be Addrressed," United States Government Accountability Office, October 2006.

French, Peter. "U.S. Railroad Safety Statistics and Trends, 1980–2010," Association of American Railroads, May 17, 2011.

Friedlaender, Ann F., Ernst R. Berndt, and Gerard McCullough. "Governance Structure, Managerial Characteristics and Firm Performance in the Deregulated Rail Industry," *Brookings Papers on Economic Activity. Microeconomics,* Vol. 1992. www.jstor.org/stable/2534763.

"Frisco's Good Connections, Confidence Help Maintain High Level of Service," *BN News,* May 1978.

"Frisco Vice Presidents Appointed to BN Posts," BN press release, May 19, 1980.

"Frogs: More Than What Meets the Eye." 1991. *Railway Track and Structures,* September, 1991.

"From the Desk of Jerry Grinstein: A New CEO's Perspective," *Mgr.* First Quarter, 1989.

"Frontiersmen of the Powder River Basin," *TrainsTalk,* September 25, 2009.

Fruin, Jerry and Robert Crnkovich, "Western Coal Transportation Rates for Minnesota Users," Staff Paper P78-3, Department of Agricultural and Applied Economics, University of Minnesota, February 1978.

Gaertner, John T. 1990. *North Bank Road: The Spokane, Portland and Seattle Railway.* Pullman: Washington State University Press.

Gaertner, John, "The Twin Cities Joint Terminal," *The Mainstreeter,* Spring 2009.

"Galesburg Gears for Growth," *Railway Age,* April 1985.

"Galesburg Yard Dedicated," *BN News,* October 1984.

Gallagher, John, "Reconsider This: Some Ex-Railroaders Question Benefit of a Scheduled Railroad," *Traffic World,* July 12, 1999.

Gallagher, John, "UP Eating BNSF's Lunch," *Traffic World,* July 19, 1999.

Gardner, A. Dudley, and Verla R. Flores. *Forgotten Frontier: A History of Wyoming Coal Mining.* Boulder, Colo.: Westview Press.

Gallamore, Robert E. and John R. Meyer. 2014. *American Railroads: Decline and Renaissance in the Twentieth Century.* Cambridge, Mass., and London: Harvard University Press.

Gaskins, Darius W., "Regulation of Freight Railroads in the Modern Era: 1970–2010," *Review of Network Economics,* December 2008.

Gayer, Dave and Mike Danneman, "Montana Rail Link Turns 20," *Railroads Illustrated,* November 2017.

"Gerald Grinstein Speaks to BN Union Leaders," *Mgr.* Fourth Quarter, 1988.

"Gerald Grinstein Succeeded Darius Gaskins . . . ," *Burlington Northern Railroad News Update,* February 1989.

"George Lamphier: Super Gandy," *BN News,* April 1980.

Gerking, Shelby, and Stephen F. Hamilton, "What Explains the Increased Utilization of Powder River Basin Coal in Electric Power Generation?" *American Journal of Agricultural Economics,* Vol. 90, Issue 4, November 1, 2008.

Gibson, Paul, "A Railroad for the Long Haul," *Forbes,* April 27, 1981.

Gillespie, M.E., "Housing Lack Stymies Boom," *The News-Record* (Gillette, Wyo.), October 31, 1974.

Gilpin, Kenneth H., and Eric Schmitt. "Business People: Burlington Northern Promotes 2 Executives," *New York Times,* December 18, 1985.

Gibson, Paul. 1981. "A Railroad for the Long Haul." *Forbes,* April 27, 1981.

Glavin, William E. 1989. "BN Becomes Roadway Innovator," *Progressive Railroading,* March 1989.

Glavin, William E. 1991. "Quality Now Railroads' 'Religion,' " *Progressive Railroading,* March 1991.

Glischinski, Steve. 1992. *Burlington Northern and Its Heritage.* Andover, N.J.: Andover Junction Publications.

Glischinski, Steve. "Crawford Hill: Mountain Railroading in Nebraska?" *CTC Board Railroads Illustrated,* July 1999.

Glischinski, Steve, "Main Street of Big Sky Country," *Trains,* April 1993.

Glischinski, Steve. "The Train That Started It All," *Trains,* March 2011.

Gohmann, John W. 1984. *The Railway Labor Act Primer.* Chicago: Gohmann and Associates, Inc.

Golden, Erin. "High-tech Derailment Defense," Omaha *World – Herald,* 2012.

"A Good Start . . . A Lot Left To Do," *Railway Age,* February 22, 1971.

"Got a Question? Cut Through the Red Tape: DIAL BOSS," *BN News,* January 1973.

"Gov. Rounds and BNSF Announce Settlement on South Dakota & #39; s Core Railroad Line," *BNSF Today,* May 2, 2005.

Graham-White, Shean, "BN's SD70MAC," *CTC Board Railroads Illustrated,* November 2004.

Grande, Walter R. 1992. *The Northwest's Own Railway: Spokane, Portland and Seattle Railway and Its Subsidiaries.* Portland, Ore.: Grande Press.

Grant, H. Roger. 1996. *The North Western: A History of the Chicago and North Western Railway System.* DeKalb: Northern Illinois University Press.

"The Grapevine: Is it Really the Root of Unfounded Rumors?" *BN News,* April 1985.

Gratitude—Stewardship—Opportunity: 25 Years of Plum Creek, 2014. Publication by Plum Creek, LLC.

Gray, Kristy, "Explosion in Gillette," *Casper Star-Tribune,* June 27, 2007.

"Grayson: BN in Rail Business to Stay," *BN News,* April 1982.

"Grayson Named to New Post, Drexel New Rail President," *BN News,* March 1982.

Grayson, Richard G., Sr., <www.ancestry.com/boards/localities.northam.usa.states.missouri.counties.crawford/950/mb.ashx>.

"Grayson to Competitors: 'Welcome, But Bring Money.' " *BN News.*

"Grayson to Leave Frisco RR for New Post at BN Jan. 1," BN press release November 14, 1980.

Great Northern Pacific & Burlington Lines, Inc., letter to all employees, September 29, 1967.

Great Northern Pacific & Burlington Lines, Inc.—Merger, Etc.—Great Northern Railway Company, et al, Interstate Commerce Commission, Finance Docket No. 21478.

Great Northern Pacific & Burlington Lines, Inc., Merger, Etc.,

Applicants' Petition for Reconsideration, Finance Docket Nos. 21478, 21479 and 21480, July 27, 1966.

Great Northern Pacific & Burlington Lines Inc.—Merger—Great Northern Ry., et al., 348 ICC 821, 829 (1977).

Great Northern Railway Co., *Annual Reports, 1960–1968.*

"Grinstein Succeeds Bressler as Chairman of BN's Board." 1990. *Traffic World,* October 22, 1990.

"Growing the Company: A Labor Relations View," BN *News,* August–September 1985.

Gruber, John. "Huge Chance, Big Payoff," *Trains* web exclusive, February 19, 2010.

Grunwald, Michael, "Back on Tracks," *Time,"* July 9, 2012.

"Guernsey Tunnel No. 2 Loses Its Top," *Alliance Times-Herald,* September 17, 1998.

Guillen, Tomas, "Killing Time in 'Rubber Room': Surplus Workers Are Idle, But Paid," *Seattle Times,* November 30, 1986.

Hackett, Jason, "Burlington Northern Complex Houses Operations," *Alliance Times-Herald,* September 21, 1995.

Haeg, Larry. 2013. *Harriman vs. Hill: Wall Street's Great Railroad War.* Minneapolis: University of Minnesota Press.

Hage, Dave, "BN's Court Defeat Won't End Labor-Cost Battles," *Minneapolis Star-Tirbune,* November 29, 1988.

Haines, Henry S., *Efficient Railway Operation,* New York: Macmillian, 1919.

Haines, Joan, "BN Closure No Shock to Man Who Fought Railroad Merger," *Bozeman Daily Chronical,* November 7, 1985.

Haines, Joan, "BN, Labor Split on Court Drug Testing," *Bozeman Daily Chronical,* February 2, 1988.

Haines, Joan, "Judge Says Winona Can't Start," *Bozeman Daily Chronical,* June 10, 1988.

Haines, Joan, "Unions Disagree with Railroads on Winona Cost," *Bozeman Daily Chronical,* April 21, 1988.

Haines, Leslie, "Former Burlington Executives Discuss Management," *Oil and Gas Investor,* October 1, 2014.

Hamberger, Edward R., "The Anniversary of Our American Success Story," <huffingtonpost.com/Edward-r-hamberger/the-anniversary-of-our-am_b_75626>.

Hansen, Peter A., 2016. "Coal: A Twisted Future," *Trains,* March 2016.

Harball, Elizabeth and Brittany Patterson, "Coal's Western Stronghold Faces Precarious Future," <www.eenews.net/stories/106003534/>.

Harmen, Robert R., "Pueblo Junction, Colorado," *Trains,* November 1992.

"Havelock Wheel Shop Opens Soon," BN *News,* October 1976.

Hayes, Thomas C. "Taking Control At Burlington," *New York Times,* April 16, 1982.

Head-End Collision of Nine Burlington Northern Locomotive Units with a Standing Freight Train, Angora, Nebraska, February 16, 1980. National Transportation Safety Board Report No. NTSB-RAR-80-7.

Head-on Collision Between Burlington Northern Railroad Freight Trains 602 and 603 Near Ledger, Montana, on August 30, 1991, National Transportation Safety Board Report NTSB/RAR-93/01, May 25, 1993.

Head-on Collision of Burlington Northern Railroad Company Freight Trains Extra 6311 West and Extra 6575 East Near Westminster, Colorado, August 2, 1985. National Transportation Safety Board Railroad Accident Report NTSB/RAR-86/02.

Head-on Collision of Burlington Northern Railroad Company Freight Trains Extra 6760 West and Extra 7907 East Near Motley, Minnesota, June 14, 1984, National Transportation Safety Board Report No. RAR-85/06.

Healey, Judith Koll. 1998. *Frederick Weyerhaeuser and the American West.* Reprinted from William Stafford, "The Way It Is." St. Paul: Minnesota Historical Society Press.

Healy, Kent T. *The Effects of Scale in the Railroad Industry,* Yale University, 1961.

"Heavy Haul Balances the Conflicts," *Progressive Railroading,* July 1991.

Hein, Rebecca. 2014. "Campell County, Wyoming," WhoHistory.org.

Helge, Doris, "Employee Motivation: Rip Off the BandAids," www.selfgrowth.com/articles/Employee_Motivation_-_Rip_OFF_The_BandAids.

Henderson, Dean, "Burlington Northern: On the Wrong Track," www.multinationalmonitor.org/hyper/issues/1993/03/mm0393_11.

"Henry Crown and Company," www.answers.com/topic/henry-crown-and-company.

Hertog, John H. "Railroads as a Coal Transportation System," International Right of Way Association, August 1982.

Hertog, John H. "Remarks of John H. Hertog, Vice President–Operations Before Annual Executive Seminar," Electro-Motive Division, General Motors Corp., May 4, 1977.

Hertog, John H. "Remarks of John H. Hertog, Vice President, Operations, Burlington Northern, Inc., Financial Analysts Trip, September 27, 1977.

Hertog, J. H. "Coal Presentation," Western Coal Transportation Association, May 4, 1976.

Hidy, Ralph W., Muriel E. Hidy, and Roy V. Scott with Don L. Hofsommer. 1988. *The Great Northern Railway: A History.* Boston: Harvard Business School Press.

"High-Speed Turnouts," BN *Technical Bulletin,* September 1990.

Hillard, John, correspondence with the author, 2013.

Hill, James J. 1910. *Highways of Progress.* New York: Doubleday, Page and Company.

Hill, James J., letter to stockholders upon retiring from Chairmanship of the Board of Directors, Great Northern Railway Company, July 1, 1912.

Hill, James J., on the occasion of a frank talk to a group of employees on strike at the Jackson Street Roundhouse in St. Paul, Minnesota, August 1880, *St. Paul Globe,* August 23, 1880, quoted in Albro Martin's *James J. Hill and the Opening of the Northwest.*

History of Plum Creek, nhcs.wikispaces.com/History +of+Plum+Creek.

"Historical Background Information: Major Burlington Northern Predecessor Companies," Burlington Northern Railroad, 1981.

History of the Clean Air Act, <www.epa.gov/air/caa/caa_history>.

"History of TOPS," www.trainweb.org/rews/tops/history.

"HL&P, Railroad Settle Coal Transport Suit," Associated Press, November 30, 1988.

Hofsommer, Don L., "Hill's Dream Realized: The Burlington Northern's Eight-Decade Gestation," *Pacific Northwest Quarterly,* October 1988.

Hofsommer, Don L., "Robert W. Downing: A Life Worth Remembering," *Railroad History,* Fall–Winter 2010.

Hofsommer, Don L., "What It Takes: Portraits of Three Successful Managers," *Railroad History,* Fall–Winter 2002.

Holding, Brian. "Railroaders View Conditions at Galesburg BN Yards," Galesburg *Register-Mail,* August 29, 1974.

Holley, Paul J., "Labor Unions Dispute BN Spinoffs," *Billings Gazette,* December 28, 1986.

Holley, Paul J. and Jim Ludwick, "Railroad Deal Expands to Huntley," *Billings Gazette,* July 24, 1987.

Hon, D. V., "The Rising Cost of Car Maintenance," *Progressive Railroading,* February 1978.

"House Units Shelve Bill to Let Pipelines Carry Coal Slurry," *The Wall Street Journal,* June 28, 1977.

"How BN Manages M/W," *Railway Age,* July 1990.

"How BN Put the Pieces Together," *Railway Age,* March 29, 1976.

"How Milwaukee's New Train Speeds Freight," *Railway Age,* November 25, 1963.

"How Mr. Henry Crown Derailed the Rock Island," 1979. *Executive Intelligence Review,* October 2–8, 1979.

"Hub Program Successful, Expanding," *BN News,* July 1983.

Hwang Suein L., "Loss Is Expected by Burlington Northern Inc." *Wall Street Journal,* June 24, 1991.

Hwang, Suein L., "Stock of Burlington Northern Declines on Write-Off Plans," *Wall Street Journal,* June 28, 1991.

"IBM 360 Computer Is Newest Addition to GN's Data Processing Department," *Talking It Over,* March 1968.

"ICC Approves C&NW Plan to Haul Coal Out of Powder River Basin," *Traffic World,* August 3, 1981.

"ICC Proposes Coal Rate Standards Aimed at Adequate Rail Revenues," *Traffic World,* February 26, 1983.

"ICC Releases Decision Outlining Rationale of Coal Rate Guidelines," *Traffic World,* September 9, 1985.

"ICC Removes Conditions Imposed on 1967 Merger of BN's Predecessors," *Traffic World,* March 30, 1987.

"Images for GAO Report on Coal Rates 2005," a series of graphs showing trends in coal rates and other aspects of coal usage.

Implementing Agreement No. 1 (1970) Between Burlington Northern Inc. 1 (Formerly the Great Northern Pacific & Burlington Lines, Inc. 1) and Its Employees Represented by The Brotherhood of Locomotive Engineers.

Implementing Agreement No. 5 Between Burlington Northern, Inc. and Its Employees Represented by United Transportation Union, October 15, 1971.

"Improving Crosstie Materials and Design," *Railway Track and Structures,* September 1991.

"Industry Seeks Smaller Train Crews," *BN News,* July 1983.

"The Information Evolution: Computers and the Systems They Process Assume a Greater Importance at Burlington Northern," *BN News,* August/September 1985.

Information Statement Covering the Distribution of Approximately 87% of the Outstanding Shares of Burlington Resources, Inc., November 14, 1988.

Ingles, J. David, "Burlington Northern: Window to the Future? Dedicated to exploring alternatives to diesel fuel," *Trains,* November 1983.

"Injuries, Accidents Down in 1982," *BN News,* August-September 1982.

"In Memoriam: Bob Binger, WMCA *Camp Widjiwagan.*

"In Memoriam: Robert Downing, the Driving Force Behind the Burlington Northern Railroad Merger," *Progressive Railroading,* August 6, 2010.

"In Memoriam: Two Former Top Execs Who Helped Shape the Class I Landscape," *Progressive Railroading,* September 2010.

Inmon, Cliff, data on line changes constructed on Colorado and Southern Railway addressed to author, July 27, 2014.

"In Search of Quality," *BN News,* August/September 1983.

"Intermodal Piggybacking Key to Railroading's Future," *BN News,* October 1980.

"Innovative Intermodal . . . A Matter of Attitude," *Progressive Railroading,* November 1986.

International Heavy Haul Association. 2001. *Guidelines to Best Practices for Heavy Haul Railway Operations: Wheel and Rail Interface Issues.* Virginia Beach, Va.: International Heavy Haul Association.

Interstate Commerce Commission, "ICC Authorizes Merger of Northern Lines Railroads," November 30, 1967.

An Interview with Al Egbers, "Across the Bargaining Table," *BN News,* April 1984.

An Interview with Chairman and CEO Walter A. Drexel, "Becoming No. 1," *BN News,* July 1984.

An Interview with D. W. Gaskins, "The Long View," *BN News,* December 1982.

An Interview with D. W. Gaskins, *BN News,* January/February 1986.

An Interview with D. W. Gaskins, "Marketing in a Challenging Environment," *BN News,* November 1984.

An Interview with D. W. Gaskins, Jr., "Together We Have Ironed Out Most of the Problems, *BN News,* January/February 1986.

An Interview with Eli Wackenstein, *BN News,* August/September 1985.

An Interview with Gerald Grinstein, Charting a Steady Course, *Railway Age,* January 1, 1991. Interviewed by Gus Welty

An Interview with Jim Dagnon, Working Across the Table, *BN News,* Spring 1988.

An Interview with John Hertog, *BN News,* November 1985.

An Interview with Raymond D. Burton, Jr. (name of interviewer not given), Railroad Executive Oral History Program for The John W. Barriger III National Railroad Library, December 2000.

An Interview with Richard M. Bressler, Chairman, President and Chief Executive Officer, Burlington Northern, Inc., June 15, 1984, conducted by John F. Kawa, First Vice President, Dean Whitter Reynolds, Inc.

An Interview with Robert Downing by Professor Donald Hofsommer, Railroad Executive Oral History Program #3, John W. Barriger III National Railroad Library, January 12, 2000.

An Interview with Robert D. Krebs by Professor Donald Hofsommer at Lake Forest, Ill., Railroad Executive Oral History Program, John W. Barriger III National Railroad Library, June 7, 2008.

An Interview with Thomas J. Lamphier by Professor Donald Hofsommer, Railroad Executive Oral History Program, John W. Barriger III National Railroad Library, University of Missouri-St. Louis.

An Interview with Tom Matthews, "Building Teamwork," *BN News,* March 1984.

An Interview with the Top Rail Officers (Richard Grayson and Walter Drexel), "What's in Store," *BN News,* May/June 1982.

An Interview with William Quinn by John W. Barriger IV and Paul Cruikshank, John W. Barriger III National Railroad Library, St. Louis Mercantile Library, University of Missouri–St. Louis.

An Interview with Worthington Smith by John W. Barriger IV, Railroad Executive Oral History Interview, John W. Barriger III National Railroad Library, July 31 and August 1, 2001.

"In the Last Half of '70, We Really Got Rolling," Interview with Louis W. Menk, *Railway Age,* February 22, 1971.

"Investing in the Future: 1984 Track Program Sharpens BN's Competitive Edge," *BN News,* April 1984.

"Is Clear-cutting Bad? How About Select Cutting?" The Pennsylvania Forestry Association, <www.paforestry.org/is-clear-cutting/bad/how/about/select/c...>.

"Issues and Comments: Addressing Employee Concerns," *BN News,* March 1985.

"I've Been Working on the Railway," *BN News,* September 1974.

"Jackson Finds New Home in Newcastle," *Powder River Reflection.*

Jamison, Michael, "Plum Creek to Close Pablo Mill," *Missoulian,* April 28, 2009.

Jenkins, Bess, "Mountains (of Paper) Waiting in Burlington Northern Inc, Merger," *Sunday Journal and Star,* February 8, 1970.

Jensen, Derrick, and George Draffan with John Osborn, M.D. 1995. *Railroads and Clearcuts: Legacy of Congress's 1864 Northern Pacific Railroad Land Grant.* Spokane: Inland Empire Public Lands Council.

"Jimmy Little and Art Mattila Cleared of Motley Accident Charges," *Straight Track,* Inter Craft Association of Minnesota Newsletter, August 1984.

Johnson, Gregory S., "BN Rail Begins Selling Grain Transport Futures," *Journal of Commerce,* January 13, 1988.

Johnson, Stephen S., "A Tycoon in the Old Mold," *Forbes,* May 20, 1996.

Jones, Todd, "Milwaukee Road in the '70's: What Really Happened?" <www.trainweb.org/milwaukee/article>.

Joplin, William N. 1990. "Burlington Northern Railroad: A Centennial Review," pamphlet prepared by BN's Corporate Communications Department.

"Jury Awards ETSI $345 Million in Slurry Suit," Associated Press, November 10, 1989.

"Justice Clears BN in Milwaukee Case," *Railway Age,* April 26, 1982.

Kalbach, John, "Crossing a Creek in Style," *Trains,* March 1984.

Kametz, Tom, "A Layman's Analysis of Plum Creek Timber," seekingalpha.com/article/507561-a-laymans-analysis-of-plum-creek-timber.

Karr, Albert R., "Coal-Slurry Pipelines Face Key Hurdle This Week in Fight for Eminent Domain," *Wall Street Journal,* May 18, 1976.

Karr, Albert R., "Nation's Railroads to Seek Big Reduction in Federal Regulation of Their Industry," *Wall Street Journal,* March 7, 1979.

Katzenbach, Jon R. and Douglas K. Smith. 2002. *The Wisdom of Teams.* New York: HarperCollins.

Kaufman, Jonathan, "Burlington Northern Taps Arco Executive as President and Chief, *Wall Street Journal,* May 20, 1980.

Kaufman, Larry, "BN Chief Executive Apologizes for Crew-Calling Snafus," *Journal of Commerce,* December 15, 1993.

Kaufman, Larry, "Focus on Bottom Line Drove BN to Abandon 'Sun Belt Strategy,'" *Journal of Commerce,* February 23, 1994.

Kaufman, Larry, "Old Union Foe of BN Bows Out in Severance Deal," *The Journal of Commerce,* July 14, 1893.

Kaufman, Lawrence H., "Air Force General to Join Burlington Northern," *Journal of Commerce,* December 7, 1990.

Kaufman, Lawrence H., "BN Seeks Talks with UTU on Northern Tier Crew Sizes," *Journal of Commerce,* March 12, 1992.

Kaufman, Lawrence H., "'Duopoly' Grates on Grain Shippers," *Railway Age,* May 1999.

Kaufman, Lawrence H., "Focus on Bottom Line Drove BN to Abandon 'Sun Belt Strategy,'" *The Journal of Commerce*, February 23, 1994.

Kaufman, Lawrence H. 2005. *Leaders Count: The Story of BNSF Railway*. Austin, Texas: Texas Monthly Custom Publishing.

Keeney, Kathleen R., and Allen R. Wastler, "Winona Transaction Mishandled, ICC Official, Rail Labor Claim," *Traffic World*, December 14, 1987.

Keeney, Kathleen R., "Picketing, Taxation Top List of High Court Transport Acts," *Traffic World*, July 6, 1987.

Keeney, Kathleen R., "Senators Seek Investigation of BN Line Sale in Montana," *Traffic World*, October 25, 1987.

Kelly, Bruce E., "BNSF Greenlights Second Idaho Bridge," *Railway Age*, May 2017.

Kelly, Bruce E., "BNSF Plans Second Bridge over Idaho Chokepoint," web.mail.comcast.net/zimbra/h/printmessage?id= 231240&tz=...

Kelly, Bruce, "Stampede Pass Comes Back to Life," *Trains*, January 1997.

Kelly, Bruce E., 2014. "The Crude Oil Challenge," *Railway Age*, February, 2014.

Kelly, Bruce E., "Three-sided Traffic Solution," *Railway Age*, November 2012.

Kipling, Rudyard, "The Sons of Martha," www.online-literature.com.

Kirby, Jeff, letter re elimination of EMD-assigned jobs and elimination of EMD supervisory roles, *Glendive Diesel Shop* (newsletter), February 2011.

Klauser, P. E. "Assessing the Benefits of Tangential-Geometry Turnouts," *Railway Track and Structures*, January 1991.

Klein, Maury. 2011. *Union Pacific: The Reconfiguration: America's Greatest Railroad from 1969 to the Present*. Oxford University Press.

Klinger, Sue, "Alliance Division's Maintenance Complex," *BN News*.

Klobuchar, Jim, "A Railway Threatens to Leave," *Minneapolis Star*, April 7, 1978.

Knutson, Rick and Karl Rasmussen, "Dispatching BN's Dakota Division," *Trains*, October 1992.

Kochan, Thomas A., "A Jobs Compact for America's Future," *Harvard Business Review*, March 2012.

Kohrs, ElDean V., 1974. "Social Consequences of Boom Growth in Wyoming," paper presented at the Rocky Mountain American Association of the Advancement of Science Meeting, April 24–26, 1974; Laramie, Wyoming.

Kolpach, Dave, "BNSF President Says Company Will Improve Service," Associated Press, January 9, 2004.

Kolstad, Toby, "Next Game Changer: The Containerization of Domestic Intermodal Traffic," *Progressive Railroading*, May 2012.

Kroll, Luisa, "Billionaire Highway Man: Life Lessons Dennis Washington Learned When He Was 25," *ForbesLife*, June, 2013.

Kursar, Robert J., "BN President Says Political Enmity Between Rails, Truckers Is Waning," *Traffic World*, March 17, 1986.

Kursar, Robert J., "Rail 'Renaissance' Still Unfolding, Shippers Told by Industry Leaders," *Traffic World*, May 20, 1985.

Kursar, Robert J., "Rail Secondary Picketing Upheld by Supreme Court," *Traffic World*, May 4, 1987.

"Labor Loses in Attempt to Halt FRA Drug Rules," *Traffic World*, December 9, 1985.

Lalonde, James E., "Paid to Do Nothing—Suit Says BN Broke Promise to Workers," *Seattle Times*, December 5, 1988.

Lamphier, George K., track inspection and evaluation reports for BN system, 1970–1981, Northern Pacific Railway Historical Association archives.

Lamphier, Thomas J., "Track Building No Job for Amateurs; Each Component Has Role in Final Product," *BN News*, April 1978.

Lane, Polly, "Burlington Chairman Named," *Seattle Times*, October 11, 1990.

Lang, A. Scheffer, "Electrification: A Dissenting View," *Railway Age*, January 12, 1961.

Lasher, Steve, "'Booming' in Modern Times," *Classic Trains*, Spring 2017.

"Larson Leaves BN Truck Lines After Leveraged Buyout Fails," *Traffic World*, March 21, 1988.

Larsen, Michael E., "Extension of Second Main Line on C&NW Western Coal," Proceedings of the 104th Annual Conference of the Roadmasters and Maintenance of Way Association of America, September 21–23, 1992.

Lavallee, Omer. 1998. *Van Horne's Road*. Toronto: Railfare Enterprises Limited.

Leachman, Rob. 1998. *Northwest Passage*. Mukilteo, Wash.: Hundman Publishing Company.

Lee, Peter M. "Consolidation of Great Northern Pacific and Burlington Lines, Inc." October 2012.

Lenzen, Mike. 1997. "A C&NW Coal Line Story." *North Western Lines*, Summer 1997.

"Les Vaughan Ends BN Social Counseling Career" *BN News*, June 1978.

Levine, Jonathan B., "Will a Takeover Derail Burlington Northern's Makeover?" *Business Week*, August 3, 1987.

Lewis, Eugene M. 2005. *12,000 Days on the North Western Line*. Chicago and North Western Historical Society.

Lindblom, Mike, "Feds Give $16M for Slide Control by Rail Tracks," *Seattle Times*, September 21, 2011.

Lindeman, Eric D., "Burlington Northern Truck Unit Moves Ahead on Strategic Plan," *Traffic World*, April 29, 1985.

Lindquist, John, "Burlington Northern in Good Shape," *Associated Press*, May 18, 1972.

Linn, Linda, "Feds Oppose Slurry Plan," *Billings Gazette*, August 2, 1977.

"Livingston, Mont., Likes BN—And Shows It Annually," *Railway Age*.

"Livingston Shows Appreciation of BN," *BN News,* July 1977.

"Locomotive 'Power-by-the-Hour' Largely Stalled at the Station," *Traffic World,* July 27, 1991.

"Logs to Lumber," *BN News,* January 1971.

Loidolt, Jared, presentation at the University of Wisconsin – Superior, April 2012.

"The Long and Winding Road: Unraveling the Lengthy, Complicated Bargaining Process," *BN News,* April 1984.

"Long – Standing Rate Fight May Be Over After City Pays BN, SP $26 Million," *Traffic World,* January 28, 1985.

"Lorentzsen: 'Gamesmanship' Not His Style," *BN News,* January 1978.

"Lorentzsen of the BN," *Inside Story International—Stanley Corporation,* Winter 1978–1979.

"Lorentzsen Reflects on New CEO Role," *BN News,* October 1978.

Lorsch, Jay W. 1989. *Pawns or Potentates: The Reality of America's Corporate Boards.* Boston: Harvard Business School Press.

Loving, Rush, Jr., "The Merger That Worked," *Trains,* June 2016.

LSU College of Engineering, "Reflections of Donald W. Clayton, "Hall of Distinction Class of 1992–1993.

Lustig, David, "AC vs. DC: What's the Difference?" *Trains,* May 2010.

Lustig, David, "Catching Trouble Before the Fact," *Trains,* November 2005.

Lutch, Russell H., Devin K. Harris and Theresa M. Ahlborn, "Causes and Preventative Methods for Rail Seat Abrasion in North America's Railroads," American Railway Engineering Association, October 2014.

Machalaba, Daniel, "Burlington Northern Shows Risks of Hiring an Outsider as CEO." *The Wall Street Journal,* April 6, 1993.

Machalaba, Daniel, "Burlington Names Grinstein to Head Its Railroad Unit," *The Wall Street Journal,* January 20, 1989.

Machalaba, Daniel, "Burlington Says Two Executives Leave Rail Firm," *The Wall Street Journal,* June 2, 1994.

Machalaba, Daniel, "Railroads Confront Unions on Crew Size," *The Wall Street Journal,* November 23, 1987.

Malone, Frank, "C&NW: Flying High as Rival Roads Founder," *Railway Age,* July 28, 1980.

Malone, Frank, "Why a Federal Court Said Yes," *Railway Age,* December 2, 1968.

Malone, Michael P. 1996. *James J. Hill Empire Builder of the Northwest.* Norman and London: University of Oklahoma Press.

Mannello, Timothy A., "Problem Drinking Among Railroad Workers: Extent, Impact and Solutions," University Research Corporation, Washington, DC.

Marshall, Evelyn T., "Mary and Martha: Faithful Sisters, Devoted Disciples," *Ensign,* January 1987.

Martin, Albro. 1976. *James J. Hill and the Opening of the Northwest.* New York: Oxford University Press.

Martin, Jack, "Electrification One Answer to Rising Fuel Costs," *BN News,* January 1981.

Martin, Jack, "116-Mile Rail Line Nearing Completion in Central Wyoming," *BN News,* July 1979.

Martin, Jack, "Zestful, Zealous Don Zeiss Zooms in on Problems as Alliance Project Construction Director," *BN News.*

Martin, Mike, "The Yellowstone Division Story: From 1973 Through 1980," *BN Expediter,* October 2011.

Matzenbach, Jon R. 1993. *The Wisdom of Teams.* Harper Collins 1993.

"Maze Heads Northwest," *Nebraska Division Whistle Post,* December 2001/January 2002.

McCarty Farms, Inc. v. Burlington Northern, Inc., 787 F. Supp. 937 (decided Montana 1992).

McCue, Dan, "Intermodal: A New Level of Reliability," *World Trade WT100,* January 3, 2012.

McDonnell, Greg. 2008. *Locomotives: The Modern Diesel and Electric Reference.* Erie, Ontario, Canada: Boston Mills Press.

McFarlen, G. L., presentation, "Burlington Northern and Western Coal," to the Montana Water Development Association, September 27, 1976.

"McGovern: Federal Loans for UP/C&NW Unacceptable," *BN News,* October 1980.

McKay, Floyd, "How Great Corporate Power Shadows Gregoire on Coal Shipments to China," *Crosscut,* January 7, 2011.

McPhee, John, "Coal Train—I and II," *New Yorker,* October 3 and October 10, 2005.

McPhee, John. 2006. *Uncommon Carriers.* New York: Farrar, Straus and Giroux.

"Meet CAPMAC: A Sophisticated Computer That Helps BN Make the Most of Its Locomotive Fleet," *BN News,* October 1983.

"Meeting an M/W Challenge—Burlington Northern," *Railway Age,* June, 1991.

Memorandum of Agreement Between BNSF Railway and Brotherhood of Railway Carmen Division/TCIU and International Association of Machinists and Aerospace Workers and International Brotherhood of Electrical Workers/System Council No. 16, October 12, 2010.

Memorandum of Agreement Between Burlington Northern Railroad and the United Transportation Union, May 20, 1993, Article I, Crew Consist.

"Memorable Moment in Time," *Powder River Reflection,* December 2009.

"Memories of Merger Day—and the Days After," *Mainstreeter,* Summer 2010.

"Menk Happy with BN's 'First Hundred Days,' " *Railway Age,* July 13, 1970.

Menk, Louis W., *A Railroad Man Looks At America; Excerpts from the Speeches of Louis W. Menk.* No place: privately published, n.d. [circa 1974].

Menk, Louis W., Remarks Before the Committee on Interior and Insular Affairs, U.S. House of Representatives, November 7, 1975.

"Merger History," *BN News*, March 3, 1970.

"Merger Memories, Part 3: The View from the West End," *Mainstreeter*, Winter 2010.

Merger Protective Agreement and Implementing Agreement No. 1 and Implementing Agreement No. 2 for Employees of Burlington Northern, Inc. Represented by Brotherhood of Locomotive Engineers.

"A Merger That Started on the Right Track," *Business Week*, June 12, 1971.

"Merle, W. Geiger, Jr.," *BLET Journal*, Winter 2010.

Merrild Augspurger et al., Appellants, v. Brotherhood of Locomotive Engineers, Appellee. 510 F.2d 853 No. 74-1363. February 4, 1975.

"A Message to All Employees from Burlington Northern's Top Officers," *BN News*, March 3, 1970.

"Method for Computation of Frequency and Severity Rates for Industrial Injuries and Classification of Industrial Accident?" <wiki.asnswers.com/Q/Method_for_computation_of_ frequency_and_severity_rates_fo . . .>.

"Metra: BNSF Line Puts on Daily Parades," *Railway*, Fall 2010.

Mickelson, Sig. 1993. *The Northern Pacific Railroad and the Selling of the West*. Sioux Falls, S.D.: The Center for Western Studies.

Middleton, William D. 1999. *Landmarks on the Iron Road: Two Centuries of North American Railroad Engineering*. Bloomington, Ind., and Indianapolis: Indiana University Press.

"Milwaukee Asks Inclusion in BN Merger," *BN News*, April 1973.

Miotek, Bill, "BN's Chicago Commuter Operations," *The BN Expediter*, January 1999.

M. M. Winter, Petitioner, v. Interstate Commerce Commission and United States of America, Respondents. No. 88-1250. United States Court of Appeals, Eighth Circuit. Decided July 12, 1988.

Moin, Ebrahim, "The Current Status of Field Welding of Rail," *Railway Track and Structures*, October 1988.

"Montana Line Up for Sale by Burlington Northern," *Minneapolis Tribune*, November 27, 1986.

Moody's Transportation Manual, annual issues 1965–95.

Morgan, David P., "Burlington Northern: A Railroad Worth Waiting for," Burlington Northern advertising supplement, 1970.

Morgan, David P., "Farewell, CB&Q, GN, NP and SP&S; Hello, BN," *Trains*, June 1970.

Morrell, Peter, "Airlines Within Airlines: An Analysis of U.S. Network Airline Responses to Low Cost Carriers," *Journal of Air Transport Management*, September 2005.

"Mount St. Helens—From the 1980 Eruption to 2000," U.S. Geological Survey Fact Sheet 036-00.

Muchnic, Suzanne, *Norton Simon and the Pursuit of Culture: Odd Man In*, Berkeley: University of California Press, 1998.

Murray, David. "Back on Track: State and Feds Give $4 Million to Fix Railroad Trestle," *Great Falls Tribune*, May 31, 2013.

Murray, David. 2014. "Right on Track: Flood-damaged Trestle Set for Repair," *Great Falls Tribune*, February 13, 2014.

Murray, Tom. 2005. *The Milwaukee Road*. St. Paul: MBI Publishing, 2005.

Myers, Edward T., "GN Sleds Ballast Under Track," *Modern Railroads*, June 1964.

Nagarajan, Ravi, "Buffet's Plans for Burlington Northeren May Involve Rapid Expansion," *Seeking Alpha*, January 26, 2010.

"Natural Resources: A Proper Balance," *BN News*, November 1974.

Neal, Roger, "Why Merge If You Can Share?" *Forbes*, May 20, 1985.

"Nebraska Division Develops Programs to Make Working on Railroad Safer for Track Employees," *BN News*, June 1978.

"The Neighborhood Store: Materials Stores Give Better, Less – Costly Service," *BN News*, May/June 1983.

Nelson, Roger H., Letter to Earl J. Currie, July 25, 2012.

"New BN Grain Train Operations to Benefit Shippers," *BN News*, August/September 1980.

"New BN President Richard Bressler Sees Exciting, Challenging Future," *BN News*, August/September 1980.

"New Drug and Alcohol Program Initiated," *BN News*, November 1984.

"New EPA Rules to Devastate Coal Industry," <www.scam.com/ showthread.php?t=140153>.

"New Record for BN Coal," *Railway Age*, October 11, 1982.

"New Strategy for Labor," *Railway Age*, June 1991.

"New Technology to Find Problems Before They Occur," *Powder River Reflections*, November–December 2000.

Nickerson, Gregory, "Clean Power Plan May Cut Wyoming Coal Revenue 31–63 Percent," <wyofile.com/clean-power-plan-may-cut-mining-coal-revenue-by-46-percent>.

Nisselius, Jack, "Campbell County Ushers in Coal Age," *News-Record* (Gillette, Wyo.), December 23, 1976.

Nixon, Ron. 2015. "Rail Industry Had Safety Technology Decades Ago," *The New York Times*, November 3, 2015.

"NMB Rules That BN Is One Company, Not 11. Will Unions Agree?" *Railway Age*, June 1991.

"No Appeal on C&NW Ruling," *BN News*, November 1981.

Norman, James R., "Divide and Prosper," *Forbes*, March 30, 1992.

Norman M. Lorentzsen, Narrator, James E. Fogerty, Inverviewer. Interviewed on August 16 and September 23, 2005, at the Minnesota Historical Society, St. Paul, Minnesota.

"Northern Lines Merger Approved," *Railway Age*, February 2/9, 1970.

Norton Simon Biography, <www.nortonsimon.org/norton-simmon-biography>.

Norton Simon, letter to BN shareholders following annual meeting of shareholders held in 1974.

"Norton Simon Quits BN Board Position," *St. Paul Sunday Pioneer Press*, January 27, 1974.

North, Irene, "Inside Nebraska's Only Rail Tunnel the Belmont Tunnel," *Scottsbluff, Nebr., Star Herald*, May 3, 1982.

"North Western Gets Ready to Absorb and Expand," *Railway Age*, May 1983.

Northern Pacific Railway Co., *Annual Reports, 1960–1968.*

"No We're Not Moving," *Mgr.,* April 1982.

O'Brien, Gerard, "Tech a Family Tradition," *The Montana Standard,* <www.mtech.edu/alumni/Profiles/dyk.php>.

Odden, Anne S., "The Powder River Basin, Chapter 1: The Coming Conflict," Research Note, Dean Witter Reynolds, Inc., June 21, 1983.

"Oil, Timber, Coal, Minerals Loom Big in BN's Future," *Railway Age,* February 22, 1971.

"The Old Order Changeth, Yielding Place to New," *BN News,* March 3, 1970.

Olson, John, "Early Burlington Northern Fuel Tenders," <jimsjunction.com/sidetrack/st02/ftl>.

"$175 Million Settlement in BN Antitrust Lawsuit," *Traffic World,* December 12, 1988.

O'Neil, Martha, "Does BN Want Out of Railroading?" *BN News,* October 1981.

OOIDA vs, Burlington Motor Carriers, Inc.

O'Reilly, Joseph, "Rail Intermodal: Where Rail Meets Road," *Inbound Logistics,* October 2012.

"Osage Refinery," *BN News,* March 1976.

"Osage Refinery in Production, Among Cleanest, *BN News,* May 1977.

Oslund, John J., "BN Railroad Receives Approval to Operate Subsidiary Railway," *Minneapolis Star-Tribune,* December 2, 1987.

Oslund, John J., "Different Rulings Could Send 'Short Rails' to High Court," *Minneapolis Star-Tribune,* June 1, 1988.

"Over the Goal Line," *BN News,* January 1983.

Overton, Richard C. 1965. *Burlington Route: A History of the Burlington Lines.* New York: Reprint by Alfred A. Knopf, Inc.

Overton, Richard C. 1953. *Gulf to Rockies: The Heritage of the Fort Worth and Denver-Colorado and Southern Railways,1861-1898.* Austin: University of Texas Press.

"Pace of Litigation Speeds Up in BN Coal Contract Rate Disputes," *Traffic World,* August 2, 1982.

Page, Paul, "Intermodal's Switching Perceptions," *Traffic World,"* June 7, 2004.

Palley, Joel, "Freight Railroads Background," Office of Rail Policy and Development, Federal Railroad Administration, March 2012.

Palmeri, Christopher, "Serving Two (Station) Masters," *Business Week,* July 24, 2006.

Park, Donald K. II, "The Hastings Rail Relocation Project," *The Streamliner,* Vol. 16 No. 2.

Park, Donald K. II. 1990. *Powder River Coal: A Guide to Facilities and Operations.* Fort Collins, Colo.: PARKRAIL.

Partlow, Bob, and Barbara Winslow, "Coal May Be Sent to Cherry Point," *Bellingham Herald,* December 30, 1980.

Patterson, Brittany, "Creating a New Future in Wyoming's Biggest Coal Town," *EE News,* April 12, 2016.

Patterson, Steve, "A Route Caught in Traffic," *Trains,* April 1999.

"People, Not Planes, Deliver" *Traffic World,* May 11, 1981.

Peterson, Chris, "Coal Creek: More Tax Base But More Services," *News-Record* (Gillette, Wyo.), April 27, 1978.

"Pertinent Personnel." 1980. *Trains,* August, 1980.

Petersen, Jim, "Burlington Northern Rolls into the South," *Oregon and Washington Industrialist,* March/April 1981.

Petery, Andras R., Investment Perspectives Excerpt, Burlington Northern, September 14, 1981.

"Petroleum Important in BN's Resource Program," *BN News,* June 1977.

Phillips, Don, "Bob Downing Deserves His Place in History Books," *Trains,* January 2011.

Phillips, J. A. III, "From A to Z with Warren McGee," *Mainstreeter,* Winter 1999.

Piller, Dan, "BN Inc. Returns to Railroading," *Modern Railroads,* November 1989.

Pitts, Jerry, "The BN Fuel Tender Story," *The BN Expediter,* October 1993.

Ploss, Thomas H. 1991. *The Nation Pays Again: The Demise of the Milwaukee Road, 1928–1986.* Thomas H. Ploss, 1991.

Plough, Bob, "From the Foothills to No Man's Land: BNSF's Colorado to Texas Coal Routes," *Railroads Illustrated,* May 2012.

Plous, F. K., Jr., "BN's Extraordinary Expediters," *Railway Age,* November 1987.

"Plum Creek History," <plumcreek.com/about/history>.

"Plum Creek Timber Company, Inc.—Company Profile, Information, Business Description, History, Background Information on Plum Creek Timber Company, Inc." <referenceforbusiness.com/history2/35/Plum-Creek-Timber-Company-Inc>.

"Plum Creek Timber Announces Montana Plant Closures," *Associated Press,* June 5, 2009.

Plum Creek: Timber Conversion with Emphasis on Efficiency," *BN News,* August/September 1980.

Plume, Janet, "BN, UP Look to Liquified Natural Gas As Alternative Locomotive Fuel," *Traffic World,* May 17, 1993.

"Pole Line to Be Eliminated," *Whistle Post,* December 2001/January 2002.

"Pooling of Grain Cars, Elimination of Private Ownership Eyed by BN," *Traffic World,* January 3, 1983.

Poplawski, Dave, "Locomotive Chronicles: 1986, Part I," *The BN Expediter,* October 2012.

"Portable Rewelder Cuts Costs," *Progressive Railroading,* November 1990.

"Positive Train Control Ran Successfully in Years-Long Burlington Northern Railroad Trial," *DesignNews,* October 21, 2008.

Potarf, E. L., "Statement," Management-Labor Conference, St. Paul, October 27, 1960.

Powder River Basin, <www.sourcewatch.org/index.php?title=Powder_River_Basin>.

"Powder River Basin Coal Production." 2016. Bureau of Land Management, May 6, 2016.

Powder River Basin Coal Resource and Cost Study Prepared for Xcel Energy by John T. Boyd Company, Report No. 3155.001, September 2011.

"Powder River 'Blowout' Sends BNSF, UP Reeling," *Trains,* September 2005.

"Powder River Country: America's Last Rail Frontier." 1989. *Trains,* November, 1989.

Powell, Warren B., "Real-Time Dispatching for Truckload Motor Carriers," 2007, http://thetenneygroup.com "BN Transport," *BN News,* October 1972.

Powers, Bob P. E., and Ron Poulsen, "Stampede Pass Reopens," American Railway Engineering Association, Bulletin 61.

"Preventive Maintenance Keeps BN Coal Cars Rolling," *Railway Age,* October 9, 1978.

Price, Donna, "Merry Rachetts: A Strong Voice for Unionism," *Alliance Times-Herald,* September 23, 1993.

"Production Tie Renewal vs. Quick-Removal Operations," *Railway Track and Structures,* October 1991.

"Productivity: From a Union View," *BN News,* October 1983.

Projected Retirements of Coal-Fired Power Plants, U.S. Energy Information Administration, July 31, 2012.

"Project Tames Curves in Canyon", *BN News,* October 1984.

"Project Yellow," *The Mixed Train,* September 1984.

"A Proper Balance: Natural Resources," *BN News,* September 1974.

Proposed Revisions to AFE Accounting Procedures for Ratable Depreciation on Property Accounts (including track structures), discussion draft, 49 CFR 1201, Docket No. 36988.

Pulliam, John R., "Galesburg BNSF Yards Poised for Growth," *The Register-Mail,* October 20, 2008.

"Putting Advanced Turnouts to the Test," *Railway Track and Structures,* March 1991.

Pyle, Joseph Gilpin. 1936. *The Life of James J. Hill.* New York. Peter Smith.

"Quarterly Coal Production Lowest Since the Early 1980s." 2016. U.S. Energy Information Administration, June 10, 2016.

"A Rail Giant Is Born," *Railway Age,* March 30, 1970.

"Rail Grinding," *BN Technical Bulletin,* August 1989.

"Rail Link Reveals Data on BN Deal," *Journal of Commerce,* November 17, 1987.

Rail Transportation of Coal to Power Plants: Reliability Issues, Every CRS Report, September 26, 2007.

Railroad Accidents Report, Head-on Collision of Burlington Northern Railroad Freight Trains Extra 6714 West and Extra 7820 East, Wiggins, Colorado, April 13, 1984 and Rear-end Collision of Burlington Northern Railroad Freight Trains Extra 7843 East and Extra ATSF 8112 East Near Newcastle, Wyoming, April 22, 1984, NTSB Report No. NTSB/RAR-85/04 (note NTSB made this a combined report, covering both accidents).

"Railroad Asks Damages Against Union," *Gillette News-Record,* July 16, 1975.

"Railroad Regulation—Shipper Experiences and Current Issues in ICC Regulation of Rail Rates," United States General Accounting Office, Report to the Chairman, Subcommittee on Oversight and Investigations, Committee on Energy and Commerce, House of Representatives, September 1987.

"Railroads Enhancing Turnouts," *Progressive Railroading,* December 1991.

"Railroads: Safer Than Most Other Industries," <www.montanarail.com/images/safer-than-most-industries.jpg>.

"Railroads Progress Report, Burlington Northern," Morgan Stanley Investment Research, December 20, 1984.

"A Railtown Booms," *Fortune,* August 25, 1980.

Ramsey, Bruce, "BN Shuns Ideas of Nuclear Foes," *Seattle Post-Intellingencer,* April 4, 1987.

"RAND Corp. Study Says Railroad Capacity Is Public Concern," <www.rtands.com/breaking_news.shtml#Feature3-8-21>.

"Rankings: Coal Production, 2014, Montana" U.S. Energy Information Administration.

"Rapid Response Ready for Orin Work," *Alliance Mechanical Employee Newsletter,* second quarter 2005.

Rattner, Steven. 1977. "The Battle for Western Coal," *New York Times,* January 1, 1977.

Rebello-Rees, Kathy, "Coal Boom Heating Up the West," *USA Today,* January 26, 1983.

"The Rebirth of Stampede Pass," *Railway,* November/December 1996.

"Reconstructive Surgery: BN's Laurel Yard Undergoes a Face-Lift," *BN News,* September 1984.

"Recovering Alcoholics Help Co-workers—Treatment Program 80% Successful," *BN News,* October 1980.

"Refund of $40 Million to San Antonio from BN, SP Ordered by ICC," *Traffic World,* April 21, 1986.

"Regulation of Freight Railroads in the Modern Era: 1970–2010," *Review of Network Economics,* December 2008.

Remarks of Thomas J. Lamphier, President, Transportations Division, Burlington Northern, Inc., Financial Analysts Trip, September 27, 1977.

Remarks to Security Analysts by Walter A. Drexel, September 29, 1982.

"Removing-All-Doubt Department." *Trains,* 1967.

Report to the President by Emergency Board No. 219, January 15, 1991.

Responsibilities of Railroads and Employees, 49 CFR 219.203.

"Retired Railroad Employee Calls BN 'Corporate Terrorists,'" 1983. *Star-Tribune of the Twin Cities,* March 31, 1983.

Rhodes, Michael. 2003. *North American Railyards.* St. Paul: MRI Publishing Company.

Richards, Bill, "How Dennis Washington Won a Beaut of a Gamble on His Butte Copper Mine," *The Wall Street Journal,* May 3, 1988.

Richards, Bill, "Worried States Enter the Railroad Business to Save Branch Lines," *The Wall Street Journal,"* September 22, 1983.

Riley, John H., Testimony Before the Committee on the Judiciary United States Senate, March 4, 1986.

Roberts, Phil, *A New History of Wyoming,* <uwacadweb.uwyo.edu/robertshistory/New_History_of_Wyoming_chapter_19>.

Robl, Ernest H., "The Powder River Basin: A Guide to America's Power Railroading Base," second edition, self-published, 2000.

Robl, Ernest H., "Powder River Country," *Railfan and Railroad,* March 2002.

Robl, Ernest H., "Stampede Pass Revival," *Pacific RAILNEWS,* October 1996.

Rockwell ARES: Advanced Railroad Electronics System, brochure prepared by Rockwell International in 1983.

Rodgers, David H., 2006, "The Permanent Legacy of Expo '74: An Insider's Story of How Spokane Acquired Riverfront Park, The Opera House and the Convention Center," *The Pacific Northwesterner,* Volume 50, Issue 2, October, 2006.

Rosenfeld, Ira, "Burlington Northern and Unions Dance, But It's Not Clear Which Is Leading," *Traffic World,* December 11, 1989.

Rosenfeld, Ira, "Burlington Northern, Labor Union Reach Tentative Acord on Crew Size," *Traffic World,* September 9, 1991.

Rosenfeld, Ira, "BN to Rely More on COTS Program As Part of Its Strategy for the Future," *Traffic World,* October 29, 1990.

Rosenfeld, Ira, "Holes in C&NW Buy-Out Blanket Sending Shivers Through Analysts," *Traffic World,* April 9, 1990.

Rosenfeld, Ira, "Locomotive 'Power by the Hour' Largely Stalled," *Traffic World,* July 22, 1991.

Rosenfeld, Ira, "NMB Decertifies Union Negotiators, Handing BN Big Victory over Labor," *Traffic World,* April 29, 1991.

Rosenfeld, Ira, "Rail Labor Leaders Feeling Wrath as Member Backlash Claims First Victim," *Traffic World,* July 29, 1991.

Rosenfeld, Ira, "Rail Labor Takes Uncle Sam to Court Over Law Ending April Rail Strike," *Traffic World,* August 5, 1991.

Rosenfeld, Ira, "Railroad Stocks Claiming Higher Despite Recession-Hit Earnings in Quarter," *Traffic World,* July 29, 1981.

Rosenfeld, Ira, "SP Names Tough Union Negotiator to Key Labor-Relations Position," *Traffic World,* March 4, 1991.

Rosenfeld, Ira, "Santa Fe Settles ETSI Suit, Agrees to Pay $350 Million by 1997," *Traffic World,* April 30, 1990.

Rosenfeld, Ira, "3,500 Burlington Northern Workers Boycott National Rail Labor Talks," *Traffic World,* April 15, 1991.

Ruff, Joe, "Why Buy Burlington?" *Omaha World-Herald,* January 31, 2010.

Rung, Albert M., Remarks at Isaac Walton League Annual Dinner, Lincoln, Nebr., December 23, 1975.

Runte, Alfred, *Burlington Northern and the Dedication of Mount St. Helens: New Legacy of a Proud Tradition,* 1982.

Sachar, Laura, "Rail Stripper," *Financial World,* January 10, 1989.

"A Safer Work Environment," *BN News,* December 1984.

"Safety Issues: FRA Gives BN a Whitewash," *Labor,* July 17, 1985.

Salpukas, Agis, "Burlington Northern's Spinoff Plan," *The New York Times,* June 3, 1986.

Sanders, Dale. 2002. *The Northern Pacific.* Mukilteo, Wash.: Hundman Publishing Company.

Sanzillo, Tom, "PRB Coal Fast Becoming a Stranded Asset," *High Country News,* August 2, 2015.

Sartore, D. V., "BN's Taconite Transfer and Storage Facility at Superior, Wisconsin," *Bulletin 653—American Railway Engineering Association.*

Sauer, Stacy J., "An Overview of Burlington Northern's Advanced Turnout Designs Including Recommendations for Future Turnout Technology," Association of American Railroads Advanced Turnout Evaluation Program, January 17, 1991.

Sauer, Stacy J., "Tackling the Turnout Problem," *Railway Age,* March, 1991.

Saunders, Richard, Jr. 2003. *Main Lines: Rebirth of the North American Railroads, 1970–2002.* DeKalb: Northern Illinois University Press.

Scates, Shelby, "Town Gets Derailed," Seattle *Post-Intelligencer,* March 1, 1984.

Schmeling, Richard L., "BNSF Construction Update," *The Diamond Newsletter,* 2016.

Schmollinger, Steve, "Public/Private Cooperation Means Upgrades for BNSF's Ottumwa Subdivision," http://trn.trains.com/Interactive/Web%20Exclusives/2011/01/Incoming%20improvements.

Schneider, Lewis M., Peter F. Rousselot, Paul L. Joffe and George W. Mayo, Jr., "Rail Service Contracts—The New Frontier," *Traffic World,* August 3, 1981.

Schneider, Paul D. 1993. *Burlington Northern Diesel Locomotives: Three Decades of BN Power.* Waukesha, Wisc.: Kalmbach Publishing Company.

Schulz, John D., "BN Truck Subsidiary Buy-out by Investor Group Completed," *Traffic World,* September 5, 1988.

Schuster, T. G. 2016. *History of the West Suburban Mass Transit District.* Compiled and self-published by Earl J. Currie.

Scribbins, Jim. 2001. *The Milwaukee Road: 1928–1985.* Forest Park, Ill.: Heimburger House Publishing Company.

Securities and Exchange Commission, Form 10-K, Burlington Northern, Inc., December 31, 1980.

"Segment of Third Main Line Opens," *UP Online,"* January 8, 1998.

Senate Bill 315 Rail Freight Competition Study, prepared by R. L. Banks and Associates for the Governor's (Montana) Office of Economic Opportunity, 2004.

Senior, Jeannie, "Becky McMahon Makes Way in BN Train World," Correspondent, *The Oregonian,* quoted in *BN News,* June 1978.

Shaffer, David, "Utilities to BNSF: More Coal," *Star Tribune,* August 5, 2014.

Shaffer, Frank E., "Burlington Northern Off to a Smooth Start," *Modern Railroads,* May 1970.

Shedd, Tom, "Burlington Northern: Aggressive, Innovator – and Thoroughly Non-Traditional," *Modern Railroads,* November 1986.

Shedd, Tom, "In MoW, BN Seeks Long-Term Solutions," *Modern Railroads,* April 1991.

Shedd, Tom, "MoW Trends at BN: The Sky's the Limit," *Modern Railroads,* March 1991.

Shine, Joseph W. 1991. *Burlington Northern Into the 90's: Motive Power Pictorial.* La Mirada, Calif.: Four Ways West Publications.

"(Shop) Closings Announced," BN *News,* January/February 1986.

"Shop Craft Unions Signed an Agreement with BNRR . . . ," *Burlington Northern Railroad News Update,* October 1987.

"Shops: Less Can Be More—Railroad Repair Facilities," *Railway Age,* March 1990.

Shands, Tom, "BN Sells Southern Line to Missoula Entrepreneur," *Livingston Enterprise,* July 23, 1987.

Sheet Metal Workers' International Association, Appellant, v. Burlington Northern Railroad Company, Appellee, 893 F.2d 199, 1990.

Shuster, Rep. Bill, "Rail Re-regulation May Be Catastrophic Public Policy," *The Hill,* March 11, 2010.

Shuster, Rep. Bill, "Shuster: Don't Regulate Freight Rail System," <rollcall.com/features/Transportation_Energy/tande/47012-1.html>.

"Signal Group Marks Ten Years, Zero Injuries," *Northern Light.*

"Sky Eyes: BN Becomes the First Railroad in the United States to Test Space-age Technology in Its Dispatching Operations by Using Satellites to Track Trains," BN *News,* March 1985.

Sloane, Donald, "BUSINESS PEOPLE; Burlington Names a Chief Executive," February 17, 1982.

Smalley, E. V. 1883. *History of the Northern Pacific Railroad.* Reprinted by David S. Coster in 2002.

"SMART-TD Members Reject One-Man Crew Agreement with BNSF," *Progressive Railroading,* September 12, 2014.

Smith, Jeffery, "Sensory Deprivation on the High Plains," *High Country News,* www.hen.ord/issues/105/3285/print_view.

Smith, Rebecca, "The Coal Age Nears Its End," *Wall Street Journal,* December 23, 2011.

Smith, Rebecca and Cameron McWhirter, "Mississippi Plant Shows the Cost of 'Clean Coal,'" *Wall Street Journal,* October 14, 2013.

Snider, Joe, "The Denver Railroad Scene," *The BN Expediter,* April 2006.

"Sniffer Dog Case Feb. 15," *Mgr.* January 30, 1985.

Soendker, Sandi, "Burlington Motor Carriers Files for Voluntary Reorganization Under Chapter 11," <ooida.com/issues&actions/Judicial/Burlington/Burlingtonbankrupt.shtml>.

Sol, Michael, "The Milwaukee Road's Demise, Notes from Michael Sol," <online.fliphtml5.com/trun/ahvm>.

Solomon, Brian. 2000. *The American Diesel Locomotive.* Osceola, Wisc.: MBI Publishing Company.

Solomon, Brian. 2005. *Burlington Northern Santa Fe Railway.* St. Paul: MBI Publishing Company.

Solomon, Brian. 2007. *Intermodal Railroading.* St. Paul: MBI Publishing Company.

Solomon, Brian. 2001. *Railway Maintenance: The Men and the Machines That Keep the Railroads Running.* St. Paul: MBI Publishing.

Solomon, Brian, and Patrick Yough. 2009. *Coal Trains: The History of Railroading and Coal in the United States."* Minneapolis: MBI Publishing Co.

"South Dakota Completes Track Sale," BN *News,* August/September 1982.

Southern Pacific Transportation Company, et al., v. San Antonio, Texas, Acting by an Through Its City Service Board, 748 F.2d 266, 1984.

"Speaking Out: Employees Voice Opinions on Survey Results," BN *News,* April 1984.

"Special Bridge Designs Win Approval for a BN Line Change," *Railway Age,* March 8, 1971.

Speroff, Leon, M.D. 2007. *The Deschutes River Railroad War.* Portland: Arnica Publishing.

Sprau, Dave, Milwaukee lines in Washington, correspondence with the author.

"Staggers Act Handcuffs U.S. Grain in World Markets, Gargill Says," *Traffic World,* December 6, 1982.

"The Staggers Act: Under Attack; New Threats Surface to Reimpose Old Constraints," BN *News,* October 1984.

Staggers, Dr. Margaret, "Staggers Act Saved Railroads," *Charleston Gazette,* October 4, 2010.

Stagl, Jeff. 2006. "What Will It Take for the Powder River Basin's 20 Mines to Meet the Nation's Insatiable Energy Needs? *Progressive Railroading,* August, 2016.

Staroski, Nick, PE. 2013. "Design Challenges of BNSF Bridge 3.8 Over the Missouri River Near Plattsmouth," American Railway Engineering and Maintenance Association, 2013.

Statement by N. S. Westergard, Vice President and General Manager, SP&S Railway, Management-Labor Conference, St. Paul, Minn., October 27, 1960.

Stephens, Bill, "Union Pacific Sees Coal Traffic Holding Steady," *Trans Industry Newsletter,* June 8, 2017.

Stephens, Bill, "Why Coal's Rebound Might Be Just a Bounce Down the Stairs, *Trains Industry Newsletter,* September 19, 2017.

Stern, Richard L., "Denny's Always the Low-Cost Producer," *Forbes,* May 15, 1989.

Stack, Don (compiled by), "Western Railroad Properties, Inc. (WRPI), <utahrails.net/up/wrpi.php>.

"Straight Bids Division Farewell," *Powder River Reflection,* 4th Space ed., 2012.

Strack, Don, "Western Railroad Properties, Inc. (WRPI)," <utahrails.net/up/wrpi.php>.

Strauss, John F. Ph.D. 1998. *The Burlington Northern: An Operational Chronology 1970–1995*. West Bend, Wisc.: Friends of Burlington Northern Railroad.

"STB Denies AAR's Petition to Consider Indirect Competition as a Coal Rate Factor," *Progressive Railroading Daily News*, March 21, 2013.

"STB Proposes New Standards to Railroad Rates for Nonj-Coal Shippers," *infor@thompsonhine.com*, August 18, 2006.

Stern, Richard L., "Denny's Always the Low-Cost Producer," *Forbes*, May 15, 1989.

"Steve Straight of Alliance," *Powder River Reflection*, October 2009.

Stone, Jeffrey B., Wertheim Schroder report on Burlington Northern, Inc., June 1, 1987.

Stucke, John, "Montana Rail Link Wins National Safety Award," *The Missoulian*, May 21, 1999.

"Study Says Gillette Will Adjust to Coal Development," *The News-Record*, September 12, 1974.

"Study Team Lists Benefits of BN-Frisco Merger Plan," *BN News*, February/March 1978.

"Suburban Equipment Grand Is Announced," *Chicago Region News*, March/April 1972.

Sullivan, John L., "Engineering Expedites Third Main Line," *UP Online*, September 2, 1998.

Sullivan, Patricia, "Robert Harris; Helped Push D.C. Home Rule," *Washington Post*, October 7, 2007.

Summary of Proceedings, 1985 Burlington Northern Energy Seminar.

"Summary of the Clean Air Act," 42 U.S.C. 7401 et seq. (1970).

Summary of Proposed Labor Agreements, Burlington Northern Railroad, June 1987.

"'Super System' for Ballast Undercutting/Cleaning," *Railway Track and Structures*, October 1987.

"A Superior Coal Facility," *BN News*, May 1976.

"Survey to Tell More Than How 'Happy' We Are," *Mgr.,* third quarter 1985.

Swenson, Ty, "Port Completes $52 Million E. Marginal Way Grade Separation: Expected to Improve Freight Movement, Ease Gridlock," *West Seattle Herald*, April 12, 2012.

"Swing Nose Frogs," *BN Technical Bulletin*, October 1988.

"Swing-Nose Frogs, Tangential Geometry Extend Turnout Life," *Progressive Railroading*, August 1990.

"Switch 'Em Once and Go" *Modern Railroads*, October 1973.

"Tab for Signing Bonuses at BN Put at $11.6 Million and Counting," *Traffic World*, June 24, 1991.

"Taconite Facility Dedicated at Superior," *BN News*, October 1977.

"Taming the Curves: Crawford Line Changes Completed," *BN News*, August–September 1982.

Tarbox, Gary, "Robert W. Downing, 1913–2010," *The Mainstreeter*, Fall 2010.

Taylor, Jeremy. 1997. *Powder River Coal Trains*. Telford, Penn.: Silver Brook Junction Publishing.

"Teamwork: Management, Labor Must Unite at BN," *BN News*, May/June 1983.

"Technology Overtakes the Caboose," <www.uprr.com/aboutup/history/caboose/cabo002.shtml>.

"Tenders Return," *BN News*, August–September 1983.

Ten Year Accident-Incident Overview by Railroad, Montana Rail Link, Federal Railroad Administration Office of Safety Analysis, 1998–2006.

"The Land Bridge: Myth or Method?" *Railway Age*, July 7, 1969.

"The Truck Question: Radial Or Not?" *Trains*, May 2007.

"Three Are Killed in Minnesota Head-on Collision," *Mgs.,* June 20, 1984.

"Three Railroads Join Forces to Rediscover Northwest Passage." 1990. *Traffic World*, July 30, 1990.

"Timber and Western Lands: 'A Keen Sense of Awareness Guides Their Land Management Objectives,'" *BN News*, November 1970.

"Tipping the Scales: BN's Coal Contracts Outweigh the Competition's," *BN News*, September-October 1986.

"Track Work" (to tie gangs from Montana sent to work on lines in Illinois), *BN News*, December 1974.

"Tractors Speed Up Car Inspections," *BN News*, April 1982.

"Train Accidents Claim Lives of Seven Employees," *BN News*, June 1984.

"Trainmaster: Attitude More Important Than Age," *BN News*, February 1982.

"Treatment Program 80% Successful," *BN News*, October 1980.

"Trucking Subsidiaries Sold," *BN News*.

"Tulsa Yard Betterments Circa 1974–1976," *Frisco All Aboard*, January 18, 2007.

Turley, Mike, "Personal Railroad History Page," <bearstoc.com/MRL/hist/career>.

Turney, D.E., "Advantages of #132 RE as a Heavy-Duty Rail Section," *Bulletin 711-American Railway Engineering Association*.

"Turnout Technology Takes a New Direction." 1990. *Railway Track and Structures*, July, 1990.

Tutumluer, Erol, William (Zach) Dombrow and Hai Huang, "Effect of Coal Dust on Railroad Ballast Strength and Stability," Railroad Engineering Program, University of Illinois, October 10, 2008.

Tuzik, Robert E., "Fast-on, Fast-off Machines," *Railway Age*, September 1991.

"12,000 Jobs in Jeopardy Railroad Says," *Chicago Tribune*, June 17, 1987.

"Two Railroads Assail Critics of Montana Track Transfer," *Journal of Commerce*, November 6, 1987.

Udland, Myles, "Warren Buffett Made a Deal in 2009 That Was So Good You Could Say He Stole It," *Business Insider*, November 12, 2014.

"Understanding the Clean Air Act," <epa.gov/air/peg/understand>.

"Union Leader Assails U.S. Justice Dept. Action in Northern Lines Merger Case." 1968. *Talking It Over*, June, 1968.

"Union Opposes Rail Sale Despite Signing Contract," *Livingston (Mont.) Enterprise,* September 15, 1987.

"Union Pacific Marks Silver Anniversary of Its First Southern Powder River Basin Coal Train, <uprr.com/newsinfo/releases/service/2009/0817_silver.shtml?print>.

"Union Pacific Moves Its 200,000th Loaded Coal Train from Wyoming's Southern Powder River Basis, <uprr.som/newsinfo/relesases/service/2009/0520_200000-coal.shtml>.

Union Pacific Railroad North Platte Area Timetable #1, October 25, 1998.

"Unions Win Fight to Stop New BN Line," *Associated Press,* June 11, 1988.

"Unit Train Car Shop," *Progressive Railroading,* September 1977.

United States, Appellant, v. Interstate Commerce Commission et al. Charles E. Brundage et al., Appellants, v. United States et al. City of Auburn, Appellant, v. United States et al. Livingston Anti-Merger Committee, Appellant, v. Interstate Commerce Commission 396 U.S. 491 (90 SCt 708, 24 L.Ed.2d 700), February 2, 1970.

United States Code, Section 10901 (e) of Title 49

United States v. ICC 396 U.S. 491(1970).

"Unprofitable Truck Lines Sold by BN's Parent Firm," *Traffic World,* September 3, 1984.

"UP, BNSF Announce Southern Powder River Basin Joint Line $100 Million Capacity Expansion Plan," <uprr.com/newsinfo/releases/capital_investment/2006/0508_sprb.shtml>, June 12, 2010.

U. S. Energy Information Administration, "Coal Transportation Issues," 2007.

"Use of Satellites Signals New Era in Train Control," BN *News,* Winter 1987.

Vance, Catherine A., "Secondary Picketing In Railway Labor Disputes: A Right Preserved Under the Norris-LaGuardia Act," *Fordham Law Review,* Volume 55, Issue 2

"Vandalism Mars Takeover of Lines in Two States by Montana Rail Link, Inc.," *Traffic World,* November 9, 1987.

Van Hattem, Matt, "Sweeping Aside a Dusty Problem," *Trains,* November 2007.

Van Hattem, Matt, "Wyoming Coal Line Expansion," *Trains.*

Vantuono, William C., "How Staggers 'Saved the Freight Railroad Industry from Socialism,'" *Railway Age,* November 2005.

Vantuono, William C., "Maximizing Productivity in 'The Heart of Coal,'" *Railway Age,* August 2012.

Vantuono, William C., "Regional Railroad of the Year: Montana Rail Link," *Railway Age,* April 2013.

Vartabedian, Ralph, "Bressler: Engineer on BN's Train Out of St. Paul," *The Minneapolis Star, April 1,* 1981.

Vartabedian, Ralph, "Bressler Has BN Making Tracks," *Minneapolis Star,* April 1, 1981.

Verstraete, Jim and Kim Forman, "Mt. Saint Helen's: A Thundering, Crashing Symphony of Awesome Power," BN *News,* August/September 1980.

Vietor, Richard H. K. 1994. *Contrived Competition: Regulation and Deregulation in America.* Cambridge, Mass., and London: Belknap Press of Harvard University Press.

"Volcano Damage Limited," BN *Westwords,* June 1980.

Voorhees, Josh, "Railroads, Utilities Clash Over Dust from Coal Trains," *EE News,* January 25, 2010.

"Walking Away From Clean Coal." 2015. *Bloomberg,* February 16–February 22, 2015.

"Walter Drexel: An Interview with the New President," BN *News,* April 1982.

"The Washington Companies—Company Profile, Information, Business Description, History, Background Information on the Washington Companies," <referenceforbusiness.com/history2/47/The-Washington-Companies>.

"Washington Star Exposes Coal Slurry Tactics, Sees Conflict of Interest Issue," *Burlington Northern Grass Roots Report,* November 1975.

Wastler, Allen R., "BN Hopes That Class III Railroads Will Develop Traffic for Main Line," *Traffic World,* October 27, 1986.

Wastler, Allen R. and Kathleen R. Keeney, "Labor Feathers Ruffled over BN Subsidiary Move," *Traffic World,* December 7, 1967.

Wastler, Allen R., "Peer Drug Intervention Effective, Riley Says, But Drug Tests Needed," *Traffic World.*

Wastler, Allen r., "Transport Labor Unions Attack Random Drug, Alcohol Testing," *Traffic World,* November 9, 1987.

"Wayside Wheel Inspection," www.dapcondt.com/?page_id=101.

Webb, H. G., "Advantages of 136/# RE Rail as a Heavy Duty Rail Section," *Bulleton 711—American Railway Engineering Association.*

Weinstein, William W., Philip, S. Babcock IV and Frank Leong, "Safety Analysis of ARES," Charles Stark Draper Laboratory, Inc., October 1987.

Welty, Gus, "BN America: Quality Counts," *Railway Age,* November 1991.

Welty, Gus, "BN and ARES: 'Control' in a New Dimension," *Railway Age,* May 1988.

Welty, Gus, "BN Confronts That Old Problem: Too Many Unions," October 1987.

Welty, Gus, "BN Employees Help Make the Merger Work," *Railway Age,* July 27, 1970.

Welty, Gus, "BN/Frisco: A Good Fit of Properties and Personalities," *Railway Age,* January 23, 1981.

Welty, Gus, "BN Gets Flexibility—And a 'Spirit of Cooperation,'" *Railway Age,* December 14, 1970.

Welty, Gus, "BN Maps Power Strategies," *Railway Age,* October 1989.

Welty, Gus, "BN Objects to Lines on Labor Column," *Railway Age,* September 1993.

Welty, Gus. 1977. "For Burlington Northern, It's Coal—And Whole Lot More." *Railway Age,* December 28, 1977.

Welty, Gus, "BN America: Quality Counts—Burlington Northern's Domestic Container Business," *Railway Age,* November 1991.

Welty, Gus. "BN—High-Tech Training," *Railway Age, July 1989.*

Welty, Gus. "BN's New Alliance Shop: Big, Capable, Expandable," *Railway Age,* August 3, 1979.

Welty, Gus. "BN's New Control Center: 22,000 Miles Under One Roof," *Railway Age,* June 1995.

Welty, Gus, "C&NW, BN, UP Play Catch-up: Capacity Problems in the Powder River Basin," *Railway Age,* October 1994.

Welty, Gus, "Crew Consist: The New Pacts Are Paying Off," *Railway Age,* March 31, 1980.

Welty, Gus, "From BN to Its Employees: 250 Questions," *Railway Age,* April 1984.

Welty, Gus., "What's So Good About Holding Companies?" reprint from *Railway Age* in *BN News,* July 1981.

Welty, Gus, "Is All of This BN-Bashing Really Necessary? Burlington Northern Railroad Co. CEO Jerry Grinstein's Management," *Railway Age,* May 1993.

Welty, Gus, "Look, I've Got a Problem," story from *Railway Age* quoted in *BN News,* November 1973.

Welty, Gus, "Management, Labor, Philosophies and 'A Major Thrust,'" *Railway Age,* June 1983.

Welty, Gus, "'Preventive' Payoff Is Big," *Railway Age,* June 1990.

Welty, Gus. "The BN at 10: 'The Amazing Thing Was, It Worked,'" *Railway Age,* February 25, 1980.

Welty, Gus, "The Era of the Giants," *Railway Age,* Fegruary 23, 1981.

Welty, Gus, "The Strange Case of the BN and the UTU," *Railway Age,* April 1992.

Welty, Gus, "Tracking Heavy-Car Impact," *Railway Age,* March 1989.

Welty, Gus, "Two Accidents That Didn't Have to Happen," *Railway Age,* August 1985.

Welty, Gus. 1989. "What's New in Turnouts," *Railway Age,* April, 1989.

Wendt, Darrell, "Bridging the 'Mighty Mo,'" *Railroads Illustrated,* April 2013.

Wermiel, Stephen, "High Court to Decide Whether Railroads Must Negotiate with Workers Over Sales," *Wall Street Journal,* November 29, 1988.

Wernick, Robert, "Norton Simon's Zeal for the Best and Its Results," *Smithsonian,* September 1979.

Wesselmann, Carl H., "Ray Burton: Piggyback Leadeer/Corporate Healer," *Modern Railroads,* January 1986.

Wesselmann, Carl, "The Case for Concrete Tie Turnouts," *Modern Railroads,* October 1987.

"Western Coal to Meet Energy Crisis," *BN News,* April 1973.

"Western Railroad Properties, Inc. (WRPI)," <utahrails.net/up/wrpi.php>.

"Weyerhaeuser Is Buying Plum Creek for $8.4B to Form Timber Giant," *Seattle Times,* November 9, 2015.

"What Price Railroad Safety?" *Baseline,* <www.baselinemag.com/c/a/Projects-Management/What-Price-Railroad-Safety>.

"What the Merger Taught BN About Its Employees," *Railway Age,* July 14, 1975.

Wiedrich, Bob, "12,000 Jobs in Danger, Railroad Tells Workers," *Chicago Tribune,* June 26, 1987.

Wiedrich, Bob, 1989. "Power-by-the-Hour: A Rail Success Story," *Chicago Tribune,* April 16, 1989.

Wiedrich, Bob, "Railroad Women Write a New Saga," *Chicago Tribune,*

Wiedrick, Bob, "She's a Railroader—Simple as That," *Chicago Tribune,* re-published in the *Alliance Times-Herald,* March 5, 1977.

"Will Crews Really Benefit from UTU's Crew Consist Fight?" *Railway Age,* September 1991.

"Will a Takeover Derail Burlington Northern's Makeover?" *Business Week,* August 3, 1987.

Will, George F., "China Has Seen the Future, and It Is Coal," *Washington Post,* December 30, 2010.

Williams, Winston, "Burlington Road Limps Faster," *The New York Times,* February 12, 1978.

Wilner, Frank N., "Coal Shippers in Revolt," *Railway Age,* March 1999.

Wilner, Frank N., "STB Expert: Consider PTC's Business Benefits," *Railway Age,* November 2008.

Wilner, Frank N., *Understanding the Railway Labor Act,* Omaha: Simmons-Boardman Books, Inc., 2009.

Wilner, Frank N., "When You Get a Good Deal, Take It," *Railway Age,* July 28, 2014.

"With BN Unit Grain Trains, Rates Reduced 5-15% from Single Car Rates," *Traffic World,* August 24, 1981.

Withers, Paul K., "Oakway SD60s," *Diesel Era,* vol. 7, no. 4.

Wong, Brad, "Transportation Officials Seek Solution to Bottlenecks on Rails," *Seattle Post-Intelligencer,* April 18, 2005.

Wood, Charles R. 1968. *The Northern Pacific: Main Street of the Northwest."* New York: Bonanza Books.

Wood, Charles and Dorothy. 1974. *Spokane, Portland and Seattle Railway.* Seattle: Superior Publishing Company.

Wood, Charles, and Dorothy, 1978. *The Great Northern Railway: A Pictorial Study by Charles and Dorothy Wood.*

"Working: Maintainer Puts Training to Work," *BN News,* January 1983.

"Working the Line," *BN News,* May/June 1985.

"Working Women: The Challenges of the Workplace," *BN News,* November 1985.

Woxland, M. O., "New Facilities for Increased Traffic at Alliance," American Railway Engineering Association, Bulletin 678.

Wright, Robert A. "The Way West Looks Bright for Merged Railway," *The New York Times,* March 25, 1973.

"WRPI Buys into BN Coal Line," *Railway Age,* February 1987.

Wyer, Dick and Co., *Burlington—Great Northern—Northern Pacific—SS&S Consolidation Study,* Study I, II, III, VI, X, X-L, X-F and XI, Common Points, Duplicate Lines, Car Load

Freight—Shortest or Most Economical Joint Routes, General Repair Facilities—Diesel Locomotives, General Repair Facilities—Freight Train Cars.

"Wyoming Coal Line Expansion," *Trains*, November 2008.

Wyoming Coal Production, 1969–2015. <wyomingmining.org>.

Yenne, Bill. 1991. *A History of the Burlington Northern.* New York: Bonanza Books.

Young, David, "How Do Our Commuter Lines Rate?" *Chicago Tribune Magazine*, January 11, 1981.

Zarembski, Allan M., "Effect of Material Quality on Ballast Life," *Railway Track and Structures*, March 1989.

Zarembski, Allan M., "Tracking R&D: Field Welding Rail," *Railway Track and Structures*, June 1987.

Zarembski, Allan M., "Track Replacement Needs as a Function of Traffic Densities," *Railway Track and Structures*, May 1990.

"Zestful, Zealous Don Zeiss Zooms in on Problems as Alliance Project Construction Director," *BN News*, December 1978.

Zeutschel, D. Larry, "Mountain Country Regional," *Pacific Rail News*, February 1991.

INDEX